Methods in Molecular Biology

Series Editor
John M. Walker
School of Life and Medical Sciences
University of Hertfordshire
Hatfield, Hertfordshire, UK

For further volumes:
http://www.springer.com/series/7651

For over 35 years, biological scientists have come to rely on the research protocols and methodologies in the critically acclaimed Methods in Molecular Biology series. The series was the first to introduce the step-by-step protocols approach that has become the standard in all biomedical protocol publishing. Each protocol is provided in readily-reproducible step-by step fashion, opening with an introductory overview, a list of the materials and reagents needed to complete the experiment, and followed by a detailed procedure that is supported with a helpful notes section offering tips and tricks of the trade as well as troubleshooting advice. These hallmark features were introduced by series editor Dr. John Walker and constitute the key ingredient in each and every volume of the Methods in Molecular Biology series. Tested and trusted, comprehensive and reliable, all protocols from the series are indexed in PubMed.

Manuscript Instructions https://drive.google.com/file/d/1K4Az5vpAFVzYadJ3jKx2-QSo7Dqa_SNA/view?usp=sharing

The Preimplantation Embryo

Methods and Protocols

Edited by

Alexandra J. Harvey

Melbourne IVF, Collingwood, VIC, Australia

David K. Gardner

Melbourne IVF, Collingwood, VIC, Australia

Humana Press

Editors
Alexandra J. Harvey
Melbourne IVF
Collingwood, VIC, Australia

David K. Gardner
Melbourne IVF
Collingwood, VIC, Australia

ISSN 1064-3745 ISSN 1940-6029 (electronic)
Methods in Molecular Biology
ISBN 978-1-0716-5291-6 ISBN 978-1-0716-5292-3 (eBook)
https://doi.org/10.1007/978-1-0716-5292-3

This Humana imprint is published by the registered company Springer Science+Business Media, LLC, part of Springer Nature.
The registered company address is: 1 New York Plaza, New York, NY 10004, U.S.A.

Foreword

Anybody who has read the book *Lessons in Chemistry* (by Bonnie Garmus) will appreciate that knowing how to perform a task (like following a recipe) is far more likely to succeed if one has an underlying knowledge of how to undertake the task. To be able to successfully conduct a specific technique is important in all fields of scientific endeavor, particularly in reproductive science and technology where one is focused on the earliest beginnings of life.

In *Lessons in Chemistry*, the fictional Elizabeth Zott encouraged her television audience to adopt the thinking that methodological knowledge leads to greater success. The same principle is on show in this book, edited by Alexandra (Alex) Harvey and David Gardner. In this project, they have assembled authors from around the world, all of whom have developed and mastered specific techniques, spanning the field of reproductive biology and medicine. It is the most up-to-date offering about methods I have seen. This sharing of such "recipes" and methodological insights and why they matter, within a single publication, is rare. All the contributors are well known in their fields, many of whom I personally know and admire for what they have contributed.

Understanding the regulation of gamete and embryo development remains one of the central challenges in reproductive biology. Capturing the complexity of human embryo development has historically been difficult, limited by the availability of clinical material and the associated ethical considerations that are at odds with the need to optimize the process. This is combined with technical challenges in supporting gametes and the embryo in vitro, which remains a poor representation of the dynamic environment in vivo, the analysis of discrete stages of development, and the use of methods requiring large numbers of gametes/embryos that mask cellular heterogeneity. Animal models have been invaluable for exploring molecular regulators and signaling pathways, and for establishing efficacy and safety against which all assisted reproductive technologies (ARTs) have been benchmarked; yet they have been similarly limited by the amount of biological material required and methods available to analyze samples.

Consequently, the development of single-embryo and single-cell methods, as well as non-invasive approaches, to assess cell viability and regulation has revolutionized the way in which we can interrogate the earliest stages of life. Such technologies provide unparalleled resolution and enhance our ability to improve each step in the context of ARTs—whether that is to simplify the process or to limit external stressors and better support gametes and embryos prior to utilization. Indeed, harnessing these technologies is critical to understand the mechanisms regulating gamete and preimplantation embryo development given that these early stages of development set the stage for the health and viability of all tissues and organs of the resultant individual.

Likewise, as the field embraces embryo-like models, the technologies outlined in this edition will be of significance not only in establishing whether these models accurately replicate normal developmental trajectories, but also for their use as models to explore genomic, environmental, and drug interactions on cellular signaling, heterogeneity, lineage specification, and spatial organization.

A personal note about the editors: Alex began her PhD with me in 2000, quickly growing her interest in regulators of embryo development and an understanding in molecular biology; we were the first to examine the role of *Hypoxia Inducible Factors* in the preimplantation embryo. She then spent time as a post-doc in the USA with Barry Bavister and Carol Brenner examining mitochondrial properties in stem cells before returning to Australia to join David Gardner at The University of Melbourne. One of her strengths that both David and I recognized is her ability to understand and interrogate the methodology used in the laboratory. David has been in our field for over 40 years, and I know him not only as a brilliant researcher and influential thought-leader in reproductive technologies, but also as a great friend. He is genuinely passionate about developing solutions to alleviate infertility and driven to make our field more responsive to the plight of the infertile.

I am therefore confident that this book will have considerable value to both students and those experienced in the field, as it provides insight into why the best science will always depend on a full appreciation of the latest methods.

Adelaide, SA, Australia ***Jeremy Thompson***

Preface

Since the birth of Louise Brown in 1978, significant advances have been made in our understanding of mammalian preimplantation embryo development, the development and contribution of gametes, and how to manipulate such reproductive cells in the laboratory. This has been achieved through ever-evolving technologies with increasing sensitivity, combined with the analysis of embryos from a range of species. Consequently, it is now possible to analyze life at single-cell resolution. Ongoing advances are likewise setting the stage to open a veritable Pandora's box on the regulation of embryo development. The intention of this book is therefore to detail essential and emerging technologies. This compendium of methods, written by experts in their field, provides detailed step-by-step guides and protocols. Technologies covered include gamete preparation, in vitro embryo culture and cryopreservation, physiological and molecular methods for embryo assessment, including single-cell sequencing, and emerging methods for modeling key developmental processes encompassing gamete and embryo manipulation through implantation.

This work serves as a comprehensive and practical toolbox and will be of immense value to all scientists and laboratories working on the truly fascinating and dynamic preimplantation embryo and the gametes from which it originates.

Collingwood, VIC, Australia *Alexandra J. Harvey*
Collingwood, VIC, Australia *David K. Gardner*

Contents

Contributors

GEORGIA E. BLACKLEY • *School of BioSciences, Faculty of Science, Bio21 Institute, The University of Melbourne, Parkville, VIC, Australia*

RENATA BLOCHER • *Department of Animal, Dairy and Veterinary Sciences, Utah State University, Logan, UT, USA*

MICHELE BOIANI • *Max Planck Institute for Molecular Biomedicine, Muenster, Germany*

VESNA BOJOVIC • *Centre of Reproductive Medicine and Andrology (CeRA), University Münster, Münster, Germany*

CHARLES L. BORMANN • *Massachusetts General Hospital Fertility Center, Department of Obstetrics, Gynecology, and Reproductive Endocrinology and Infertility, Massachusetts General Hospital and Harvard Medical School, Boston, MA, USA*

ELIZABETH G. BROMFIELD • *School of BioSciences, Faculty of Science, Bio21 Institute, The University of Melbourne, Parkville, VIC, Australia; Infertility and Reproduction Research Program, Hunter Medical Research Institute, New Lambton, NSW, Australia; Centre for Reproductive Science, School of Environmental and Life Sciences, The University of Newcastle, Callaghan, NSW, Australia*

CAMILA BRUNA DE LIMA • *Centre de Recherche en Reproduction, Développement et Santé Intergénérationnelle (CRDSI), Département des Sciences Animales, Université Laval, Québec City, QC, Canada*

NATHAN D. BURKE • *School of BioSciences, Faculty of Science, Bio21 Institute, The University of Melbourne, Parkville, VIC, Australia; Infertility and Reproduction Research Program, Hunter Medical Research Institute, New Lambton, NSW, Australia; Centre for Reproductive Science, School of Environmental and Life Sciences, The University of Newcastle, Callaghan, NSW, Australia*

ANTONIO CAPALBO • *Juno Genetics, Rome, Italy; Unit of Molecular Genetics, Center for Advanced Studies and Technology (CAST), "G. d'Annunzio" University of Chieti-Pescara, Chieti, Italy; Department of Psychological Health and Territorial Sciences, School of Medicine and Health Sciences, "G. d'Annunzio" University of Chieti-Pescara, Chieti, Italy*

JOHN CARROLL • *Monash Biomedicine Discovery Institute, Monash University, Clayton, VIC, Australia*

DARREN J. X. CHOW • *Robinson Research Institute, College of Health, Adelaide University, Adelaide, Australia; Institute for Photonics and Advanced Sensing, Adelaide University, Adelaide, Australia; Centre of Light for Life, Adelaide University, Adelaide, Australia*

PIERRE COMIZZOLI • *Smithsonian's National Zoo and Conservation Biology Institute, Washington, DC, USA*

RAUL DA COSTA • *Centre of Reproductive Medicine and Andrology (CeRA), University Münster, Münster, Germany; Münster Flow Cytometry Facility (MFlow), University Münster, Münster, Germany*

GABRIEL DALVIT • *Advanced Institute of Reproductive Technologies Inc., Tokyo, Japan, Advanced Fertility Centre, Cancun, Mexico*

KISHAN DHOLAKIA • *Centre of Light for Life, Adelaide University, Adelaide, Australia; School of Biological Sciences, Adelaide University, Adelaide, Australia; School of Physics and Astronomy, University of St Andrews, North Haugh, Fife, Scotland*
EVDOKIA DIMITRIADIS • *Department of Obstetrics, Gynecology and Newborn Health, University of Melbourne, Parkville, VIC, Australia; Gynaecology Research Centre, Royal Women's Hospital, Parkville, VIC, Australia*
BASTIEN DUCREUX • *Université de Paris Cité, Faculty of Medicine, INSERM 1016, Paris, France*
KYLIE R. DUNNING • *Robinson Research Institute, College of Health, Adelaide University, Adelaide, Australia; Institute for Photonics and Advanced Sensing, Adelaide University, Adelaide, Australia; Centre of Light for Life, Adelaide University, Adelaide, Australia*
PATRICIA FAUQUE • *Université de Paris Cité, Faculty of Medicine, INSERM 1016, Paris, France; Department of Reproductive Biology-CECOS, APHP Centre-Université Paris Cité, Cochin, Paris, France*
LUCILE FERREUX • *Université de Paris Cité, Faculty of Medicine, INSERM 1016, Paris, France; Department of Reproductive Biology-CECOS, APHP Centre-Université Paris Cité, Cochin, Paris, France*
JULIE FIRMIN • *Université de Paris Cité, Faculty of Medicine, INSERM 1016, Paris, France; Department of Reproductive Biology-CECOS, APHP Centre-Université Paris Cité, Cochin, Paris, France*
DAVID K. GARDNER • *Melbourne IVF, Collingwood, VIC, Australia; School of Biosciences, University of Melbourne, Parkville, VIC, Australia*
ROBERT B. GILCHRIST • *Fertility & Research Centre, Discipline of Women's Health, School of Clinical Medicine, University of New South Wales, Sydney, NSW, Australia*
JURI GNAGNARELLI • *Department of Internal Medicine, Reproduction and Population Medicine, Faculty of Veterinary Medicine, Ghent University, Merelbeke, Belgium*
KATHRYN H. GURNER • *Melbourne IVF, Collingwood, VIC, Australia*
TUONG M. HO • *HOPE Research Center, My Duc Hospital, Ho Chi Minh City, Vietnam; IVFMD, My Duc Hospital, Ho Chi Minh City, Vietnam*
SEAN J. HUMPHREY • *Charles Perkins Centre, School of Life and Environmental Sciences, University of Sydney, Sydney, NSW, Australia; Murdoch Children's Research Institute, Royal Children's Hospital, Melbourne, Australia; Murdoch Children's Research Institute, Royal Children's Hospital, Melbourne, VIC, Australia*
JESSICA ISPADA • *Animal Biotechnology, AgResearch, Hamilton, New Zealand*
STEFFEN ISRAEL • *Max Planck Institute for Molecular Biomedicine, Muenster, Germany*
HONGBIN JIN • *Australian Regenerative Medicine Institute, Monash University, Clayton, VIC, Australia*
MASASHIGE KUWAYAMA • *Advanced Institute of Reproductive Technologies Inc., Tokyo, Japan; Advanced Fertility Centre, Cancun, Mexico*
ANH H. LE • *HOPE Research Center, My Duc Hospital, Ho Chi Minh City, Vietnam; IVFMD, My Duc Hospital, Ho Chi Minh City, Vietnam*
PEI-CHIH LEE • *Smithsonian's National Zoo and Conservation Biology Institute, Washington, DC, USA*
YING LIU • *Department of Animal, Dairy and Veterinary Sciences, Utah State University, Logan, UT, USA*
JEFFREY R. MANN • *Monash Biomedicine Discovery Institute, Monash University, Clayton, VIC, Australia*

Marcella Pecora Milazzotto • *Laboratory of Embryonic Metabolism and Epigenetics, Center of Natural and Human Science, Federal University of ABC, Santo Andre, Brazil*

Jess E. Mulhall • *Infertility and Reproduction Research Program, Hunter Medical Research Institute, New Lambton, NSW, Australia; Centre for Reproductive Science, School of Environmental and Life Sciences, The University of Newcastle, Callaghan, NSW, Australia*

Zsolt Peter Nagy • *Reproductive Biology Associates - Prelude, Atlanta, GA, USA*

Sushmita Nath • *Centre of Reproductive Medicine and Andrology (CeRA), University Münster, Münster, Germany*

Luis Navarro-Sánchez • *Research and Development Department, Igenomix - part of Vitrolife Group, Ronda Narcís Monturiol, Paterna, Spain*

Brett Nixon • *Infertility and Reproduction Research Program, Hunter Medical Research Institute, New Lambton, NSW, Australia; Centre for Reproductive Science, School of Environmental and Life Sciences, The University of Newcastle, Callaghan, NSW, Australia*

Thomas Nolte • *Max Planck Institute for Molecular Biomedicine, Muenster, Germany*

Yvonne L. Odey • *Centre for Reproductive Health, Institute of Regeneration and Repair, University of Edinburgh, Edinburgh, UK; Biomedical Teaching Organisation (BMTO) Biomedical Sciences, Hugh Robson Building, George Square, University of Edinburgh, Edinburgh, UK*

Christian Ottolini • *Juno Genetics, Rome, Italy; Department of Maternal and Fetal Medicine, UCL Institute for Women's Health, University College London, London, UK*

Catherine Patrat • *Université de Paris Cité, Faculty of Medicine, INSERM 1016, Paris, France; Department of Reproductive Biology-CECOS, APHP Centre-Université Paris Cité, Cochin, Paris, France*

Tayler Patrick • *Department of Animal, Dairy and Veterinary Sciences, Utah State University, Logan, UT, USA*

Krishna Chaitanya Pavani • *Department of Internal Medicine, Reproduction and Population Medicine, Faculty of Veterinary Medicine, Ghent University, Merelbeke, Belgium; Department for Reproductive Medicine, Ghent University Hospital, Gent, Belgium*

Luc Peelman • *Department of Veterinary and Biosciences, Faculty of Veterinary Medicine, Ghent University, Merelbeke, Belgium*

Irina A. Polejaeva • *Department of Animal, Dairy and Veterinary Sciences, Utah State University, Logan, UT, USA*

Maurizio Poli • *Juno Genetics, Rome, Italy*

Rebecca L. Robker • *Robinson Research Institute, College of Health, Adelaide University, Adelaide, SA, Australia*

Carmen Rubio Lluesa • *Research and Development Department, Igenomix - part of Vitrolife Group, Ronda Narcís Monturiol, Paterna, Spain*

Darryl L. Russell • *Robinson Research Institute, College of Health, Adelaide University, Adelaide, SA, Australia*

Michaela Sacco • *Department of Obstetrics, Gynecology and Newborn Health, University of Melbourne, Parkville, VIC, Australia; Gynaecology Research Centre, Royal Women's Hospital, Parkville, VIC, Australia*

MUNUSE C SAVASH ISHANZADEH • *University of Oxford, Nuffield Department of Women's and Reproductive Health, John Radcliffe Hospital, Women's Centre, Oxford, UK; Life Research Oxford, Hayakawa Building, Oxford, UK*
STEFAN SCHLATT • *Centre of Reproductive Medicine and Andrology (CeRA), University Münster, Münster, Germany*
MARC-ANDRÉ SIRARD • *Centre de Recherche en Reproduction, Développement et Santé Intergénérationnelle (CRDSI), Département des Sciences Animales, Université Laval, Québec City, QC, Canada*
DAVID A. SKERRETT-BYRNE • *Infertility and Reproduction Research Program, Hunter Medical Research Institute, New Lambton, NSW, Australia; School of Biomedical Sciences and Pharmacy, College of Health, Medicine and Wellbeing, The University of Newcastle, Callaghan, NSW, Australia; Institute of Experimental Genetics, Helmholtz Zentrum München, German Research Center for Environmental Health, Neuherberg, Germany; German Center for Diabetes Research (DZD), Neuherberg, Germany*
KATRIEN SMITS • *Department of Internal Medicine, Reproduction and Population Medicine, Faculty of Veterinary Medicine, Ghent University, Merelbeke, Belgium*
SHANNON P. SMYTH • *School of BioSciences, Faculty of Science, Bio21 Institute, The University of Melbourne, Parkville, VIC, Australia*
ROSS STOCKER • *Monash Biomedicine Discovery Institute, Monash University, Clayton, VIC, Australia*
TIFFANY C. Y. TAN • *Robinson Research Institute, College of Health, Adelaide University, Adelaide, Australia; Institute for Photonics and Advanced Sensing, Adelaide University, Adelaide, Australia; Centre of Light for Life, Adelaide University, Adelaide, Australia*
EVELYN E. TELFER • *Institute of Cell Biology, School of Biological Sciences, Hugh Robson Building, George Square, University of Edinburgh, Edinburgh, UK; Centre for Reproductive Health, Institute of Regeneration and Repair, University of Edinburgh, Edinburgh, UK*
ANN VAN SOOM • *Department of Internal Medicine, Reproduction and Population Medicine, Faculty of Veterinary Medicine, Ghent University, Merelbeke, Belgium*
CONSTANTINE S. VELMAHOS • *Department of Urology, Massachusetts General Hospital and Harvard Medical School, Boston, MA, USA*
LAN N. VUONG • *University of Medicine and Pharmacy at Ho Chi Minh City, Ho Chi Minh City, Vietnam; HOPE Research Center, My Duc Hospital, Ho Chi Minh City, Vietnam*
DAGAN WELLS • *University of Oxford, Nuffield Department of Women's and Reproductive Health, John Radcliffe Hospital, Women's Centre, Oxford, UK; Juno Genetics, Hayakawa Building, Oxford, UK*
EMMA G. WHATLEY • *Melbourne IVF, Collingwood, VIC, Australia*
YASMYN E. WINSTANLEY • *Robinson Research Institute, College of Health, Adelaide University, Adelaide, SA, Australia*
JENNIFER ZENKER • *Australian Regenerative Medicine Institute, Monash University, Clayton, VIC, Australia*
WEI ZHOU • *Department of Obstetrics, Gynecology and Newborn Health, University of Melbourne, Parkville, VIC, Australia; Gynaecology Research Centre, Royal Women's Hospital, Parkville, VIC, Australia*

Chapter 1

In Vitro Development of Primordial Follicles from Human Ovarian Tissue

Evelyn E. Telfer and Yvonne L. Odey

Abstract

The ability to develop human oocytes from the primordial stage to full maturation entirely in vitro presents profound opportunities. Beyond advancing basic science in oogenesis and meiosis, in vitro growth systems hold immense potential for novel assisted reproductive technologies (ARTs). Advances in deriving oocytes from stem cells combined with in vitro growth and in vitro maturation hold the promise of revolutionizing fertility preservation. Should these stem cell-derived oocytes prove functionally competent, in combination with IVG systems they could ultimately reduce reliance on donor gametes and expand reproductive options for a wide range of patients. Looking ahead, refining culture conditions to more closely mimic the in vivo ovarian niche will be essential for improving oocyte quality and developmental competence. Ethical and regulatory frameworks will also need to evolve to address the unique challenges posed by in vitro-derived gametes. As genomic, epigenetic, and metabolic profiling techniques advance, they will enable rigorous assessment of the safety and fidelity of IVG derived oocytes. In the long term, these technologies could not only transform fertility treatment but also serve as powerful models for studying human germline development, inheritance, and disease. This chapter outlines a methodology for developing human oocytes from primordial follicles to maturity.

Keywords Ovarian cortex, Follicle culture, Oocyte maturation, Ovary, In vitro growth (IVG), In vitro maturation (IVM)

1 Introduction

The capacity of mammalian oocytes to develop outside the body has long intrigued reproductive biologists. In 1934, Pincus and Enzmann posed the seminal question of whether mammalian oocytes could complete their development in vitro. Their pioneering studies demonstrated that oocytes isolated from large antral follicles could spontaneously resume meiosis, marking a pivotal moment in the emergence of systems designed to support oocyte maturation outside the ovarian environment [1].

By 1965, Edwards had successfully achieved in vitro maturation (IVM) of oocytes from several species, including mice, cows, pigs,

Alexandra J. Harvey and David K. Gardner (eds.), *The Preimplantation Embryo*, Methods in Molecular Biology, vol. 3038, https://doi.org/10.1007/978-1-0716-5292-3_1,

and rhesus monkeys [2]. However, it would take years of sustained research before scientists could reliably produce embryos and viable offspring from oocytes matured in vitro. A significant breakthrough came in 1984, when Schroeder and Eppig reported the first successful fertilization and embryo development from in vitro matured mouse oocytes, validating decades of foundational work and setting the stage for the eventual clinical application of IVM [3]. In parallel, efforts have focused on developing methods to support the maturation of even earlier oocyte stages, such as primordial and preantral follicles (Fig. 1).

Efforts to culture mouse oocytes from preantral stages in vitro began over 40 years ago. Eppig pioneered techniques that enabled mouse oocytes grown for extended periods in vitro to subsequently undergo meiotic maturation [4]. Continued improvements in these culture systems culminated in the production of embryos and live offspring from oocytes derived from in vitro grown preantral follicles [5]. This foundational work provided critical models for understanding oocyte development, particularly emphasizing the importance of oocyte–somatic cell interactions and the regulatory influence of oocyte-secreted factors [6, 7]. As understanding deepened, significant

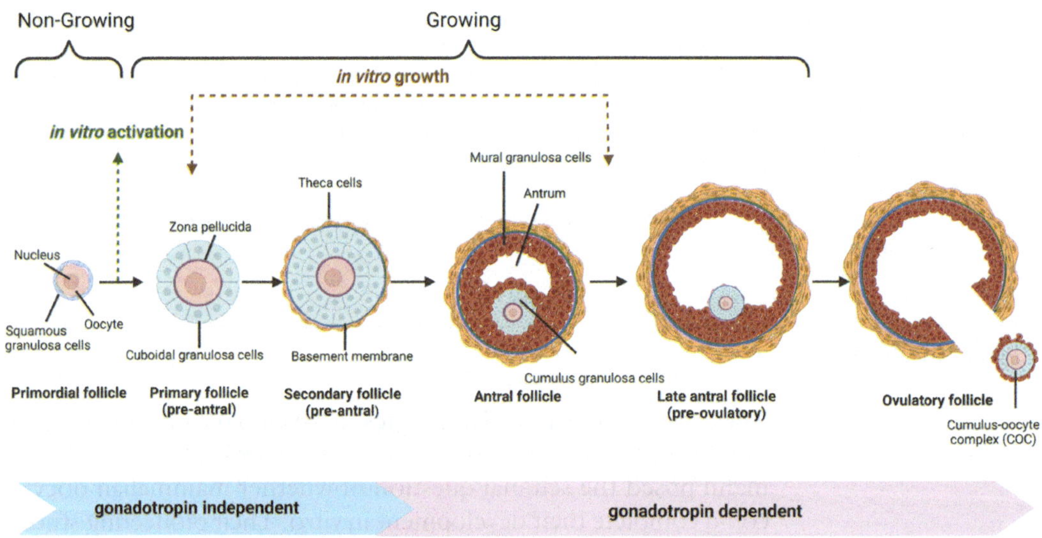

Fig. 1 Stages of follicle growth (primordial to preovulatory). Primordial follicles, once activated, progress to the primary stage, characterized by an oocyte enveloped by a single, complete layer of cuboidal granulosa cells. Under the influence of paracrine signaling, granulosa cells proliferate, forming multilayered (preantral) follicles, which are surrounded by differentiated thecal cells external to the basement membrane. As development continues, follicles form a fluid-filled cavity (antrum), resulting in antral follicles with mural granulosa cells lining the follicle wall and cumulus granulosa cells directly surrounding the oocyte. These antral follicles undergo rapid growth, culminating in the preovulatory stage, where the oocyte-cumulus complex is released at ovulation in response to luteinizing hormone (LH) signaling. While early follicular stages develop independently of follicle-stimulating hormone (FSH), progression through the multilaminar stages is acutely dependent on FSH for continued growth

advancements in culture systems followed throughout the 1980s and 1990s (for review [8]).

A major milestone was reached when the earliest stages of follicles, i.e., primordial follicles, were successfully activated and grown in vitro. Using a carefully staged, two-step culture system, primordial follicles were first activated to grow to the preantral stage, after which the enclosed oocytes could be matured and fertilized, producing viable embryos and live offspring [9, 10]. While initial experiments yielded only a single mouse exhibiting adult abnormalities [9], subsequent refinements to the culture medium enabled the production of multiple embryos and healthy live-born mice from fully in vitro grown and matured oocytes [10]. These studies provided conclusive proof that complete oocyte development could be achieved entirely in vitro, sparking efforts to extend these methods to human and large animal systems.

Adapting in vitro growth (IVG) technologies from rodents to humans and domestic species has proven challenging. Fundamental biological differences, including the significantly larger size of human follicles and the prolonged duration of human folliculogenesis, complicate the direct application of murine protocols to human tissue. Nevertheless, significant progress has been made and advances in culturing human, nonhuman primate, and domestic animal follicles have brought the goal of achieving complete human oocyte development in vitro closer to reality [8, 11–13].

Supporting the entire multiphased process of human oocyte development in vitro presents a formidable technical challenge. Isolated preantral follicles from human tissue have developed to maturity within an alginate matrix [14], and more recent studies have demonstrated that complete development from primordial or unilaminar follicles to metaphase II oocytes is now achievable [15, 16]. Culture systems that start with activation of primordial follicles utilize ovarian cortical biopsies obtained from healthy women or individuals undergoing fertility preservation procedures [11]. For a comprehensive review on development of systems for human application see [13] and Fig. 2A.

The methodology outlined in this chapter describes a multistep culture system to support the development of human oocytes within primordial/unilaminar follicles to maturity from healthy adult ovarian tissue (Figs. 1 and 2A). The emphasis is on oocyte development rather than supporting the development of large antral follicles. The first step is to support primordial follicle activation (PFA), and this involves the precise coordination of stimulatory and inhibitory signals within an intricate intrafollicular network [17]. Initial efforts to culture isolated human primordial follicles were largely unsuccessful, with follicles failing to activate and grow [18–20]. Later studies revealed that maintaining the structural integrity of ovarian cortical tissue and preserving interactions between primordial follicles and surrounding stromal cells are essential for sustaining follicle viability and promoting activation [21–25].

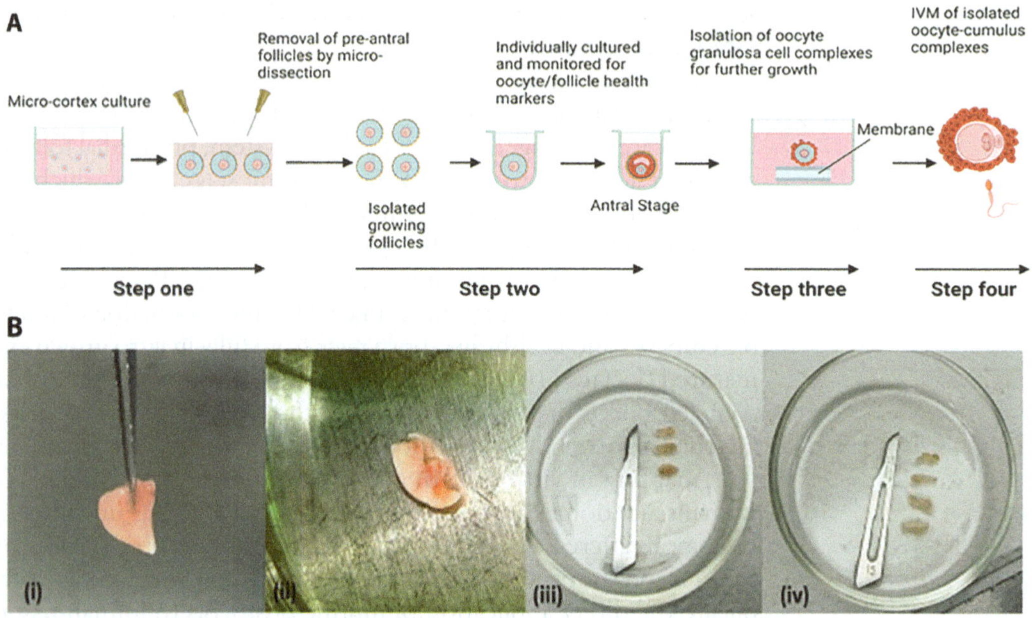

Fig. 2 (**A**) Multistep culture system for human oocyte development from primordial to maturity (adapted from [15]). Step 1: Ovarian tissue containing primordial or unilaminar follicles is prepared for culture. Step 2: Once follicles reach the multilaminar stage, they are mechanically isolated using fine needles. Isolated follicles are individually cultured from the preantral to antral stages. Step 3: Cumulus-oocyte complexes (COCs) are retrieved from the antral follicles and cultured until the oocyte diameter exceeds 100 μm. Step 4: COCs are transferred to maturation medium for in vitro maturation (IVM), where they are assessed for cumulus cell expansion, metaphase II spindle formation, and the presence of a polar body. (**B**) Preparation of Microcortex fragments (i) Biopsied human ovarian tissue prior to preparation, (ii) Stromal aspect of human ovarian tissue uppermost; stretching allows removal of excess stromal tissue, (iii) Fragments of human ovarian tissue, excess stroma removed, cortical aspect uppermost, (iv) Post fully stretched and loosened fragments, surface area increased by >10% but <20%

2 Materials

2.1 Media and Reagents

Media should be prepared no more than 24 h before use. All supplements are made up to the appropriate concentration and stored as aliquots at −20°C and thawed at 37°C before adding to media. All media prepared with supplements should be stored in the refrigerator and used within 24 h.

1. Transport and Dissection medium; Leibovitz medium Supplemented with 2 mM sodium pyruvate, 2 mM glutamine, 3 mg/mL Human Serum Albumin (HSA), 75 μg/mL penicillin G, 50 μg/mL streptomycin.
2. Microcortex culture medium; McCoy's 5 A medium with bicarbonate. Supplemented with 20 mM HEPES, 3 mM glutamine, 0.1% HAS, 0.1 mg/mL penicillin G, 0.1 mg/mL streptomycin, 2.5 μg/mL transferrin, 4 ng/mL selenium, 10 ng/mL

human insulin, 1 ng/mL recombinant human FSH (rhFSH), 50 µg/mL ascorbic acid.

3. Isolated follicle culture medium; microcortex culture medium additionally supplemented with 100 ng/mL human recombinant activin A and 5 ng/mL rhFSH.
4. OGC holding medium [a holding medium for oocyte–granulosa cell complexes (OGCs) after isolation from antral follicles and prior to IVM]; microcortex culture medium with the addition of 50 ng/mL human recombinant activin A and 10 ng/mL rhFSH.
5. Maturation medium; SAGE IVM medium supplemented with 75 mIU/mL FSH and 75 mIU/mL LH [15].
6. Sterile mineral oil.
7. Bouin's solution (for fixing tissue fragments for histological processing on days 0 and 8 of culture).

2.2 Consumables

1. Thermos flask (to hold tubes to transport tissue and maintain tissue temperature at 37°C during transport) or portable incubator (for transport >30 min).
2. Sterile 20 mL tubes/universal containers for media.
3. Glass Petri dishes (for dissection, culture handling).
4. Fine forceps (for dissecting cortex strips, follicles).
5. Scalpel blades (Nos. 10, 24, 25).
6. 24-well cell culture plates (for fragment culture).
7. 96-well V bottomed cell culture plates (for isolated follicles).
8. Fine-bore Pasteur pipettes (pastettes; for gentle handling of follicles/COCs).
9. 25 G needles (for microdissection, follicle handling).
10. Track-etched nucleopore membranes (for membrane-based culture of COCs in the final step).
11. 4-well cell culture plates (for final culture steps, IVM).

2.3 Equipment

1. Dissecting microscope equipped with a heated stage, capable of magnification 40–100× for microdissection of cortex and follicles.
2. CO_2 incubator set at 5% CO_2 in air (for standard culture).
3. Tri-gas/multigas incubator (to control O_2 and CO_2, e.g., reduced O_2 environment for culture of complexes at 5:5:90).
4. Temperature-controlled transport container (to maintain 37°C during handling).
5. Laminar flow hood/sterile biosafety cabinet (for aseptic culture setup).
6. Microscope with inverted stage and an eyepiece graticule for monitoring follicles/COCs during culture.

3 Methods

Ovarian cortex tissue, which predominantly contains primordial follicles, serves as the primary substrate for the culture system described here. Two systems that have developed human oocytes from this material to obtain mature oocytes are [15, 16]. This chapter outlines a methodology for developing human follicles from the primordial and preantral stages to obtain mature oocytes based on McLaughlin et al. [15].

3.1 Culture of Human Primordial/Unilaminar Follicles Within Cortical Tissue

3.1.1 Preparation of Human Ovarian Cortical Tissue

Biopsies of human ovarian cortical tissue obtained from healthy donors or patients undergoing fertility preservation, with appropriate ethical approvals and informed consent, should be used (Fig. 2Bi).

1. Warm 50 mL of dissection medium to 37°C for a minimum of 1 h.
2. Transport ovarian tissue to the laboratory in prewarmed Transport/dissection medium (10 mL) within a sterile tube. If transportation time is less than 30 min, the sterile container should be placed in a small Thermos to maintain temperature at 37°C during transportation. For longer transportation times, a portable incubator maintained at 37°C without gas should be used.
3. On arrival to the laboratory, transfer the ovarian tissue to a sterile glass petri dish with fresh warmed dissection medium. Tissue should be examined under light microscopy and any damaged, hemorrhagic, or excess stromal tissue removed using a scalpel and fine needles (*see* **Note 1**; Fig. 3Bi).
4. Remove large follicles (>100 μm) using two 25G needles (*see* **Note 2**).
5. With the medullary surface facing up, hold the tissue using fine flat toothed forceps. Remove excess medullary tissue using a No. 24 scalpel blade. The tissue must be held firmly while removing excess medulla to prevent cortical tearing. Excess medulla is defined as tissue exceeding 3 mm above the cortical surface. Once the excess tissue has been removed, the biopsied tissue is turned over to expose the cortical surface uppermost that should now be flattened (Fig. 2Bii).
6. Anchor the tissue firmly to the base of the petri dish using fine forceps to hold a small edge or corner of the tissue piece. To facilitate primordial follicle activation (PFA), manually dissect

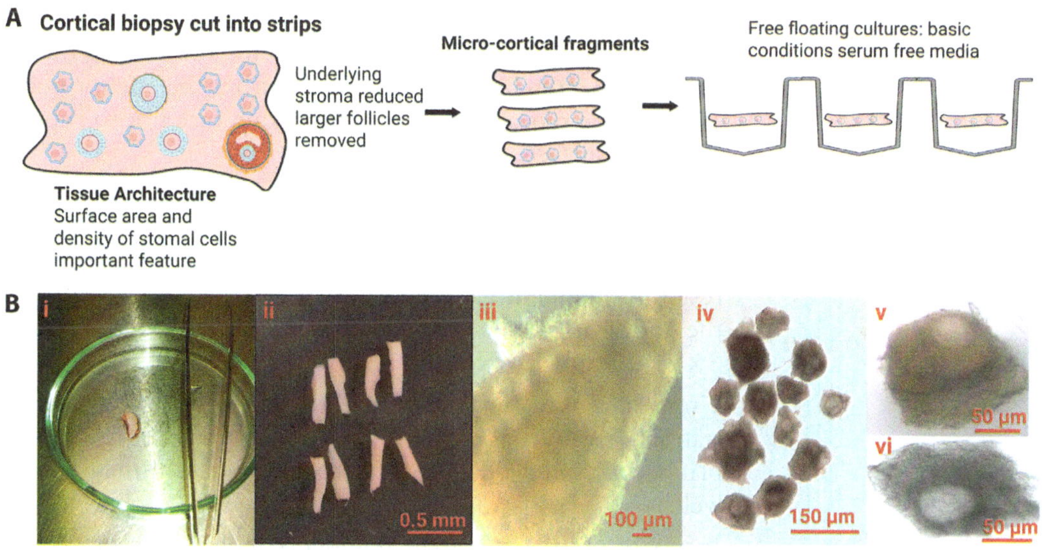

Fig. 3 Preparation of tissue and Step 1 of culture: **(A)** illustrates the preparation of the ovarian tissue sample to produce the "microcortex". The Biopsy is prepared by removing stromal cells and growing follicles then cutting the tissue into microcortex fragments before stretching. (**B**) Photomicrographs of an ovarian cortical biopsy before preparation of microcortex pieces (i) Prepared microcortex pieces, (ii) In vitro activated follicles within a cultured microcortex, (iii) Preantral follicles with theca layers dissected from cultured microcortex, (iv) range of follicles dissected from microcortex of culture. (v, vi) Individual follicles selected for Step 2 showing surrounding stromal cells

the cortical biopsy into "microcortex" fragments by slicing the tissue with a scalpel blade into fragments of approximately 3×1×0.5 mm for incubation (Fig. 3Bii).

7. Reserve two cortical fragments from each biopsy for processing and histological evaluation by placing them in Bouin's solution.
8. Reexamine the fragments under light microscopy and measure any visible follicles (Fig. 2Biii). Follicles with a mean diameter >40 μm should be excised from the tissue fragments using fine needles to pull the stromal cells from the follicle and remove with fine forceps, to ensure that the presumptive population of follicles contained within the fragments are unilaminar, i.e., primordial, transitory, or primary (*see* **Note 3**).

3.1.2 Preparation of Stretched Loosened Microcortex and Fragment Culture

Microcortex fragments should undergo a loosening procedure that slackens the fibrous cortex, relaxing the cortical surface. Cortical loosening is accomplished by tethering each fragment using a 25 G or finer needle at a corner point and firmly drawing the blunt end of a small scalpel blade evenly over the cortical surface of the tissue piece. This action is repeated until the cortical surface area of each tissue fragment has increased by more than 10% but not greater than 20% (Fig. 2Biv).

1. Using the blunt edge of a No. 10 scalpel blade, angled at 45° to the cortical surface, and gentle pressure, draw along the tissue surface stretching the tissue away from the anchor point. This may be repeated several times taking care not to rupture the tissue surface.
2. With the cortical surface still facing up, cut the tissue with a No. 24 scalpel blade using an angled incision. The blade is positioned at 45° to the cortical surface to ensure that each resulting fragment has a larger cortical surface area relative to the underlying tissue (ratio of 3:2). Fragments should measure 3 × 1 × 0.5 mm (*see* **Note 4**).
3. Using fine-bore sterile pastettes to minimize tissue damage, transfer tissue fragments individually intO24-well culture plates containing 300 μL of microcortex culture medium per well.
4. Incubate prepared ovarian fragments individually in 24-well cell culture plates. Culture fragments at 37°C in humidified air with 5% CO_2 for 6–10 days (*see* **Note 5**). Replace half the volume of medium every 48 h.
5. At the end of culture, reserve two cortical fragments from each biopsy for processing for histological evaluation by placing in Bouin's solution.

3.2 Isolation and Growth of Secondary (Multilaminar) Follicles

Visible follicles ranging in size from 90 to 150 μm can be observed between 6 and 10 days of culture (Fig. 3Biii). Within 6–10 days activated primordial follicles progress to secondary (multilaminar) stages [15, 22] (Fig. 3Biv). However, if retained within cortical tissue, secondary follicles typically degenerate, necessitating their isolation for continued culture [26, 27]. The decision to isolate follicles is based on observing the cultured microcortex and measuring growing follicles that are obvious on the surface.

1. To isolate follicles, place fragments with visible growing follicles (100–150 μm) in dissection media warmed to 37°C. Mechanically dissect growing follicles from the cultured cortical fragments using 25-gauge needles (*see* **Note 6**).
2. Follicles with intact basement membranes are selected for further culture (Fig. 3Biv).
3. Individually transfer isolated secondary follicles to 96-well V-bottomed culture plates containing 150 μL of isolated follicle culture medium (*see* **Note 7**). Culture individual follicles at 37°C in humidified air with 5% CO2. Refresh half of the medium every 48 h. Follicle diameter measurements should be taken at each medium change using a dissecting microscope with a graticule eyepiece. Formation of antral cavities should be observed within 10 days of culture.

3.3 Extraction and Culture of Oocyte-Granulosa Cell Complexes (OGCs)

1. Transfer in vitro grown follicles from each culture well into a glass petri dish containing prewarmed holding medium using a plastic pastette. Hold each follicle using fine forceps and examine under light microscopy to ascertain the presence of an antral cavity (Fig. 4A, B).
2. If a cavity is present, while holding the follicle with fine forceps, apply light pressure to puncture the follicle with a 25 G needle to release the oocyte-granulosa cell complex (OGC) into the petri dish. Care must be taken to puncture the basal lamina only and not damage the underlying OGC.
3. On release, the OGCs must be immediately removed from the holding medium into prewarmed OGC holding medium using pulled glass pipettes.
4. If no antral cavity is visible, puncture the basal lamina gently several times using 2 × 25-gauge needles and gently stretch the basal lamina until it ruptures and the inner cells are exposed. Using a pulled glass pipette, gently draw the inner cells back and forth until the OGC is detached from the follicle. Remove the complex from the holding medium into prewarmed OGC holding medium using a pulled glass pipette.
5. Using pulled glass pipettes, transfer OGCs from holding medium onto track-etched nucleopore membranes (Fig. 4C) in 300 μL of OGC culture medium in 4-well culture plate. A maximum of six complexes can be matured per membrane (*see* **Note 8**).
6. Note the position of every complex on each membrane and measure each individual oocyte diameter and cumulus spread.

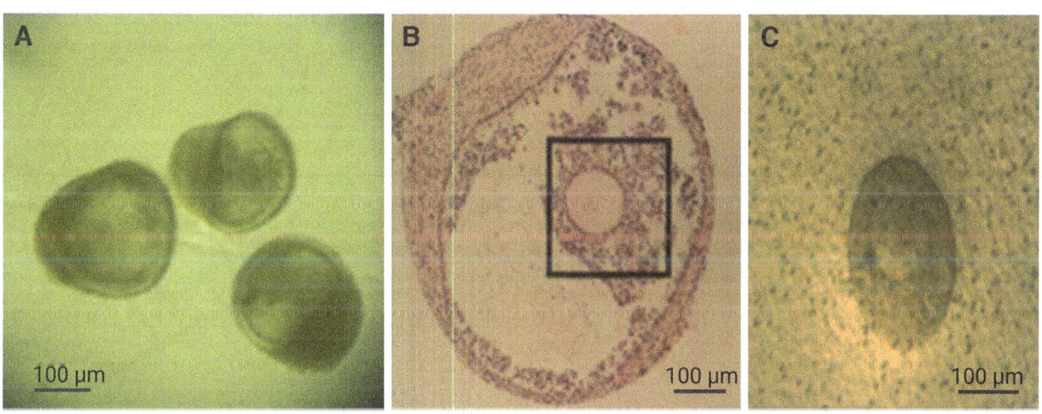

Fig. 4 Removal of oocyte granulosa cell complex following Step 2 of culture: Photomicrographs of (**A**) an in vitro grown antral follicle following Step 1 and 2 of the multistep culture process (**B**) a histological section of an IVG antral follicle showing the antral cavity and oocyte-granulosa cell complex (**C**) the isolated OGC aspirated from an IVG antral follicle and placed on membrane for a further culture period (Step 3) before undergoing in vitro maturation (IVM)

7. Incubate complexes at 37°C under low oxygen (5:5:90 gas mix). Monitor oocyte diameter and cumulus expansion daily. Replace half the OGC medium every 48 h, concomitant with oocyte diameter and cumulus expansion measurements.
8. Growth is considered complete when oocytes reach at least 100 μm in diameter, indicative of readiness for subsequent maturation. The duration of complex culture is 2–4 days depending on rate of oocyte growth but no longer than 4 days (*see* **Note 9**).

3.4 In Vitro Maturation

When oocyte diameters are large enough (>100 microns) to undergo in vitro maturation, the membranes are removed from the OGC medium using fine forceps and placed into 24-well cultures plates, one membrane per well. This protocol has previously used SAGE IVM medium containing 75 mIU/mL of FSH and 75 mIU/mL of LH [15], however, there are several alternative IVM Protocols that can be used once oocytes >100 microns have been obtained (*see* **Note 10**).

1. Remove each membrane from the OGC medium using sterile fine forceps and place into a well of a 4-well culture plate, allowing the membranes to settle on the base of the well before carefully adding maturation medium.
2. Gently remove complexes from the membranes by agitating the membrane within the culture well and removing the complexes with a fine bore micropipette before transferring to preequilibrated microwells containing 25 μL of maturation media under mineral oil.
3. Incubate for 22–28 h at 37°C in humidified air with 5% CO_2.
4. Following IVM, expulsion of the first polar body can be visualized by light microscopy and used to confirm meiotic resumption/maturation.

4 Notes

1. It is critical that the cortical surface be distinguished from the underlying tissue. Failure to correctly identify the cortex will lead to tissue damage and loss of follicles.
2. Any follicles larger than 100 μm are removed as the presence of large developing follicles will impede the activation and growth of others. Where large follicles are inaccessible or the tissue is particularly fibrous a no. 10 scalpel blade and fine forceps are used to prevent the tissue tearing while follicles are removed. Any torn, damaged or hemorrhagic areas of tissue are excised using the tip of a no. 10 scalpel blade.

3. Primordial follicles are contained within the cortex and 98% of primordial follicles are found within 350 μm of the surface. These follicles are densely packed and require to pulled apart (but not isolated) to release them from inhibitory factors in order to activate their growth. Following the trimming of stromal tissue and removal of growing follicles, the microcortex fragments are ready to be mechanically loosened to stretch the fibers of the microcortex pieces. This is an important step to support activation of primordial follicles within the cultured fragments.
4. Fragmentation of the ovarian cortex disrupts the Hippo signaling pathway, crucial in regulating tissue growth. Alterations in actin dynamics shift the balance of growth regulators, leading to nuclear translocation of YAP in granulosa cells and the transcription of growth-promoting factors [28, 29]. This disruption, initiated during tissue preparation, is a key trigger for follicular awakening. Recent studies have highlighted the critical influence of biomechanical forces, including tissue pressure and stromal support, in modulating follicle activation and growth [30–32]. These findings suggest that the physical environment of the follicle, in addition to biochemical signals, plays a crucial role in regulating the transition from dormancy to active growth.
5. When to dissect out follicles for Step 2 is determined by monitoring follicle growth and observing follicle structures measuring 100–150 μ. These can appear as early as 6 days of culture but should be monitored daily up to 10 days of culture.
6. In this system, mechanical isolation is preferred over enzymatic digestion to preserve the integrity of the follicular basal lamina and thecal layers. Follicle isolation may be achieved enzymatically, using collagenase, DNase, or Liberase [33–35], though we have found that enzymatic methods risk damage to the critical thecal layers. Mechanical isolation using fine needles preserves follicle integrity, albeit with lower yield and requiring considerable expertise [15, 22, 36, 37].
7. Maintaining follicular structure during in vitro growth (IVG) is challenging. Encapsulation within biomatrices, particularly alginate, has been utilized to provide structural support [38–40]. However, alginate's rigidity may impede gene expression, ECM remodeling, and cellular interactions unless optimized or supplemented with bioactive molecules [41, 42]. Emerging tissue-engineered scaffolds, including decellularized ovarian matrices and three-dimensional porous scaffolds [43, 44], offer promising alternatives, particularly for future applications in artificial ovary development [45]. We have found that when the thecal layer is intact, a supporting matrix is not essential [15], enabling more physiological follicle development and ease of recovering mature oocytes.

8. It is critical that complexes containing oocytes of a similar diameter are cultured together on the same membrane. If complexes with disparate sized oocytes are placed together on the membrane, oocyte development, cumulus cell expansion, and subsequent in vitro maturation can be comprised.
9. Oocytes within the complexes should measure between 85 and 95 μm before being placed on membranes, The majority of complexes will reach a diameter of greater than 100 μm after 2 days in membrane culture but some may take up to 4 days to reach the optimum diameter. Complexes should not be cultured on membranes for longer than 4 days.
10. IVM protocols have improved over recent years, including the CAPA-IVM protocol, which has been particularly successful [46].

References

1. Pincus G, Enzmann EV (1934) Can mammalian eggs undergo normal development in vitro?. Proc Natl Acad Sci U S A 20(2):121–122. https://doi.org/10.1073/pnas.20.2.121
2. Edwards RG (1965) Maturation in vitro of mouse, sheep, cow, pig, rhesus monkey and human ovarian oocytes. Nature 208(5008):349–351. https://doi.org/10.1038/208349a0
3. Schroeder AC, Eppig JJ (1984) The developmental capacity of mouse oocytes that matured spontaneously in vitro is normal. Dev Biol 102(2):493–497. https://doi.org/10.1016/0012-1606(84)90215-x
4. Eppig JJ (1979) A comparison between oocyte growth in coculture with granulosa cells and oocytes with granulosa cell-oocyte junctional contact maintained in vitro. J Exp Zool 209(2):345–353. https://doi.org/10.1002/jez.1402090216
5. Eppig JJ, Schroeder AC (1989) Capacity of mouse oocytes from preantral follicles to undergo embryogenesis and development to live young after growth, maturation, and fertilization in vitro. Biol Reprod 41(2):268–276. https://doi.org/10.1095/biolreprod41.2.268
6. Eppig JJ, Peters AH, Telfer EE, Wigglesworth K (1993) Production of cumulus expansion enabling factor by mouse oocytes grown in vitro: preliminary characterization of the factor. Mol Reprod Dev 34(4):450–456. https://doi.org/10.1002/mrd.1080340415
7. Eppig JJ, Telfer EE (1993) Isolation and culture of oocytes. Methods Enzymol 225:77–84. https://doi.org/10.1016/0076-6879(93)25008-p
8. Telfer EE, Grosbois J, Odey YL, Rosario R, Anderson RA (2023) Making a good egg: human oocyte health, aging and in vitro development. Physiol Rev. 103(4):2623–2677. https://doi.org/10.1152/physrev.00032.2022
9. Eppig JJ, O'Brien MJ (1996) Development in vitro of mouse oocytes from primordial follicles. Biol Reprod 54(1):197–207
10. O'Brien MJ, Pendola JK, Eppig JJ (2003) A revised protocol for in vitro development of mouse oocytes from primordial follicles dramatically improves their developmental competence. Biol Reprod 68(5):1682–1686. https://doi.org/10.1095/biolreprod.102.013029
11. Telfer EE, Andersen CY (2021) In vitro growth and maturation of primordial follicles and immature oocytes. Fertil Steril 115(5):1116–1125. https://doi.org/10.1016/j.fertnstert.2021.03.004
12. Telfer EE, Zelinski MB (2013) Ovarian follicle culture: advances and challenges for human and nonhuman primates. Fertil Steril 99(6):1523–1533. https://doi.org/10.1016/j.fertnstert.2013.03.043
13. Vitale F, Dolmans MM (2024) Comprehensive review of in vitro human follicle development for fertility restoration: recent achievements, current challenges, and future optimization strategies. J Clin Med 13(6):1791. https://doi.org/10.3390/jcm13061791
14. Xiao S, Zhang J, Romero MM, Smith KN, Shea LD, Woodruff TK (2015) In vitro follicle growth supports human oocyte meiotic maturation. Sci Rep 5:17323. https://doi.org/10.1038/srep17323

15. McLaughlin M, Albertini DF, Wallace WHB, Anderson RA, Telfer EE (2018) Metaphase II oocytes from human unilaminar follicles grown in a multi-step culture system. Mol Hum Reprod 24(3):135–142. https://doi.org/10.1093/molehr/gay002
16. Xu F, Lawson MS, Bean Y, Ting AY, Pejovic T, De Geest K, Moffitt M, Mitalipov SM, Xu J (2021) Matrix-free 3D culture supports human follicular development from the unilaminar to the antral stage in vitro yielding morphologically normal metaphase II oocytes. Hum Reprod (Oxford, England) 36(5):1326–1338. https://doi.org/10.1093/humrep/deab003
17. Grosbois J, Devos M, Demeestere I (2020) Implications of nonphysiological ovarian primordial follicle activation for fertility preservation. Endocr Rev 41(6):847–872. https://doi.org/10.1210/endrev/bnaa020
18. Abir R, Roizman P, Fisch B, Nitke S, Okon E, Orvieto R, Ben Rafael Z (1999) Pilot study of isolated early human follicles cultured in collagen gels for 24 hours. Hum Reprod 14(5):1299–1301. https://doi.org/10.1093/humrep/14.5.1299
19. Hovatta O, Wright C, Krausz T, Hardy K, Winston RM (1999) Human primordial, primary and secondary ovarian follicles in long-term culture: effect of partial isolation. Hum Reprod 14(10):2519–2524. https://doi.org/10.1093/humrep/14.10.2519
20. Abir R, Fisch B, Nitke S, Okon E, Raz A, Ben Rafael Z (2001) Morphological study of fully and partially isolated early human follicles. Fertil Steril 75(1):141–146. https://doi.org/10.1016/s0015-0282(00)01668-x
21. Hovatta O, Silye R, Abir R, Krausz T, Winston RM (1997) Extracellular matrix improves survival of both stored and fresh human primordial and primary ovarian follicles in long-term culture. Hum Reprod 12(5):1032–1036. https://doi.org/10.1093/humrep/12.5.1032
22. Telfer EE, McLaughlin M, Ding C, Thong KJ (2008) A two-step serum-free culture system supports development of human oocytes from primordial follicles in the presence of activin. Hum Reprod 23(5):1151–1158. https://doi.org/10.1093/humrep/den070
23. Wright CS, Hovatta O, Margara R, Trew G, Winston RM, Franks S, Hardy K (1999) Effects of follicle-stimulating hormone and serum substitution on the in-vitro growth of human ovarian follicles. Hum Reprod 14(6):1555–1562
24. Garor R, Abir R, Erman A, Felz C, Nitke S, Fisch B (2009) Effects of basic fibroblast growth factor on in vitro development of human ovarian primordial follicles. Fertil Steril 91(5 Suppl):1967–1975. https://doi.org/10.1016/j.fertnstert.2008.04.075
25. Grosbois J, Bailie EC, Kelsey TW, Anderson RA, Telfer EE (2023) Spatio-temporal remodelling of the composition and architecture of the human ovarian cortical extracellular matrix during in vitro culture. Hum Reprod 38(3):444–458. https://doi.org/10.1093/humrep/dead008
26. Baker SJ, Spears N (1999) The role of intra-ovarian interactions in the regulation of follicle dominance. Hum Reprod Update 5(2):153–165. https://doi.org/10.1093/humupd/5.2.153
27. Da Silva-Buttkus P, Marcelli G, Franks S, Stark J, Hardy K (2009) Inferring biological mechanisms from spatial analysis: prediction of a local inhibitor in the ovary. Proc Natl Acad Sci U S A 106(2):456–461. https://doi.org/10.1073/pnas.0810012106
28. Kawamura K, Cheng Y, Suzuki N, Deguchi M, Sato Y, Takae S, Ho CH, Kawamura N, Tamura M, Hashimoto S, Sugishita Y, Morimoto Y, Hosoi Y, Yoshioka N, Ishizuka B, Hsueh AJ (2013) Hippo signaling disruption and Akt stimulation of ovarian follicles for infertility treatment. Proc Natl Acad Sci U S A 110(43):17474–17479. https://doi.org/10.1073/pnas.1312830110
29. Grosbois J, Demeestere I (2018) Dynamics of PI3K and Hippo signaling pathways during in vitro human follicle activation. Hum Reprod 33(9):1705–1714. https://doi.org/10.1093/humrep/dey250
30. Fiorentino G, Cimadomo D, Innocenti F, Soscia D, Vaiarelli A, Ubaldi FM, Gennarelli G, Garagna S, Rienzi L, Zuccotti M (2022) Biomechanical forces and signals operating in the ovary during folliculogenesis and their dysregulation: implications for fertility. Hum Reprod Update 29(1):1–23. https://doi.org/10.1093/humupd/dmac031
31. Nagamatsu G, Shimamoto S, Hamazaki N, Nishimura Y, Hayashi K (2019) Mechanical stress accompanied with nuclear rotation is involved in the dormant state of mouse oocytes. Sci Adv 5(6):eaav9960. https://doi.org/10.1126/sciadv.aav9960
32. Shah JS, Sabouni R, Cayton Vaught KC, Owen CM, Albertini DF, Segars JH (2018) Biomechanics and mechanical signaling in the ovary: a systematic review. J Assist Reprod Genet 35(7):1135–1148. https://doi.org/10.1007/s10815-018-1180-y
33. Dolmans MM, Michaux N, Camboni A, Martinez Madrid B, Van Langendonckt A, Nottola SA, Donnez J (2006) Evaluation of Liberase, a purified enzyme blend, for the isolation of human primordial and primary ovarian follicles. Hum Reprod 21(2):413–420. https://doi.org/10.1093/humrep/dei320

34. Telfer EE, Binnie JP, McCaffery FH, Campbell BK (2000) In vitro development of oocytes from porcine and bovine primary follicles. Mol Cell Endocrinol 163(1-2):117–123
35. Kim EJ, Lee J, Youm HW, Kim SK, Lee JR, Suh CS, Kim SH (2018) Comparison of follicle isolation methods for mouse ovarian follicle culture in vitro. Reprod Sci (Thousand Oaks, Calif) 25(8):1270–1278. https://doi.org/10.1177/1933719117737851
36. McLaughlin M, Kinnell HL, Anderson RA, Telfer EE (2014) Inhibition of phosphatase and tensin homologue (PTEN) in human ovary in vitro results in increased activation of primordial follicles but compromises development of growing follicles. Mol Hum Reprod 20(8):736–744. https://doi.org/10.1093/molehr/gau037
37. Anderson RA, McLaughlin M, Wallace WH, Albertini DF, Telfer EE (2014) The immature human ovary shows loss of abnormal follicles and increasing follicle developmental competence through childhood and adolescence. Hum Reprod 29(1):97–106. https://doi.org/10.1093/humrep/det388
38. Shea LD, Woodruff TK, Shikanov A (2014) Bioengineering the ovarian follicle microenvironment. Annu Rev Biomed Eng 16:29–52. https://doi.org/10.1146/annurev-bioeng-071813-105131
39. Xu M, Barrett SL, West-Farrell E, Kondapalli LA, Kiesewetter SE, Shea LD, Woodruff TK (2009) In vitro grown human ovarian follicles from cancer patients support oocyte growth. Hum Reprod 24(10):2531–2540. https://doi.org/10.1093/humrep/dep228
40. Jones ASK, Shikanov A (2019) Follicle development as an orchestrated signaling network in a 3D organoid. J Biol Eng 13:2. https://doi.org/10.1186/s13036-018-0134-3
41. Jiao ZX, Woodruff TK (2013) Follicle microenvironment-associated alterations in gene expression in the mouse oocyte and its polar body. Fertil Steril 99(5):1453–1459.e1451. https://doi.org/10.1016/j.fertnstert.2012.12.009
42. Vanacker J, Amorim CA (2017) Alginate: a versatile biomaterial to encapsulate isolated ovarian follicles. Ann Biomed Eng 45(7):1633–1649. https://doi.org/10.1007/s10439-017-1816-6
43. Laronda MM, Jakus AE, Whelan KA, Wertheim JA, Shah RN, Woodruff TK (2015) Initiation of puberty in mice following decellularized ovary transplant. Biomaterials 50:20–29. https://doi.org/10.1016/j.biomaterials.2015.01.051
44. Liverani L, Raffel N, Fattahi A, Preis A, Hoffmann I, Boccaccini AR, Beckmann MW, Dittrich R (2019) Electrospun patterned porous scaffolds for the support of ovarian follicles growth: a feasibility study. Sci Rep 9(1):1150. https://doi.org/10.1038/s41598-018-37640-1
45. Dadashzadeh A, Moghassemi S, Shavandi A, Amorim CA (2021) A review on biomaterials for ovarian tissue engineering. Acta Biomater 135:48–63. https://doi.org/10.1016/j.actbio.2021.08.026
46. Smitz J, Sánchez F, Romero S, Van Ranst H, Anckaert E, Gilchrist RB, Ho TM, Vuong LN, Morimoto Y (2025) Human oocyte capacitation culture: essential step toward hormone-free assisted reproductive technology. Reprod Med Biol 24(1):e12640. https://doi.org/10.1002/rmb2.12640

Check for updates

Chapter 2

Capacitation In Vitro Maturation (CAPA-IVM) of Human Oocytes

Lan N. Vuong, Anh H. Le, Tuong M. Ho and Robert B. Gilchrist

Abstract

Oocyte in vitro maturation (IVM) is a patient-friendly technique that generates mature oocytes in vitro from patients who receive minimal or no gonadotrophin stimulation. Although it was discovered a long time ago, the clinical application of IVM has been constrained by its lower success rates compared with conventional IVF, principally due to suboptimal oocyte competency. Recent advancements in the culture system for immature oocytes have significantly enhanced both maturation efficiency and embryo quality. Consequently, this technique is experiencing a resurgence in interest among clinicians and embryologists globally. This chapter details the laboratory protocols employed in a novel form of IVM, called capacitation (CAPA) IVM.

Keywords In vitro maturation of oocyte, Biphasic IVM, Standard IVM

1 Introduction

In vitro maturation (IVM) was first performed on oocytes from rabbits in 1935 and on human oocytes in 1944 [1, 2]. Robert G. Edwards was the first to note that human oocytes could mature after being removed from the body [1]. Although IVM has been known for a long time, its application in human infertility treatment has been quite cautious and slow. It wasn't until 2021 that IVM was declared a routine and viable fertility treatment, rather than an experimental technique [3].

IVM involves collecting immature, germinal vesicle (GV)-stage cumulus-oocyte complexes (COCs) from antral follicles obtained from unstimulated or minimally stimulated ovaries. These intact COCs are then cultured in vitro until they reach the metaphase II (MII) stage [1]. Subsequently, IVM oocytes and any resulting embryos are treated the same as those generated by conventional ovarian stimulation. Thus, the main differences between IVM and conventional and in vitro fertilization (IVF) are the precollection

Alexandra J. Harvey and David K. Gardner (eds.), *The Preimplantation Embryo*, Methods in Molecular Biology, vol. 3038,
https://doi.org/10.1007/978-1-0716-5292-3_2,

preparation of the patient and the fact that oocyte meiotic maturation occurs in vitro rather than in vivo.

For IVM, immature COCs can be obtained from any stage of antral follicle development, ranging from 1 mm to approximately 20 mm. The ovarian follicular environment maintains oocyte meiotic arrest, so removing a COC and culturing it in vitro will lead to most oocytes spontaneously resuming meiosis, with approximately 50% of human oocytes reaching the MII stage without notable meiotic inducers [1].

Using standard IVM procedures, COCs are cultured in tissue culture-like medium supplemented with a protein source such as albumin or serum follicle-stimulating hormone (FSH), and possibly other additives like human chorionic gonadotropin (hCG), epidermal growth factor (EGF), estradiol, and/or cysteamine, under atmospheric oxygen for 1–2 days. Unfortunately, progress in IVM has been slow because these IVM culture conditions have largely remained unchanged from those originally described by Edwards over 50 years ago [1]. Many clinics still use this same method today. As a result, pregnancy rates after use of the simple standard IVM protocol are modest [4]. This is partly due to the fact that this type of IVM is non-physiological because human oocytes from follicles larger than ~6 mm mature spontaneously in vitro without the appropriate maternal signals that normally induce oocyte maturation in vivo [5]. Therefore, there has been a need for a more elaborate IVM system designed based on the latest knowledge about oocyte maturation physiology in vivo [6, 7].

The integrity of the COC as an intact functional unit is fundamental to the success of IVM. Cumulus cells are highly specialized ovarian somatic cells that support the growth, development, and maturation of the oocyte. Compared with the in vivo generation of mature oocytes by superovulation for IVF, the success of using oocytes matured by IVM is fundamentally limited by three core biological challenges: 1) oocytes from small antral follicles are not fully developmentally competent; 2) oocytes are prone to spontaneous precocious maturation once removed from the follicle; and 3) oocytes matured in vitro are not exposed to the full ovulatory cascade that induces oocyte maturation in vivo [8].

Recently, significant advances in our understanding about fundamental aspects of oocyte biology and ovulation from animal studies have led to novel approaches to IVM [9]. One notable development in IVM technology is the use of biphasic oocyte maturation approaches, such as IVM with a pre-maturation (pre-IVM) step, exemplified by capacitation IVM (CAPA-IVM) [10–12]. Biphasic IVM involves using the C-type natriuretic peptide (CNP) molecule to inhibit the maturation-promoting factor, maintaining oocyte meiotic arrest at the GV stage, for the pre-IVM stage that is the first phase of biphasic IVM. Pre-IVM allows for the synchronization of oocyte nuclear and cytoplasmic maturation, thereby capacitating the oocyte for development.

In this chapter, we will present the laboratory protocol for CAPA-IVM. Other variations on IVM also exist, most notably standard IVM and hCG-triggered IVM, but these will not be discussed in this chapter.

2 Materials

2.1 Equipment

1. IVF chamber (Crib).
2. IVF workstation equipped with two stereomicroscopes and a heated stage.
3. Test tube warmer.
4. Benchtop incubator.
5. Non-CO_2 dry incubator.
6. Incubator: mixed-gas incubator (37°C, 7% CO_2, 20% O_2) or CO_2 incubator (6% CO_2, air oxygen and nitrogen).

2.2 Consumables

1. Serological pipettor.
2. Pipettors.
3. 35-mm Petri dish.
4. 60-mm Petri dish.
5. 90-mm Petri dish.
6. 4-well dish.
7. Pasteur pipette.
8. 70-μm cell strainer.
9. Round-bottom tubes (14 mL).
10. 10-mL syringe.
11. Serological pipettes.
12. Sterile single-use tips (200 μL).
13. Sterile pipette tips (5000 μL).
14. Sterile metal rack.
15. Paraffin film.

2.3 Media

1. Basal buffer: HEPES medium such as Global Collect. Store at 2–8°C, avoid light, and use within 7 days after opening.
2. Medicult IVM medium (*see* **Note 1**). Store at 2–8°C.
3. Human serum albumin (HSA) stock solution (100 mg/mL; *see* **Note 2**).

4. C-type natriuretic peptide (CNP; *see* **Note 3**).
5. Estradiol (*see* **Note 4**).
6. Amphiregulin (rhAREG; *see* **Note 5**).
7. Insulin.
8. FSH (*see* **Note 6**).
9. Hyaluronidase stock solution (5 mg/mL).
10. Polyvinylpyrrolidone (PVP) stock solution (5 mg/mL).
11. IVF and embryo culture medium.
12. Liquid paraffin. Store at 2–8°C, avoid light. Purified and sterile.

2.4 Media Preparation for Oocyte Collection, Searching, and IVM

Medium preparation is undertaken in an IVF workstation or biosafety cabinet to maintain sterility. The volumes outlined below represent requirements per patient.

1. Oocyte collection base medium: basal buffer supplemented with 5 mg/mL HSA. In a 90-mm Petri dish, add 2.5 mL HSA to 47.5 mL of basal buffer and mix well (*see* **Note 7**).
2. Retrieval medium: oocyte collection base medium supplemented with 20 nM estradiol and 50 nM CNP. Transfer all the retrieval medium into 3 × 14-mL round-bottom tubes labeled "Retrieval" and equilibrate at 37°C.
3. Searching medium: oocyte collection base with 10 nM estradiol and 25 nM CNP. Transfer all the searching medium into two 14-mL round-bottom tubes labeled "Searching".
4. Store the rinse dishes, the searching tube, the retrieval tubes, and the tubes marked at the 2 mL level in the non-CO_2 incubator.
5. IVM basal medium: Medicult supplemented with 5 ng/mL insulin, 10 nM estradiol, 5 mg/mL HSA. Store at 4°C (*see* **Notes 2 and 4**).
6. CAPA medium: basal IVM medium with 1 mIU/mL rFSH and 25 nM CNP (*see* **Notes 3 and 6**).
7. IVM medium: basal IVM medium with 100 mIU/mL rFSH and 100 ng/mL rhAREG (*see* **Notes 5 and 6**).

3 Methods

CAPA-IVM is a 2-day biphasic IVM procedure consisting of two steps: (1) a pre-IVM culture step of 22–26 h, and (2) an IVM culture step of 30–32 h [11]. A flowchart of the CAPA-IVM process by day is shown in Fig. 1 (*see* **Note 8**).

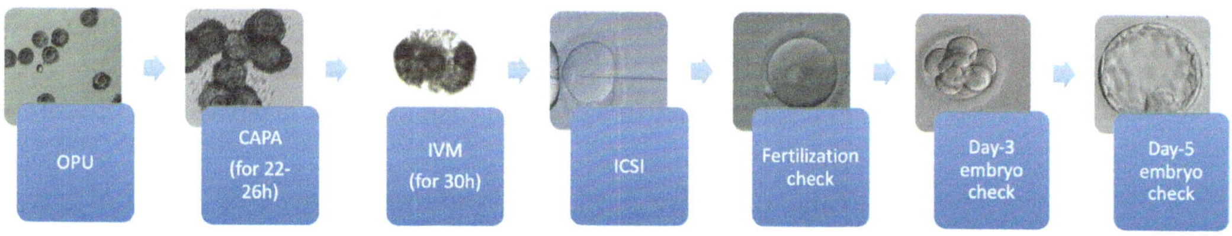

Fig. 1 Flow chart of the capacitation in vitro maturation (CAPA-IVM) process

3.1 Medium Preparation and Immature Oocyte Pick-Up

1. The day before oocyte pick-up, prepare retrieval, searching, and CAPA media. Transfer remaining IVM basal medium into a 14-mL round-bottom tube (labeled "Basal"). Wrap paraffin film around the tube cap of the remaining IVM basal medium and store it in a refrigerator at 2–8°C for preparation of IVM the following day (*see* **Note 9**).
2. Prepare two plates of CAPA medium. Using a 1-mL pipette, add 500 μL of the freshly prepared CAPA medium to each well of a pre prepared 4-well plate (CAPA culture dish). Cover the medium with 0.5 mL oil immediately. In a second 4-well plate, pipette 700 μL of CAPA medium to each well (CAPA washing dish). Cover the medium with 0.5 mL oil immediately. Place the two CAPA IVM plates in the mini incubator for at least 12 h or overnight.
3. On the morning of the aspiration, transfer 1 mL of pre-warmed retrieval medium into 10 × 14-mL tubes marked at the 2 mL line and place in a non-CO_2 dry incubator (*see* **Note 10**).
4. Transfer 2.5 mL of pre-warmed searching medium into four 35-mm dishes and place them in a non-CO_2 incubator. The remaining searching medium in the tube will be used to rinse the follicular fluid tubes and the strainer when searching for immature oocytes.
5. Provide the midwife with one 14-mL tube containing 10 mL Global Collect medium for the doctor to rinse the needle and 10 × 14-mL tubes marked at 2 mL for oocyte retrieval.
6. In the theater, oocytes are aspirated from all visible antral follicles (*see* **Note 11**), with aspirates collected into 14-mL tubes containing 1 mL retrieval medium (*see* **Note 12**). Once the 14-mL tube containing follicular fluid aspirate is received from the midwife, perform identification according to the patient identification procedure.

3.2 Searching for Immature Oocytes

COC searching and processing is performed at the IVF workstation equipped with a heated stage (*see* **Note 13**).

1. Place a cell strainer into a 90-mm dish, pour the searching medium into a 60-mm dish, rinse a 10-mL syringe with the searching medium, then create an air gap before drawing the

medium into the syringe (*see* **Note 14**). Place the syringe into a clean 60-mm dish until required to maintain sterility.

2. Place a strainer in a 60-mm dish and wet it with 1 mL of searching medium.
3. Take a 35-mm petri dish containing searching medium out of the incubator and confirm the patient's identity according to the patient identification procedure.
4. Evenly divide the follicular fluid from the 14-mL tube into the four prepared 60-mm petri dishes. Rinse the follicular fluid tube with the syringe of 0.3 mL searching medium.
5. Working quickly, use a stereomicroscope to identify the presence of the COC masses (Fig. 2). Being careful not to damage the oocyte's cumulus cells, use a micropipette to aspirate the COC mass and transfer it to an oocyte retrieval dish, then wash the COC mass into another oocyte retrieval dish. After the search is complete, aspirate the follicular fluid from the 60-mm Petri dishes and transfer it to the centre of the cell strainer. Repeat the searching process for each follicular fluid tube, using the same cell strainer for all aspirates for one patient.
6. After all follicular fluid tubes have been searched, rinse the inner surface of the strainer, then invert the strainer and flush the inner and outer walls of the strainer with medium. Transfer the strainer to another dish and flush the remaining medium into the centre of the strainer. Check the medium after rinsing the strainer to identify any further immature oocytes that may have been missed during the initial search (*see* **Note 15**).
7. Transfer any remaining COC masses to a 35-mm dish containing search medium.

3.3 Pre-maturation Culture (Pre-IVM)

Sanitize hands and clean the IVF chamber area before starting and verify the patient's identity according to the patient identification

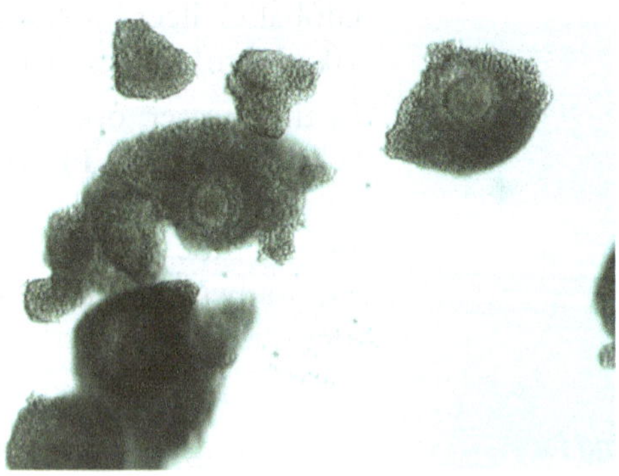

Fig. 2 Cumulus-oocyte complexes (COCs) at oocyte pick-up before capacitation in vitro maturation. Integrity of the COC is important to the success of in vitro maturation

protocol. Record the start time of CAPA culture on relevant paperwork.

1. Aspirate the COCs using a 10-μL pipette and transfer them into a well of one of the CAPA wash dishes (used to wash the oocytes after collection; *see* **Note 16**). The oocytes should be gently aspirated up and down two to three times at different positions in the well, then washed several times in three different wash wells before being placed in the culture well of the CAPA dish in order from well one onwards (*see* **Note 17**). Each culture well should contain up to ten COCs (Fig. 2).
2. Check the search plates and the tips used for washing to ensure no COCs are left behind.
3. Place the CAPA pre-IVM dish in the mini incubator in the correct case position.
4. The CAPA culture time is 22–26 h in the mini incubator at 37°C, 7% CO_2, and 20% O_2.

3.4 IVM Culture

Sanitize hands and clean the IVF chamber area before starting. Verify the patient's identity according to the patient identification protocol.

1. On the day of oocyte pick-up, prepare IVM medium, then prepare IVM culture plates by pipetting 500 μL of the freshly prepared IVM medium into each well of a 4-well plate (IVM culture dish). Cover the medium with 0.5 mL oil per well immediately. In a second 4-well plate, pipette 700 μL of IVM medium to each well (IVM washing dish). Cover the medium with oil immediately. Equilibrate the IVM medium plates in the mini incubator for at least 12 h before use.
2. After 22–26 h of CAPA pre-IVM culture, take the CAPA dish, the IVM dish, and the IVM wash dish (used to wash the COCs when transferring them from CAPA dish to IVM dish) set from the mini incubator and place them in the IVF chamber/crib.
3. Using a 10 μL pipette, transfer the COCs from the CAPA culture dish to the first well of the IVM wash dish (*see* **Note 18**, Fig. 3). The oocytes should be gently aspirated up and down two to three times at different positions in the well, then washed several times in three different wells. Then, transfer the COCs to the IVM culture dish corresponding to the COC clusters cultured in the CAPA culture well.
4. After culturing, before disposal, check the CAPA and IVM wash dishes to ensure that no COCs are left behind.
5. Culture the COCs for 30–32 h in the mini incubator at 37°C, 7% CO_2, and 20% O_2.
6. After 30 h of oocyte maturation culture, MII oocytes generated by CAPA-IVM can be frozen or fertilized by IVF

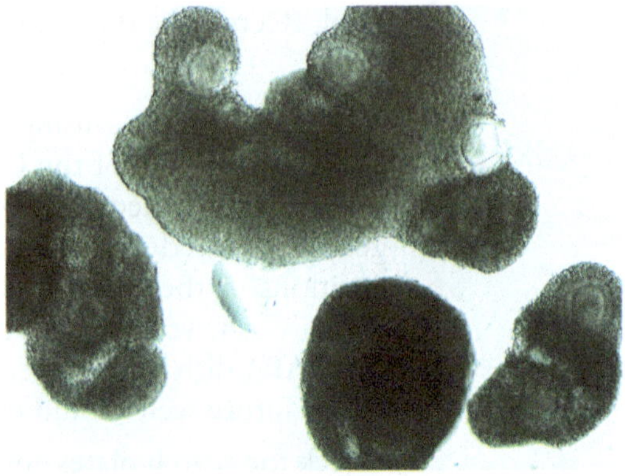

Fig. 3 Cumulus-oocyte complexes (COCs) after pre-maturation (capacitation) culture during capacitation in vitro maturation. Note, there is no cumulus expansion after the pre-IVM phase

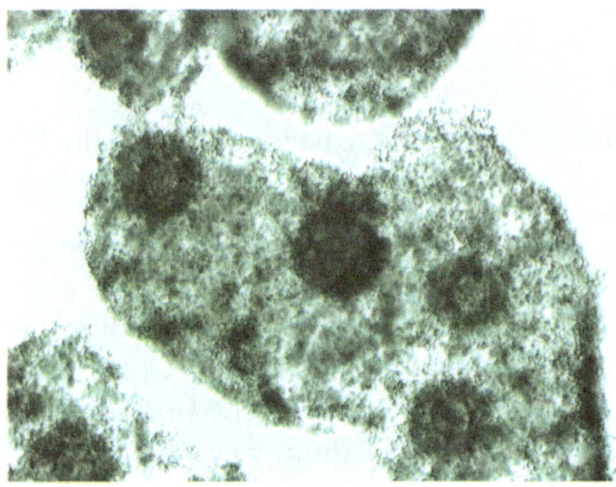

Fig. 4 Cumulus-oocyte complexes (COCs) after the in vitro maturation phase of capacitation in vitro maturation

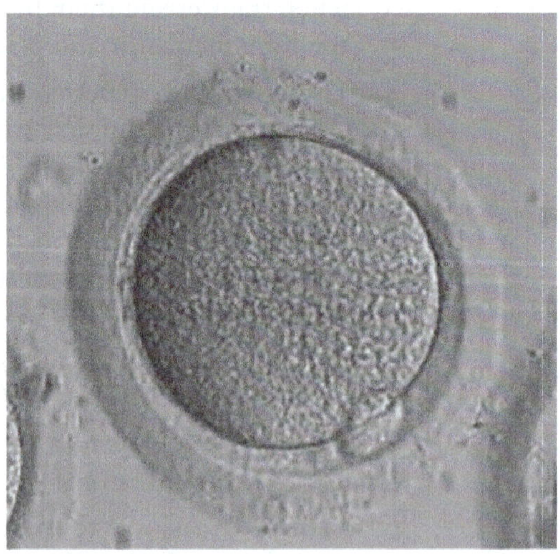

Fig. 5 Denuded mature oocyte after capacitation in vitro maturation

or ICSI (*see* **Note 11**, Figs. 4 and 5), using the same decision criteria as used in conventional IVF procedures (*see* **Note 19**).

4 Notes

1. Medicult IVM medium is currently the preferred IVM base medium when using the CAPA-IVM protocol. This is because all preclinical safety studies and clinical efficacy studies have been conducted using this medium. No other IVM base medium has yet been tested in the CAPA-IVM system. Although, in theory, other IVM base medias may be suitable, there is currently no safety data for these.
2. HSA quality: an ART grade HSA should be used.
3. CNP is a central component of the CAPA-IVM protocol because it is the meiotic inhibitor that is used to intentionally keep the oocyte meiotically arrested in an immature (GV) state during the pre-IVM culture step [10, 11] (Fig. 1). CNP is included in the retrieval, search and CAPA (pre-IVM) media.
4. Estradiol is an important additive used to maintain cumulus cell expression of natriuretic peptide receptor 2 (NPR2), which is the receptor for CNP. Hence, CNP is ineffective as a meiotic inhibitor without the addition of estradiol [13]. Estradiol is included in the retrieval, search, and CAPA (pre-IVM) media.
5. Amphiregulin (AREG) is a natural meiotic inducer because it is one of the three EGF-like peptides expressed by granulosa cells and cumulus cells in response the LH-induced ovulatory cascade [14]. Hence, AREG is added to the second IVM step as the principal inducer of oocyte maturation.
6. FSH is a vital hormone for normal granulosa and cumulus cell function. FSH is used at a low dose of 1 mIU/mL in the retrieval, search, and CAPA (pre-IVM) media to maintain cumulus cell viability and proliferation, whereas FSH is used a high dose of 100 mIU/mL in the IVM medium to promote oocyte meiotic resumption and cumulus expansion in conjunction with AREG.
7. Oocyte retrieval and searching media, along with pre-IVM media, should be prepared the day before oocyte collection and allowed to equilibrate overnight. IVM medium should be prepared on the day of oocyte pick-up and equilibrated overnight.

8. Note that CAPA-IVM differs considerably from traditional standard IVM, in that standard IVM is a 1-day monophasic IVM procedure consisting of a single culture step of 30–32 h, whereas CAPA-IVM is a two-step protocol requiring 2 full days of oocyte culture before egg freezing or fertilization.
9. Basal medium is common to both pre-IVM and IVM culture steps, so both steps of the CAPA system contain HSA, insulin, and estradiol. Sufficient basal medium is prepared to ensure both pre-IVM and IVM culture use the same batch of medium.
10. Tubes are marked to indicate 2 mL to ensure that 1 mL of follicular fluid is mixed with 1 mL of medium containing CNP. This shortens the time from oocyte aspiration to searching and washing oocytes. Therefore, while tubes containing the retrieval medium are changed more frequently, this reduces the time the oocytes are exposed to the external environment after aspiration.
11. Although all visible antral follicles are aspirated, the CAPA-IVM system is particularly well suited to oocytes in smaller, more numerous follicles in the 4–12 mm range. The less numerous follicles with a diameter of >12 mm are also aspirated, but the COC recovery rate/follicle aspirated is lower from these follicles when patients do not receive an hCG-trigger.
12. It is important to use supplemented retrieval medium during OPU to prevent spontaneous oocyte meiotic resumption initiating in theater.
13. Given that searching for immature oocytes (compact COCs) takes longer than searching for IVF eggs, it is important that a heated stage is used to prevent oocyte cooling.
14. Creating a gap in the syringe ensures that all medium is pushed out and fully utilized.
15. Use a micropipette to gently press the bottom of the strainer to allow the medium and any remaining oocytes to rise to the surface and/or gradually aspirate the medium in the strainer and transfer it to the medium surrounding the strainer to check for the presence of any remaining COC masses.
16. Only culture oocytes that have cumulus cells surrounding them because the CAPA-IVM system relies on the presence of cumulus cells to modulate oocyte maturation in response to the signaling molecules present in the CAPA-IVM medium. Discard denuded and degenerated oocytes.
17. Handle COCs gently to avoid breaking the surrounding cumulus cells because it is important that the cumulus remain intact to maintain signaling to the oocyte to prevent spontaneous nuclear maturation.

18. At the end of CAPA pre-IVM culture (i.e., Day 1), COCs are expected to be compacted (Fig. 3), whereas after IVM culture (Day 2) they are expanded (Fig. 4), although less expanded than oocytes matured in vivo in conventional IVF cycles.
19. Maturing oocytes (metaphase I stage) can be cultured further and rechecked after 2 h (if the maturation rate is below 30%) (Fig. 5).

References

1. Edwards RG (1965) Maturation in vitro of mouse, sheep, cow, pig, rhesus monkey and human ovarian oocytes. Nature 208(5008):349–351. https://doi.org/10.1038/208349a0
2. Rock J, Menkin MF (1944) In vitro fertilization and cleavage of human ovarian eggs. Science 100:105–107
3. Practice Committees of the American Society for Reproductive Medicine, the Society of Reproductive Biologists and Technologists, and the Society for Assisted Reproductive Technology (2021) In vitro maturation: a committee opinion. Fertil Steril 115(2):298–304. https://doi.org/10.1016/j.fertnstert.2020.11.018
4. Zheng X, Guo W, Zeng L, Zheng D, Yang S, Xu Y, Wang L, Wang R, Mol BW, Li R, Qiao J (2022) In vitro maturation without gonadotropins versus in vitro fertilization with hyperstimulation in women with polycystic ovary syndrome: a non-inferiority randomized controlled trial. Hum Reprod 37(2):242–253. https://doi.org/10.1093/humrep/deab243
5. Richani D, Gilchrist RB (2018) The epidermal growth factor network: role in oocyte growth, maturation and developmental competence. Hum Reprod Update 24(1):1–14. https://doi.org/10.1093/humupd/dmx029
6. Gilchrist RB (2011) Recent insights into oocyte-follicle cell interactions provide opportunities for the development of new approaches to in vitro maturation. Reprod Fertil Dev 23(1):23–31. https://doi.org/10.1071/rd10225
7. Norris RP, Ratzan WJ, Freudzon M, Mehlmann LM, Krall J, Movsesian MA, Wang H, Ke H, Nikolaev VO, Jaffe LA (2009) Cyclic GMP from the surrounding somatic cells regulates cyclic AMP and meiosis in the mouse oocyte. Development 136(11):1869–1878. https://doi.org/10.1242/dev.035238
8. Gilchrist RB, Smitz J (2023) Oocyte in vitro maturation: physiological basis and application to clinical practice. Fertil Steril 119(4):524–539. https://doi.org/10.1016/j.fertnstert.2023.02.010
9. Gilchrist RB, Ho TM, De Vos M, Sanchez F, Romero S, Ledger WL, Anckaert E, Vuong LN, Smitz J (2024) A fresh start for IVM: capacitating the oocyte for development using pre-IVM. Hum Reprod Update 30(1):3–25. https://doi.org/10.1093/humupd/dmad023
10. Romero S, Sánchez F, Lolicato F, Van Ranst H, Smitz J (2016) Immature oocytes from unprimed juvenile mice become a valuable source for embryo production when using C-type natriuretic peptide as essential component of culture medium. Biol Reprod 95(3):64. https://doi.org/10.1095/biolreprod.116.139808
11. Sánchez F, Lolicato F, Romero S, De Vos M, Van Ranst H, Verheyen G, Anckaert E, Smitz JEJ (2017) An improved IVM method for cumulus-oocyte complexes from small follicles in polycystic ovary syndrome patients enhances oocyte competence and embryo yield. Hum Reprod 32(10):2056–2068. https://doi.org/10.1093/humrep/dex262
12. Vuong LN, Ho VNA, Ho TM, Dang VQ, Phung TH, Giang NH, Le AH, Pham TD, Wang R, Smitz J, Gilchrist RB, Norman RJ, Mol BW (2020) In-vitro maturation of oocytes versus conventional IVF in women with infertility and a high antral follicle count: a randomized non-inferiority controlled trial. Hum Reprod 35(11):2537–2547. https://doi.org/10.1093/humrep/deaa240
13. Zhang M, Su YQ, Sugiura K, Wigglesworth K, Xia G, Eppig JJ (2011) Estradiol promotes and maintains cumulus cell expression of natriuretic peptide receptor 2 (NPR2) and meiotic arrest in mouse oocytes in vitro. Endocrinology 152(11):4377–4385. https://doi.org/10.1210/en.2011-1118
14. Park JY, Su YQ, Ariga M, Law E, Jin SL, Conti M (2004) EGF-like growth factors as mediators of LH action in the ovulatory follicle. Science 303(5658):682–684. https://doi.org/10.1126/science.1092463

Check for updates

Chapter 3

Sperm Preparation in ART: Traditional Methods and Emerging Technologies

Constantine S. Velmahos and Charles L. Bormann

Abstract

Sperm preparation techniques are essential for enhancing fertilization potential by isolating motile, viable sperm while removing harmful components such as debris, reactive oxygen species (ROS), and immotile sperm. These methods are fundamental to assisted reproductive technologies (ART), optimizing outcomes in procedures like intrauterine insemination (IUI) and in vitro fertilization (IVF). Traditionally, sperm processing has been categorized into three primary techniques: density gradient centrifugation, swim-up, and simple wash. This chapter provides detailed protocols for these classic methods while exploring advanced techniques such as microfluidics, which offer improved precision and efficiency in sperm selection for ART applications.

Keywords Sperm processing, Density gradient centrifugation, Swim-up technique, Simple wash, Microfluidics, Assisted reproductive technologies (ART), Sperm DNA integrity

1 Introduction

As sperm traverse the female reproductive tract during natural selection, motile sperm are naturally separated from immotile sperm, cell debris, reactive oxygen species, and seminal plasma. This process ensures the optimal preparation of sperm for oocyte penetration and fertilization by inducing a series of morphological changes, collectively referred to as "capacitation" [1, 2].

The advent of assisted reproductive technologies (ART) has revolutionized fertility treatments by introducing numerous in vitro techniques that replicate in vivo fertilization conditions. One critical component of ART is sperm preparation, which isolates motile sperm from immotile sperm and removes other detrimental components of semen that can impair fertilization potential.

Sperm processing techniques have traditionally been grouped into three primary categories: swim-up, simple wash, and density gradient centrifugation. These methods differ in their approach

Alexandra J. Harvey and David K. Gardner (eds.), *The Preimplantation Embryo*, Methods in Molecular Biology, vol. 3038,
https://doi.org/10.1007/978-1-0716-5292-3_3,

to migration, centrifugation, medium utilization, and the yield of viable sperm. Emerging biomimetic techniques like filtration columns, microfluidics, and barrier-based methods are reshaping reproductive medicine by offering more precise, efficient, and selective sperm isolation strategies. Microfluidic approaches, for e.g., utilize microscale fluid channels to replicate natural reproductive environments, enabling the isolation of motile, morphologically normal sperm.

This chapter will detail the laboratory protocols and methodologies for established and novel sperm processing techniques, providing a comprehensive guide to current best practices in ART.

2 Materials

2.1 Equipment and Consumables

1. Microscope (equipped with phase contrast optics: 10×, 20×, and 40×).
2. Centrifuge with swing bucket rotors.
3. CO_2 incubator at 37°C.
4. Heated dry block.
5. Sterile serological pipettor and pipettes.
6. Glass Pasteur pipettes and bulb.
7. Glass Pasteur pipettes with syringe and tubing to allow for volumetric control.
8. Mechanical pipettors and pipette tips.
9. 15 mL conical centrifuge tubes.
10. Round-bottom polystyrene 7-mL and 14-mL culture tubes.
11. Conical-bottom polystyrene 15-mL centrifuge tubes.
12. Test tube rack.

2.2 Density Gradient, Swim-Up, and Combined DGC+ Swim-Up Media

1. Sperm processing medium, such as ISolate, prepared to 50% and 90% concentrations (*see* **Note 1**).
2. Commercially available synthetic Human Tubal Fluid (HTF), such as InVitroCare® HTF Culture medium with EDTA and glutamine, or Multipurpose Handling Medium (MHM), such as Fujifilm MHM-c dual-buffered medium with HEPES, MOPS, and gentamicin.
3. Sperm Washing Medium (SWM), such as Fujifilm Modified HEPES-buffered HTF with human serum albumin (5.0 mg/mL).
4. Bicarbonate-buffered salt medium, such as G-RINSE (*see* **Note 2**).
5. IVF media, such as Gx-IVF, pre-equilibrated overnight at 37°C in a 6% CO_2 incubator.

2.3 Simple Wash Media

1. Embryo culture medium, such as Biggers, Whitten, and Whittingham (BWW), Earle's balanced salt solution (EBSS), human tubal factor or any other tested and appropriate media supplemented with human serum albumin (HSA) or serum.
2. Highly purified HSA, which is free from viral, bacterial, prion contaminants, and endotoxins.
3. Heat inactivated patient serum: following blood collection and centrifugation for serum isolation, transfer the serum to a new vial. Heat to 56°C for 20 min.
4. Sperm wash medium, (6 mg/mL HSA, 0.27 mM Na pyruvate, 19 mM Na lactate, 23.81 mM NaHCO3): add 300 mg of HSA, 1.5 mg of sodium pyruvate, 180 µL of 60% (v/v) sodium lactate syrup, and 100 mg of sodium bicarbonate to 50 mL of culture medium (e.g., human tubal fluid or multipurpose handling medium). If a serum supplement is required (for nonproprietary media), prepare by adding 4 mL of heat-inactivated patient serum to 46 mL of medium, along with 1.5 mg of sodium pyruvate, 0.18 mL of sodium lactate, and 100 mg of sodium bicarbonate [3].

2.4 Microfluidic Sperm Separation

1. Microfluidic sperm separation device. This method outlines the procedure for separating sperm using the single-use Zymot ICSI Sperm Separation Device.
2. Sperm Washing Medium (SWM), such as Fujifilm Modified HEPES-buffered HTF with human serum albumin (5.0 mg/mL).

3 Methods

The choice of sperm preparation technique depends largely on the quality of the semen sample and the intended application [4, 5]. For ART procedures such as in vitro fertilization (IVF) or intracytoplasmic sperm injection (ICSI), the technique must yield sperm with minimal DNA damage [6]. An ideal preparation method isolates functional sperm while maintaining DNA integrity and minimizing oxidative stress from ROS produced by sperm or leukocytes [7–9]. For intrauterine insemination (IUI) with normozoospermic samples, diluting semen with culture media followed by centrifugation is often effective [10]. For instance, the direct swim up method is highly suitable for selecting motile sperm due to its reliance on sperm's intrinsic motility. Both direct and pellet swim-up methods have been shown to reduce DNA fragmentation, although the centrifugation step in pellet swim-up may affect other sperm functions [11]. Direct swim-up is typically reserved for samples with normal parameters, while density gradient centrifugation is favored for samples with significant abnormalities such as severe oligozoospermia, teratozoospermia, or asthenozoospermia due to its higher yield of motile sperm. Modifications to density gradient parameters—such as adjusting gradient volume or centrifugation duration—can further optimize recovery for specific sample characteristics.

To accommodate the inherent variation in semen quality and sperm characteristics between patient samples, this chapter outlines each of these methods.

3.1 Density Gradient Centrifugation (DGC)

Density gradient centrifugation is a highly effective and versatile technique for isolating high-quality sperm for assisted reproductive technologies (ART). By selecting motile sperm while efficiently removing debris and nonsperm cells, this method provides consistent and reliable results, making it a preferred choice for preparing sperm for in vitro fertilization (IVF) and intracytoplasmic sperm injection (ICSI). Consequently, the density gradient technique is ideal for processing poor-quality semen samples with high debris, abnormal morphology, or multiple pathologies, as it effectively isolates motile, morphologically normal sperm while removing debris and nonsperm cells. Furthermore, density gradient centrifugation is easier to standardize, ensuring reproducibility across different samples and laboratories.

This technique involves layering liquefied semen over density gradients composed of colloidal silica coated with silane, which separates cells based on their density. The most widely used approach is a two-step discontinuous density-gradient centrifugation, where sperm with optimal motility and morphology migrate to the bottom layer, free from contaminants such as debris, leukocytes, nongerm cells, and degenerating sperm cells.

Density gradient centrifugation offers several advantages, particularly for patients with suboptimal semen parameters, such as oligozoospermia or asthenozoospermia. By processing the entire ejaculate, this method achieves a significantly higher yield of motile sperm compared to alternative techniques. Moreover, the sperm retrieved through density gradient centrifugation tend to have a higher percentage of normal morphology, superior DNA integrity, and improved chromatin packaging [12, 13]. This process also generates cleaner samples with minimal cellular debris and leukocyte contamination, thereby reducing the levels of reactive oxygen species and associated complications [9].

Despite its benefits, density gradient centrifugation has limitations. The technique can be more costly than simpler methods, such as swim-up, and often requires multiple centrifugation steps, which may increase the handling time and complexity of the procedure. Nonetheless, its ability to consistently produce high-quality sperm samples with improved genetic and functional attributes makes it an indispensable tool in ART.

1. Check that the 50% and 90% gradient ISolate and SWM (or MHM) are all at 37°C or room temperature.
2. Once collected, semen samples should be allowed to liquify for 20–30 min by incubating them at 37°C or room temperature.
3. Using a sterile glass Pasteur pipette, transfer 1.5–2 mL of the 90% gradient ISolate into a sterile disposable 15 mL conical centrifuge tube (*see* **Note 3**).
   i. If pursuing one-layer gradient separation, proceed directly to Step 4 without overlaying the second gradient solution (*see* **Note 4**).
4. With a new sterile glass Pasteur pipette, gradually transfer 1.5–2 mL of the 50% gradient ISolate on top of the 90% gradient Isolate (*see* **Note 5**).
5. Using a 5 or 10 mL serological pipette, mix the liquified semen by gently pipetting up and down, taking great care to avoid producing any bubbles.
6. Using a new, sterile glass Pasteur pipette, carefully release 1.5–2.0 mL of liquified semen on top of the 50% gradient layer (*see* **Note 6**).
7. Place the tubes in the centrifuge and spin for 20 min at 200–300 × *g* (*see* **Note 7**). Avoid using the brake and confirm that the centrifuge is appropriately calibrated before use.
8. Using a clean 5 mL serological pipette, insert the tip below the surface of each layer to remove each individual layer in sequence (from top to bottom), without disturbing the pellet (*see* **Note 7**).
9. Wash the sperm pellet by transferring the sperm pellet to a new sterile tube using a new glass Pasteur pipette. Resuspend the pellet in 2–3 mL of MHM or SWM.
10. Centrifuge at 200–300 × *g* for 10 min. If the pellet is not visible, continue to centrifuge the sample at 500 × *g* for 10 min.
11. Once the pellet is visible, aspirate the entirety of the supernatant utilizing a new, sterile glass Pasteur pipette being careful not to disturb the pellet. Discard the supernatant.
12. Repeat the wash step once more.
13. If the sperm will be utilized for IUI, remove the supernatant and resuspend the pellet in 0.25–0.5 mL in SWM or MHM, and keep the tube at 37°C. If the sperm will be utilized for IVF or ICSI, resuspend the pellet in Human Tubal Fluid and prepare a dilution of 75,000–175,000/mL motile sperm, and keep the tube in a CO_2 incubator until it is ready for use.

3.2 Simple Wash

The simple wash method is a widely used sperm preparation technique that is particularly valued for its cost-effectiveness, simplicity, and ability to quickly process samples (*see* **Note 8**).

Recent research, including the findings of a retrospective single-center cohort study by Vagios et al., highlights that the simple wash technique can be effectively utilized for intrauterine insemination (IUI) cycles without compromising clinical pregnancy or live birth outcomes when compared to density gradient centrifugation. This study demonstrated equivalent clinical efficacy between the two techniques, even for patients with varying semen qualities and ovulation induction regimens [14].

Given its efficiency and lower cost, the simple wash method remains an appealing option for processing high-quality semen samples, particularly in resource-constrained settings or situations where quick processing is necessary. However, its limitations, including the retention of contaminants, must be carefully considered, and the method should be selected based on the specific requirements of each clinical case.

1. Warm sperm washing medium to 37°C prior to use.
2. Once collected, semen samples should be allowed to liquify for 20–30 min by incubating them at 37°C or room temperature.
3. Add sperm washing medium directly to the liquefied semen in the specimen cup to achieve a minimum 2:1 ratio of wash medium to the sample volume. Utilizing a pipettor and a pipette tip, pipette the mixture up and down several times, gently while thoroughly mixing the sample and medium.
4. Transfer the mixed sample into labeled 15 mL conical centrifuge tubes, ensuring no tube is filled beyond 75% capacity.
5. Allow the sample to settle for approximately 30 s. Using a glass Pasteur pipette, carefully remove any sediment (e.g., gelatinous clumps) from the bottom of the tube to prevent clogging during centrifugation.
6. Centrifuge the sample at 600 × *g* for 10 min. Ensure the centrifuge is balanced and calibrated before processing.
7. After centrifugation, separate the sperm pellet from the supernatant by carefully aspirating the supernatant using a sterile glass Pasteur pipette without disturbing the pellet (*see* **Note 9**).
8. Transfer the sperm pellet to a clean, sterile tube. Resuspend it in prewarmed SWM to a final volume of 0.3–0.5 mL, depending on the intended application (*see* **Note 10**).
9. Place the processed sample into a room-temperature holding incubator until use.

3.3 Swim-Up

The swim-up technique is a widely used method to isolate motile spermatozoa based on their ability to swim into a culture medium, away from seminal plasma and contaminants. This approach is particularly effective for selecting functional sperm with progressive motility, a critical factor in assisted reproductive technologies (*see* **Note 11**).

The swim-up technique is best suited for high-quality samples with good motility and low debris, particularly in cases without male factor infertility. While density gradient centrifugation offers higher recovery rates and works well with challenging samples, the swim-up method is simpler, faster, and more cost-effective but less effective for compromised semen quality.

The swim-up procedure involves layering liquefied semen with culture medium to enable highly motile sperm to swim into the upper layer, with the sample incubated at an angle to maximize the interface surface area. This technique eliminates the need for centrifugation, thereby reducing ROS production and preserving sperm quality. It yields a population of motile and morphologically normal sperm free from nongerm cells or debris. However, this method may recover fewer motile sperm than other approaches and requires careful monitoring to prevent prolonged exposure to seminal plasma, which can negatively affect sperm quality [15, 16]. The swim-up technique can also be performed on pelleted samples (*see* **Note 12**). While this method has been successfully utilized in many clinical IVF programs, particularly in cases without male factor infertility, it is less favored in modern practice. This shift is due to the established risk of impairing sperm fertilizing potential, which is particularly concerning for couples facing significant male factor infertility. Employing techniques that involve steps known to cause irreversible damage to sperm function is not consistent with responsible laboratory management, as such damage could compromise the likelihood of a successful IVF outcome. Consequently, more refined methods that minimize harm to spermatozoa are recommended to ensure optimal results. However, familiarity with both variations of the swim-up technique remains essential in ART laboratories.

3.3.1 Swim-Up from a Non-pelleted Sample

1. Once collected, semen samples should be allowed to liquify for 20–30 min by incubating them at 37°C or room temperature.
2. Add 2 mL of culture sperm medium (Gx-IVF) to a round-bottomed tube.
3. Carefully add 1 mL of the semen sample underneath the culture medium using a sterile glass pasteur pipette without mixing (*see* **Note 13**).
4. Tilt the tube at a 45° angle to increase the surface area between the semen sample and the culture medium (*see* **Note 14**).
5. Incubate the tube at 37°C with 6% CO_2 for 30–60 min to allow motile sperm to swim into the upper medium layer.
6. After incubation, carefully collect 1 mL of the supernatant (containing motile sperm) into a new tube for further experimentation or analysis (*see* **Note 15**).

3.3.2 Swim-Up from a Pelleted Sample

1. Once collected, semen samples should be allowed to liquify for 20–30 min by incubating them at 37°C or room temperature.

2. Transfer 1 mL of semen into a new tube (*see* **Note 2**) using a sterile serological pipette. If the semen sample is greater than 1 mL, pipette the remaining sample into multiple new, rinsed tubes, with no more than 1 mL in each tube.
3. With care not to disturb the layer underneath, pipette 2 mL of pre-equilibrated IVF medium (e.g., Gx-IVF) over the semen sample (*see* **Note 13**). Repeat this step for every tube and then incubate the tubes at 37°C at 6% CO_2 for 30–60 min.
4. Taking care not to disturb the lower semen sample layer, transfer the top layer of IVF medium with a clean glass Pasteur pipette into a new, rinsed tube. In the new tube, add 5 mL of pre-equilibrated IVF medium and mix the samples by gentle pipetting. Proceed to centrifuge the tubes at 300–600 × *g* for 10 min. Repeat for all tubes that have been layered with IVF medium (*see* **Note 16**).
5. Carefully remove the supernatant without disturbing the pellet. Resuspend the pellet in 5 mL of pre-equilibrated IVF medium and centrifuge at 300–600 × *g* for 10 min, once more. Repeat the step for all tubes.
6. Remove the supernatants, combine the pellets, and resuspend them in 0.5–1.0 mL. Analyze motility and concentration of spermatozoa in the washed sample.
7. Dilute to a final concentration of 75,000–200,000 motile sperm/mL using equilibrated IVF medium.
8. Add 0.5–1.0 mL of sperm medium to the center wells of insemination dishes. When using an oil overlay, ensure each droplet is at least 100 μL. Place the dishes in an incubator at 37°C with 6% CO_2 for a minimum of 2 h to equilibrate, until required for use.

3.4 Combined Density Gradient and Swim-Up

The combined density gradient and swim-up technique offers an optimized approach to sperm selection by leveraging the advantages of both methods. Initially, a semen sample undergoes density gradient centrifugation to separate sperm based on motility and morphology, yielding a highly concentrated population of motile and morphologically normal sperm. This step effectively reduces contaminants such as cellular debris, leukocytes, and nongerm cells, which are known to produce reactive oxygen species and impair sperm function. Following this, the swim-up method is employed on the density gradient pellet to further isolate motile sperm, ensuring minimal DNA fragmentation and improving overall sperm quality.

This combined approach has demonstrated significant benefits. The sequential use of density gradient and swim-up methods results in a sperm sample with low levels of DNA fragmentation and enhanced motility, as shown by Jackson et al., where the combined method exhibited superior outcomes compared to either technique alone. Specifically, DNA fragmentation rates following the combined method were reduced to 6%, compared to 12% for density gradient alone and 20% for swim-up alone, and sperm motility increased to 84% versus 67% and 53%, respectively [17].

However, the combined method also has drawbacks. It is more labor-intensive, involves higher costs, and requires multiple centrifugation steps, which may not be suitable for all laboratory settings or patient populations. Additionally, longer swim-up durations may be needed to recover adequate sperm for procedures like IVF or ICSI. Despite these limitations, the combined method's ability to yield high-quality sperm with minimal DNA damage makes it a preferred choice in cases requiring optimal sperm selection for assisted reproductive technologies.

1. Check that the 50% and 90% gradient Isolate, SWM or MHM are all at 37°C or room temperature.
2. Once collected, semen samples should be allowed to liquify for 20–30 min at 37°C.
3. Using a sterile glass Pasteur pipette, transfer 1.5–2 mL of the 90% gradient ISolate into a sterile disposable 15 mL conical centrifuge tube (*see* **Note 3**).
   i. If pursuing one-layer gradient separation, proceed directly to Step 4 without overlaying the second gradient solution (*see* **Note 4**).
4. With a new sterile disposable pipette, gradually transfer 1.5–2 mL of the 50% gradient ISolate on top of the 90% gradient Isolate (*see* **Note 5**).
5. Using a 5 or 10 mL serological pipette, mix the liquified semen by gently pipetting up and down, taking great care to avoid producing any bubbles.
6. Using a new, sterile glass Pasteur disposable pipette, carefully release 1.5–2.0 mL of liquified semen on top of the 50% gradient layer (*see* **Note 6**).
7. Place the tubes in the centrifuge and spin for 20 min at 200–300 × *g* (*see* **Note 7**). Avoid using the brake and confirm that the centrifuge is appropriately calibrated before use.
8. Using a clean 5 mL pipette tip, insert the tip below the surface of each layer to remove each individual layer in sequence (from top to bottom), without disturbing the pellet (*see* **Note 7**).
9. Wash the sperm pellet again by transferring the sperm pellet to a new sterile tube using a new glass Pasteur pipette. Resuspend the pellet in 2–3 mL of MHM or SWM. Centrifuge at 200–300 × *g* for 10 min. Aspirate the supernatant utilizing a new, sterile glass Pasteur pipette and discard it.
10. Add 2 mL of IVF medium (e.g., Gx-IVF) to a fresh round-bottomed tube.

11. Carefully add 1 mL of the centrifuged semen sample to the bottom of the tube without mixing (*see* **Note 13**).
12. Tilt the tube at a 45° angle to increase the surface area between the semen sample and the culture medium (*see* **Note 14**).
13. Incubate the tube at 37°C with 6% CO2 for 3–4 h to allow motile sperm to swim into the upper culture medium layer.
14. After incubation, carefully collect 1 mL of the upper layer into a new tube for further experimentation or analysis using a glass Pasteur Pipette (*see* **Note 15**).

3.5 Microfluidic Sperm Separation

Microfluidic sperm selection techniques leverage microscale environments to isolate high-quality sperm efficiently by mimicking the natural in vivo conditions of the female reproductive tract (*see* **Note 17**). These devices employ physical principles such as fluid dynamics, hydrodynamic forces, and dielectrophoresis to selectively separate motile sperm from nonviable cells and debris. Notably, this process often avoids the need for centrifugation, which has been associated with increased reactive oxygen species and subsequent DNA damage [18–21]. Devices such as the Zymot ICSI sperm separation chip demonstrate the potential of microfluidics to produce sperm populations with significantly lower DNA fragmentation compared to traditional methods like density-gradient centrifugation [22, 23].

Microfluidic approaches may be particularly valuable for specific patient populations, such as those with unexplained infertility or high levels of sperm DNA fragmentation. These cases highlight the potential of microfluidics to offer tailored solutions that enhance laboratory outcomes while reducing oxidative stress. However, challenges such as higher costs and the technical expertise required to operate these devices limit their widespread adoption in clinical practice.

Here, we detail the methods for sperm processing using the Zymot ICSI Sperm Separation Device, a leading microfluidic product on the market. The Zymot ICSI Sperm Separation Device features five individual microfluidic channels, each with a sample input capacity of 2 mL. The microchannels are designed to create a hydrodynamic environment that restricts the movement of compromised sperm, allowing only motile sperm to progress to the outlet for collection [18, 19].

1. Incubate the semen sample to allow for liquefaction. Using a 2–20 μL pipette with the appropriate tip, aspirate 15 μL of sperm wash medium.
2. Insert the pipette tip gently into the inlet port (the smallest of the ports), keeping the micropipette vertical and applying light pressure to form a seal. Avoid pressing the pipette tip too far, as this may block the flow. Slowly dispense the sperm wash medium through the inlet until it fills both the channel and outlet port, taking care to prevent any air bubbles from forming in the device.

3. Using a 1–10 μL micropipette with the appropriate tip, carefully add 2 μL of the liquefied semen sample to the inlet port. As before, avoid letting the pipette tip touch the bottom to ensure an unobstructed flow of the sample. Disregard any excess medium that may accumulate at the inlet port.
4. Turn a 100 mm Petri dish upside down so that its lid acts as the base. Fill a 35 mm Petri dish halfway with water and place both the uncovered 35 mm dish and the device into the inverted lid of the larger Petri dish. Then, use the original base of the 100 mm Petri dish as a cover.
5. Incubate the device at 37°C in a 6% CO_2 incubator for 30 min (*see* **Note 18**).
6. From each of the outlet ports, aspirate 2 μL of the sorted sperm-containing sample and transfer the collected samples to a capped 15 mL conical tube, until required for use.

4 Notes

1. This method employs two standard isolates, a 50% and 90% gradient, to separate sperm cells effectively. However, in cases where custom gradients are required, specific concentrations can be prepared by diluting the gradient material with a compatible medium such as modified Human Tubal Fluid (HTF) or Multipurpose Handling Medium (MHM). Custom gradients are particularly beneficial for managing atypical semen samples. For instance, with highly viscous samples, diluting the semen with sperm buffer and gently mixing it before gradient layering can enhance processing. "Dirty" specimens may require three-step gradients with layers like 40%, 60%, and 80%. Poor-quality samples may benefit from less dense gradients (e.g., 36% and 72%) to increase yield. Retrograde ejaculate urine specimens can be processed using standard 40%/80% gradients after concentrating and resuspending the material. For surgically retrieved sperm, a single 40% colloid gradient is typically sufficient to isolate spermatozoa from other cells. If custom gradients are prepared, the layers can be mixed in advance, stored at 4°C, and warmed to room temperature before use to ensure consistency and efficiency.
2. Warm G-RINSE overnight at 37°C in a 6% CO_2 environment. To prepare utensils such as tubes and dishes for a sperm processing procedure using G-Rinse, begin by ensuring all items are clean and dry before handling. Dispense an appropriate volume of G-Rinse medium into a sterile container, ensuring that the working environment remains contamination-free, such that each utensil is completely immersed in the G-Rinse

medium, allowing it to coat all surfaces thoroughly. Gently swirl or rotate items like tubes to ensure the medium contacts the inner surfaces evenly. Once the utensils have been adequately rinsed, carefully remove them from the medium, allowing any excess to drain off. Place the rinsed utensils on a sterile, non-absorbent surface or in a laminar flow hood to air dry completely before use. Discard of the rinsed fluid. Ensure that all utensils are dry and free from residual G-Rinse before proceeding with the sperm processing procedure.

3. These density gradient media consist of colloidal silica particles, characterized by low viscosity, which covalently bound silane molecules making them nontoxic and approved for human in vivo use. This design minimizes resistance to the sedimentation of spermatozoa during centrifugation. During the process, highly motile sperm migrate more efficiently through the gradient, penetrating the interface between layers faster than their less motile counterparts. This results in a clean fraction enriched with highly motile and viable sperm. ISolate® is one of several products developed specifically for clinical IVF and ICSI use. Similar products include PureSperm®, IxaPrep®, and Enhance®. To ensure isosmotic conditions, these gradients often include polysucrose, which aids in maintaining physiological osmolarity.

4. A single-layer density gradient is a simple and effective way to process sperm, especially when the goal is to separate motile, viable sperm from debris and immotile sperm without the need for a more complex setup. This method works well for cases with normal or near-normal semen parameters and is often used in procedures like IUI. Research shows that for fresh, extended semen, single-layer and two-layer density gradient methods yield similar results in terms of motile sperm recovery [24]. Both significantly improve sperm motility and survival compared to unprocessed samples, with no noticeable difference in quality between the two techniques. However, when working with cold-stored semen, the yield from either centrifugation technique tends to be 18%–20% lower than with fresh semen. Overall, single-layer density gradient centrifugation remains appealing given its simplicity and convenience, making it a practical option for use in the field while still delivering high-quality results.

5. This process involves lowering the pipette tip down the side of the tube, until it touches the surface of the 90% gradient ISolate. Delicately release the 50% gradient ISolate by moving the pipette tip in a spiral motion along the tube's circumference upward as the 50% gradient ISolate's level increases. It's crucial to maintain a distinct boundary between the 50% and 90% gradient layers without disturbing them, ensuring a

sharp interface is preserved. Additionally, it is important to layer the gradients immediately before use. If custom dilutions are employed, the ISolate layers may be prepared in advance, stored at 4°C, and then warmed to room temperature prior to use.

6. Pipette sperm carefully onto the surface of the gradient by moving the pipette tip in a spiral motion along the tube's circumference. If the total volume of semen is greater than 2 mL, the operator should use multiple tubes and distribute the volumes evenly. Each tube with 50% and 90% gradient layers can process 1–3 mL of semen.
7. It is important to stop aspirating when 0.5 mL of the lower (90%) gradient layer remains, whether the pellet is visible or not. If the pellet exceeds 0.5 mL, aspirate as much liquid as possible, without disturbing the pellet. Additionally, to avoid pellet contamination, the following actions should be strictly avoided: removing the entire bottom gradient layer to expose the pellet, using the same pipette for pellet recovery that was previously used to remove the upper layers, and resuspending the pellet directly within the gradient tube. Avoiding these methods will greatly reduce the risk of contamination as each layer is removed.
8. The simple wash method involves a single centrifugation step to separate spermatozoa from seminal plasma. However, this method also retains cellular debris, leukocytes, and immotile sperm, which can generate reactive oxygen species. These ROS may lead to DNA damage in sperm, potentially compromising embryo development, reducing pregnancy rates, and increasing miscarriage risk [25–31]. Repeated centrifugation during this process may also impair sperm motility and fertilization capacity, which is why more advanced methods, such as swim-up and density gradient centrifugation, are often preferred in clinical scenarios requiring optimal sperm quality.
9. Alternatively, traverse the supernatant with a glass Pasteur pipette and directly withdraw the pellet
10. For intrauterine insemination, a final resuspension volume of 0.5 mL is preferred in our lab. This volume provides a concentrated sperm dose while minimizing uterine cramping and reflux, which may occur with larger volumes. It reflects a balance between efficacy and patient comfort based on clinical experience.
11. The direct swim-up method involves carefully layering culture medium over liquefied semen in a conical tube. Ideally, semen should not be diluted or centrifuged before performing the swim-up technique to minimize oxidative damage and preserve sperm integrity. Motile sperm actively migrate

upward into the medium, which is then collected. While the direct swim-up method yields fewer sperm than techniques like washing or density gradient centrifugation, it effectively isolates high-quality motile sperm for cases such as IVF or ICSI, where motility is low [12].

12. In this variation, semen is first centrifuged to concentrate sperm into a pellet. Culture medium is added over the pellet, allowing motile sperm to swim upward into the medium. This approach isolates sperm populations with over 90% motility, free of cellular debris, and enriched with sperm exhibiting superior morphology, swimming speeds, acrosome reaction rates, and nuclear maturity. However, this method has limitations. Centrifugation required to form the pellet can increase exposure to reactive oxygen species produced by immotile sperm, leukocytes, or debris, potentially impairing fertilization capacity. Additionally, the recovery of motile sperm may be low due to the limited interface between the pellet and the overlying medium [32].

13. A sterile glass Pasteur pipette should be used to gently layer well-mixed, liquefied semen beneath the "sperm buffer" at the bottom of each tube. Ensure that the interface is distinct and clearly defined. Around 300 μL is sufficient to fill the hemispherical base of the tube, as adding more semen will not enhance the yield. However, if desired, the sperm pellet can either be disrupted and suspended in media or left intact and overlaid with media. If the pellet is disrupted, care must be taken when adding media to avoid mixing and contamination with immotile sperm, debris, and other cell types. Disrupted or resuspended pellets can be placed in 4-well dishes and overlaid with culture media to increase the surface area between the pellet and the media.

14. To maximize the interface area, tubes can be positioned at a 45° angle for incubation (approximately 30–60 min). This can be done by placing the tube at an angle within a standard lab test tube rack. Tilting the test tube at a 45° angle can increase the interface surface area between the pellet and the overlying culture media, allowing more motile sperm to migrate into the media. However, they should remain at this angle until after harvesting to prevent the risk of contaminating the medium layer with seminal plasma when returning the tube to an upright position. Additionally, a visible change in the media's turbidity indicates successful migration. If the media remains clear after 60 min, allow additional time for motile sperm to swim up from the pellet.

15. If desired, after aspirating the supernatant, it may be transferred to a clean conical tube, centrifuged at 300–600 × *g* for 4–10 min, and then resuspended in 0.5–1.0 mL for analysis or centrifugation.

16. Centrifugation speed should be kept low to avoid overly compacting the final pellet. This can be checked by tilting the test tube to see if the pellet shifts accordingly. Each laboratory should adjust centrifugation speeds as needed to achieve the desired pellet formation.
17. A key advantage of microfluidics is its ability to simulate the microanatomy of the female reproductive tract, employing mechanisms like chemotaxis and rheotaxis [33]. This biomimicry enhances the functional quality of selected sperm, as evidenced by improvements in motility and morphology. Laboratory studies have shown that microfluidic sperm selection leads to better DNA integrity, with reductions in fragmentation levels of up to 80% compared to conventional techniques [21, 23]. Despite promising laboratory outcomes, clinical applications of microfluidics have yielded mixed results. Recent randomized controlled trials and meta-analyses suggest that while microfluidic devices reduce oxidative stress and DNA damage, they do not consistently improve clinical outcomes such as fertilization rates, embryo quality, or live birth rates when compared to standard methods like density-gradient centrifugation or swim-up [22, 23]. For example, a systematic review found no statistically significant differences in clinical pregnancy rates or miscarriage rates between microfluidic sperm selection and traditional techniques, emphasizing the need for further large-scale, multicenter trials to validate these findings [23].
18. For an alternate incubation procedure, using a 1–10 μL micropipette and the appropriate tip, carefully apply a 2 μL oil overlay to both the inlet and outlet ports.

References

1. Bedford JM (1983) Significance of the need for sperm capacitation before fertilization in eutherian mammals. Biol Reprod 28(1):108–120. https://doi.org/10.1095/biolreprod28.1.108
2. Mortimer D, Templeton AA (1982) Sperm transport in the human female reproductive tract in relation to semen analysis characteristics and time of ovulation. J Reprod Fertil 64(2):401–408. https://doi.org/10.1530/jrf.0.0640401
3. Organization WH (2021) WHO laboratory manual for the examination and processing of human sperm. 6th edn. World Health Organization, Geneva
4. Henkel R (2012) Sperm preparation: state-of-the-art-physiological aspects and application of advanced sperm preparation methods. Asian J Androl 14(2):260–269. https://doi.org/10.1038/aja.2011.133
5. Canale D, Giorgi PM, Gasperini M, Pucci E, Barletta D, Gasperi M, Martino E (1994) Inter and intra-individual variability of sperm morphology after selection with three different techniques: layering, swimup from pellet and percoll. J Endocrinol Invest 17(9):729–732. https://doi.org/10.1007/BF03347769
6. Volpes A, Sammartano F, Rizzari S, Gullo S, Marino A, Allegra A (2016) The pellet swim-up is the best technique for sperm preparation during in vitro fertilization procedures. J Assist Reprod Genet 33(6):765–770. https://doi.org/10.1007/s10815-016-0696-2
7. Iwasaki A, Gagnon C (1992) Formation of reactive oxygen species in spermatozoa of infertile patients. Fertil Steril 57(2):409–416. https://doi.org/10.1016/s0015-0282(16)54855-9

8. Aitken RJ, Clarkson JS (1988) Significance of reactive oxygen species and antioxidants in defining the efficacy of sperm preparation techniques. J Androl 9(6):367–376. https://doi.org/10.1002/j.1939-4640.1988.tb01067.x
9. Henkel RR, Schill WB (2003) Sperm preparation for ART. Reprod Biol Endocrinol 1:108. https://doi.org/10.1186/1477-7827-1-108
10. Boomsma CM, Heineman MJ, Cohlen BJ, Farquhar C (2007) Semen preparation techniques for intrauterine insemination. Cochrane Database Syst Rev 4:CD004507. https://doi.org/10.1002/14651858.CD004507.pub3
11. Simon L, Emery B, Carrell DT (2019) Sperm DNA fragmentation: consequences for reproduction. Adv Exp Med Biol 1166:87–105. https://doi.org/10.1007/978-3-030-21664-1_6
12. Gabriel LK, Vawda AI (1993) Preparation of human sperm for assisted conception: a comparative study. Arch Androl 30(1):1–6. https://doi.org/10.3109/01485019308988361
13. Hammadeh ME, Kuhnen A, Amer AS, Rosenbaum P, Schmidt W (2001) Comparison of sperm preparation methods: effect on chromatin and morphology recovery rates and their consequences on the clinical outcome after in vitro fertilization embryo transfer. Int J Androl 24(6):360–368. https://doi.org/10.1046/j.1365-2605.2001.0317a.x
14. Vagios S, Velmahos CS, Cherouveim P, Dimitriadis I, Bormann CL (2023) The impact of different sperm preparation methods on clinical pregnancy and live birth rates in intrauterine insemination cycles: a retrospective single-center cohort study. Fertil Steril 120(3 Pt 2):617–625. https://doi.org/10.1016/j.fertnstert.2023.05.153
15. Erel CT, Senturk LM, Irez T, Ercan L, Elter K, Colgar U, Ertungealp E (2000) Sperm-preparation techniques for men with normal and abnormal semen analysis. A comparison. J Reprod Med 45(11):917–922
16. Oehninger S, Acosta R, Morshedi M, Philput C, Swanson RJ, Acosta AA (1990) Relationship between morphology and motion characteristics of human spermatozoa in seme and in the swim-up sperm fractions. J Androl 11(5):446–452
17. Jackson RE, Bormann CL, Hassun PA, Rocha AM, Motta EL, Serafini PC, Smith GD (2010) Effects of semen storage and separation techniques on sperm DNA fragmentation. Fertil Steril 94(7):2626–2630. https://doi.org/10.1016/j.fertnstert.2010.04.049
18. Zhang X, Khimji I, Gurkan UA, Safaee H, Catalano PN, Keles HO, Kayaalp E, Demirci U (2011) Lensless imaging for simultaneous microfluidic sperm monitoring and sorting. Lab Chip 11(15):2535–2540. https://doi.org/10.1039/c1lc20236g
19. Tasoglu S, Safaee H, Zhang X, Kingsley JL, Catalano PN, Gurkan UA, Nureddin A, Kayaalp E, Anchan RM, Maas RL, Tuzel E, Demirci U (2013) Exhaustion of racing sperm in nature-mimicking microfluidic channels during sorting. Small 9(20):3374–3384. https://doi.org/10.1002/smll.201300020
20. Asghar W, Velasco V, Kingsley JL, Shoukat MS, Shafiee H, Anchan RM, Mutter GL, Tuzel E, Demirci U (2014) Selection of functional human sperm with higher DNA integrity and fewer reactive oxygen species. Adv Healthc Mater 3(10):1671–1679. https://doi.org/10.1002/adhm.201400058
21. Cho BS, Schuster TG, Zhu X, Chang D, Smith GD, Takayama S (2003) Passively driven integrated microfluidic system for separation of motile sperm. Anal Chem 75(7):1671–1675. https://doi.org/10.1021/ac020579e
22. Prien S, Moseley T, Singh-Sharma N, Liebermann J, VerMilyea MD, Penrose LL (2023) A simple one-step system enhances the availability of high=quality sperm for assisted reproductive procedures. Open J Obstet Gynecol 13:1676–1687. https://doi.org/10.4236/ojog.2023.1310141
23. Nosrati R, Vollmer M, Eamer L, San Gabriel MC, Zeidan K, Zini A, Sinton D (2014) Rapid selection of sperm with high DNA integrity. Lab Chip 14(6):1142–1150. https://doi.org/10.1039/c3lc51254a
24. Morrell JM, Dalin AM, Rodriguez-Martinez H (2009) Comparison of density gradient and single layer centrifugation of stallion spermatozoa: yield, motility and survival. Equine Vet J 41(1):53–58. https://doi.org/10.2746/042516408x322139
25. Tesarik J, Greco E, Mendoza C (2004) Late, but not early, paternal effect on human embryo development is related to sperm DNA fragmentation. Hum Reprod 19(3):611–615. https://doi.org/10.1093/humrep/deh127
26. Henkel R, Hajimohammad M, Stalf T, Hoogendijk C, Mehnert C, Menkveld R, Gips H, Schill WB, Kruger TF (2004) Influence of deoxyribonucleic acid damage on fertilization and pregnancy. Fertil Steril 81(4):965–972. https://doi.org/10.1016/j.fertnstert.2003.09.044
27. Bungum M, Humaidan P, Spano M, Jepson K, Bungum L, Giwercman A (2004) The predictive value of sperm chromatin structure assay (SCSA) parameters for the outcome of intrauterine insemination, IVF and ICSI. Hum Reprod 19(6):1401–1408. https://doi.org/10.1093/humrep/deh280
28. Seli E, Gardner DK, Schoolcraft WB, Moffatt O, Sakkas D (2004) Extent of nuclear DNA damage in ejaculated spermatozoa impacts on blastocyst development after in vitro fertilization.

Fertil Steril 82(2):378–383. https://doi.org/10.1016/j.fertnstert.2003.12.039
29. Benchaib M, Braun V, Lornage J, Hadj S, Salle B, Lejeune H, Guerin JF (2003) Sperm DNA fragmentation decreases the pregnancy rate in an assisted reproductive technique. Hum Reprod 18(5):1023–1028. https://doi.org/10.1093/humrep/deg228
30. Virant-Klun I, Tomazevic T, Meden-Vrtovec H (2002) Sperm single-stranded DNA, detected by acridine orange staining, reduces fertilization and quality of ICSI-derived embryos. J Assist Reprod Genet 19(7):319–328. https://doi.org/10.1023/a:1016006509036
31. Lopes S, Jurisicova A, Sun JG, Casper RF (1998) Reactive oxygen species: potential cause for DNA fragmentation in human spermatozoa. Hum Reprod 13(4):896–900. https://doi.org/10.1093/humrep/13.4.896
32. Ford WC, McLaughlin EA, Prior SM, Rees JM, Wardle PG, Hull MG (1992) The yield, motility and performance in the hamster egg test of human spermatozoa prepared from cryopreserved semen by four different methods. Hum Reprod 7(5):654–659. https://doi.org/10.1093/oxfordjournals.humrep.a137714
33. Li Z, Liu W, Qiu T, Xie L, Chen W, Liu R, Lu Y, Mitchelson K, Wang J, Qiao J, Cheng J (2014) The construction of an interfacial valve-based microfluidic chip for thermotaxis evaluation of human sperm. Biomicrofluidics 8(2):024102. https://doi.org/10.1063/1.4866851

Chapter 4

Evaluation of Sperm Quality by Flow Cytometry: Sequential Protocols for DNA Fragmentation and Acrosome Reaction in Membrane-Intact Spermatozoa

Sushmita Nath, Raul da Costa, Vesna Bojovic and Stefan Schlatt

Abstract

Evaluation of sperm DNA fragmentation and membrane integrity plays a critical role in understanding male fertility potential. Flow cytometry serves as a powerful and precise technique for this purpose. Our group has developed a convenient, reliable, and reproducible flow cytometry-based strategy for simultaneous evaluation of DNA and Membrane integrity (DMI). Building on this, we are currently conducting in vitro research on acrosome reaction analysis in live spermatozoa, an additional sperm membrane integrity-related parameter, using flow cytometry. This chapter presents both approaches as complementary protocols that can be run side by side: simultaneous evaluation of DNA and membrane integrity (DMI), and assessment of in vitro acrosome reaction and membrane integrity.

Keywords Sperm, DNA fragmentation, DMI, Acrosome reaction, Membrane integrity, Flow cytometry

1 Introduction

Standard semen analysis in current practice focuses on the volume, concentration, motility, morphology, and pH of the semen sample. However, when it comes to more detailed insights into male fertility potential, basic semen analysis needs additional parameters to be taken into consideration. This is particularly important in cases of unexplained infertility (even when the standard semen analysis indicates normal results), men with recurrent miscarriages in their partner, and men with a history of lifestyle risks and/or failed IVF/ICSI cycles. Recent studies have revealed that sperm DNA fragmentation (SDF) and sperm membrane integrity (SMI) are two vital indicators of sperm functionality. For successful fertilization, and a healthy live birth, these two parameters play a crucial role [1, 2]. Insights from these valuable parameters are useful for assessing testicular function, diagnosing male infertility, and guidance for treatment options [2].

The authors Sushmita Nath and Raul da Costa contributed equally.

Alexandra J. Harvey and David K. Gardner (eds.), *The Preimplantation Embryo*, Methods in Molecular Biology, vol. 3038, https://doi.org/10.1007/978-1-0716-5292-3_4,

Sperm DNA integrity (SDI) and membrane integrity (SMI) are decisive factors of the functional competence of sperm. SDI, determined by the quality and stability of the sperm's genetic material, is its DNA integrity. Any damage, whether in the form of breakage or lesions in the DNA strand(s), is referred to as SDF. Likewise, SMI reflects the structural state of the outer membrane, an essential condition for sperm motility and fertilization following the physiological events of capacitation, acrosome reaction, and zona binding [3–6].

Despite its status as a logical biomarker of sperm quality, SDI is not included in the existing standard routine human semen analysis. Nevertheless, a number of techniques have been developed for the purpose of detecting the different aspects of SDF. These include sperm chromatin structure assay (SCSA), terminal deoxynucleotidyl transferase dUTP nick end labeling or TUNEL, comet assay or single cellgel electrophoresis (SCGE), sperm chromatin dispersion (SCD) [2, 7, 8]. Among these, the SCSA assay utilizes flow cytometry for precise and efficient analysis of the susceptibility of sperm DNA to an acid-induced denaturation [8, 9].

In SCSA, sperm are exposed to an acidic environment to test their chromatin integrity, thereby denaturing and exposing sites of DNA fragmentation. Subsequently, a metachromatic fluorescent dye, acridine orange (AO), is used to quantify DNA fragmentation in terms of DNA Fragmentation Index (DFI) and high DNA stainability (HDS). DFI therefore denotes the percentage of sperm exhibiting DNA damage in a semen sample, while HDS quantifies the amount of sperm in a semen sample with incomplete chromatin condensation associated with an increased amount of retained histones resulting from lack of protamination [10–12]. DFI is further categorized into <15%, 15%–30%, and >30% designated as low, moderate, and high DFI values respectively. Infertile males are more likely to have low sperm DNA integrity, typically represented by high DFI, compared to fertile males. Likewise, an elevated level of HDS indicates compromised sperm maturation [9].

Interestingly, none of the above-mentioned flow cytometric approaches for analyzing SDF considers sperm viability in terms of SMI; consequently, there is no distinction made between live and dead spermatozoa. Quantification of only live and membrane-intact spermatozoa can actually predict the fertility potential of the individual. In this regard, a convenient, reliable and reproducible multiparametric flow cytometric method for simultaneous evaluation of DNA and Membrane integrity (DMI), has been developed by our group [2]. This approach blends the principle of the SCSA to evaluate SDF with the evaluation of SMI, thereby allowing the classification of spermatozoa into four distinct categories based on their membrane integrity and degree of DNA fragmentation (Fig. 1). Furthermore, it also enables the evaluation of High DNA Stainability (HDS) in

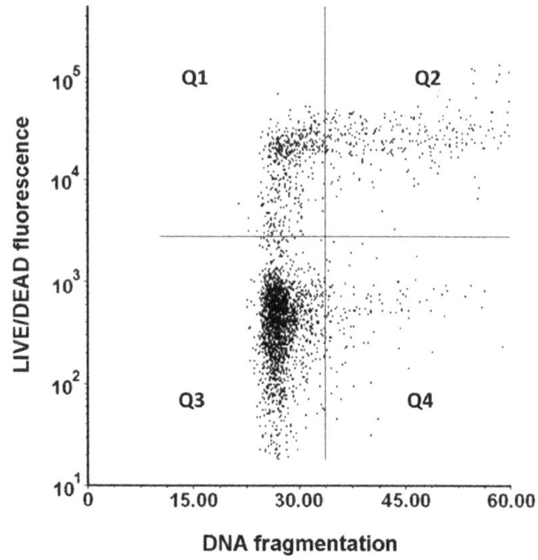

Fig. 1 Simultaneous evaluation of DNA and membrane Integrity. The *x*-axis and *y*-axis represent values of DNA fragmentation and the fluorescence values of the membrane integrity dye, LIVE/DEAD™ Fixable Blue, respectively. Four sperm subpopulations can be identified (Q1–4). Q1: membrane non-intact + no DNA fragmentation, Q2: membrane non-intact + DNA fragmentation, Q3: membrane intact + no DNA fragmentation, and Q4: membrane intact + DNA fragmentation. (Reprinted from [2])

membrane intact spermatozoa, which reflects the proportion of viable spermatozoa with immature chromatin structure.

As a continuation of this established method, our group is currently conducting in vitro research on acrosome reaction (AR) analysis, an additional SMI-related parameter and key event in the process of fertilization. In vitro AR analysis is a complement to standard semen analysis, offering more precise insights into sperm functionality. This test is able to detect subtle defects undetectable by standard semen analysis [13] by analyzing the in vitro acrosome reaction and membrane integrity in live spermatozoa by flow cytometry using the same semen sample.

In this chapter, we summarize methods for the simultaneous evaluation of DNA and membrane integrity (DMI), and parallel assessment of in vitro acrosome reaction and membrane integrity, carried out using the same sample, as indicators of sperm function.

2 Materials

Ensure that all listed reagents and materials are properly prepared before beginning the experimental procedures. Storage conditions and duration vary depending on the composition of each solution.

Photosensitive solutions should be protected from light during preparation, either by placing them in amber-colored containers or by covering them with aluminum foil for additional shielding.

2.1 Equipment

1. Plastic wide-mouthed specimen container for semen sample collection.
2. 12×75 mm flow cytometry tubes or equivalent.
3. Micropipettes and tips.
4. Centrifuge.
5. Vortex mixer.
6. Incubator at 37°C with 5% CO_2 supply.
7. Aluminum foil or amber colored container.
8. Magnetic stirrer and stirring bar.
9. Ice bath.
10. Timer.
11. Gloves.
12. Cell counter.
13. Water bath.
14. Hot plate with temperature control.
15. 1.5 mL Eppendorf tubes.
16. 15 mL Falcon tubes.
17. 2 L glass bottle.
18. 2 L volumetric flask.
19. pH meter.
20. 0.22 μm filter.
21. A flow cytometer equipped with the appropriate lasers and filters to meet the minimum requirements (*see* **Note 1**).
22. Flow cytometry data analysis software, e.g., FlowJo or FCS Express.

2.2 Reference Sample

Collect a human reference semen sample in a specimen collection container following an abstinence period of at least 72 h. The sample should be classified as normozoospermic according to World Health Organization (WHO) guidelines.

2.3 Reagents

1. 1 M NaOH.
2. 2 N NaOH.
3. *4 mM human tubal fluid (HTF) buffer*: Prepare stock solutions as outlined in Table 1. Fill a 2 L glass bottle to 90% of its capacity with deionized water equipped with a magnetic stirrer. While stirring, add each weighed component in the order listed in Table 1, making sure that each is completely

Table 1
List of components with concentration for preparing 4 mM HTF buffer

Components		Stock	For 2 L	Final concentration
Sodium chloride	NaCl	5 M	37.52 mL	93.8 mM
Potassium chloride	KCl	2.5 M	3.76 mL	4.7 mM
Magnesium sulfate	$MgSO_4$	1 M	0.4 mL	0.2 mM
Monopotassium phosphate	KH_2PO_4	1.5 M	492 µL	0.369 mM
Calcium chloride	$CaCl_2$	1 M	4.08 mL	2.04 mM
2-[4-(2-hydroxyethyl) piperazin-1-yl] ethanesulfonic acid	HEPES	1 M	41.96 mL	20.98 mM
Glucose	$C_6H_{12}O_6$	0.278 M	20 mL	2.78 mM
Sodium lactate	$NaC_3H_5O_3$	60% (w/w)	6.12 mL	21.4 mM
Sodium bicarbonate	$NaHCO_3$	–	0.672 g	4 mM

dissolved before adding the next. Glucose and HEPES must be warmed to 37°C in a water bath prior to addition. Transfer the solution to a volumetric flask, and measure the pH of the solution using a calibrated pH meter. Adjust the pH to 7.35 with 1 M NaOH at 37°C. Rinse the bottle with deionized water and add the rinses to the flask. Make up to the desired volume mark with deionized water. Filter the solution through a 0.22 µm filter, prepare 50 mL aliquots, and store at 4°C for up to 1 month (*see* **Note 2**).

4. *Swim upbuffer*: Add 100 mM sodium pyruvate to 4 mM HTF buffer in a 3.3 µL/mL concentration on the day of the procedure. Prewarm this mixture at 37°C 5% CO_2 incubator before using (*see* **Note 3**).
5. *Fixable viability dye*: Amine reactive dye, e.g., LIVE/DEAD™ Fixable Blue Dead Cell Stain kit (*see* **Note 4**). Allow the fixable viability dye to equilibrate to room temperature before opening. Reconstitute by adding 50–100 µL of DMSO (depending on the kit size) to one vial of dye and mix until fully dissolved. Aliquot into 1 µL volumes and store at −20°C. Avoid freeze-thaw cycles.
6. *Dimethyl sulfoxide (DMSO)*: Store at −20°C.
7. *1× Tris-NaCl–EDTA (TNE) buffer, pH 7.4*: Prepare a 10× TNE stock solution (0.1 M Tris.Cl, 1.50 M NaCl, 10 mM disodium EDTA; pH 7.4). To make 1 L, place approximately 800 mL of distilled water in a suitable beaker or bottle equipped with a magnetic stirring bar, add 9.48 g of Tris-HCl,

52.6 g of NaCl, and 2.23 g of disodium EDTA to the water. Stir until all solids are dissolved. Adjust pH to 7.4 with HCl or NaOH 1 M as needed. Add distilled water to make the final volume 1 L and store up to 1 year at 4°C. To make a 1× TNE solution, dilute the 10× TNE buffer 1:10 with distilled water and check the pH.

8. *Acid detergent solution, pH 1.2*: 0.08 N HCl, 0.15 M NaCl, 1% Triton X-100; pH 1.2 with 5 N HCl. To make 1 L, place approximately 800 mL of distilled water in a suitable beaker or bottle equipped with a magnetic stirring bar, add 40 mL 2.0 N HCl, 8.77 g NaCl, and 1 mL Triton X-100. Adjust pH to 1.2 with 5 N HCl or 10 N NaOH. Add distilled water to make the final volume 1 L and store up to 6 months at 4°C.
9. *Acridine orange (AO) stock solution*: In a glass scintillation vial, dissolve AO in distilled water to obtain a solution of 1 mg/mL. For example, dissolve 3 mg Acridine Orange powder in 3 mL distilled water. Protect the vial with aluminum foil and store at 4°C for up to 6 months.
10. *Staining Buffer, pH 6.0*: Prepare 500 mL (0.1 M citric acid, 0.2 M Na2PO4, 1 mM EDTA, 0.15 mM NaCl). Place about 300 mL of distilled water in a suitable beaker or bottle equipped with a magnetic stirring bar, add 3.554 g of citric acid monohydrate, 8.943 g of Na_2HPO_4, 186 mg of disodium EDTA (wt 372.24), and 4.385 g of NaCl. Stir overnight and adjust to pH 6.0. Store at 4°C for up to 6 months.
11. *AO staining solution (0.0015%)*: To prepare 100 mL add 100 mL of staining buffer and 600 μL of AO stock solution. Store at 4°C, for up to 2 weeks.
12. *25 mM HTF buffer*: The entire preparation procedure is the same as that outlined for 4 mM HTF buffer, with the exception of variations in component concentrations as given in Table 2.
13. *Capacitation buffer*: Add 100 mM sodium pyruvate and HSA to a 25 mM HTF buffer in a 3.3 μL/mL and 30 μL/mL concentration ratio, respectively, and prewarm this mixture at 37°C 5% CO_2 incubator before using (*see* **Note 5**).
14. *Ionomycin*: Prepare a stock concentration of 20 mM of ionomycin (as per manufacturer instructions). Store in 1-5 μL aliquots at −20°C.
15. *APC-conjugated CD46 antibody*: Ready to use at a concentration of 200 μg/mL. Store at 4°C.
16. *Propidium iodide (PI)*: Prepare a stock concentration of 2.5 mg/mL in deionized water, store at 4°C.

Table 2
List of components with concentration for preparing 25 mM HTF buffer

Compounds		Stock	For 2 L	Final concentration
Sodium chloride	NaCl	5 M	29.12 mL	72.8 mM
Potassium chloride	KCl	2.5 M	3.76 mL	4.7 mM
Magnesium sulfate	$MgSO_4$	1 M	0.4 mL	0.2 mM
Monopotassium phosphate	KH_2PO_4	1.5 M	492 µL	0.369 mM
Calcium chloride	$CaCl_2$	1 M	4.08 mL	2.04 mM
2-[4-(2-hydroxyethyl)piperazin-1-yl] ethanesulfonic acid	HEPES	1 M	41.96 mL	20.98 mM
Glucose	$C_6H_{12}O_6$	0.278 M	20 mL	2.78 mM
Sodium lactate	$NaC_3H_5O_3$	60% (w/w)	6.12 mL	21.4 mM
Sodium bicarbonate	$NaHCO_3$	–	4.200 g	25 mM

3 Methods

The method comprises two complementary procedures that can be carried out on the same semen sample.

3.1 Simultaneous Evaluation of DNA and Membrane Integrity (DMI)

The DMI protocol outlined below consists of a two-step co-staining process for simultaneous assessment of DNA fragmentation and membrane integrity. In the first step, spermatozoa are washed to remove seminal plasma and then stained with a viability dye, in this case LIVE/DEAD™ Fixable Blue Dead Cell Stain Kit (LD). After removal of the excess viability dye, the second step consists of staining the cells with AO, following the protocol similar to SCSA described by Evenson et al. [8]. This consists of denaturing sperm chromatin with an acidic solution to access breaks in the nuclear DNA, followed by staining with AO. When excited at 488 nm, AO associated with double-stranded DNA emits green fluorescence, and AO associated with single-stranded (damaged) DNA emits red fluorescence.

3.1.1 Semen Sample Collection and Staining

1. Collect the human semen samples to be assessed in a clinical specimen container following an abstinence period of at least 72 h.
2. Allow the samples to liquefy for at least 30 min in an incubator at 37°C with 5% CO_2 supply.
3. Perform a standard semen analysis within 1 h of collection according to WHO guidelines, including the determination of sperm concentration.
4. Centrifuge the semen in a 15 mL Falcon tube (300 × *g*, 5 min) to remove seminal plasma, then carefully remove the supernatant using a 1 mL pipettor without disturbing the sperm pellet.

5. Resuspend the pellet in 4 mM HTF to obtain a sperm concentration of 1–10 × 10^6 spermatozoa/mL.

3.1.2 Preparation of Reference Control

For calibration of the flow cytometer to perform DFI determinations, a frozen reference sperm sample from a normozoospermic donor is used during each experiment.

1. Collect a human semen sample from a healthy donor with an abstinence of at least 72 h in a clinical specimen jar.
2. Allow the sample to liquefy for 30 min in an incubator at 37°C with 5% CO_2 supply.
3. Perform the standard semen analysis according to WHO guidelines. The sample needs to satisfy all WHO criteria for normal sperm samples. Otherwise, discard the sample and choose another donor.
4. Dilute the sample with 1× TNE buffer to a working concentration of 1.1 × 10^6 sperm/mL.
5. Aliquot the reference sample into 1.5 mL Eppendorf tubes, each containing 300 μL of this working sample. Store in a −80 °C freezer. Take out one aliquot and thaw it each time you run a DNA fragmentation assessment. Repeated freezing is not recommended, and a new aliquot should be thawed for each experiment.

3.1.3 Viability Staining

1. Add 1 μL/mL viability dye to the sperm suspension (including the reference sample when prepared using a fresh sample).
2. Incubate in the dark for 30 min at 25°C.
3. Centrifuge the sperm sample (300 × *g*, 5 min) to remove excess dye.
4. Resuspend the pellet in 1× TNE Buffer to achieve a concentration of 1–2 × 10^6 sperm/mL.
5. Keep viability-stained samples covered from light on ice until analysis, ensuring a maximum holding time of up to 3 h (*see* **Note 6**).

3.1.4 DNA Staining

1. If not using a fresh reference sample, thaw an aliquot of the frozen reference sample at RT for a few minutes and continue with the following steps on ice (*see* **Note 7**).
2. To 300 μL of sperm suspension, including the reference sample, add 400 μL of acid detergent solution and manually shake the tube(s) constantly for exactly 30 s.
3. Add 1.2 mL of AO staining solution to each tube.
4. Incubate for 3 min in the dark on ice before analysis.
5. During the incubation time, prepare a blank buffer tube by combining 400 μL of acid detergent solution with 1.2 mL of AO staining solution.

3.2 In Vitro Acrosome Reaction and Membrane Integrity

The steps outlined below assess both the acrosome reaction (AR) and membrane integrity, adapted and modified based on the methodology described by Young and his group [12]. To assess the acrosome reaction, removal of seminal plasma is first undertaken by washing the semen sample(s), followed by the collection of highly motile sperm via swim-up. Thereafter, spermatozoa are stained with APC-conjugated CD46 for in vitro acrosome reaction, followed by staining with PI for membrane integrity. As a control, capacitated but not ionomycin-induced samples should be included for each donor.

3.2.1 Semen Sample Preparation

1. Following collection, liquefaction, and standard assessment, gently mix the neat sample and remove 300–500 μL into an Eppendorf tube.
2. Add a 1:1 volume of swimup buffer and centrifuge at 390 × *g* for 10 min.
3. Remove and discard the supernatant.
4. Carefully overlay the pellet with a 1:2 volume (neat sample) of swimup buffer and incubate for 1 h at 37°C, 5% CO_2 to allow motile sperm to swim up.
5. Gently collect the upper 50% fraction and place it into a fresh Eppendorf tube. Discard the bottom fraction.
6. Calculate the concentration of swim-up sperm and adjust to a final concentration of 1×10^6 sperm/mL using swim-up buffer. Use 1 mL of this sample for capacitation and acrosome reaction staining.

3.2.2 Capacitation, Induction of Acrosome Reaction, and Staining of Acrosome-Reacted Sperm

1. Spin down 1 mL of sample at 390 x g for 10 min at room temperature.
2. Discard the supernatant and resuspend the pellet in 1 mL capacitation buffer.
3. Incubate this suspension for at least 2 h at 37°C in an incubator with 5% CO_2 supply to capacitate.
4. After 2 h, carefully mix and split the sample into two aliquots of 500 μL (one for ionomycin exposure, one as a control).
5. Dilute the capacitated sample with capacitation buffer to obtain 5×10^5 sperm/mL in both aliquots.
6. To induce the acrosome reaction, add ionomycin stock to one of the aliquots to obtain a final concentration of 5 μM (*see* **Note 8**).
7. Incubate samples for 30 min at 37°C in a 5% CO_2 incubator.
8. To both tubes, add 0.05 μg/mL of APC-conjugated CD46–antibody (*see* **Note 9**).
9. Incubate this mixture at 37°C with 5% CO_2 for another 30 min.

3.2.3 Co-Staining with PI

1. Co-stain samples with PI at a final concentration of 0.8 mg/mL to assess membrane integrity.
2. Incubate the PI labeled sample in the dark for 10 min at room temperature.
3. At the end of the entire co-staining process, cover the sample tubes with aluminum foil and gently vortex the sample tube(s) for a fraction of a second to thoroughly mix.
4. Keep the sample tube(s) wrapped in aluminum foil at room temperature.
5. Just before starting flow cytometry analysis, transfer each sample with a minimum volume of 0.5 mL one by one to flow cytometry tubes. Use separate tubes and pipette tips for each individual sample.

3.3 Flow Cytometer Analysis Settings

The protocol outlined below has been optimized for use on a CytoFlex S benchtop flow cytometer from Beckman Colter equipped with four diode lasers and 15 optical filters. This flow cytometer has the capacity for reading samples loaded into either tubes or microplates. Alternative flow cytometers can be used, provided they have the appropriate lasers and filters for the dyes used in this protocol (*see* **Note 1**).

3.3.1 General Instrument Settings and Calibration

1. Check sheath fluid and waste tanks; ensure adequate sheath fluid and empty waste container.
2. Turn on the flow cytometer and computer, allowing sufficient warm-up time (typically 30 min).
3. Open flow cytometry software.
4. Run a quality control program of the instrument using the recommended manufacturer's beads to verify instrument performance.
5. Create a new experiment template specifically for each protocol by following the gating scheme outlined below.
6. Set the following parameters in the detector settings: (a) Set FSC and SSC to linear scale, (b) Set FL1 and FL3 to logarithmic scale.
7. Before running samples, allow the flow cytometer to equilibrate by running the blank buffer tube.
8. Place the sample tube on the sample injection port (SIP).
9. Set the flow cytometer to "Run" to adjust the parameters.
10. Run each sample for 15 s before starting data acquisition (to clear the dead volume).
11. Adjust voltages to obtain mean red and green fluorescence values on respective channels.

12. Set the threshold on FSC to exclude debris (typically around 52 for the primary parameter).
13. Start data acquisition once all the parameters are adjusted.
14. From the acquired data, gate the sperm based on their FSC and SSC properties.
15. Select the singlets population from the gated cells on a bidimensional plot, FSC-A vs. FSC-H
16. Analyze the singlets by plotting red vs. green fluorescence.
17. Process the obtained data in FCS file format using suitable flow cytometry data analysisdata analysis software, e.g., FlowJo, FCS Express.

3.3.2 Setup for DMI

1. Gating strategy:
   a) Configure dot plots for visualizing and gating cell populations: (a) FSC vs. SSC to identify sperm, (b) FSC-A vs. FSC-H to eliminate doublets or sperm aggregates, (c) red fluorescence (FL3) vs. green fluorescence (FL1) (Fig. 2A).
2. Data acquisition:
   a) Adjust the flow rate to achieve a throughput of approximately 500 events per second.
   b) Using the reference sample, adjust voltages to obtain mean red and green fluorescence values of 130 ± 5 and 500 ± 5, respectively (*see* **Note 10**).
   c) Set the threshold on FSC to exclude cell debris (typically around 52 for the primary parameter).
   d) Acquire data as 5000–10,000 sperm cells/event.
3. Data analysis:
   a) Define the HDS region as sperm with >2 standard deviations above the mean green fluorescence (FL1) of the main sperm population.
   b) Calculate High DNA Stainability (%HDS) by %HDS = (number of sperm in HDS gate/Total gated sperm) × 100. To measure chromatin immaturity, defined as sperm with excess nuclear histones (Fig. 2B).
   c) Calculate the percentage of sperm with DNA fragmentation (%DFI) by DFI = (red fluorescence (FL3)/total red + green fluorescence (FL1 + FL3)) × 100% (Fig. 2B, C).
   d) Plot the DFI vs viability to obtain the bidimensional plot (Fig. 2D).

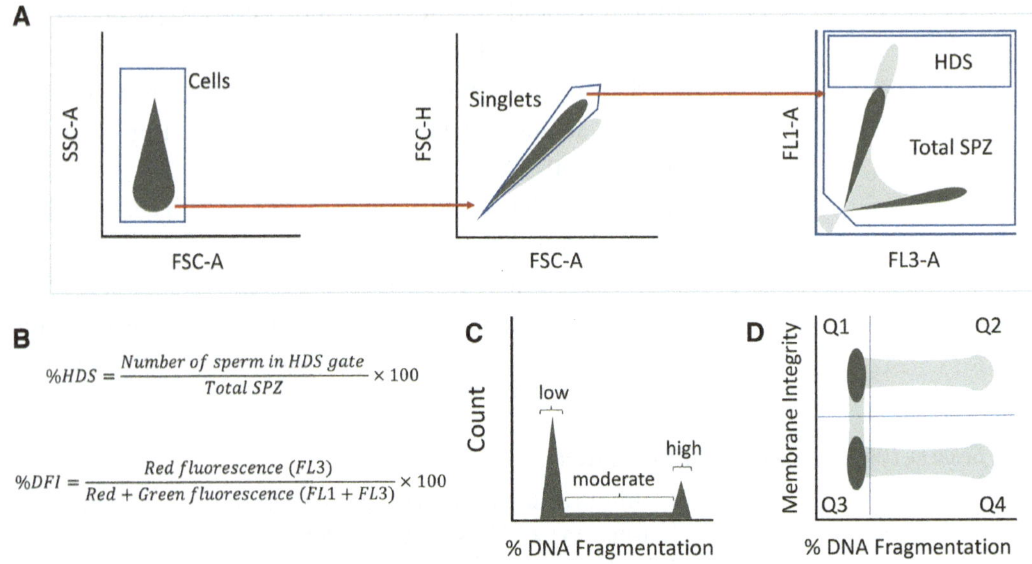

Fig. 2 Gating strategy and calculation of DNA fragmentation and membrane integrity. (**A**) Gating Strategy: Cells are initially selected based on forward scatter (FSC) and side scatter (SSC) properties, followed by identification of single cells using FSC-area versus FSC-height. Red (FL3) versus green (FL1) fluorescence is then plotted. High green fluorescence events are gated to quantify %HDS (high DNA stainability), while the total spermatozoa (total SPZ) population is used for %DFI (DNA Fragmentation Index) calculation. (**B**) Equations: Formulas for %HDS and %DFI are presented. (**C**) %DFI Categorization: %DFI values are categorized into low, moderate, or high levels. (**D**) Fragmentation in live and dead cells: DNA fragmentation is detected in spermatozoa based on their membrane integrity (live vs. dead)

3.3.3 Setup for Assessment of In Vitro Acrosome Reaction and Membrane Integrity

1. Gating strategy:
   a) Configure dot plots for visualizing and gating cell populations by FSC-A vs. SSC-A to identify sperm, FSC-A vs. FSC-H dot plot to discriminate doublets, and histogram plot of red fluorescence (FL3) vs. green fluorescence (FL1) to distinguish acrosome reacted (APC for CD46 positive) only in live (PI negative) sperm (Fig. 3).
2. Data acquisition:
   a) Adjust the gain in acquisition settings for APC to 2000, and for PI to 500.
   b) Set the flow rate to 60 μL/min.
   c) Acquire data as 10,000 sperm/event.
3. Data analysis:
   a) Analyze the Live cells in the histogram, where the APC-conjugated CD46 negative population will appear as a peak of low fluorescence (left side), and the APC-conjugated CD46 positive population will appear as a peak of high fluorescence (right side) (Fig. 3).

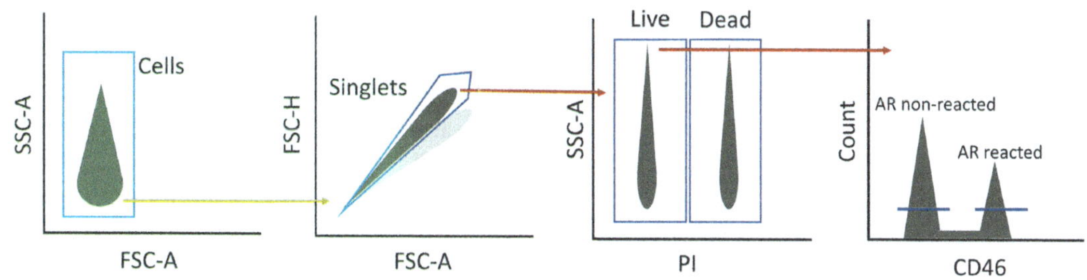

Fig. 3 Gating scheme for co-staining of acrosome reaction and vitality test. Start by gating out the area of interest (sperm) from all events, followed by the selection of singlets. Then identify live and dead cells in singlets. Determination of CD46-negative (AR-non-reacted) and CD46-positive (AR-reacted) cells within the live cell population is shown in a histogram plot

4 Notes

1. The Flow Cytometer should include a 488 nm blue laser along with the necessary filter sets to detect green fluorescence (515–530 nm band pass) and red fluorescence (630 nm long pass). In addition, an additional laser line, UV (355 nm), violet (405 nm), and/or red (638 nm) should be integrated with appropriate filters to allow detection of viability dye fluorescence without relying on the blue laser.
2. HTF is an azide-free and protein acid-free buffer solution, which minimizes staining artifacts, ensuring the dye accurately reflects cell viability. Alternatively, washing with protein-containing buffers (e.g., FBS) can help quench unreacted dye and prevent nonspecific binding. While this step is optional, it is recommended, as protein-containing buffers neutralize residual dye, thereby improving specificity.
3. For example, in a 50 mL 4 mM HTF buffer, you would require 165 μL of sodium pyruvate. Do not use left over swim-up buffer; it must be prepared fresh daily.
4. Fixable viability dyes are typically amine-reactive dyes that covalently react with exposed protein amines, external amines in live cells, and external and internal amines in dead cells, to reflect membrane integrity by differential fluorescence intensity. There are several commercially available fixable viability dyes. To avoid compensation issues, it is critical that the selected dye does not rely on the flow cytometer's blue laser and does not emit fluorescence within the AO filters, 525/40 BP (FL1) for green fluorescence and 690/50 BP (FL3) for red fluorescence. Alternative dyes include LIVE/DEAD Fixable Violet, LIVE/DEAD Fixable Lime, LIVE/DEAD Fixable Aqua, Zombie UV, Zombie Violet, Zombie Aqua, and others.

5. For example, in 2 mL 25 mM HTF buffer, you would require to add 6.6 μL Na-Py and 60 μL HSA. Capacitation buffer needs to be prepared fresh on the day of use.
6. If stained but unfixed samples are held for too long before analysis, ongoing cell death and DNA damage can introduce artefacts, resulting in data that does not accurately represent the sample at the time of staining. To lock in an accurate "snapshot" of cell viability and DNA status, immediate processing after staining is recommended. We have held the samples for up to 3 h without observing any negative effects.
7. Frozen reference sperm samples with normal parameters in accordance with WHO guidelines need to be used to calibrate the flow cytometer and subsequently set the thresholds for DNA fragmentation determination. Although the levels of DNA fragmentation remain the same during storage, there may be a loss of viability during storage. Therefore, the DFI reference samples should not be used to set viability gates, as such gates are sample-dependent and must be adjusted accordingly in each case based on the negative population (viable cells).
8. If you are using a sample volume of 1 mL, then directly add 2.5 μL of 20 mM ionomycin stock solution (prepared as per manufacturer instructions).
9. Check the concentration supplied by the vendor and dilute with capacitation if required. The working concentration of APC-conjugated CD46–antibody can be obtained by diluting its actual concentration of 200 μg/mL in a 1:9 ratio with capacitation buffer. The fluorescent marker APC-conjugated CD46 determines the percentage of the parent population of cells that have undergone the acrosome reaction by detecting the exposure of CD46, a membrane cofactor protein present in the inner acrosomal membrane of sperm.
10. All reference-sample mean red and green fluorescence values must remain within ±5 channels of the established laboratory standard. These reference values should be strictly maintained within this range throughout the experiment to ensure accuracy and consistency.

Acknowledgments

The protocols described in this chapter are part of projects supported by the DAAD and the Deutsche Forschungsgemeinschaft (DFG). R.D.C. during the development of the DMI protocol was supported by a DAAD Research Grant doctoral program in Germany (91590556) and additional funding from DFG (Clinical Research Unit, CRU 326). S.N. and V.B. are supported by the DFG-funded project (BD605447) for the ongoing investigation. The authors extend their gratitude to Prof. Strünker and his group for providing donor samples for optimization experiments.

References

1. Agarwal A, Majzoub A, Esteves SC, Ko E, Ramasamy R, Zini A (2016) Clinical utility of sperm DNA fragmentation testing: practice recommendations based on clinical scenarios. Transl Androl Urol 5(6):935–950. https://doi.org/10.21037/tau.2016.10.03
2. Da Costa R, Redmann K, Schlatt S (2021) Simultaneous detection of sperm membrane integrity and DNA fragmentation by flow cytometry: a novel and rapid tool for sperm analysis. Andrology 9(4):1254–1263. https://doi.org/10.1111/andr.13017
3. Ramirez JP, Carreras A, Mendoza C (1992) Sperm plasma membrane integrity in fertile and infertile men. Andrologia 24(3):141–144. https://doi.org/10.1111/j.1439-0272.1992.tb02627.x
4. McCallum C, Riordon J, Wang Y, Kong T, You JB, Sanner S, Lagunov A, Hannam TG, Jarvi K, Sinton D (2019) Deep learning-based selection of human sperm with high DNA integrity. Commun Biol 2:250. https://doi.org/10.1038/s42003-019-0491-6
5. Shamsi MB, Imam SN, Dada R (2011) Sperm DNA integrity assays: diagnostic and prognostic challenges and implications in management of infertility. J Assist Reprod Genet 28(11):1073–1085. https://doi.org/10.1007/s10815-011-9631-8
6. Tanga BM, Qamar AY, Raza S, Bang S, Fang X, Yoon K, Cho J (2021) Semen evaluation: methodological advancements in sperm quality-specific fertility assessment—a review. Anim Biosci 34(8):1253–1270. https://doi.org/10.5713/ab.21.0072
7. Mitchell LA, De Iuliis GN, Aitken RJ (2011) The TUNEL assay consistently underestimates DNA damage in human spermatozoa and is influenced by DNA compaction and cell vitality: development of an improved methodology. Int J Androl 34(1):2–13. https://doi.org/10.1111/j.1365-2605.2009.01042.x
8. Evenson DP (2022) Sperm chromatin structure assay (SCSA(R)) for fertility assessment. Curr Protoc 2(8):e508. https://doi.org/10.1002/cpz1.508
9. Sharma R, Iovine C, Agarwal A, Henkel R (2021) TUNEL assay—standardized method for testing sperm DNA fragmentation. Andrologia 53(2):e13738. https://doi.org/10.1111/and.13738
10. Blachman-Braun R, Best JC, Sandoval V, Lokeshwar SD, Patel P, Kohn T, Jacobs M, Ramasamy R (2020) Sperm DNA fragmentation index and high DNA stainability do not influence pregnancy success after intracytoplasmic sperm injection. F S Rep 1(3):233–238. https://doi.org/10.1016/j.xfre.2020.08.005
11. Esteves SC, Zini A, Coward RM, Evenson DP, Gosalvez J, Lewis SEM, Sharma R, Humaidan P (2021) Sperm DNA fragmentation testing: summary evidence and clinical practice recommendations. Andrologia 53(2):e13874. https://doi.org/10.1111/and.13874
12. Young S, Schiffer C, Wagner A, Patz J, Potapenko A, Herrmann L, Nordhoff V, Pock T, Krallmann C, Stallmeyer B, Ropke A, Kierzek M, Biagioni C, Wang T, Haalck L, Deuster D, Hansen JN, Wachten D, Risse B, Behre HM, Schlatt S, Kliesch S, Tuttelmann F, Brenker C, Strunker T (2024) Human fertilization in vivo and in vitro requires the CatSper channel to initiate sperm hyperactivation. J Clin Invest 134(1):e173564. https://doi.org/10.1172/JCI173564
13. Franken DR, Oehninger S (2012) Semen analysis and sperm function testing. Asian J Androl 14(1):6–13. https://doi.org/10.1038/aja.2011.58

Chapter 5

Sperm Proteomics and Phosphoproteomics: High-Throughput Methods Tailored to Gain Insight into Sperm Cell Signaling Pathways

Elizabeth G. Bromfield, Nathan D. Burke, Shannon P. Smyth, Georgia E. Blackley, Jess E. Mulhall, Brett Nixon, Sean J. Humphrey and David A. Skerrett-Byrne

Abstract

Protein phosphorylation is critical to successful sperm cell maturation and oocyte interaction. However, despite longstanding and largely global use of protein tyrosine phosphorylation as a robust marker of "readiness for fertilization," a deeper understanding of sperm-specific phosphorylation events and phosphoproteins is still lacking. To address this critical knowledge gap, here we present a complete guide to produce and analyze highly comprehensive phosphoproteomes from human spermatozoa. This chapter provides a bespoke guide to in vitro human sperm capacitation, phosphopeptide enrichment, and high-throughput sample preparation for mass spectrometry based on the renowned EasyPhos method. Replete with hints for successful application, we intend this guide to be of equal use in improving our understanding of male infertility, as well as establishing new insight for the development of non-hormonal, sperm-targeted contraceptives.

Keywords Germ cells, Kinase, Mass spectrometry, Phosphatase, Phosphorylation, Phosphoproteome, Proteomics, Spermatogenesis, Sperm capacitation, Spermatozoa

1 Introduction

Due to their transcriptionally and translationally suppressed state, mammalian spermatozoa are reliant on post-translational modifications to drive the generation of new proteoforms, such as phosphorylated proteins, that support their function and survival [1, 2]. Critically, during their transit of the cervix, uterus, and oviduct within the female reproductive tract, the establishment of sperm fertilization competency occurs via a phosphorylation-mediated signaling process called "capacitation" [3–5]. Despite the essential nature of phosphorylation to sperm function and fertility, only a limited repository of phosphorylated substrates and kinases have been

Alexandra J. Harvey and David K. Gardner (eds.), *The Preimplantation Embryo*, Methods in Molecular Biology, vol. 3038,
https://doi.org/10.1007/978-1-0716-5292-3_5,

identified from sperm undergoing capacitation. This is largely due to the limited adoption of high-resolution proteomic strategies in the field of reproductive biology [6], as well as a lack of high-throughput approaches to yield comprehensive phosphoproteomes. In humans, our limited knowledge of the proteins involved in fertilization signaling is prohibitive to the development of sperm-targeted non-hormonal contraceptives that prevent sperm–egg interaction [7], but also a limiting factor for the establishment of novel treatments for sperm dysfunction and infertility [8].

To increase the resolution, reproducibility and uptake of proteomics performed on sperm cells, here we provide a tailored guide to high-throughput phosphopeptide enrichment and high-resolution, label-free tandem mass spectrometry based on the EasyPhos method developed by Humphrey et al. [9]. This method has been used by our team to produce high resolution proteomes and/or phosphoproteomes of human and mouse spermatozoa [10], epididymal epithelial cells [11, 12], as well as developing germ cells, and seminal vesicle fluid [13]. This protocol allows for the precise and concurrent quantification of both the sperm proteome and phosphoproteome, greatly streamlining the workflow for sample preparation for mass spectrometry. From sperm isolation and preparation through to mass spectrometry and bioinformatics (Fig. 1), this protocol will serve as a step-by-step guide to yield high resolution proteomes and phosphoproteomes from sperm cells that can be used to deepen our understanding of sperm cell biology and fertilization.

2 Materials

2.1 Equipment

1. Megafuge with 15 mL tube adaptors.
2. Dry block thermal mixer (e.g., Eppendorf ThermoMixer C).
3. 0.1–2.5 μL, 0.5–10 μL, 2–20 μL, 20–200 μL, and 100–500 μL 1-channel, variable volume pipettors (optional—5–50 μL and 30–300 μL multichannel pipettors to decrease pipetting time).
4. Deep-well plate adapter for orbital shaker.
5. 2 mL tube adapter for orbital shaker.
6. ThermoTop accessory for orbital shaker.
7. Tip probe sonicator (e.g., Hielscher ultrasonic processor).
8. Plate reader compatible with BCA assay.
9. Vacuum liquid aspiration line and disposable borosilicate Pasteur pipettes, 5¾ in.
10. Custom 3D-printed StageTip adapter [9] (*see* **Note 1**).
11. Acclaim PepMap 100 C18 75 μm × 320 μm trap column.

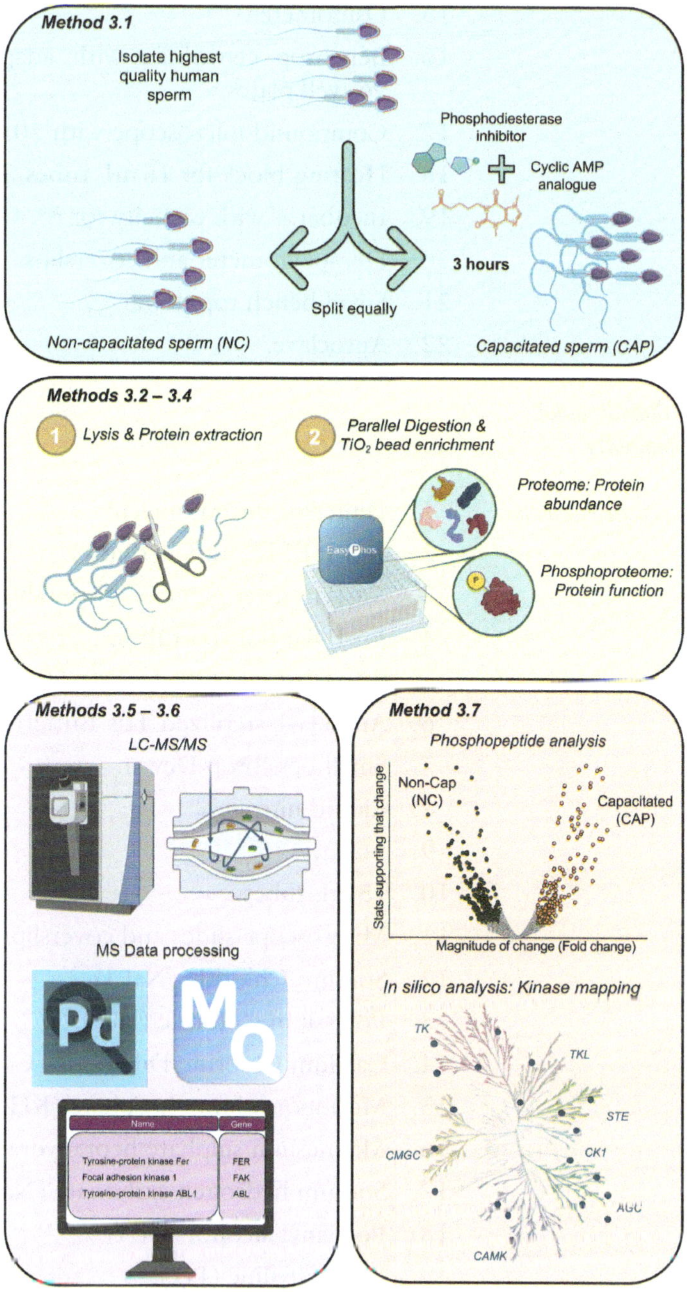

Fig. 1 A visual summary of the workflow provided in this chapter starting from human sperm isolation and capacitation through to mass spectrometry, data processing, and phosphopeptide analysis

12. Evaporative concentrator/Speedvac, with 96-well plate rotor.
13. EASY-Spray PepMap C18 75 μm × 250 μm column.
14. Mass spectrometer and an Ultra-High-Performance Liquid Chromatography (UHPLC) system (*see* **Note 2**).

15. Osmometer.
16. Benchtop centrifuge with adaptor for 1.5 mL tubes and 96-well plates.
17. Compound microscope with 20× and 40× objectives.
18. Heating block for 15 mL tubes, 37°C.
19. Incubator with capacity for 5% CO_2, 37°C.
20. Hemocytometer and coverslips.
21. Small bench top dewar.
22. Autoclave.

2.2 Consumables and Reagents

2.2.1 Sperm Preparation

1. Penicillin/Streptomycin.
2. 1 M HEPES buffer solution.
3. 3 mL transfer pipettes, disposable and flexible.
4. 5 mL flat bottom tubes.
5. 0.1–10 μL, 2–200 μL, 100–500 μL pipette tips.
6. Autoclave sterilized Tris-Buffered Saline.
7. Small benchtop Dewar.
8. Liquid nitrogen.
9. Eosin.
10. 15 mL tubes.
11. Miscroscope slides and coverslips.
12. Sodium Chloride (NaCl).
13. Potassium Chloride (KCl).
14. Calcium Chloride Dihydrate (CaCl22H2O).
15. Monopotassium phosphate (KH2PO4).
16. Magnesium sulphate heptahydrate (MgSO47H2O).
17. Sodium hydrogen carbonate (NaHCO3).
18. Polyvinyl alcohol (PVA).
19. Pentoxifylline (Ptx).
20. N6,2′-O-Dibutyryladenosine 3′,5′-cyclic monophosphate sodium salt (dbcAMP).
21. D-glucose.
22. Sodium pyruvate.
23. Sodium lactate.
24. Percoll.
25. Earle's Balanced Salt Solution 10x.

2.2.2 Mass Spectrometry and Proteomic Sample Preparation

For consumables and reagents, use of mass spectrometry grade reagents for the preparation of all mass spectrometry buffers is recommended.

1. Deep-well plates, 96 wells (2 mL square wells).
2. Adhesive PCR Plate Foils.
3. 2 mL Protein LoBind Tubes.
4. 5 mL flat bottom tubes.
5. 14-gauge syringes.
6. Bicinchoninic acid (BCA) Protein Assay Kit including BCA reagent A, BCA reagent B, and 2 mg/mL Albumin Standard Ampules.
7. 96-well plates for BCA assay (suitable for your plate reader).
8. MS-grade trypsin/Lys-C mixture.
9. Titanium dioxide (TiO_2) beads, Titansphere Phos-TiO bulk, 10 μm.
10. Styrenedivinylbenzene-reverse phase sulfonated (SDB-RPS) StageTips can be commercially obtained, but we make ours in house with solid-phase extraction disks from Empore SDP-RPS (*see* **Note 3**).
11. Solid-phase extraction disks for C8 StageTips (e.g., Empore C8) (*see* **Note 3**).
12. PCR strip tubes.
13. Disposable Pasteur pipettes.
14. HPLC grade water.
15. Sodium deoxycholate.
16. Tris(2-carboxyethyl)phosphine hydrochloride (TCEP-HCl).
17. 2-Chloroacetamide (CAM).
18. Potassium hydroxide (KOH).
19. Calcium chloride (CaCl2).
20. Potassium dihydrogen phosphate (KH2PO4).
21. Tris(hydroxymethyl)aminomethane hydrochloride (Tris-HCl).
22. Acetonitrile (ACN).
23. Methanol (MeOH).
24. Isopropanol (ISO).
25. Acetic acid (AcOH).
26. Trifluoroacetic acid (TFA).
27. Ammonia solution (25% w/v; NH4OH).
28. Ethylacetate.

2.3 Buffers

2.3.1 Sperm Preparation and Capacitation

1. Non-capacitating BWW (make fresh on day of experiment): 120 mM NaCl, 4.6 mM KCl, 1.7 mM $CaCl_22H_2O$, 1.2 mM KH_2PO_4, 1.2 mM $MgSO_47H_2O$, 5.6 mM D-glucose, 0.27 mM sodium pyruvate, 44 mM sodium lactate, 5 U/mL penicillin, 5 mg/mL streptomycin, 20 mM HEPES buffer, and 1 mg/mL PVA (osmolarity of 290–310 mOsm/kg, prewarm to 37°C).
2. Capacitating BWW (make fresh on day of experiment): 25 mM $NaHCO_3$, 4.6 mM KCl, 1.7 mM $CaCl_22H_2O$, 1.2 mM KH_2PO_4, 1.2 mM $MgSO_47H_2O$, 5.6 mM D-glucose, 0.27 mM sodium pyruvate, 44 mM sodium lactate, 5 U/mL penicillin, 5 mg/mL streptomycin, 20 mM HEPES buffer, 1 mg/mL PVA, 3 mM pentoxifylline, and 5 mM dibutyryl cyclic adenosine monophosphate (osmolarity of 290–310 mOsm/kg, prewarm to 37°C).
3. Sodium pyruvate solution: 300 mg sodium pyruvate, 5 mL Milli-Q-H_2O. Store at 4°C, can be stored for approximately 3 months.
4. 100% Percoll: 100 mg PVA, 210 mg $NaHCO_3$, 50 μL Sodium pyruvate solution, 370 μL sodium lactate, 10 mL 10× Earle's Balanced Salt Solution, and 90 mL Percoll. Store <1 week at 4°C.
5. 100% Percoll diluted to 80% and 40% solutions with non-capacitating BWW (make fresh on the day of experiment).
6. Eosin solution: 0.5% (w/v) eosin B and 0.9% (w/v) NaCl, in Milli-Q-H_2O. Can be stored <6 months at room temperature (RT).

2.3.2 Proteomic Buffers

1. 1 M Tris-HCl (pH 8.5). Store at RT for >1 year.
2. 5M KOH. Store at RT for >1 year.
3. 100 mM KH2PO4. Store at RT for >1 year.

All buffers are prepared with HPLC grade water (purity >99%). The storage requirements are listed in parentheses. All buffers lasting longer than the day of use, should be stored in amber bottles. Take caution with 2-Chloroacetamide (CAM; toxic) and trifluoroacetic acid (TFA; corrosive), both must be prepared in a fume hood and handled with gloves:

1. SDC lysis buffer (make fresh on day of experiment): 4% (w/v) SDC in 100 mM Tris-HCl (pH 8.5).
2. Reduction/Alkylation buffer (make fresh on day of experiment): 100 mM Tris(2-carboxyethyl)phosphine hydrochloride and 400 mM CAM. Adjust to between pH 7 and 8 with 5 M KOH.

3. EP loading buffer: 6% (v/v) TFA and 80% (v/v) acetonitrile (ACN). Store at RT for <3 months.
4. EP enrichment buffer: 48% (v/v) TFA and 8 mM KH_2PO_4. Store at RT for <3 months.
5. EP wash buffer: 5% (v/v) TFA and 60% (v/v) isopropanol. Store at RT for <3 months.
6. EP transfer buffer: 0.1% (v/v) TFA and 60% (v/v) isopropanol. Store at RT for <3 months.
7. EP elution buffer (time sensitive—make fresh on day of experiment, within 1 h of intended use): 5% (v/v) NH_4OH in 40% (v/v) ACN.
8. SDB-RPS StageTip loading buffer: 1% (v/v) TFA in isopropanol. Store at RT for <3 months.
9. SDB-RPS StageTip wash buffer 1: 1% (v/v) TFA in isopropanol. Store at RT for <3 months.
10. SDB-RPS StageTip wash buffer 2: 0.2% (v/v) TFA and 5% (v/v) ACN. Store at RT for <3 months.
11. SDB-RPS StageTip elution buffer (time sensitive - make fresh on day of experiment, within 1 h of intended use): 5% (v/v) NH_4OH in 60% (v/v) ACN. Store at RT for <3 months.
12. MS loading buffer : 0.3% (v/v) TFA in 2% (v/v) ACN. Store at RT for <6 months.
13. Proteome cleanup buffer: 99% (v/v) Ethylacetate and 1% (v/v) TFA. Store at RT for <3 months.

2.4 Software

1. Xcalibur (Thermo Fisher Scientific; version 4.2.47) is the interface of the mass spectrometers (MS) we utilize, but there will be MS-specific software with any MS you use.
2. Proteome Discover (Thermo Fisher Scientific; version 3.2) or MaxQaunt [14] are the most widely used MS processing software.
3. For postprocessing analysis of your proteomic data, we utilize a mixture of bioinformatic tools including UniProt (https://www.uniprot.org/), Perseus [15], Ingenuity Pathway Analysis (https://digitalinsights.qiagen.com/products-overview/discovery-insights-portfolio/analysis-and-visualization/qiagen-ipa/), Cytoscape [16], Phosphomatics [17], and KinMap [18].

3 Methods

The complete process described here takes several days of laboratory work to complete sperm preparation/capacitation and phosphopeptide enrichment and proteome clean up (Fig. 1).

3.1 Human Sperm Preparation and Capacitation

1. Following collection, incubate semen samples for 30–60 min at RT to allow for liquefaction before semen volume and raw sperm motility are recorded via light microscopy (*see* **Note 4**).
2. Determine motility by placing 5 μL of the sample on a clean microscope slide and cover with a regular coverslip.
3. View the slide immediately using a 40× objective on a light microscope.
4. Count 100 cells as either motile or nonmotile. This is usually expressed as the percentage of motile cells.
5. To assess sperm viability, mix Eosin solution 1:1 with semen on a microscope slide by gentle swirling of the liquid with a clean pipette tip.
6. Place a coverslip over the sample and view immediately using the 40× objective on a light microscope.
7. Count 100 cells as either viable (cells that appear clear), or nonviable (cells that appear pink). This count is usually expressed as the percentage of live cells (*see* **Note 5**).
8. Prepare a discontinuous Percoll gradient (one gradient per sample) by carefully underlaying 3 mL of 80% Percoll under 3 mL of 40% Percoll in a 15 mL Falcon tube (total volume 6 mL) using a disposable 3 mL flexible transfer pipette.
9. Layer each semen sample over its own Percoll gradient gently with a clean transfer pipette. Be careful not to overload the gradient. Aim for 2–4 mL of semen in total.
10. Centrifuge gradients at 500 × *g* for 30 min at RT in a megafuge. After centrifugation, three fractions should now be visible: seminal plasma and debris at the top, the lower quality sperm at the interface between the two Percoll layers, and the high-quality sperm pelleted at the bottom.
11. Isolate the highest quality sperm in the sample by removing the supernatant, disposing of it in a bleach solution (i.e., 12.5% sodium hypochlorite) and resuspend the pellet in non-capacitating BWW.
12. Wash the resuspended pellet via centrifugation at 500 × *g* for 15 min at RT, discard the supernatant and resuspend in 1 mL of non-capacitating BWW.
13. Assess the processed sperm for motility and viability again as described above (steps 2 - 7) and determine sperm concentration using a hemocytometer.
14. Take half of the resuspended sample and place it into a 2 mL LoBind tube and centrifuge at 500 × *g* for 5 min at RT.
15. Wash the sample three times via centrifugation at 500 × *g* for 3 min at room temperature, resuspending each time in autoclaved, sterile Tris-Buffered Saline (TBS) (*see* **Note 6**).

16. Remove as much TBS as possible prior to snap freezing the pellet in liquid nitrogen and storing at −80°C. This sample is the non-capacitated (NC) sperm sample.
17. Centrifuge the remaining sperm suspension for 3 min at 500 × *g* to remove non-capacitating BWW and then resuspend in capacitating BWW in a 5 mL flat bottom tubes to a concentration of 10×10^6 cells/mL.
18. To capacitate the sperm, place samples with their lids off in a 37°C incubator under a 5% CO_2/95% air atmosphere for 3 h and gently invert manually every 30 min, capping the tube for the inversions each time.
19. Following capacitation, assess sperm for motility and viability as described previously (steps 2-7), then centrifuge at 500 × *g* and wash three times in sterile TBS (*see* **Note 6**). This sample can now be snap frozen in liquid nitrogen and stored at −80°C. This is the "capacitated" (CAP) sperm sample.

3.2 Proteomic Sample Preparation

1. To each non-capacitated and capacitated sample, add 250 μL SDC lysis buffer and boil immediately (95°C, 8 min) to inactive endogenous phosphatases (*see* **Notes 7 and 8**).
2. Place samples on ice and sonicate with the following settings: maximum output (100%), 10 s pulse followed by 20 s rest on ice.
3. Repeat sonication at least four times, to ensure sperm cells are completely lysed (*see* **Note 9**).
4. Gently mix samples thoroughly by pipetting up and down. These samples are ready to quantify using a protein concentration assay (BCA assay; *see* **Note 10**).
5. To perform the BCA assay, the preprepared BSA standards are needed (*see* **Note 10**) Make up the standards by making a 9-point dilution series from a the 2 mg/mL stock. Using clean Eppendorf tubes, first pipette 300 μL into tube "A."
6. Next, carry out the series of dilutions as outlined in Table 1 using SDC lysis buffer.
7. Mix gently and use immediately or store at −20°C for later use (*see* **Note 10**).
8. Next, the working buffer needs to be prepared. To either a 50 or 15 mL falcon tube, add BCA Reagent A and B in a 50:1 ratio based upon the following calculation:

   (#Standards + #Samples being tested) × (#Replicates) × (200 μL) = Total working reagent volume.
9. Add Reagent B to Reagent A, mix thoroughly and gently until a clear green solution is achieved.
10. Use immediately or store working reagent sealed at RT where it is stable for several days.

Table 1
BSA standard dilutions for BCA protein quantification

Tube	Volume of SDC lysis buffer	Volume and source of BSA	Final BSA concentration
A	0	300 µL stock	2000 µg/mL
B	125 µL	375 µL stock	1500 µg/mL
C	325 µL	325 µL stock	1000 µg/mL
D	175 µL	175 µL of tube B	750 µg/mL
E	325 µL	325 µL of tube C	500 µg/mL
F	325 µL	325 µL of tube E	250 µg/mL
G	325 µL	325 µL of tube F	125 µg/mL
H	400 µL	100 µL of tube G	25 µg/mL
I	400 µL	0	Blank

11. To a 96-well plate, add 10 µL of each standard into wells.
12. For samples add 10 µL into wells. These raw samples can be diluted further in SDC lysis buffer if the concentration is suspected to be high (> 2000 µg/µL).
13. Once all the standards and samples are in the wells, add 200 µ L of working reagent to each well.
14. Shake plate gently for 30 s to mix, then cover with aluminium foil.
15. Incubate the plate for 30 min at 37°C.
16. Allow the plate to cool to RT.
17. Using a plate reader, measure the absorbance at 562 nm (540–590 nm acceptable).
18. Using the software available, subtract the average of the blank wells from all standard and sample wells (i.e., blank correct the raw absorbance values of the standard and sample wells).
19. Plot standard curve (average blank-corrected absorbance vs µ g/mL BSA).
20. Interpolate sample concentrations from standard curve.
21. From the protein concentrations calculated, dilute equal amounts (e.g., 50 µg *see* **Note 11**) from each of your samples. Dilute these samples to a total volume of 270 µL in 2 mL deep well plates (*see* **Note 12**).
22. Next, to each well add 1:10 volume (30 µL) of reduction and alkylation buffer (*see* **Note 13**).

23. Place samples on the thermal mixer, seal the plate with adhesive plate foil, and incubate for 5 min at 45°C with shaking (1500 rpm).
24. Remove samples from the thermal mixer, add 1:30 (10 μL) Trypsin/Lys-C mixture and seal the plate once again.
25. Allow samples to incubate overnight (16–20 h, 37°C, 1500 rpm) with a thermal lid attached (*see* **Notes 14 and 15**).

3.3 Proteomic Sample Preparation

EasyPhos was developed as a high-throughput method for phosphopeptide enrichment in biological samples. The specific protocol below describes the steps required for spermphosphopeptide enrichment, permitting the analysis of sperm signaling changes regulated by phosphorylation that are required for physiological capacitation. This protocol could also be used to determine whether specific conditions, such as the pharmacological inhibition of sperm kinases, have an impact on sperm signaling. While the phosphopeptide enrichment protocol is specified below, with minor modifications, this high throughput protocol can also be used to analyze the complete cellular proteomeProteome/proteomics in addition (or instead of) the phosphoproteomePhosphoproteome/phosphoproteomics (*see* **Note 16**). This modified proteome only approach is outlined below in Section 3.5.

1. To the ~300 μL digested peptides in the 2 mL DeepWell plates (DWP hereafter), add 400 μL Isopropanol and replace the thermal lid.
2. Mix using the thermal mixer (30 s, RT, 1500 rpm).
3. Add 100 μL EP Enrichment buffer to each sample and replace the thermal mixer lid.
4. Mix again using the thermal mixer (30 s, RT, 1500 rpm) (*see* **Note 17**).
5. Weigh out the titanium dioxide beads (TiO_2). In the DWP you will need a 12:1 (w/w) beads to protein ratio, i.e., 500 μg of peptide requires 6 mg of beads (*see* **Note 18**). Add up the total beads required across the experiment and account for an additional ~5% extra to avoid running out mid-experiment. Resuspend this final number of beads in EP loading buffer at a concentration of 1 mg/μL.
6. Vortex the prepared TiO_2 bead suspension to ensure the beads are evenly dispersed (*see* **Notes 19 and 20**).
7. Transfer 5 μL (i.e., 5 mg) of beads into DWP wells and ensure a tight seal on the plate (*see* **Note 18**).
8. Weigh the DWP plate containing the samples, including plate seal, and prepare a balance DWP plate for the subsequent centrifugation steps. Achieve this by weighing your plate,

zeroing the balance and placing a fresh DWP on the same scales. Use water to fill the wells until the scales reach zero, matching the mass of the sample plate.

9. Incubate the bead and peptide mixture and shake on the thermal mixer (5 min, 40°C, 2000 rpm).
10. Pellet the TiO_2 beads by centrifugation (1 min, RT, 2000×*g*) and transfer the supernatant (the proteome fraction) to a fresh 2 mL LoBind tubes/DWP for desalting in the following section (*see* **Note 21**).
11. Pipette 1 mL EP wash buffer to the beads in the DWP and mix briefly (30 s) on the thermal mixer.
12. Centrifuge samples to pellet beads (1 min, RT, 2000×*g*) (*see* **Notes 22 and 23**).
13. Aspirate and discard the nonphosphopeptide supernatant using a glass aspirator tip attached to a vacuum hose, or alternatively you can pipette one by one using a manual pipettor (*see* **Note 24**).
14. Wash a further four times (five in total) with 1 mL EP wash buffer and discard the supernatant.
15. Resuspend the TiO_2 beads in 75 μL of EP transfer buffer and transfer to the top of C8 StageTips (*see* **Notes 25 and 26**).
16. Repeat this with another 75 μL of EP transfer buffer to capture any remaining beads from the wells (*see* **Note 27**).
17. Place the StageTips into the centrifuge adapters and spin through to dryness (8 min, RT, 1500×*g*) (*see* **Notes 27 and 28**). You will notice the TiO_2 beads remain on top of the C8 material.
18. During this spin time make the EP elution buffer. Do not make it in advance as it is a time sensitive buffer. Preheat the SpeedVac to 45°C before commencing the next steps (*see* **Notes 29**).
19. Discard the flow through.
20. Elute phosphopeptides into clean labeled 0.2 mL PCR strip tubes with 30 μL of EP Elution buffer, centrifuge (3 min, RT, 1500×*g*) to dryness.
21. Repeat by again placing the StageTips into the centrifuge adapters and spinning through to dryness (8 min, RT, 1500×*g*). Collect the eluate into the same PCR tube.
22. Immediately place the PCR tubes into an evaporative concentrator and concentrate under vacuum (SpeedVac) at 45°C for approximately 30 min (*see* **Notes 30 and 31**).
23. Remove phosphopeptide samples from the SpeedVac and immediately add 100 μL SDB-RPS Loading buffer to each sample.

24. Load phosphopeptides onto nonequilibrated SDB-RPS StageTips with a pipettor or multichannel pipettor set to 110 μL.
25. Centrifuge (8 min, RT, 1500 × *g*) to dryness.
26. Wash StageTips with 100 μL of SDB-RPS Wash buffer 1 and centrifuge (8 min, RT, 1500 × *g*) to dryness.
27. Add 100 μL SDB-RPS Wash buffer 2 and centrifuge (5 min, RT, 1500 × *g*) to dryness.
28. Elute with 60 μL SDB-RPS Elution buffer into clean 0.2 mL PCR tubes by centrifugation (3 min, RT, 1500 × *g*).
29. SpeedVac for 35 min (or until dry) at 45 °C
30. Immediately, reconstitute the phosphopeptides in 5 μL of MS loading buffer and place either on a thermal mixer for 2 min (RT, 2000 rpm).
31. Centrifuge the samples at 2000 × *g* for 1 min at RT.
32. Phosphoproteome samples are now ready for MS analyses (go to Sect. 3.5).

3.4 Proteomic Sample Preparation—Proteome Cleanup

1. Using the proteome fraction collected previously, measure and pipette sample into a fresh 2 mL DWP (*see* **Note 32**).
2. Dilute each sample 1:1 with mass spectrometry grade H_2O to bring the SDC concentration below 1%.
3. Vortex briefly.
4. Add 1:1 volume of proteome cleanup buffer.
5. Mix on a thermal mixer (2 min, RT, 2000 rpm).
6. Transfer 200 μL to SDB-RPS StageTips, the maximum volume limit of the StageTips.
7. Centrifuge (3 min, RT, 1500 × *g*) to dryness.
8. If the sample volume exceeds 200 μL, repeat application of samples to the SDB-RPS StageTips and centrifugation until the entire sample volume has been passed through the StageTip.
9. Wash with 100 μL Proteome cleanup buffer.
10. Centrifuge (3 min, RT, 1500 × *g*) to dryness.
11. Add 100 μL of SDB-RPS Loading buffer to each and centrifuge (8 min, RT, 1500 × *g*) to dryness.
12. Wash StageTips with 100 μL of SDB-RPS Wash buffer 1 and centrifuge (8 min, RT, 1500 × *g*) to dryness.
13. Wash with 100 μL SDB-RPS Wash buffer 2 and centrifuge (5 min, RT, 1500 × *g*) to dryness.

14. Elute into clean 0.2 mL PCR tubes with 60 μL of SDB-RPS StageTip elution buffer and centrifuge (3 min, RT, 1500 × *g*) to dryness.
15. SpeedVac for 35 min (or until dry) at 45°C (*see* **Note 30**).
16. Proteome samples are ready for MS and can be stored in the meantime (*see* **Note 31**).

3.5 Alternative Proteome Sample Preparation - Proteome only

1. Following overnight digestion, 10 μg of sample was transferred into a fresh 2mL DWP (*see* **Note 32**).
2. Dilute each proteome sample at least 4-fold with 100mM Tris-HCl (pH 8.5) to bring the SDC concentration below 1%.
3. Add a 1:1 volume of proteome clean up buffer.
4. Mix on Thermal mixer (2 min, RT, 2,000 rpm).
5. Prepare one SDB-RPS StageTip per sample for peptide clean up:
   (i) Prime with 100 uL of 100% acetonitrile (1 min, RT, 1000 × g).
   (ii) Equilibrate with 100 uL of 0.1% TFA in isopropanol (3 min, RT, 1000 × g).
   (iii) Activate with 100 uL of 30% methanol/1% TFA (3 min, RT, 1000 × g).
6. Transfer 200 μL of sample to the SDB-RPS StageTips, the maximum volume limit of the StageTips.
7. Centrifuge (3 min, RT, 1000 × g) to dryness.
8. If the sample volume exceeds 200 μL, repeat application of samples to the SDB-RPS StageTips and centrifugation until the entire samples volume has passed through the StageTip.
9. Wash with 100 μL Proteome cleanup buffer.
10. Centrifuge (3 min, RT, 1000 × g) to dryness.
11. Repeat steps 7-8 for a total of two washes.
12. Wash StageTips with 100 μL of 0.2% TFA in isopropanol and centrifuge (3 min, RT, 1000 × g) to dryness.
13. Elute into clean 0.2 mL PCR tubes with SDB-RPS elution buffer and centrifuge (5 min, RT, 1000 × g) to dryness.
14. SpeedVac for 35 min (or until dry) at 45°C (*see* **Note 30**).

3.6 High-Performance Liquid Chromatography–Mass Spectrometry (LC–MS/MS)—Phosphoproteome and Proteome

1. Estimate total mass spectrometry time prior to analysis. Allow 90 min/sample for phosphoproteome MS and 90 min/fractionated sample and 130 min/unfractionated sample for proteome MS.
2. Once ready to analyze samples on the LC–MS/MS, place the samples into the LC autosampler, which should be cooled to 4–8°C. Samples are typically dispensed directly from the PCR

tubes in which they were dried, but samples can be transferred to 96-well plates or vials, depending on user preference.

3. Typically for the proteome cleanup, you should have ~7.5 μg of peptide dried. Reconstitute in 10 μL of MS Loading buffer, giving you ~750 ng / 1 μL injection
4. Analyze the phosphopeptides and peptides (proteome) using the LC and MS settings outlined in Tables 2 and 3 (*see* **Note 33**).

3.7 MS Data Processing

We typically process RAW (phospho)proteomics data using Proteome Discoverer [10, 11, 13] and conduct downstream bioinformatics analysis with the Perseus platform [15], as well as UniProt (https://www.uniprot.org/), Ingenuity Pathway Analysis [10, 11, 13], Cytoscape [16, 19], Phosphomatics [17], and KinMap [18] (detailed below). Nonetheless, there are various other suitable software options, both free and commercial, available for phospho(proteomics) data processing (e.g., MaxQuant [14], https://www.maxquant.org/) and Spectronaut, https://biognosys.com/software/spectronaut/) [20].

1. Download protein FASTA files from UniProt (https://www.uniprot.org/), with a focus on reviewed proteins, including isoform data.
2. The default settings on Proteome Discoverer will suit 99% of one's needs (Table 4), but slight adjustments can be made to include additional post-translational modifications of interest (*see* **Note 34**).

3.8 Data Analysis—UniProt

UniProt represents the world's most comprehensive catalogue of information on proteins. It can be accessed via https://www.uniprot.org/. The UniProt Knowledgebase (UniProtKB) is the central access point for extensive curated protein information (*see* **Note 35**).

Table 2
LC setups for samples

Phosphoproteome		Proteome (unfractionated)		Proteome (fractionated: *see* Note 44)	
Time interval (min)	Gradient (% Buffer B)	Time interval (min)	Gradient (% Buffer B)	Time interval (min)	Gradient (% Buffer B)
0	3	0	2	0	2
52	20	100	35	46	35
72	40	120	95	50	95
87	98	130	2	60	2
90	2				

Table 3
MS key settings for proteome and phosphoproteome analyses

QExactive HF-X (phospho)		Exploris 480 (proteome)	
Settings	**Value**	**Settings**	**Value**
Full MS		**Full MS**	
Resolution	60,000	Resolution	60,000
Automatic gain control target	3×10^6	Normalized automatic gain control target	300%
Maximum fill time	120 ms	Maximum fill time	120 ms
Scan range	350–1500 m/z	Scan range	350–1500 m/z
dd-MS^2		**dd-MS^2**	
Resolution	15,000	Resolution	15,000
Automatic gain control target	1×10^5	Automatic gain control target	1×10^5
Maximum fill time	50 ms	Maximum fill time	50 ms
Loop count	10	Loop count	20
Normalized collisional energy	27	Normalized collisional energy	30

1. Select *ID Mapping* under tools.
2. Copy and paste the accession numbers from your protein or phosphopeptide lists (generated following *MS Data Processing*) into the text box provided.
3. By default, the settings should be "From database: UniProtKB AC/ID" and "To database UniProtKB", ensure this is the case and select "Map IDs."
4. Once the job is complete, click "completed" to access the results. This process generates gene names and protein names from your list of accessions but can also be used to access further information about every protein in your dataset.
5. Customize the results view to include information of interest related to your proteins by firstly selecting "Customize columns."
6. We recommend ticking, at minimum, the following boxes: *Reviewed, Entry Name, Protein Name, Gene Name, Protein existence, Subcellular location, GO (cellular component), and keywords.*
7. Click on the download button, choose the Excel format and ensure the output is not compressed.

Table 4
Proteome Discoverer key settings

Tab (nodes)	Note(s)
Study definition	
Study factors	Here, you can implement conditions/treatments/mutations, etc. for quantification analysis, such as capacitation and non-capacitated sperm.
Grouping & Quantification	
Study variables	Building from the study factors, you can decide what ratios (fold change) comparisons you want calculated in the final protein list.
Analysis Results	
Processing step	
Minora Feature Detector	This is required as for label-free quantification.
Sequest HT	This is the default search engine of Proteome Discoverer. The key settings to focus on are:
	(1) *FASTA File*: select the latest version downloaded via UniProt, (2) *Enzyme*: Trypsin (Full), (3) *Maximum missed cleavage*: 2, (4) *Precursor mass tolerance*: 10 ppm, (5) *Fragment mass tolerance*: 0.02 Da, (6) *Fix modifications*: Carbamidomethyl (C), (7) *Variable modifications*: Oxidation (M), Acetyl (protein N-term), and Phospho (STY).
IMP-ptmRS	This is a key node for phospho analysis, to gain confident localization of modification sites within validated peptide sequences [21].
Consensus step	
Feature Mapper & Precursor Ions Quantifier	These must be included as here we are analyzing label-free samples.
Peptide and Protein Filter	Ensure "*Peptide Confidence At Least*" is set to High.

8. From this Excel spreadsheet, you can sort or filter your protein data based on the column headings. For example, you could assess how many of your identified sperm proteins currently have annotated protein evidence or whether there is any known sperm cell localization associated with those proteins. To filter, highlight all columns in your spreadsheet and click "filter" under the "Data" menu, At the top of each column there will now be a drop-down menu demarked by an arrow. Click the "reviewed" drop-down menu, for e.g., and you will see options to tick or untick "reviewed" proteins. Click ok and your column will be reduced to only the proteins that have been reviewed by Uniprot.
9. To complement this, the Human Protein Atlas now has a subcellular resource for sperm (https://www.proteinatlas.org/

humanproteome/subcellular/sperm), where they have localized 654 proteins to specific sperm subcellular structures in human sperm. You can query your proteins individually or upload a list of proteins for bulk analysis.

3.9 Data Analysis—Perseus

There are several platforms that can be utilized to normalize data, perform statistical analysis, and interpret protein quantification. We recommend the Perseus software platform [15]; a free, user-friendly, interactive workflow environment that provides complete documentation of computational method. They also provide YouTube videos of their summer schools.

1. To import into Perseus, you first need your files of proteins to be a CSV (or txt) file. To generate this file from Excel click file > save as > Text (tab delimited) > save.
2. Once you have installed and opened Perseus, use the green arrow at the top left corner under "matrix" to load your file.
3. Click *Select* and in the pop-up box, choose your newly generated CSV or txt file.
4. Next, you need to tell Perseus what types of data it will be handling. There are typically two important data types in Perseus named:
   - *Main*: The data of interest the statistical analysis is carried out on (usually your protein or peptide abundance data).
   - *Text*: Columns containing text-based information about the proteins but not included in the raw analysis. Examples include protein accession, gene name, protein name, and description.

   Perseus may automatically categorize your data but you can also click the column names in the left hand panel and then click the arrows to move those columns into "main" or "text" and click ok.
5. Now Perseus has processed your data generating a matrix. Save your Perseus session by *File* > *Save as* and name it (*see* **Note 36**).
6. Right click the blue matrix box (center panel) and rename it "*Raw analysis.*"
7. Next, categorize your samples according to type. For this, select the *Processing* panel > *Annot. Rows* > *Categorical annotation rows.*
8. In the pop-up box, rename all replicates to their appropriate collective name, such as "NC" for all non-capacitated sperm samples and "CAP" for all capacitation sperm samples. To rename, first click the name under "Keys" and then click the tick box and enter the new label (*see* **Note 37**).

9. Now transform the data; *Basic*> *Transform*> Leave default settings for log2 (x)> *OK* (*see* **Note 38**).
10. Next, assess the consistency and tightness between replicates though correlation plots. To do this select in the *Analysis panel*> *Visualization*> *Multi scatter plot.*
11. Choose that replicates will be compared. Select each group separately (i.e., as per this example, NC and then CAP).
12. Click onto the generated plot and widen your panel by clicking on the *Fit width icon.*
13. At the top of the plot, click on the drop-down menu where <*None*> is listed and choose *Pearson correlation.* Scores will appear in the top-left hand corner of each plot (*see* **Note 39**).
14. Next to the drop-down menu click on the equal (=) sign to ensure all graphs have equal ranges on all axes.
15. Export this image by clicking on the PDF icon.
16. Carry out statistical testing on your two groups by first filtering out proteins that have less than three quantitative values for both groups (three being the minimum required for statistical testing, depending on your study design). *Filter rows*> *Filter based on valid values* and then click *OK.*
17. Next click *Tests*> *Two-sample tests* (*see* **Note 40**).
18. Ensure Perseus has selected your two groups for analysis, leave default with the exception of:
    - *Valid value filter* > *Min number of valid values*= **3**
      Min. number mode= ***In both groups***
19. Click *OK.*
20. The resulting new matrix will have two additional columns:
    - First Student's *t*-test significance—the plus symbol indicates significance
    - Second Student's *t*-test significance—states the comparison carried out
21. Right click on the matrix and rename as *Tested Matrix.*
22. Create a volcano plot by clicking on *Analysis* panel> *Look for volcano plot icon* and leave settings on default.
23. Click on the newly generated square in the bottom right corner of your matrix box to view the volcano plot. All data can be exported via the export panel, and you have the capacity to alter the aesthetics of the graph by clicking on the *Configure* icon (a wrench and hammer symbol).
24. Principal component analysis (PCA), is a computational technique for reducing the dimensionality of large datasets, increasing interpretability while simultaneously minimizing

information loss. It does so by creating new, uncorrelated variables that maximize variance. For a PCA, a quantitative value is required for all replicates. To adjust for this, select *Filter rows> Filter based on valid values> min. = 100%.*

25. *Next, select Analysis panel> Clustering/PCA > Principal component analysis.* Leave at default and click *OK.*
26. Click on the newly generated square in the bottom right corner of your matrix box to see the PCA. The top panel is your 2-dimensional PCA. To expand the space on the screen, drag the edges of the PCA panel to the bottom; just below the x-axis.
27. Click on the *Configure* icon to adjust the style of the PCA plot.
28. On the right-hand side panel, in the top box, click on *Categories.* Select your group/category of interest to adjust the symbol color, type, size, and labels.
29. When finished, right click the group in the box and select *Clear Selection.*
30. Export your image by clicking on the PDF icon and save.
31. Finally, heatmaps are an efficient and aesthetically pleasing visualization of complex data, displaying how your data groups together on a large scale or highlighting the key changes in your dataset. Click back on *Tested Matrix* and apply a *Z*-normalization by selecting *Processing panel> Normalization> Z-Score.* Change *Matrix access* to *columns* and click *OK* (*see* **Note 41**).
32. To plot only the significantly different proteins: *Filter rows> Filter based on categorical column.* Change mode to *Keep matching rows.*
33. *Next, select Analysis panel> Clustering/PCA > Hierarchical clustering.* Leave the parameters at default, click *OK.* Export this image by clicking on the PDF icon and save.

    Alter the colors of the heat map by clicking on the *Change color gradient icon* (the color wheel symbol). When you have the gradient open, right click above the numbers and adjust the minimal and maximum values. Click and drag markers to the numbers you wish (e.g., −2 and 2) and then left click to change colors.
34. Back on the heatmap plot, you can highlight and analyze clusters of interest by hovering over the intersection of a cluster until a red box appears, click once it does. This gives that cluster a unique color and cluster number. Each cluster can be exported for further analysis using Ingenuity Pathway Analysis or UniProt, by clicking on the *Export row clusters* icon. This will give you a new matrix with the same information as your

previous matrix but with one new column, *Cluster* with the unique cluster number.

35. Once the Cluster information has been generated select *Export panel> save icon> Select* and name it (*see* **Note 42**).
36. There are many bioinformatical tools available for further analysis of these data. We have provided a brief list of recommendations for further analysis (*see* **Note 43**).

3.10 PRIDE Upload

Once all data have been processed and the authors wish to proceed to manuscript submission, it is critical authors make the proteomic data available to reviewers, and subsequently, the field upon publication. This can be achieved through PRIDE [22] using their submission tool (https://www.ebi.ac.uk/pride/markdownpage/pridesubmissiontool), which requires the following files: RAW, peak (.mzML), result (mzIdentML). The RAW files are those that were generated by the LC–MS/MS.

1. For the mzIdentML and peak files, these can be obtained through Proteome Discoverer. Open your final results file of your processed data and select *File> Export.*
2. From the drop-down menu, select mzIdentML and *Export.*
3. For "peak", within Proteome Discoverer click "spectra.". Export from peptide spectrum match "PSM" and select mzML.
4. Lastly, using software such as MSConvertGUI/Proteowizard (https://proteowizard.sourceforge.io/), convert each of your RAW files to mzML files by clicking browse, selecting your RAW file and then selecting mzML as the output format. Leave the settings as they are and then click start.
5. In total, for each sample a mzIdentML and mzML file, generated from the Proteome Discover results, as well as the RAW files for each processed sample and the corresponding mzML files, are required to complete a submission to PRIDE.
6. Within 2 working days, you will receive private login details for your dataset, which can be easily placed in the manuscript. All data remain private until you select to make it public upon publication.

4 Notes

1. The centrifuge adapter is an in-house printed 96-well StageTip centrifuge device. 3D printers using fused-deposition modeling, such as the Zortax M200 or Ultimaker 3

instruments, enable the rapid (~12 h) and inexpensive fabrication of devices capable of holding StageTips and collection tubes for centrifugal-based processing of up to 96 samples in parallel. If such a device is not available, a pipette-tip box may serve as a suitable StageTip holder. Care should be taken when centrifuging to ensure items are fully contained within centrifuge buckets, that they are equally balanced and that centrifugation speeds used do not result in damage.

2. Any contemporary mass spectrometer capable of MS1 and MS2 scans with high resolution and fast scanning speed can be used. This includes various instruments such as the Q Exactive Plus, HF-X, Eclipse and Exploris 480, LTQ-Orbitrap Elite, as well as several time-of-flight (TOF) instruments. Any Ultra-High-Performance Liquid Chromatography (UHPLC) system can be used. An example is the Dionex Ultimate 3000SRLC nanoflow system). While non-UHPLC systems are also compatible, especially when using shorter columns, we have observed a reduction in phosphopeptide identifications under these conditions.
3. StageTips should be prepared in advance to ensure smooth timing. C8 and SDB-RPS StageTips are prepared as described [23], using Empore solid-phase extraction disks. We prepare a double plug layer for C8 StageTips and a triple plug layer for SDB-RPS StageTips. We achieve this by punching out the material using a blunt-end 14-gauge syringe and firmly placing them in the ends of clean 200 μL pipette tips. You will require one of each per sample. Once prepared, they can be stored for several months at RT.
4. Following liquefaction, all semen samples should remain in the 37°C incubator when not being analyzed.
5. Eosin is toxic to cells over time. Hence, cell viability must be read immediately after the addition of eosin to ensure a true representation.
6. Avoid using Phosphate-Buffered Saline (PBS) as the final wash of cells/tissues as excessive residual PBS during lysis can interfere with enrichment of phosphorylated proteins. TBS can be used as an alternative.
7. Prior to commencing the protocol, preheat the thermal mixer to 95°C as immediate boiling after addition of lysis buffer to cells/tissues is crucial to ensure complete inactivation of phosphatases and proteases. Following boiling, be sure to set the thermal mixer to 45°C.
8. Throughout the protocol there are a number of settings required for the thermal mixer. Where possible, utilize the preprograming capacity so each step's requirements are easily achieved by a flick of a button.

9. A small aliquot (e.g., 5 µL) can be taken to confirm complete lysis on a light microscope. If complete lysis has been achieved, there should be no intact sperm cells. If intact cells are present, repeat the sonication and additional time.
10. It is important to check the BCA Assay instruction manual after purchasing your kit in case of any minor changes to the protocol. BSA standards with a higher concentration can also be prepared, aliquoted, and stored at −20°C. Be sure to place standards on ice on the morning before sample prep to ensure they are fully thawed by the time they are required.
11. We find that 50 µg of protein is an ideal starting amount when working with sperm samples, and we recommend a minimum of 20 µg and a maximum of 120 µg.
12. If you require very few samples, you could consider using 2 mL LoBind tubes rather than a DeepWell plate (DWP), but this will require slightly more time overall due to the manual handling.
13. This step helps to linearize the proteins, breaking down tertiary structures by reducing disulfide bonds and carbamidomethylate cysteine residues, preventing any novel disulfides from forming.
14. A stepped program feature can be used, where available on your thermal mixer, to keep samples at 4°C following digestion.
15. Following digestion, peptide populations can be stored for up to 4 weeks at −20°C or for >12 months at −80°C.
16. If you are only interested in the proteome and wish to utilize, the high-throughput EasyPhos approach for proteomics without phosphopeptide enrichment, please refer to Smyth et al. (2022) for slight modifications to achieve this [13].
17. If you observe some cloudiness, first mix once again for an additional 5 min. If not improved and you observe a precipitate-like deposit in the bottom of the well, this could indicate the presence of lipids. It is critical this is removed as it can interfere with the phospho-enrichment. To achieve this, spin the DWP plate at 2000 × *g* (15 min, RT) and carefully transfer the clear supernatant to a fresh DWP before proceeding. Manual and electronic multichannel pipettes can help to expedite this process.
18. If using <500 µg protein the rule of thumb is 5 mg beads minimum, or 12:1 beads:protein. The latter is more likely when working with sperm samples within the suggested limits of starting material above.

19. For any experiment with a large number of samples, it is advised to weigh out the titanium dioxide beads before commencing day 2 steps.
20. TiO_2 beads settle out of solution rapidly and as such must be mixed thoroughly immediately before pipetting to ensure that the beads are evenly distributed.
21. Make labeled 2 mL LoBind tubes in advance for the proteome fraction. Samples can be stored at −80°C or keep at 4°C for desalting on the same day.
22. Adjust the weight of the balance plate as required and double check that the sample plate and balance plate are equal by weighing them.
23. Always check the plate from the bottom after every shake and spin. Each time you should notice an even distribution of beads in the wash buffer.
24. A trick for removing the supernatant is to gently slide the glass aspirator tip down along the corner of the sample well, stopping only upon touching the base of the well. This is well represented in Humphrey et al. (2018), Supplementary Fig. 4 [9].
25. StageTips should be prepared in advance to ensure smooth timing. C8 and SDB-RPS StageTips are prepared as described [23], using Empore solid-phase extraction disks. We prepare a double plug layer for C8 StageTips and a triple plug layer for SDB-RPS StageTips. We achieve this by punching out the Empore solid-phase extraction disk material using a blunt-end 14-gauge syringe and firmly placing the small punch outs in the ends of clean 200 μL pipette tips. You will require one of each per sample. Once prepared they can be stored for several months at RT.
26. When dispensing samples into the C8 StageTip, after the first point of resistance wait a moment as the residual volume pools in the tip. Then push past the second point of resistance.
27. The first transfer captures ~90% of beads, the second is to ensure maximum recovery. Using a multichannel pipette, you can transfer many samples at once.
28. Ensure that all transfer buffer has passed through the StageTips before proceeding to elution. In the case that not all buffer has passed through the StageTips, centrifuge once again (5 min, 1500 × *g*).
29. Be careful not to confuse the EP Elution buffer with SDB-RPS Elution buffers. Additionally, note that NH_4OH must be stored at 4°C and pipetted rapidly to ensure its equilibrium is maintained. Prepare sufficient EP elution buffer for

the number of samples being desalted, plus make extra to ensure you do not run out

30. Do not allow the samples to fully dry, aiming for about ~20 μL remaining. A PCR tube with 20 μL of water can be prepared as a helpful gauge of volume remaining.
31. Phosphopeptides and peptides (proteome) can be stored in MS loading buffer for several days at 4°C, several weeks at −20°C, or several months at −80°C, provided an airtight seal is maintained on the sample tubes.
32. You will notice in the following steps in the protocol, there are a number of 1:1 volumes being added. Given the 2 mL limit of the DWP, be sure to account for all these additions before beginning. If beyond 2 mL, take the largest volume possible to maximize your recovery. Pipette the remaining volume into either 2 mL LoBind tubes or a 96-well plate sealed with a plate foil and store at −80°C.
33. MS setup will differ between technologies but presented in the tables are settings related to top-tier orbitrap mass spectrometers across, which the settings are widely applicable. Despite this, we encourage researchers to run trial samples and consult the latest literature for their particular MS.
34. Addition of modifications has a two-fold effect; (1) increase in the computation time required and (2) increase in False Discovery Rate for nonpositive modifications.
35. As a starting point, researchers should give strong consideration to deploying Shiny Applications to support accessibility and interpretability of these complex datasets, allowing for effective data-driven insights. To facilitate this, we have published a step-by-step guide with full annotated code, which can be easily repurposed by researchers without coding experience [6, 24].
36. Note how the middle panel is a flowchart, this is a great way to keep track of what you have done. It is good practice to save the session regularly as there is no autosave feature.
37. It is critical that you ensure each name for the appropriate groups are spelled exactly the same, otherwise they will be treated as separate groups.
38. Log2 transformation is common in proteomics as it aids in interpretability of fold changes and it stabilizes the variance, which is important as protein abundance data can have a wide range of values. Transformation makes the protein abundance data more suitable for common data visualization approaches like volcano plots and heat maps.

39. A Pearson correlation is a number between −1 and +1 (positive correlation) that indicates to which extent two variables are linearly related.
40. There is also the option to carry out an ANOVA when looking at more than two groups.
41. To normalize these data, Perseus takes the mean of each column and subtracts it from each value. The result is then divided by the standard deviation of the column. This helps the data to sit between a defined range of −2 and 2.
42. Every data table you produce throughout this analysis (e.g., the PCA values or volcano plot values) can be exported as a txt, csv, or tsv file, and then manipulated in excel. We usually export the data and then regraph it in software such as GraphPad Prism https://www.graphpad.com/features to improve the visual quality of the data.
43. There are many bioinformatical tools available, all with their own in-depth tutorial documentation. We have highlighted four recommendations below for proteomic and phosphoproteomic data but this list is by no means exhaustive.

    (A) Ingenuity Pathway Analysis (IPA)—here you can get a breakdown of the types (classification) and location of proteins, as well as pathways, downstream molecular functions and upstream regulators of your data. Refer to [6, 10, 11, 13] for examples. Link: https://digitalinsights.qiagen.com/ipa-getting-started-and-onboarding-matrix-quick-start/

    (B) Phosphomatics—a suite of tools for the downstream analysis of phosphoproteomics data, producing PCAs, volcano plots, upstream networks, interactive data of kinases-substrates identified and more [17]. Link: https://phosphomatics.com/

    (C) KinMap—allows users to explore their kinases of interest on an interactive Human Kinome Tree with live annotations [18]. Link: http://www.kinhub.org/kinmap/tutorial.html

    (D) String and Cytoscape can be utilized in union to understand protein–protein interaction networks coupled with functional GO enrichment analysis. Cytoscape provides the capacity to control the visual elements of the outputs, see examples [19]. Link: https://string-db.org/; see Help > Getting Started.

44. To gain the greatest depth in your sample of interest, it is encouraged to explore high pH fractionation (*see* Table 2). While this approach increases MS time, the benefit is deep proteome coverage. This can be useful particularly with sperm

samples as they are often dominated by several highly abundant proteins such as heat shock proteins [10]. If this increased depth is of interest, the samples are prepared the same way as described in this protocol, but we encourage users to consult their mass spec technician within their facility to adapt the mass spectrometry runs to include fractionation.

Acknowledgments

This research was supported by an Australian Research Council Discovery Early Career Researcher Award (DECRA; DE210100103) awarded to E.G.B. and a National Health and Medical Research Council of Australia (NHMRC) Emerging Leadership Fellowship (APP2034392) awarded to D.A.S.B. We would like to thank Dr Ben Crossett and Angela Connolly at the Charles Perkins Centre (Sydney Mass Spectrometry), The University of Sydney, Nathan Smith at The University of Newcastle and the Melbourne Mass Spectrometry and Proteomics Facility at the Bio21 Institute, University of Melbourne for their support. We also wish to extend our gratitude to all our collaborators over the years supporting the development of this work here.

References

1. Serrano R, Garcia-Marin LJ, Bragado MJ (2022) Sperm phosphoproteome: unraveling male infertility. Biology (Basel) 11(5):659. https://doi.org/10.3390/biology11050659
2. Zhang X, Tu H, Zhou X, Wang B, Guo Y, Situ C, Qi Y, Li Y, Guo X (2024) Quantitative phosphoproteomic profiling of mouse sperm maturation in epididymis revealed kinases important for sperm motility. Mol Cell Proteomics 23(8):100810. https://doi.org/10.1016/j.mcpro.2024.100810
3. Baro Graf C, Ritagliati C, Stival C, Luque GM, Gentile I, Buffone MG, Krapf D (2020) Everything you ever wanted to know about PKA regulation and its involvement in mammalian sperm capacitation. Mol Cell Endocrinol 518:110992. https://doi.org/10.1016/j.mce.2020.110992
4. Mostek-Majewska A, Majewska A, Janta A, Ciereszko A (2023) New insights into post-translational modifications of proteins during bull sperm capacitation. Cell Commun Signal 21(1):72. https://doi.org/10.1186/s12964-023-01080-w
5. Nixon B, Cafe SL, Eamens AL, De Iuliis GN, Bromfield EG, Martin JH, Skerrett-Byrne DA, Dun MD (2020) Molecular insights into the divergence and diversity of post-testicular maturation strategies. Mol Cell Endocrinol 517:110955. https://doi.org/10.1016/j.mce.2020.110955
6. Skerrett-Byrne DA, Teperino R, Nixon B (2024) ShinySperm: navigating the sperm proteome landscape. Reprod Fertil Dev 36:RD24079. https://doi.org/10.1071/rd24079
7. Nickels L, Yan W (2023) Nonhormonal male contraceptive development-strategies for progress. Pharmacol Rev 76(1):37–48. https://doi.org/10.1124/pharmrev.122.000787
8. Dey S, Brothag C, Vijayaraghavan S (2019) Signaling enzymes required for sperm maturation and fertilization in mammals. Front Cell Dev Biol 7:341. https://doi.org/10.3389/fcell.2019.00341
9. Humphrey SJ, Karayel O, James DE, Mann M (2018) High-throughput and high-sensitivity phosphoproteomics with the EasyPhos platform. Nat Protoc 13(9):1897–1916. https://doi.org/10.1038/s41596-018-0014-9
10. Skerrett-Byrne DA, Anderson AL, Bromfield EG, Bernstein IR, Mulhall JE, Schjenken JE, Dun MD, Humphrey SJ, Nixon B (2022) Global profiling of the proteomic changes associated with the post-testicular maturation of mouse

spermatozoa. Cell Rep 41(7):111655. https://doi.org/10.1016/j.celrep.2022.111655

11. Skerrett-Byrne DA, Stanger SJ, Trigg NA, Anderson AL, Sipilä P, Bernstein IR, Lord T, Schjenken JE, Murray HC, Verrills NM, Dun MD, Pang TY, Nixon B (2024) Phosphoproteomic analysis of the adaption of epididymal epithelial cells to corticosterone challenge. Andrology 12(5):1038–1057. https://doi.org/10.1111/andr.13636
12. Mulhall JE, Trigg NA, Bernstein IR, Anderson AL, Murray HC, Sipilä P, Lord T, Schjenken JE, Nixon B, Skerrett-Byrne DA (2024) Immortalized mouse caput epididymal epithelial (mECap18) cell line recapitulates the in-vivo environment. Proteomics 24(7):e2300253. https://doi.org/10.1002/pmic.202300253
13. Smyth SP, Nixon B, Anderson AL, Murray HC, Martin JH, MacDougall LA, Robertson SA, Skerrett-Byrne DA, Schjenken JE (2022) Elucidation of the protein composition of mouse seminal vesicle fluid. Proteomics 22(9):e2100227. https://doi.org/10.1002/pmic.202100227
14. Tyanova S, Temu T, Cox J (2016) The MaxQuant computational platform for mass spectrometry-based shotgun proteomics. Nat Protoc 11(12):2301–2319. https://doi.org/10.1038/nprot.2016.136
15. Tyanova S, Temu T, Sinitcyn P, Carlson A, Hein MY, Geiger T, Mann M, Cox J (2016) The Perseus computational platform for comprehensive analysis of (prote)omics data. Nat Methods 13(9):731–740. https://doi.org/10.1038/nmeth.3901
16. Shannon P, Markiel A, Ozier O, Baliga NS, Wang JT, Ramage D, Amin N, Schwikowski B, Ideker T (2003) Cytoscape: a software environment for integrated models of biomolecular interaction networks. Genome Res 13(11):2498–2504. https://doi.org/10.1101/gr.1239303
17. Leeming MG, O'Callaghan S, Licata L, Iannuccelli M, Lo Surdo P, Micarelli E, Ang CS, Nie S, Varshney S, Ameen S, Cheng HC, Williamson NA (2021) Phosphomatics: interactive interrogation of substrate-kinase networks in global phosphoproteomics datasets. Bioinformatics 37(11):1635–1636. https://doi.org/10.1093/bioinformatics/btaa916
18. Eid S, Turk S, Volkamer A, Rippmann F, Fulle S (2017) KinMap: a web-based tool for interactive navigation through human kinome data. BMC Bioinf 18(1):16. https://doi.org/10.1186/s12859-016-1433-7
19. Zhang M, Chiozzi RZ, Skerrett-Byrne DA, Veenendaal T, Klumperman J, Heck AJR, Nixon B, Helms JB, Gadella BM, Bromfield EG (2022) High resolution proteomic analysis of subcellular fractionated boar spermatozoa provides comprehensive insights into perinuclear theca-residing proteins. Front Cell Dev Biol 10:836208. https://doi.org/10.3389/fcell.2022.836208
20. Bruderer R, Bernhardt OM, Gandhi T, Miladinović SM, Cheng LY, Messner S, Ehrenberger T, Zanotelli V, Butscheid Y, Escher C, Vitek O, Rinner O, Reiter L (2015) Extending the limits of quantitative proteome profiling with data-independent acquisition and application to acetaminophen-treated three-dimensional liver microtissues. Mol Cell Proteomics 14(5):1400–1410. https://doi.org/10.1074/mcp.M114.044305
21. Taus T, Köcher T, Pichler P, Paschke C, Schmidt A, Henrich C, Mechtler K (2011) Universal and confident phosphorylation site localization using phosphoRS. J Proteome Res 10(12):5354–5362. https://doi.org/10.1021/pr200611n
22. Perez-Riverol Y, Bai J, Bandla C, García-Seisdedos D, Hewapathirana S, Kamatchinathan S, Kundu DJ, Prakash A, Frericks-Zipper A, Eisenacher M, Walzer M, Wang S, Brazma A, Vizcaíno JA (2022) The PRIDE database resources in 2022: a hub for mass spectrometry-based proteomics evidences. Nucleic Acids Res 50(D1):D543-d552. https://doi.org/10.1093/nar/gkab1038
23. Rappsilber J, Mann M, Ishihama Y (2007) Protocol for micro-purification, enrichment, pre-fractionation and storage of peptides for proteomics using StageTips. Nat Protoc 2(8):1896–1906. https://doi.org/10.1038/nprot.2007.261
24. Trigg NA, Mulhall JE, Nixon B, Laurent K, Pai S, Smyth SP, Burke ND, Beckers J, Bromfield EG, Karr TL, Lord T, Pleuger C, Schjenken JE, Hrabě de Angelis M, Teperino R, Skerrett-Byrne DA (2026) Curating fertility-proteomic remodelling of sperm during epididymal transit. Reproduction 171(2), xaag012. https://doi.org/10.1093/reprod/xaag012

Chapter 6

Conventional In Vitro Fertilization and Intracytoplasmic Sperm Injection

Zsolt Peter Nagy

Abstract

In vitro fertilization (IVF) and intracytoplasmic sperm injection (ICSI) are assisted reproductive technologies used to help individuals or couples conceive a child. IVF involves combining an oocyte and sperm outside the body in a laboratory to create embryos. Following ovarian stimulation, oocytes retrieved through a minor surgical procedure are mixed with sperm in the laboratory. If fertilization occurs, resulting embryos are cultured for several days, and one (or more) embryo(s) is (are) then transferred into the uterus (either "fresh" or after cryopreservation) in hopes of implantation and pregnancy. Conversely, ICSI is a specialized form of IVF typically used when male infertility is a concern, or in previously failed conventional insemination, whereby a single sperm is injected directly into an oocyte using a fine needle. This bypasses issues like low sperm count, poor sperm motility, or abnormal sperm shape, as well as some other potential gamete structural or physiological impairments. ICSI follows the same general steps as IVF, but it requires the use of a microscope and micro-manipulators, which makes the fertilization process more predetermined. The success rates of IVF and ICSI depend on factors such as age, fertility issues, and overall health. Importantly, it has been demonstrated that IVF and ICSI are likely not increasing risks of genetic/chromosomal abnormalities in fetuses nor are they associated with increased risks of malformation of children born. IVF/ICSI can use a partner's or a donor's oocytes and sperm. Both IVF and ICSI, outlined in this chapter, are continually advancing, improving success rates and safety.

Keywords In-vitro fertilization (IVF), Intracytoplasmic sperm injection (ICSI), Oocyte, Embryo, Sperm, Fertilization

1 Introduction

Conventional in vitro fertilization (IVF) was first reported in 1978 by Professor Robert Edwards [1]. This breakthrough inspired other the scientists and clinicians around the world to establish IVF programs and shortly after this first report a number of countries reported pregnancies, including Australia, USA, and France [2–4]. When IVF began, the main indication was to address female-infertility associated with blocked Fallopian tubes, while later indications of IVF were expanded [5]. Initially, patients natural cycles were used to

Alexandra J. Harvey and David K. Gardner (eds.), *The Preimplantation Embryo*, Methods in Molecular Biology, vol. 3038,
https://doi.org/10.1007/978-1-0716-5292-3_6,

retrieve oocytes; however, the advent of hormone containing medications made it possible to stimulate ovaries to allow more follicles to develop in a controlled manner [6]. However, it was soon realized that conventional IVF did not always work efficiently, especially when sperm parameters were impaired. To address the sperm-related inefficiency of conventional insemination, scientists experimented with various micro-manipulation techniques to assist fertilization. The most successful method, which used a single sperm injected directly into the cytoplasm of the oocyte termed intracytoplasmic sperm injection (ICSI), was first reported by Lanzendorf [7], while the first pregnancy using ICSI was reported by Palermo in 1992 [8]. It is estimated that in most countries 1%–3% of all births are related to the use of infertility treatments, with IVF treatment, now combined with ICSI increasingly available across the globe.

The IVF laboratory is a specialized, stringently controlled environment designed to replicate in-vivo physiological conditions cruical for successful gamete fertilization and subsequent embryo development. Achieving optimal outcomes necessitates precise regulation of incubation and culture parameters, including elevated carbon dioxide (CO_2) and reduced oxygen (O_2) concentrations, alongside the utilization of advanced embryo culture media. Conventional insemination is usually performed "overnight" (a period of 8–14 h) in culture drops of a Petri dish during which time hundreds to thousands (typically 100,000–250,000/mL) of sperm "race" to fertilize the oocyte. However, in the end, only one sperm penetrates the zona pellucida then fuses with the oolemma, triggering the cortical reaction, which alters the zona resulting in blocking other sperm cells to enter the perivitelline space [9]. Some studies have demonstrated that "short incubation" insemination (3–4 h) works equally well compared with the traditional "overnight incubation" to obtain similar fertilization rates (which is typically 60%–80%; however, in 5%–10% of cases have failed or low (below 30%) fertilization) [10].

Following conventional insemination, fertilization is typically evaluated at 16–20 h post-initiation of insemination, through observation of pronuclei and polar bodies after gentle removal of surrounding corona and cumulus cells. While visible signs of fertilization are microscopically evident several hours in a certain period post-insemination, the presence of two pronuclei (2PN stage) alongside two polar bodies signifies normal fertilization. The absence of pronuclei indicates failed fertilization. The rare observation of a single pronucleus often suggests spontaneous oocyte activation, which typically precludes viable embryo development. Conversely, the presence of three or more pronuclei denotes abnormal fertilization, commonly attributed to polyspermy (multiple sperm penetration) or a failure in second polar body extrusion, both potentially indicative of suboptimal oocyte quality due to impaired cytoplasmic factors. When sperm parameters are adequate, the typical fertilization rate observed with conventional IVF ranges between 60% and 80% [11]. However, when sperm parameters are suboptimal, conventional IVF is usually less efficient, and it can result in lower (or no) fertilization [12].

Despite being a relatively invasive technique, ICSI has proved to be the most successful alternative to conventional IVF. Initial preclinical trials of ICSI were limited, and the first pregnancy following ICSI was achieved in 1992 by Palermo [8]. While initial ICSI fertilization rates were modest (ranging from 30% to 40%), they consistently surpassed those achieved with other assisted fertilization methods (like subzonal insemination). Through meticulous refinement and optimization of the procedural steps over subsequent years, the fertilization success rate for ICSI has significantly improved, now routinely achieving 70% or higher. Importantly, even though ICSI is a relatively invasive technique, it appears that the technique itself is not associated with a meaningfully higher rate of chromosomal abnormalities in pregnancies following the ICSI procedure [13, 14]. Similarly, most studies have not observed a correlation between ICSI procedure and birth defects [15, 16].

The timing of cumulus-corona cell removal relative to oocyte retrieval and the subsequent ICSI procedure has been optimized in the last decades [14]. Current common practice dictates decumulation approximately 2–3 h post-oocyte retrieval, with the ICSI procedure performed approximately 4 h after oocyte collection (approximately 40 h after trigger administration). Upon denuding, oocytes undergo microscopic examination (typically at 200× or 400× magnification) to ascertain nuclear maturity and document any discernible morphological anomalies.

Achieving successful ICSI necessitates the precise delivery of sperm into the oocyte's cytoplasm. Sperm selection for microinjection necessitates the operator select a motile sperm exhibiting visually "adequate" morphology prior to immobilization and aspiration into the microinjection needle for direct cytoplasmic delivery into the oocyte. Immobilization, intentionally "breaking" the sperm tail at its midregion not only simplifies the injection process but also appears to contribute to oocyte activation and enhanced fertilization rates [14]. Successful microinjection then requires initial penetration of the zona pellucida, followed by a controlled breach of the oolemma to ensure intracellular sperm deposition. The development of ICSI was a true breakthrough in infertility treatment, as the technique can achieve normal fertilization and high success rates, as long as it is possible to obtain a single live sperm to inject the oocyte [17]. A significant advantage of ICSI lies in its applicability to surgically recovered sperm from either the epididymis or testis [14]. Epididymal sperm, typically obtained for patients with obstructive azoospermia, can be processed similarly to ejaculated samples using density gradient centrifugation, or high-speed centrifugation if the sample quality is suboptimal. Testicular sperm, harvested from both obstructive and nonobstructive cases [15], necessitates mechanical disruption of the tissue [16], and subsequent injection using ICSI.

In this chapter, we describe the methodology and critical parameters of conventional in IVF and ICSI. We additionally review the respective benefits and drawbacks of these insemination modalities, alongside relevant determinants for their selection and provide detailed description on how these procedures are typically performed.

2 Materials

2.1 Sperm Preparation for Conventional IVF and for ICSI with Adequate Sperm Parameters (No Significant Male-Factor Infertility)

1. Laminar flow hood.
2. Centrifuge with a rotor for 15 mL conical tubes.
3. Phase contrast microscope.
4. Makler (sperm) counting chamber.
5. Sperm Preparation Medium with protein supplement (e.g., Origio Sperm Preparation Medium with phenol red). Store at 2–7°C.
6. Separation gradients, e.g., PureSperm (50%, 70%, and 95%). Store at 2–7°C.
7. Sterile specimen cup.
8. Sterile Pasteur pipets.
9. 5 mL serological pipettes and pipettor.
10. Rubber bulb.
11. 15 mL conical tubes.
12. Gloves.
13. Glass slides.

2.2 Sperm Preparation for ICSI with Impaired Sperm Parameters (Significant Male-Factor Infertility)

1. Laminar flow hood.
2. Centrifuge.
3. Phase contrast microscope.
4. Makler (sperm) counting chamber.
5. Sperm Preparation Medium with protein supplement (e.g., Origio Sperm Preparation Medium with phenol red). Store at 2–7°C.
6. Sterile specimen cup.
7. Sterile Pasteur pipets.
8. Rubber bulb.
9. 5 mL serological pipette and pipettor.
10. 15 mL conical tubes.
11. Gloves.

2.3 Sperm Preparation for ICSI with Epididymal Sperm

1. Laminar flow hood.
2. Centrifuge.
3. Phase contrast microscope.
4. Makler (sperm) counting chamber.
5. Sperm Preparation Medium with protein supplement (e.g., Origio Sperm Preparation Medium with phenol red). Store at 2–7°C.
6. Sterile Pasteur pipets.
7. Rubber bulb.
8. 5 mL serological pipette and pipettor.
9. 15 mL conical tubes.
10. Gloves.

2.4 Sperm Preparation for ICSI with Testicular Sperm

1. Laminar flow hood.
2. Centrifuge.
3. Sperm Preparation Medium w/HEPES, supplemented w/ protein (e.g., Fujifilm Sperm Washing Medium). Store at 2–7°C.
4. Washed Embryo Culture Mineral Oil.
5. Sterile Pasteur pipets.
6. Rubber bulb.
7. 15 mL conical tubes.
8. Glass slides.
9. 25 G 11/2" needle.
10. Dissecting microscope.
11. Inverted microscope.
12. 60 × 15 mm embryo culture dishes.
13. Gloves.

2.5 Oocyte Preparation for Conventional IVF

1. Laminar flow workstation with warming surface.
2. Humidified 5% O_2, 5% CO_2, 90% N_2 gas supply, and gassing jar.
3. 5.5% CO_2 embryo culture incubator (with 5% O_2 preferentially), for e.g., a K-systems G210 InviCell.
4. Inverted microscope.
5. Dissecting microscope.
6. Drummond pipet aid.
7. Fertilization medium with protein supplement (e.g., Quinn's Advantage™ Fertilization (HTF) Medium). Store at 2–7°C.
8. Mineral oil for embryo culture (CO_2 equilibrated, *see* **Note 1**).

9. 60×15 mm dishes.
10. 35×10 mm Nunc petri dish.
11. 1 mL serological pipets.
12. 2 mL serological pipets.
13. 15 mL conical tubes.
14. Makler chamber.
15. 25 G 11/2" needles.
16. Diamond pen.
17. Alcohol burner with wick.

2.6 Oocyte Preparation for ICSI

1. Laminar flow workstation.
2. 5.5% CO_2 incubator (with 5% O_2 preferentially), for e.g., a K-systems G210 InviCell.
3. Inverted microscope with micro-manipulators.
4. Dissecting microscope.
5. Drummond pipet aid.
6. Pipetman P-10 pipettor.
7. Fertilization medium with protein supplement (e.g., Quinn's Advantage™ Fertilization (HTF) Medium). Store at 2–7°C.
8. Mineral oil for embryo culture (e.g., Ovoil—100% paraffin oil), equilibrated in a CO_2 incubator (*see* **Note 1**).
9. Human recombinant hyaluronidase (e.g., ICSI Cumulase) to be used at a final concentration of 10–40 U/mL. Store at <0°C.
10. 10% Polyvinylpyrrolidone (PVP; e.g., FujiFilm PVP solution). Store at 2–7°C.
11. 60×15 mm center well organ culture dish.
12. 60×15 mm cell culture dish.
13. 35×10 mm petri dish.
14. 50×9 mm ICSI dish.
15. Sterile Pasteur pipets.
16. Rubber bulb.
17. 5 mL serological pipets.
18. 15 mL conical tubes.
19. 25 G 11/2" needles.
20. Pasteur pipettes.
21. 150 μm Stripper Tips.
22. 275 μm Stripper Tips.
23. Warming block suitable for 1 mL vials.

24. ICSI needles.
25. Holding pipettes.
26. Diamond pen.
27. Wax pencil.

3 Methods

3.1 Sperm Preparation for Conventional IVF and for ICSI with Adequate Sperm Parameters (No Significant Male-Factor Infertility)

The semen sample is collected by masturbation into a sterile specimen cup. Confirm the name on the specimen cup with the laboratory notes and any other relevant paperwork (*see* **Note 2**). All samples should start to be processed within 1 h of receipt by the laboratory. Maintain the sample at room temperature after collection and during the preparation procedure. Wear gloves for the sperm preparation procedure.

1. Remove the Sperm Preparation Medium and separation gradients from the refrigerator and allow them to equilibrate to room temperature for a minimum of 2 h.
2. Evaluate the sample for acceptability (*see* **Note 3**). Samples containing white or red blood cells may be processed as per normal with separation gradients.
3. Allow the sample to liquefy at room temperature (*see* **Note 4**).
4. Label a sterile 15 mL conical tube with the patient's name, partner's name, date, and patient identification number. Mix the sample well with a sterile Pasteur pipet by gentle pipetting up and down, then transfer it to the labeled tube. Measure the volume using a sterile serological pipette.
5. Place a small drop on a Makler chamber and assess concentration, motility, forward progression (*see* **Note 5**), viscosity, agglutination, gelatinous material, and presence of red and white blood cells (*see* **Note 6**).
6. Add a few milliliters (~5) of Sperm Preparation Medium to the semen sample, mix the sample well using a Pasteur pipet then centrifuge at $500 \times g$ for 5 min.
7. While the sample is spinning, prepare a "separation" tube. Label a new 15 mL conical tube with the patient's details. Layer separation gradients with a sterile Pasteur pipet starting with 1 mL of 95%, overlayed with 1 mL 70%, and then 1 mL of 50%.
8. Using a new pipet, aspirate the supernatant from the semen sample tube taking care not to disturb the pellet. The supernatant may be expelled into the original specimen cup or

saved as "waste" in a labeled tube (*see* **Note 7**). Reconstitute the pellet in 1 mL Sperm Preparation Medium and mix well. Carefully layer the sperm suspension onto the surface of the gradient in the separation tube without disturbing the interface between each of the gradients. Spin at approximately 300 × *g* for 10 min.

9. Using another sterile pipet, remove the sperm pellet from the bottom of the conical tube, taking care not to pick up debris and nonmotile sperm as the pipet passes through the gradients. Transfer the pellet to a new, labeled 15 mL conical tube. Reconstitute the pellet in 4 mL Sperm Preparation Medium. Centrifuge at 500 × *g* for 5 min. Remove the supernatant and repeat the wash step once more.
10. After the final wash, remove the supernatant and reconstitute the pellet in 1 mL Sperm Preparation Medium.
11. Evaluate the sample using the Makler chamber and adjust the concentration using the following formula:

$$\frac{\text{Desired final concentration}}{\text{Concentration of sample}} = \text{X (the volume of sperm to add to Fertilization medium for a total volume of 1 mL)}$$

For example: if the concentration of the patient's sample is 1 million motile sperm in 1 mL of Sperm Preparation Medium and the desired concentration required is 200,000 motile sperm in 1 mL:

i. 200,000 ÷ 1,000,000 = 0.2 mL

ii. Add 0.2 mL motile sperm to 0.8 mL Fertilization Medium

12. Label a new tube with the patient's name, date, and identification number. Dilute the sample to the desired concentration to be used for insemination (approximately 100,000–250,000 sperm/mL).
13. Leave the sample tightly capped at room temperature in the laminar flow hood until required.
14. Record all relevant information according to standard procedures.

3.2 Sperm Preparation for ICSI with Impaired Sperm Parameters (Significant Male-Factor Infertility)

1. Remove the Sperm Preparation Medium from the refrigerator and allow it to equilibrate to room temperature for a minimum of 1 h.
2. Evaluate the sample for acceptability (*see* **Note 3**). Samples containing white or red blood cells may be processed as per normal with separation gradients.

3. Allow the sample to liquefy at room temperature (*see* **Note 4**).
4. Label a sterile 15 mL conical tube with the patient's name, partner's name, date, and patient identification number. Mix the sample gently with a sterile Pasteur pipet and bulb and transfer it to the labeled tube. Measure the volume using a sterile serological pipette.
5. Place a small drop on a Makler chamber and assess concentration, motility, activity, forward progression, viscosity, agglutination, gelatinous material, and presence of red and white blood cells (*see* **Notes 5 and 6**).
6. Add a few (~5) milliliters of Sperm Preparation Medium to the tube containing the semen sample, mix the sample well using a Pasteur pipet by gentle pipetting up and down then centrifuge at 600 × *g* for 10 min.
7. Using a new pipet, aspirate the supernatant from the semen sample taking care not to disturb the pellet. The supernatant may be expelled into the original specimen cup or saved as "waste" in a labeled tube. Save all supernatants until the sperm preparation is finished (*see* **Note 7**). Reconstitute the pellet in 1 mL Sperm Preparation Medium and mix well.
8. Evaluate the sample for concentration with a Makler chamber.
9. Leave the sample tightly capped at room temperature in the laminar flow hood.
10. Record all relevant information according to standard procedures.

3.3 Sperm Preparation for ICSI with Epididymal Sperm

Epididymal sperm is collected in a surgical room by a qualified urologist. The urologist collects the contents of the epididymis in a sterile conical tube containing 1–2 mL Sperm Preparation Medium that will be handed to the laboratory. Maintain the specimen at room temperature after collection and during the preparation procedure. All samples should be processed within 1–2 h of receipt by the laboratory. Wear gloves for the sperm preparation procedure.

1. Remove the Sperm Preparation Medium from the refrigerator and allow it to equilibrate to room temperature for at least 60 min prior to collection and processing.
2. Examine the specimen provided by the surgeon first visually then under the inverted microscope for tissue/cell content.
3. Mix the sample well with a sterile Pasteur pipet and bulb after adding 1–2 mL Sperm Preparation Medium. Measure the volume with a serological pipette, if possible.
4. Place a small drop on a Makler chamber and assess concentration, motility, activity, forward progression, and the presence

of red blood cells. Since this is a specimen recovered from the epididymis, there is no "reference range."

5. Centrifuge the sperm sample tube at 600 × *g* for 10 min.
6. Using a new pipet, aspirate the supernatant from the tube taking care not to disturb the pellet. The supernatant may be expelled into another labeled conical tube (*see* **Note 7**).
7. Reconstitute the pellet in 1 mL Sperm Preparation Medium and mix well. Evaluate the sample for concentration and motility using a Makler chamber.
8. Leave the sample tightly capped at room temperature in the laminar flow hood.
9. Record all relevant information according to standard procedures.

3.4 Sperm Preparation for ICSI with Testicular Sperm

Testicular tissue/sperm is collected in a surgical room by a qualified urologist. The urologist removes small pieces of the testis and puts them in a sterile conical containing 1–2 mL of Sperm Preparation Medium tube that will be handed to the laboratory. All samples should be processed within 1–2 h of receipt by the laboratory. Wear gloves for the sperm preparation procedure.

1. Remove the Sperm Preparation Medium from the refrigerator and allow it to equilibrate to room temperature for at least 60 min prior to testicular tissue collection and processing.
2. Examine the specimen provided by the surgeon using first visual inspection and then under the inverted microscope for tissue content.
3. Maintain the specimen at room temperature after collection and during the preparation procedure.
4. Evaluate the sample using an inverted microscope for tissue and cell content.
5. Add sperm preparation medium to a 60 mm culture dish and place the testicular tissue pieces in the dish.
6. Working at room temperature, use needles to disrupt the testicular tissue by tearing the tissue/tubes into smaller pieces as much as possible under a dissecting microscope.
7. Cover the content of the dish with mineral oil to prevent medium evaporation. Evaluate the sample using the inverted microscope for the presence (or absence) of sperm.
8. Leave the sample at room temperature in the laminar flow hood until required for ICSI.
9. Record all relevant information according to standard lab procedures.

3.5 Oocyte Preparation and Conventional IVF

Oocyte retrieval for conventional IVF typically occurs around 36-h post-ovulation trigger. During oocyte collection, the follicular fluid is aspirated into a tube that is passed to the embryologist who identifies oocytes that are surrounded by cumulus cells. For conventional insemination, cumulus-corona-oocyte complexes (CCOCs) are typically placed in culture drops of a Petri dish containing an adequate amount of sperm. Insemination typically occurs 40 h post trigger administration (or approximately 4 h after oocytes were collected).

3.5.1 Oocyte Preparation for Conventional IVF

All oocyte manipulations are performed on the warm surface of a laminar flow workstation. Work quickly to limit pH and temperature changes. Keep all dishes not in use under a gassing jar.

1. Prepare 60 × 15 mm cell culture dishes by dispensing 6 × 100 μL drops of Quinn's Advantage™ Fertilization (HTF) medium and cover with 5 mL mineral oil. Pre-equilibrate dishes overnight, or for at least 12 h, into an embryo culture incubator. Prepare sufficient dishes to account for the expected number of follicles to be aspirated (two to four oocytes per drop).
2. Following retrieval, dissect CCOCs using two 25 G 11/2" needles to remove degenerated cumulus cells and any cells with debris/blood in a 60 × 15 mm culture dish containing Quinn's Advantage™ Fertilization (HTF) Medium.
3. Transfer CCOCs to a 60 × 15 mm culture dish containing Quinn's Advantage™ Fertilization (HTF) Medium, covered with mineral oil to rinse. Thereafter, transfer two to four CCOCs to each prepared 100 μL drop. Transfer the dish to an embryo culture incubator until insemination time.

3.5.2 Conventional IVF Insemination

Perform insemination within 4 h post-retrieval (40 h after trigger administration).

1. Prepare a conventional IVF insemination dish by pipetting six drops of 100 μL Quinn's Advantage™ Fertilization (HTF) medium in each dish and cover with 5 mL mineral oil. Pre-equilibrate dishes overnight, or for at least 12 h, in an embryo culture incubator.
2. Remove patient dishes from the incubator. Confirm patient name on the bottom of the dish and the tube containing the sperm sample for insemination with the patient's laboratory forms.
3. Using a flame-polished Pasteur pipet and rubber bulb, underlay 10–12 sperm droplets (60–80 μL) along the edge of the conventional IVF dish. Prepare sufficient drops so that each drop can hold one to four oocytes. Check drops for correct density of sperm on an inverted microscope (*see* **Note 8**).

Number the insemination drops consecutively with a diamond pen and return the dish to the incubator for 1 h.

4. After 1 h, turn on the gas to the gassing chamber and remove the insemination and oocyte dishes from the incubator. Confirm patient name and oocyte number on the dishes and laboratory forms. Place the dishes under the gassing jar on the warming surface of the hood.
5. Using a flame-polished pipet and bulb, pick up each oocyte (cumulus-corona-oocyte complex), wash it through three of the drops. If necessary, use a pulled glass probe to gently loosen each oocyte from the bottom surface of the dish. Transfer each oocyte to one of the sperm drops along the edge of the dish. Work quickly but carefully (*see* **Note 9**). When all oocytes have been transferred to the sperm drops check the oocyte culture dish again to confirm the dish is empty. Check the insemination drops again for sperm motility and density on the inverted microscope.
6. Return dishes to the incubator for overnight culture (*see* **Note 10**).
7. Record all information on an IVF database or relevant paperwork as per standard laboratory procedures.

3.6 Oocyte Preparation and ICSI

Oocyte retrieval for ICSI employs the same procedures as conventional IVF, maintaining the typical 36-h interval post-ovulation trigger. A critical distinction for ICSI, however, necessitates the complete removal of surrounding cumulus and corona cells from the oocytes. This denuding process typically involves the synergistic application of enzymatic and mechanical techniques.

3.6.1 Oocyte Preparation for ICSI

All oocyte manipulations are performed on the warming surface of the laminar flow workstation. Work quickly to limit pH and temperature changes. Keep all dishes not in use under a gassing jar. Prepare sufficient culture dishes to accommodate the expected number of follicles to be aspirated.

1. Prepare a post-retrieval dish by pipetting 6×100 μL drops of Fertilization Medium into a 60×15 mm culture dish then overlaying with 4 mL of mineral oil. Pre-equilibrate dishes overnight, or for at least 12 h, in an embryo culture incubator.
2. Prepare 35 mm oocyte culture dishes by first pipetting approximately 3.5 mL Ovoil then underlaying 6× drops of 60–80 μL Fertilization Medium. These dishes serve as "pre-ICSI" and "post-ICSI" dishes. Pre-equilibrate dishes overnight, or for at least 12 h, in an embryo culture incubator.
3. Aliquot 5–10 mL of Fertilization Medium into a tube and pre-equilibrate overnight, or for at least 12 h, in an incubator gassed with 5.5% CO_2 (ideally in combination with 5% O_2).

4. On the day of oocyte retrieval, for each patient that requires ICSI insemination, place a Cumulase tube to a warming block (set to 37°C), at least 1 h before use.
5. At the conclusion of the retrieval, dissect CCOCs using two 25 g 11/2" needles to remove any degenerated cumulus cells and cells with debris/blood in a 60 × 15 mm culture dish containing pre-equilibrated Fertilization Medium.
6. At 30–60 min time post-retrieval, transfer CCOCs to a "post-retrieval" dish and rinse through two to three culture drops. Thereafter, two to four CCOCs are transferred to each drop of a "pre-ICSI" 35 × 10 mm oocyte culture dish labeled with the patient's name. Return the dish to the incubator until insemination time.
7. Immediately prior to stripping oocytes, prepare a 60 mm Centre Well Organ Culture Dish by adding the full content of a Cumulase tube.
8. Flame-polish a Pasteur pipet, then remove the patient's "pre-ICSI" dish from the culture incubator and confirm the name on the bottom of the dish. Using the flame-polished pipet, transfer the CCOCs to the center well dish containing the Cumulase.
9. Using a new flame-polished Pasteur pipet, slowly pipette in and out to remove the cumulus cells (1–2 min). Starting with 275 μm Stripper Tips, pick up the oocyte(s) and transfer them to a wash drop. Using consecutively smaller pipets (150 μm Stripper Tips), strip excess coronal cells from the oocyte by gentle pipetting the oocytes in and out from the pipette.
10. Wash the oocytes through at least three fresh Fertilization Media drops of a "pre-ICSI" dish then place in a holding drop.
11. When all oocytes are stripped, assess them for maturity on the inverted microscope. Gently roll them with a stripper tip (if needed) and separate those with a clear first polar body (Metaphase-II).
12. Place the mature oocytes into a fresh drop of Fertilization Medium in a "pre ICSI" dish. Return the dish to the incubator until ready for use.
13. Record the results on relevant paperwork according to standard procedures.

3.6.2 ICSI Insemination

1. Prepare "ICSI dishes" in the morning, prior to scheduled retrievals. Pipet 5 μL Fertilization medium into the middle of a 50 × 9 mm dish. Make subsequent drops of 5 μL directly above, below, to the right and to the left of the middle drops making a cross pattern. Then make four drops diagonally out

from the center drop (upper and lower right, upper and lower left) to achieve a circle. Immediately overlay the drops with 4 mL equilibrated embryo culture oil or as much as needed for the oil to cover all drops completely (*see* **Note 11**). Remove the center drop, change the pipet tip and replace the drop with 5 μL of 10% PVP. One ICSI dish should be prepared for every eight oocytes to be injected. Place the lid snugly on the dish and place into a humidified, non-CO_2 incubator for a minimum of 30 min.

2. Prepare the micromanipulation equipment. Check/adjust the following: the warming stage is on, the pipette holders and tubing for both the holding pipet and injection needle are properly attached, the holding syringe should be free of bubbles, the joysticks are adjusted midway between zero and five, the coarse control boxes are switched on and the video recording equipment is setup.
3. Confirm the patient's name, date, number of oocytes to be ICSI'd, and record the counter start time on relevant documentation.
4. Confirm the patient's name on laboratory forms, dishes, and the sperm tube containing the final specimen for insemination.
5. Adjust the left micromanipulator. Hold the back of the tool firmly and slide the holding pipette gently into the partially unscrewed microtool holder. Tighten the screw top so the holding pipet cannot move freely in the tool holder
6. Attach the tool holder containing the holding pipet to the left arm of the micromanipulator. Incline inwards under a 30° angle. Using a 4× objective, locate your tool, turn it so it will lay parallel to the bottom of the dish and enter the microscopic field at 9 o'clock. Check that coarse controls are adjusted to give you at least 1 cm range of movement up and down. Check the pipet at 40×—make sure the elbow of the tool will not drag on the bottom surface of the dish.
7. Repeat the previous steps to setup the injection needle in the right arm of the micromanipulator. This tool should enter the microscopic field at 3 o'clock. Inspect the injection needle for quality (i.e., absence of barbs on beveled edge, reasonable length of spike on needle tip, etc.).
8. Using the turbo coarse controls, move the pipets across the microscopic field to ensure range of movement. Check that neither the holding pipet nor injection needle will rest on the side of the dish. After pipets are aligned appropriately, raise them as high as the controls will allow in preparation for a dish to be placed on the stage.

9. Remove an ICSI dish from the incubator. Retrieve patient's washed sperm specimen from the culturing incubator and confirm the patient's identity with another embryologist. Confirm identity on all dishes and laboratory forms (*see* **Note 2**).
10. Using a pipetman, pipette an appropriate volume of sperm (1–2 μL) into the bottom edge of the center PVP drop. Cap the sperm tube tightly and leave it in the hood at room temperature. Starting at the top of the circle of microdrops and working counterclockwise, label drops one to eight with a wax pencil. Remove four oocytes from the oocyte dish and place each one into a labeled drop. Place the oocyte dish under the gassing jar.
11. Place the ICSI dish on the inverted microscope, first locating the sperm in the PVP microdrop. Lower the injection needle into the drop and allow the needle to fill with PVP. Scan the drop for morphologically normal, motile sperm. Immobilize the sperm with the ICSI needle by carefully rubbing the midpiece without separating the head from the tail. Ensure the sperm is immobilized by checking for a kink at the point of contact. Draw the sperm into the pipet, tail first. Draw the sperm up and down slowly in the needle to make sure the sperm will move freely within. When the sperm is stable within the needle, move to the first injection drop.
12. Refocus the ICSI needle, checking that the sperm is still in place in the pipet. Lower the holding pipet into the drop. Position the oocyte on the holding pipette (and apply sufficient suction to keep the oocytes securely in place) so that the polar body is visible in the perivitelline space at 6–8 o'clock and a fraction of the inside membrane of the zona pellucida is visualized at the 12 o'clock position. The oocyte should be barely resting on the bottom of the dish.
13. Maintain suction on the holding pipet and introduce the injection needle slowly through the zona pellucida at the 3 o'clock position. Thread the needle through the zona and move towards the 9 o'clock position. If the membrane does not break right away (which is preferable to immediate breakage), draw back the needle slightly then advance it down and back towards 9 o'clock, following the original path of the pipet.
14. Aspirate cytoplasm into the ICSI needle (until the oolemma breaks, which typically happens by aspirating half of the length of the sperm) and expel the sperm with the cytoplasm. Make sure the sperm head is seated well in a dense area of cytoplasm. If it appears to be "floating," aspirate the sperm again and expel it in another area.

15. Withdraw the ICSI needle slowly from the oocyte, then move the ICSI needle out of the microdrop. Move the oocyte to the edge of the microdrop as a way of indicating the oocyte has been injected using the holding pipet, then remove the holding pipet from the microdrop.
16. Repeat injection steps for the remaining three oocytes in their respective drops. Remove the "post-ICSI" culture dish from the incubator and confirm the patient's name. Number each drop consecutively using a diamond pen. Keeping them in numerical order, remove the oocytes from the wash droplets, and place them in the numbered drops. Place the dish under the gassing jar.
17. Repeat previous steps until all mature oocytes have been injected. Return all culture dishes to the appropriate incubator.
18. Remove pipets from the micromanipulators and discard in a sharp's container. Adjust the joysticks back to the zero position.
19. Complete relevant paperwork and input detail into patient management system as per standard procedures.

4 Notes

1. Recommended time for equilibration is a minimum of 6 h, but it typically performed overnight (i.e., 8–12 h).
2. Verifying a patient's ID at the time of gamete collection in IVF treatment is critically important to prevent the devastating consequences of misidentification. The most significant risk in an IVF lab is the mix-up of sperm, eggs, or embryos. A single mistake at the time of collection could lead to the fertilization of a woman's eggs with sperm from the wrong partner or donor. The ID check confirms that the gametes being collected—the eggs from the female partner and the sperm from the male partner or donor—belong to the correct, intended individuals. This ensures that the resulting embryo will be the biological child of the correct parents.
3. Initial assessment by visually checking for sufficient volume and presence of contamination, or the use of correct collecting cup. If there is any question regarding the acceptability of a specimen, the laboratory director is consulted.
4. You may use a stage warmer set to 37°C to liquefy the sample. It may take 30–60 min to have the sample fully liquefied.
5. ≥16 million/mL, Total Motility (any movement): ≥42%, Progressive Motility (PR): ≥30%. These values are statistical

references. A man whose semen analysis falls below these ranges may still be fertile, and a man whose results are above these ranges may still face fertility challenges. Semen parameters can vary significantly from one sample to another due to factors like sexual abstinence period, stress, and illness.

6. Viscosity, which refers to the thickness or stickiness of the semen can be assessed when pipetting the semen. Agglutination, which is clumping or sticking together of motile (moving) sperm is easy to see when checking sperm in the Makler chamber.
7. Save all supernatants until the sperm preparation is finished. The supernatant may be reprocessed by high-speed centrifugation (1000 × g) if no sperm was found in the pellet after initial processing.
8. You can adjust sperm density of the drops by adding sperm (or just fertilization media) if needed. Make sure that you have an adequate number of drops to accommodate all oocytes you intend to inseminate.
9. Make sure that you minimize the amount of medium carried over when transferring the oocytes as this can "dilute" sperm concentration.
10. The typical time period for conventional insemination is anywhere between 4 h and 18 h. Less than 4 h is not recommended because that can reduce the effectiveness of insemination and more than 18 h is also not recommended because that can lead of missing the observable signs of normal fertilization
11. It is critical that you add sufficient oil to make sure that all droplets are fully covered, otherwise some evaporation of the culture media can occur, and given the small amount of the droplets, this can result in significant changes in osmolarity.

References

1. Steptoe PC, Edwards RG (1978) Birth after the reimplantation of a human embryo. Lancet (London, England) 2(8085):366. https://doi.org/10.1016/s0140-6736(78)92957-4
2. Lopata A, Johnston IW, Hoult IJ, Speirs AI (1980) Pregnancy following intrauterine implantation of an embryo obtained by in vitro fertilization of a preovulatory egg. Fertil Steril 33(2):117–120
3. Trounson AO, Leeton JF, Wood C, Webb J, Wood J (1981) Pregnancies in humans by fertilization in vitro and embryo transfer in the controlled ovulatory cycle. Science 212(4495):681–682
4. Cohen J, Trounson A, Dawson K, Jones H, Hazekamp J, Nygren KG, Hamberger L (2005) The early days of IVF outside the UK. Hum Reprod Update 11(5):439–459. https://doi.org/10.1093/humupd/dmi016
5. Hull MG (1990) Indications for assisted conception. Br Med Bull 46(3):580–595. https://doi.org/10.1093/oxfordjournals.bmb.a072418
6. Gordon JW, Grunfeld L, Garrisi GJ, Talansky BE, Richards C, Laufer N (1988) Fertilization of human oocytes by sperm from infertile males after zona pellucida drilling. Fertil Steril 50(1):68–73. https://doi.org/10.1016/s0015-0282(16)60010-9

7. Lanzendorf SE, Maloney MK, Veeck LL, Slusser J, Hodgen GD, Rosenwaks Z (1988) A preclinical evaluation of pronuclear formation by microinjection of human spermatozoa into human oocytes. Fertil Steril 49(5):835–842. https://doi.org/10.1016/S0015-0282(16)59893-8
8. Palermo G, Joris H, Devroey P, Van Steirteghem AC (1992) Pregnancies after intracytoplasmic injection of single spermatozoon into an oocyte. Lancet (London, England) 340(8810):17–18. https://doi.org/10.1016/0140-6736(92)92425-f
9. Tsaadon A, Eliyahu E, Shtraizent N, Shalgi R (2006) When a sperm meets an egg: block to polyspermy. Mol Cell Endocrinol 252(1-2):107–114. https://doi.org/10.1016/j.mce.2006.03.037
10. Kattera S, Chen C (2003) Short coincubation of gametes in in vitro fertilization improves implantation and pregnancy rates: a prospective, randomized, controlled study. Fertil Steril 80(4):1017–1021. https://doi.org/10.1016/s0015-0282(03)01154-3
11. Wolf DP, Byrd W, Dandekar P, Quigley MM (1984) Sperm concentration and the fertilization of human eggs in vitro. Biol Reprod 31(4):837–848. https://doi.org/10.1095/biolreprod31.4.837
12. Hall J, Fishel S (1997) In vitro fertilization for male infertility: when and how?.. Baillieres Clin Obstet Gynaecol 11(4):711–724. https://doi.org/10.1016/s0950-3552(97)80008-9
13. Bonduelle M, Wilikens A, Buysse A, Van Assche E, Wisanto A, Devroey P, Van Steirteghem AC, Liebaers I (1996) Prospective follow-up study of 877 children born after intracytoplasmic sperm injection (ICSI), with ejaculated epididymal and testicular spermatozoa and after replacement of cryopreserved embryos obtained after ICSI. Hum Reprod 11(Suppl 4):131–155. https://doi.org/10.1093/humrep/11.suppl_4.131 discussion 156–139.
14. Bonduelle M, Van Assche E, Joris H, Keymolen K, Devroey P, Van Steirteghem A, Liebaers I (2002) Prenatal testing in ICSI pregnancies: incidence of chromosomal anomalies in 1586 karyotypes and relation to sperm parameters. Hum Reprod 17(10):2600–2614. https://doi.org/10.1093/humrep/17.10.2600
15. Lie RT, Lyngstadaas A, Orstavik KH, Bakketeig LS, Jacobsen G, Tanbo T (2005) Birth defects in children conceived by ICSI compared with children conceived by other IVF-methods; a meta-analysis. Int J Epidemiol 34(3):696–701. https://doi.org/10.1093/ije/dyh363
16. Yan J, Huang G, Sun Y, Zhao X, Chen S, Zou S, Hao C, Quan S, Chen ZJ (2011) Birth defects after assisted reproductive technologies in China: analysis of 15,405 offspring in seven centers (2004 to 2008. Fertil Steril 95(1):458–460. https://doi.org/10.1016/j.fertnstert.2010.08.024
17. Nagy ZP, Liu J, Joris H, Verheyen G, Tournaye H, Camus M, Derde MC, Devroey P, Van Steirteghem AC (1995) The result of intracytoplasmic sperm injection is not related to any of the three basic sperm parameters. Hum Reprod 10(5):1123–1129. https://doi.org/10.1093/oxfordjournals.humrep.a136104

Chapter 7

Culture of the Human Embryo

Kathryn H. Gurner and David K. Gardner

Abstract

Fundamental to the success of the IVF laboratory is the use of an optimized embryo culture system that facilitates the production of high-quality blastocysts. This is achieved through the establishment of a culture system that encompasses the provision of essential nutrients by use of appropriate culture media, the maintenance of a stable environment in vitro and establishment of effective quality control protocols within the laboratory. While there is considerable variability in the culture systems used in IVF laboratories worldwide, as no standardized system exists, all modern embryo culture systems could be optimized by following these fundamental principles. This chapter describes one method for human blastocyst culture using a time-lapse culture system and highlights key considerations for optimizing embryo culture protocols.

Keywords Time-lapse, Sequential medium, Low oxygen tension, Blastocyst, In vitro embryo culture

1 Introduction

The main goal of the IVF laboratory is to produce high-quality blastocysts capable of establishing successful pregnancies. Vital to this is the use of an embryo culture system that maintains the inherent viability of the embryo and helps to facilitate its development. Early culture media were simplistic and based on the simple salt solutions developed for mouse embryos [1, 2], or tissue culture media adopted from tissue culture systems for somatic cell growth. The former media lacked the complex nutritional components that we now know are essential for sustaining embryonic growth, while the latter were not designed with the specific needs of the preimplantation embryo in mind. Consequently, human IVF success rates were low, and the field faced substantial challenges. However, extensive research in the 1980s and 1990s into the nutritional requirements of mammalian gametes and embryos [3, 4], in parallel with analyses of the composition of the luminal fluid within the human fallopian tube and uterus [5, 6] facilitated the development of culture media that contained an aqueous mix of salts, glucose, lactate, pyruvate,

Alexandra J. Harvey and David K. Gardner (eds.), *The Preimplantation Embryo*, Methods in Molecular Biology, vol. 3038,
https://doi.org/10.1007/978-1-0716-5292-3_7,

vitamins, amino acids, chelators, and albumin [1, 7, 8]. Such media are capable of supporting culture of human embryos to the blastocyst stage [9], a practice that results in high implantation rates and low multiple birth rates [10] and has thus now become a standard of care worldwide.

Today, embryo culture media utilized in human IVF clinics typically take on one of two forms; sequential/two-step media, or monophasic/one-step media. Sequential media were developed to reflect the changing nutrient composition (particularly of carbohydrates and amino acids) within the female reproductive tract between the oviduct and uterus, which supports the dynamic physiological and developmental needs of the embryo [5, 6, 11, 12]. In contrast, monophasic media have been developed based on the premise of "let the embryo choose." Examples of the latter media include KSOM that was developed using computer simplex optimization [13, 14] to formulate the ions, glucose, pyruvate and lactate [13, 14], and was subsequently supplemented with the amino acids shown to improve mouse embryo development in culture [15], KSOMAA [16]. While sequential media are more physiological (i.e., reflective of the in vivo environment), interestingly to date, there is no overarching scientific agreement as to which approach is superior for human embryo culture [17, 18]. Hence, the type of culture media used varies between laboratories worldwide. Additionally, there is no consensus on whether a media renewal in monophasic media is beneficial, or at which stage it should be performed.

The production of high-quality, viable blastocysts is also influenced by factors in the culture system beyond the culture media used. Embryos can grow under various environmental conditions due to their high levels of developmental plasticity [19]. However, deviations from an in vivo-like environment, characterized by low oxygen levels (~5%) [20–22], stable 37°C temperature, and controlled pH [21, 23] can induce embryonic stress [19, 24, 25]. IVF laboratory equipment and protocols are, therefore, designed to control these environmental conditions. For instance, traditional, large box incubators originally used for embryo culture, have now predominantly been replaced by the use of benchtop-style incubators, which have improved temperature and gas control and have faster recovery times, enabling improved stability of environmental conditions [26, 27]. Further, use of time-lapse incubators, in combination with monophasic/one-step medium, facilitate undisturbed culture to the blastocyst stage by eliminating the need to remove dishes from the incubator for daily morphology checks and media changes. This minimizes fluctuations in temperature and pH and reduces exposure to atmospheric oxygen levels. However, the use of monophasic media over two-step media does pose potential risks, such as nutrient deprivation and the buildup of toxic compounds like ammonium [28, 29]. Outside of the incubator, for short-term procedures, pH is typically controlled by using HEPES or MOPS buffered media [30]. For longer procedures, the use of a gassed IVF isolette, equipped with a heated stage and microscope, can maintain pH and temperature during embryo handling before the embryo is returned to its culture incubator [31]. Further, while the osmolality of embryo culture medium is typically determined by commercial manufacturers, this can be significantly affected during culture dish preparation. Factors such as preparation time, surface temperature leading to evaporation, the size of the culture media drop, and the presence of an oil overlay can all impact osmolality [32]. Therefore, efficient protocols for dish preparation are crucial to maintaining optimal conditions.

The advent of time-lapse incubation, and the increasing desire to track embryos individually for preimplantation genetic testing has also seen individual culture of embryos become standard practice in many laboratories. Embryos were traditionally cultured in groups rather than individually due to the association of group embryo culture with better embryonic development [33–36], largely attributed to the presence of paracrine signaling factors, which are absent in individual culture. Development in individual microdrop culture can be improved by reducing the volume of culture medium used per embryo (i.e., embryo density) and thus concentrating positive autocrine factors directly around the embryo. However, reports regarding optimal density are inconsistent [19, 24] likely due to the fact that optimum embryo density depends on other culture conditions, such as oxygen levels [36, 37]. Notably, the introduction of commercially available microwell culture dishes compatible with time-lapse systems has helped to alleviate some of the challenges in individual embryo culture by utilizing designs with multiple microwells under shared media reservoirs (WOW dishes) [38] or separate media drops that help to concentrate embryo secreted factors. This advancement has made individual embryo culture a more viable and effective option in many laboratories and clinics.

Despite the clinical application of human embryo culture, there is considerable variability in protocols and techniques employed across different laboratories [39]. This variability is partly due to constraints in equipment availability and space, the desire to adopt new protocols, as well as a lack of consensus in the literature regarding best practices for embryo culture. However, general principles that aim to maintain a stable environment for the embryo throughout its development (as discussed) need to be applied in all laboratories, irrespective of which culture system is utilized. Additionally, it is crucial to ensure that all plastic consumables and supplies have been prescreened for embryo toxicity prior to use. Typically, prescreening by use of the mouse embryo assay (MEA) [40] for commercially available products is completed by manufacturers, however, additional testing by use of a sperm toxicity assay [41] prior to introduction and use of new plasticware is recommended, especially if the contact supply has not been MEA tested. Adequate quality control protocols must also be implemented to ensure that equipment is properly calibrated and functioning within acceptable ranges.

This chapter describes one method for human blastocyst culture, focusing on the use of the Embryoscope time-lapse culture system.

2 Materials

2.1 Equipment

1. Time-lapse incubator system (e.g., Embryoscope Plus).
2. Equilibration incubator (e.g., large box incubator, or small benchtop incubator).
3. Laminar flow hood equipped with stereomicroscope.
4. CO_2-O_2 Gas analyser.
5. Type K thermocouple.
6. Gassed IVF isolette equipped with stereomicroscope.
7. Heated stage/plate.
8. Round surface thermometer (RST).
9. Repeat dispense pipettor in various sizes (5 mL, 200 μL).
10. Handling micropipette and holder in various sizes (155 μm, 200 μm) (*see* **Note 1**).
11. 200 μL pipettor.
12. Thermometer.

2.2 Consumables

1. Nitrogen and CO_2 gas supply.
2. Time-lapse culture dishes.
3. Sterile Gloves.
4. Sterile single use pipette tips (5 mL, 200 μL).
5. 35 mm petri dish.
6. 5 mL round bottom tubes.

2.3 Media

1. Paraffin or mineral oil (*see* **Note 2**).
2. One-step embryo culture media (*see* **Note 3**).
3. 2 mL aliquots of media containing phenol red (*see* **Note 4**).

3 Methods

3.1 Equipment Calibration and Maintenance

All equipment used in the IVF laboratory must tested and/or calibrated prior to use. Thereafter, equipment should be monitored daily to identify major fluctuations in optimized conditions and recalibrated at least twice yearly.

3.1.1 Incubator (and Isolette) Daily Monitoring

1. Use a certified thermometer to measure and record internal incubator temperatures
2. Visually inspect indicator tubes containing phenol red inside incubators (*see* **Note 5**) to verify the absence of significant changes in incubator gas conditions (*see* **Note 6**).
3. Record the incubator set and display values.
4. Once a week, use a certified/calibrated electronic gas analyser to assess gas conditions within incubators (and isolettes) to ensure they remain properly calibrated.

3.1.2 Calibration of Heated Stages

1. Prepare a test dish containing a droplet of embryo culture media covered with oil and then warm the dish for at least 20 min (*see* **Note 7**).
2. Place test dish on heated stage and immerse thermocouple in media drop. Record temperature displayed on thermocouple and alter the set temperature of the stage until the thermocouple remains stabilized at 37°C (*see* **Note 8**).
3. Thereafter, daily monitoring of surface temperatures should be conducted with the use of a calibrated RST to ensure stage temperature does not drift outside an acceptable range (*see* **Note 9**).

3.2 Embryo Culture Dish Preparation

All culture dishes should be prepared using aseptic technique in a sterile laminar flow hood, containing a stereomicroscope on a non-heated surface.

All dishes should be prepared the day prior to use and equilibrated overnight at 5% O_2, 7% CO_2 (*see* **Note 10**). All dishes and tubes should be off-gassed prior to use (*see* **Note 11**).

The volume of embryo culture media and paraffin oil required for microwell dishes varies between dish type. As such, dishes should be prepared as per specific manufacturer's instructions (*see* **Note 12**). Here, we describe the preparation of EmbryoScope Plus dishes;

1. Fill wash wells with 40 µL of embryo culture medium using a repeat dispense pipettor (*see* **Note 13**).
2. Under the stereomicroscope, using a micropipette tip (maximum diameter of 200 µm), fill all microwells in the dish with embryo culture medium from the wash wells (*see* **Note 14**).

3. Once all microwells are filled, add 180 μL of culture medium to each reservoir of the Embryoscope plus dish using a repeat pipettor and immediately cover with 1.6 mL paraffin oil (*see* **Note 15**).
4. Ensure that oil covers all wells and then place the lid carefully on the embryo culture dish, avoiding contact with oil.
5. Following preparation, all embryo culture dishes must be transferred to an equilibration incubator (5% O_2, 7% CO_2) for overnight equilibration prior to use.

3.3 Developmental Timeline/Blastocyst Assessment

Following transfer of inseminated oocytes to pre-equilibrated embryo culture microwell dishes (developmental Day 0 (ICSI) or Day 1 (IVF) (*see* **Note 16**) embryos are cultured in the time-lapse incubator to the blastocyst stage undisturbed until developmental Day 5 (*see* **Note 17**).

1. Assess fertilization by the appearance of 2 pronuclei (2PN) at ~16–19 hpi (Day 1) utilizing the remote viewer associated with the time-lapse incubator (e.g., EmbryoViewer) (*see* **Note 18**).
2. In the morning of developmental Day 5 (~116 hpi), assess the development of each embryo individually. Firstly, reconfirm the presence of 2 pronuclei (2PN) by watching time-lapse footage (video) until 2PN fade. Then, use the Gardner Grading system system [42] to give each blastocyst a number grade based on the level of blastocoel formation and a letter grade for the quality and cohesiveness of both the ICM and the TE (*see* Fig. 1).
3. Select the embryo with the highest morphological grade for transfer (where relevant).
4. Select embryos with grade ≥3BB for vitrification (*see* **Note 19**).

3.4 Medium Replenishment

Embryos that are not at a developmental stage for utilization on Day 5 can be returned to the incubator following medium replenishment and reassessed for vitrification suitability on Day 6 and/or Day 7. The method and volume of media replenished in microwell dishes varies between dish type. As such, medium replenishment should be conducted as per specific manufacturer's instructions. For example, to replenish medium in Embryoscope Plus dishes:

1. On the day prior to blastocyst assessment (Day 4), prepare an aliquot of embryo culture medium in a loose capped 5 mL round bottom tube and pre-equilibrate overnight (*see* **Note 20**).
2. In the morning of Day 5 (*see* **Note 21**), following embryo assessment for transfer and vitrification and removal of the embryo selected for transfer, refresh the embryo culture medium by firstly taking the embryo culture dish from the

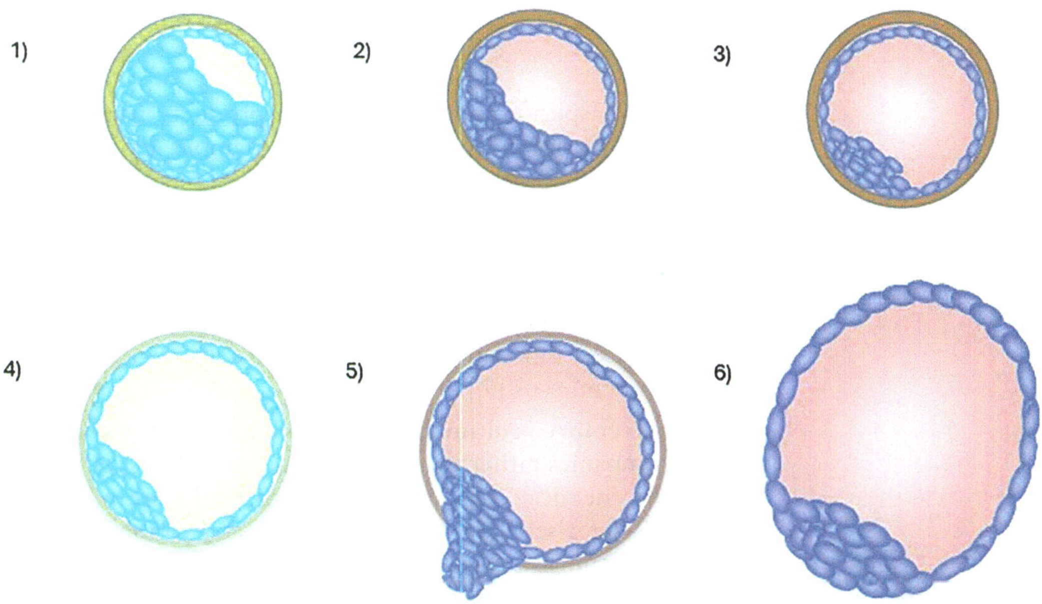

Fig. 1 Gardner Grading System for the human blastocyst. (1) An early blastocyst with a blastocoel that is less than half of the volume of the embryo; (2) a blastocyst with a blastocoel that is half of or greater than half of the volume of the embryo; (3) a full blastocyst with a blastocoel completely filling the embryo; (4) an expanded blastocyst with a blastocoel volume larger than that of the early embryo, with a thinning zona; (5) a hatching blastocyst with the trophectoderm starting to herniate through the zona; and (6) a hatched blastocyst, in which the blastocyst has completely escaped from the zona

incubator and placing it on a heated stage in a gassed isolette (*see* **Note 22**).

3. Using a 200 μL manual pipette and sterile tip, carefully remove 90 μL of medium from the top corner of one reservoir and expel the medium into a 35 mm discard petri dish. Replace the medium in the reservoir with 90 μL pre-equilibrated culture medium (*see* **Note 23**).
4. Carefully remove 90 μL of medium from the other end of the medium reservoir, expel it into a 35 mm discard petri dish and replace with 90 μL of fresh pre-equilibrated culture medium.
5. Repeat for the reservoir on the other side of the Embryoscope dish (*see* **Note 24**).
6. Ensure that all embryos are accounted for and return the culture dish to the time-lapse incubator for further culture until developmental Day 6 or 7 depending on laboratory protocols.

Any embryos that are not suitable for vitrification (i.e., <3BB) by Day 7 should be discarded as per laboratory protocols.

4 Notes

1. Many IVF specific companies produce products made for the purpose of embryo manipulation, for example, Ezy-squeeze, Vitrolife glass pipettes, or the stripper micropipettor.
2. Oil overlays are an important component of the embryo culture system, particularly where "dry incubators" are utilized to minimize fluctuations in temperature and pH [43], while also preventing evaporation of culture media and consequent increases in osmolality, all of which are detrimental to gametes and embryos [7].
3. Human embryo culture media is produced by large commercial companies rather than being prepared in-house, thus the specific chemical composition for such media are not available. Refer elsewhere [2], for example and formulations of media formulations used over the past 40 years.
4. Round bottomed 5 mL tubes containing 2 mL of phenol red containing medium can be placed in incubators to act as a visual indicator of incubator CO_2 levels. These tubes should be placed in incubators loose-capped and replaced fortnightly.
5. Phenol red solutions exhibit a gradual transition from yellow to red over the pH range 6.8–8.2. Above pH 8.2, phenol red turns a bright pink color while below a pH of 7.4, the medium tends towards yellow. Phenol red indicator tubes in the IVF laboratory should remain pale salmon/pink in color, as this indicates a pH of 7.4.
6. All readings should be taken when the incubator/isolette has been closed for an extended period of time (at least 4 h) to provide an accurate result. Additionally, in IVF labs where gassed isolettes are used, gas and temperature analyses also need to be performed on these units.
7. Stage temperature should be calibrated based on all dish types that are utilized on each particular stage in the laboratory. This is due to differences in their designs (e.g., whether the dish sits flush on the stage or has an air gap) and how this impacts temperature within the dish. Other factors such as the temperature of the lab, airflow, and use of dish lids can also impact temperature measurements and should be considered when calibrating heated stages.
8. Stage set temperature may be higher or lower than 37°C, but use of this method ensures that the drop of medium where the embryo is cultured is maintained at 37°C. Additionally, temperatures should always be set for the warmest case scenario, to ensure the temperature in the embryo culture media never exceeds 37°C.

9. Each heated stage should have an assigned round surface thermometer (RST) that is calibrated to that specific stage. RST daily values should remain consistent across daily readings. If readings change or drift, the heated stage should be removed from clinical use until validation of temperature using a thermocouple has been completed.
10. Optimal CO_2 concentrations may be different for different media systems. Therefore, laboratories need to optimize the CO_2 concentration in their incubators accordingly.
11. Off-gassing of plasticware prior to use in the IVF laboratory is important in reducing the presence of volatile organic compounds (VOCs) in the laboratory, which can be toxic to gametes and embryos. To off-gas dishes, remove dishes from packaging in a laminar flow hood and leave them to vent. [44]. It is important to maintain sterility of consumables during the off-gassing process.
12. In the absence of a time-lapse incubator system, embryos can be cultured in microdrops in a standard benchtop incubator using sequential medium. To prepare microdrop culture dishes, firstly rinse a single-use sterile tip by taking up the required volume of media and then expelling into a discard dish, before placing six 25 μL drops of phase one medium into a 60 mm petri dish. Ideally, these four drops should be placed at the 3, 6, 9, and 12 o'clock positions (to be used for embryo culture - 4 embryos per drop) and the fifth and sixth drops should be placed in the middle of the dish (wash drops). Immediately cover microdrops with 9 mL paraffin oil. Then, using a new rinsed tip for each drop, add a further 25 μL of medium to each original drop. Repeat for the required number of embryo culture dishes and ensure a maximum of two culture dishes are prepared at any given time. Following preparation, all embryo culture dishes must be transferred to an equilibration incubator for overnight equilibration prior to use.
13. Use of a repeat dispense pipette reduces the risk of RSI for embryologists and increases the speed of dish preparation. This is important from both an efficiency perspective but also for maintenance of appropriate culture medium osmolality and reduction of contamination risks.
14. It is easiest to do this by letting the tip of the pipette touch the side of the microwell and filling until a convex meniscus is created. Prefilling the microwells in time-lapse dishes is important in preventing formation of bubbles in culture wells. The presence of bubbles may hinder image acquisition in the time-lapse incubator.

15. The volume of oil required to cover culture wells may differ depending on viscosity and type of oil used. Individual laboratories will need to optimize the volume used in their dishes.
16. Ensure that IVF inseminated oocytes are transferred from insemination dishes to culture dishes by 16 h post-insemination (hpi) to ensure fertilization is not missed.
17. Alternatively, embryos cultured in standard (non-timelapse) incubators using microdrops are maintained undisturbed until Day 3, after which a media change is performed before culture is continued to the blastocyst stage. Microdrop culture dishes will additionally be removed from the incubator daily for morphology assessments under the microscope.
18. For microdrop culture/non-timelapse culture, fertilization is assessed by observing oocytes under an inverted microscope and gently rolling with a micropipette until pronuclei are visualized.
19. Some time-lapse systems create an AI-generated blastocyst grade or score. This score can be used as a selection tool to help differentiate and determine FET triage number for embryos assigned the same morphological grade. AI-based selection tools can also be purchased as additions to time-lapse systems separately (e.g., CHLOE).
20. For group microdrop culture with sequential media, a new culture dish containing microdrops of phase 2 (blastocyst stage) medium needs to be prepared (as described in **Note 12**) in the afternoon of Day 2 and equilibrated overnight. Additionally, one wash dish (per ten embryos) and one sorting dish needs to be prepared for each patient. These dishes can be prepared by placing 1 mL of second phase medium into the center of an organ well dish and 2 mL of medium into the outer well and transferring dishes to an equilibration incubator. If medium changeover is to occur outside of an isolette, HEPES/MOPS-buffered medium with amino acids should be used instead, and the dish should be warmed on a heated stage rather than equilibrating in a gassed incubator.
21. For group microdrop culture, embryos should be moved from first phase to second phase media between 9 am and midday on Day 3. To do so, use a 190 μm micropipette to firstly remove embryos from the phase 1 dish and wash thoroughly in wash dish. Next, transfer the embryos to the sorting dish and group those with similar morphology together. Move the embryos through the two wash drops in the second phase culture dish and then place 4 embryos in each drop of phase 2 medium. Return the dish to the culture incubator for further culture.

22. Use of gassed isolettes is advised where possible to reduce exposure of the embryo(s) to atmospheric air. If a gassed isolette is not available, medium replenishment must occur quickly on a heated microscope stage. This is vital to reduce the time embryos are exposed to atmospheric air and pH changes outside of the stable incubator environment.
23. To avoid creating bubbles, only aspirate or dispense to the first stop of the pipette when removing or adding media to culture dishes.
24. Medium renewal can also be performed by transferring embryos into a new, pre-equilibrated dish containing fresh media. To perform this, new culture dishes must be prepared in the afternoon the day prior to media renewal. In a gassed isolette (where possible) wash each embryo successively through two wash drops and then transfer to new culture well. Repeat until culture embryos have been transferred to new culture dish and return the dish to the incubator for further culture.

References

1. Gardner DK (2008) Dissection of culture media for embryos: the most important and less important components and characteristics. Reprod Fertil Dev 20(1):9–18. https://doi.org/10.1071/rd07160
2. Gardner DK, Lane M (2017) Embryo culture systems. In: Gardner DK, Simon C (eds) Handbook of IVF, 4th edn. pp 205–244
3. Gardner DK 1987. The nutrition and energy metabolism of the preimplantation mouse embryo.. University of York, UK
4. Gardner DK (1998) Changes in requirements and utilization of nutrients during mammalian preimplantation embryo development and their significance in embryo culture. Theriogenology 49(1):83–102. https://doi.org/10.1016/s0093-691x(97)00404-4
5. Gardner DK, Lane M, Calderon I, Leeton J (1996) Environment of the preimplantation human embryo in vivo: metabolite analysis of oviduct and uterine fluids and metabolism of cumulus cells. Fertil Steril 65(2):349–353. https://doi.org/10.1016/s0015-0282(16)58097-2
6. Gardner DK, Leese HJ (1990) Concentrations of nutrients in mouse oviduct fluid and their effects on embryo development and metabolism in vitro. J Reprod Fertil 88(1):361–368. https://doi.org/10.1530/jrf.0.0880361
7. Morbeck DE, Paczkowski M, Fredrickson JR, Krisher RL, Hoff HS, Baumann NA, Moyer T, Matern D (2014) Composition of protein supplements used for human embryo culture. J Assist Reprod Genet 31(12):1703–1711. https://doi.org/10.1007/s10815-014-0349-2
8. Chronopoulou E, Harper JC (2015) IVF culture media: past, present and future. Hum Reprod Update 21(1):39–55. https://doi.org/10.1093/humupd/dmu040
9. Gardner DK, Vella P, Lane M, Wagley L, Schlenker T, Schoolcraft WB (1998) Culture and transfer of human blastocysts increases implantation rates and reduces the need for multiple embryo transfers. Fertil Steril 69(1):84–88. https://doi.org/10.1016/s0015-0282(97)00438-x
10. Gardner DK, Schoolcraft WB, Wagley L, Schlenker T, Stevens J, Hesla J (1998) A prospective randomized trial of blastocyst culture and transfer in in-vitro fertilization. Hum Reprod 13(12):3434–3440. https://doi.org/10.1093/humrep/13.12.3434
11. Gardner DK, Lane M (1998) Culture of viable human blastocysts in defined sequential serum-free media. Hum Reprod 13(Suppl 3):148–159. https://doi.org/10.1093/humrep/13.suppl_3.148 Discussion 160.
12. Gardner DK, Lane M (1997) Culture and selection of viable blastocysts: a feasible proposition for human IVF? Hum Reprod Update 3(4):367–382. https://doi.org/10.1093/humupd/3.4.367

13. Lawitts JA, Biggers JD (1991) Optimization of mouse embryo culture media using simplex methods. J Reprod Fertil 91(2):543–556. https://doi.org/10.1530/jrf.0.0910543
14. Lawitts JA, Biggers JD (1993) Culture of preimplantation embryos. Methods Enzymol 225:153–164. https://doi.org/10.1016/0076-6879(93)25012-q
15. Gardner DK, Lane M (1993) Amino acids and ammonium regulate mouse embryo development in culture. Biol Reprod 48(2):377–385. https://doi.org/10.1095/biolreprod48.2.377
16. Ho Y, Wigglesworth K, Eppig JJ, Schultz RM (1995) Preimplantation development of mouse embryos in KSOM: augmentation by amino acids and analysis of gene expression. Mol Reprod Dev 41(2):232–238. https://doi.org/10.1002/mrd.1080410214
17. Dieamant F, Petersen CG, Mauri AL, Comar V, Mattila M, Vagnini LD, Renzi A, Petersen B, Ricci J, Oliveira JBA, Baruffi RLR, Franco JG Jr (2017) Single versus sequential culture medium: which is better at improving ongoing pregnancy rates? A systematic review and meta-analysis. JBRA Assist Reprod 21(3):240–246. https://doi.org/10.5935/1518-0557.20170045
18. Youssef MM, Mantikou E, van Wely M, Van der Veen F, Al-Inany HG, Repping S, Mastenbroek S (2015) Culture media for human pre-implantation embryos in assisted reproductive technology cycles. Cochrane Database Syst Rev 2015(11):Cd007876. https://doi.org/10.1002/14651858.CD007876.pub2
19. Gardner DK, Kelley RL (2017) Impact of the IVF laboratory environment on human preimplantation embryo phenotype. J Dev Orig Health Dis 8(4):418–435. https://doi.org/10.1017/s2040174417000368
20. Fischer B, Bavister BD (1993) Oxygen tension in the oviduct and uterus of rhesus monkeys, hamsters and rabbits. J Reprod Fertil 99(2):673–679
21. Ng KYB, Mingels R, Morgan H, Macklon N, Cheong Y (2018) In vivo oxygen, temperature and pH dynamics in the female reproductive tract and their importance in human conception: a systematic review. Human Reprod Update 24(1):15–34. https://doi.org/10.1093/humupd/dmx028
22. Mastroianni L Jr, Jones R (1965) Oxygen tensions within the rabbit fallopian tube. J Reprod Fertil 9:99–102
23. Hugentobler S, Morris DG, Kane MT, Sreenan JM (2004) In situ oviduct and uterine pH in cattle. Theriogenology 61(7–8):1419–1427. https://doi.org/10.1016/j.theriogenology.2003.08.008
24. Wale PL, Gardner DK (2016) The effects of chemical and physical factors on mammalian embryo culture and their importance for the practice of assisted human reproduction. Hum Reprod Update 22(1):2–22. https://doi.org/10.1093/humupd/dmv034
25. Sciorio R, Rinaudo P (2023) Culture conditions in the IVF laboratory: state of the ART and possible new directions. J Assist Reprod Genet 40(11):2591–2607. https://doi.org/10.1007/s10815-023-02934-5
26. Swain JE (2014) Decisions for the IVF laboratory: comparative analysis of embryo culture incubators. Reprod Biomed Online 28(5):535–547. https://doi.org/10.1016/j.rbmo.2014.01.004
27. Fujiwara M, Takahashi K, Izuno M, Duan YR, Kazono M, Kimura F, Noda Y (2007) Effect of micro-environment maintenance on embryo culture after in-vitro fertilization: comparison of top-load mini incubator and conventional front-load incubator. J Assist Reprod Genet 24(1):5–9. https://doi.org/10.1007/s10815-006-9088-3
28. Sacha CR, Gopal D, Liu CL, Cabral HR, Stern JE, Carusi DA, Racowsky C, Bormann CL (2022) The impact of single-step and sequential embryo culture systems on obstetric and perinatal outcomes in singleton pregnancies: the Massachusetts Outcomes Study of Assisted Reproductive Technology. Fertil Steril 117(6):1246–1254. https://doi.org/10.1016/j.fertnstert.2022.03.005
29. Werner MD, Hong KH, Franasiak JM, Forman EJ, Reda CV, Molinaro TA, Upham KM, Scott RT Jr (2016) Sequential versus Monophasic Media Impact Trial (SuMMIT): a paired randomized controlled trial comparing a sequential media system to a monophasic medium. Fertil Steril 105(5):1215–1221. https://doi.org/10.1016/j.fertnstert.2016.01.005
30. Swain JE, Pool TB (2009) New pH-buffering system for media utilized during gamete and embryo manipulations for assisted reproduction. Reprod Biomed Online 18(6):799–810. https://doi.org/10.1016/s1472-6483(10)60029-6
31. Swain JE, Kan MT, Moayeri SE, Schoolcraft WB (2015) Real-time assessment of pH stability within an IVF isolette chamber. Fertil Steril 103(2):e37. https://doi.org/10.1016/j.fertnstert.2014.12.071
32. Swain JE, Cabrera L, Xu X, Smith GD (2012) Microdrop preparation factors influence culture-media osmolality, which can impair mouse embryo preimplantation development. Reprod Biomed Online 24(2):142–147. https://doi.org/10.1016/j.rbmo.2011.10.008

33. Moessner J, Dodson WC (1995) The quality of human embryo growth is improved when embryos are cultured in groups rather than separately. Fertil Steril 64(5):1034–1035. https://doi.org/10.1016/s0015-0282(16)57925-4
34. Ebner T, Shebl O, Moser M, Mayer RB, Arzt W, Tews G (2010) Group culture of human zygotes is superior to individual culture in terms of blastulation, implantation and life birth. Reprod Biomed Online 21(6):762–768. https://doi.org/10.1016/j.rbmo.2010.06.038
35. Rebollar-Lazaro I, Matson P (2010) The culture of human cleavage stage embryos alone or in groups: effect upon blastocyst utilization rates and implantation. Reprod Biol 10(3):227–234
36. Kelley RL, Gardner DK (2017) In vitro culture of individual mouse preimplantation embryos: the role of embryo density, microwells, oxygen, timing and conditioned media. Reprod Biomed Online 34(5):441–454. https://doi.org/10.1016/j.rbmo.2017.02.001
37. Kelley RL, Gardner DK (2019) Individual culture and atmospheric oxygen during culture affect mouse preimplantation embryo metabolism and post-implantation development. Reprod Biomed Online 39(1):3–18. https://doi.org/10.1016/j.rbmo.2019.03.102
38. Vajta G, Korösi T, Du Y, Nakata K, Ieda S, Kuwayama M, Nagy ZP (2008) The Well-of-the-Well system: an efficient approach to improve embryo development. Reprod Biomed Online 17(1):73–81. https://doi.org/10.1016/s1472-6483(10)60296-9
39. Christianson MS, Zhao Y, Shoham G, Granot I, Safran A, Khafagy A, Leong M, Shoham Z (2014) Embryo catheter loading and embryo culture techniques: results of a worldwide web-based survey. J Assist Reprod Genet 31(8):1029–1036. https://doi.org/10.1007/s10815-014-0250-z
40. Gardner DK, Reed L, Linck D, Sheehan C, Lane M (2005) Quality control in human in vitro fertilization. Semin Reprod Med 23(4):319–324. https://doi.org/10.1055/s-2005-923389
41. Claassens OE, Wehr JB, Harrison KL (2000) Optimizing sensitivity of the human sperm motility assay for embryo toxicity testing. Hum Reprod 15(7):1586–1591. https://doi.org/10.1093/humrep/15.7.1586
42. Gardner DK, Schoolcraft WB (1999) Culture and transfer of human blastocysts. Curr Opin Obstet Gynecol 11(3):307–311. https://doi.org/10.1097/00001703-199906000-00013
43. De Munck N, Mateizel I, Panagiotidis D, Anick D, Hilde V, Céline S, Ingrid S, Danijel J, Lisbet V, Koen W (2022) The behavior of pH and osmolality by the type of embryo culture medium and the type of oil. Reprod Biomed Online 45:e22. https://doi.org/10.1016/j.rbmo.2022.08.038
44. Mortimer D, Cohen J, Mortimer ST, Fawzy M, McCulloh DH, Morbeck DE, Pollet-Villard X, Mansour RT, Brison DR, Doshi A, Harper JC, Swain JE, Gilligan AV (2018) Cairo consensus on the IVF laboratory environment and air quality: report of an expert meeting. Reprod Biomed Online 36(6):658–674. https://doi.org/10.1016/j.rbmo.2018.02.005

Chapter 8

Oocyte and Embryo Vitrification

Masashige Kuwayama and Gabriel Dalvit

Abstract

Widespread use of oocyte cryopreservation began after the introduction of vitrification, allowing women to postpone childbearing, precipitated by age-related declines in fertility. Similarly, vitrification offers poor responder patients the option of accumulating embryos. Over the last decade, several advances in vitrification technologies have improved clinical efficiency and outcome, reduced the dependence on operator skills, and raised success rates to higher and more uniform levels. As a result, vitrification has become reliable and safe, completely replacing slow cooling as the method of choice. Minimum volume methods, in particular the advanced Cryotec Ready to Use (RtU) system, as described here, ensure achieving the highest viability in oocytes and embryos at any stage can be achieved safely. A robust vitrification protocol is therefore a cornerstone of any ART program.

Keywords Vitrification, Oocytes, Embryos, Cryotec method

1 Introduction

Over the last 50 years, cryobiology has evolved to support not only sperm storage, but also oocyte and embryo viability. Slow freezing was first established for blood cells and sperm with reasonable success, supporting ~50% viability. While attempts were made to cryopreserve oocytes and embryos using the same cryoprotectants, success rates remained poor [1] until advances in vitrification technologies were made.

In 1985, Rall and Fahy were the first to describe the use of vitrification in mouse embryo cryopreservation [2]. Some years later, Kuwayama and colleagues confirmed that vitrification was suitable for human embryos, reporting that high survival rates could be obtained from pronuclear stage embryos after vitrification and warming procedures [3]. In addition to this achievement, improvements in day 3 embryo viability and blastocyst culture lead to the goal of vitrifying blastocysts being achieved, supporting clinics with strong elective single-embryo transfer policies. Consequently, the proportion of

Alexandra J. Harvey and David K. Gardner (eds.), *The Preimplantation Embryo*, Methods in Molecular Biology, vol. 3038, https://doi.org/10.1007/978-1-0716-5292-3_8,

frozen cycles being undertaken has been reported to be as high as 42.8% [4], thus reducing multiple pregnancy rates [5].

Cryopreservation media development has been critical to improvements in survival, with media developed by Kuwayama remaining the most effective in supporting high survival rates. All other cryopreservation media have followed the principles established by Kuwayama, with slight modifications of the same components. All cryopreservation media largely consist of a base culture medium, such as TCM 199 or MEM, supplemented with cryoprotectants such as ethylene glycol and dimethyl sulfoxide, sucrose or trehalose as an osmotic agent, and one of the following supplements: Serum Supplement Substitute (SSS), hydroxyprolyl cellulose, or xanthan gum. The latest medium developed by Kuwayama, RtU, has been shown to support high rates of blastocyst survival (99%–100%) and pregnancy (58%–65%) [6, 7]. In parallel, devices have evolved from vials for liquid media containing cells to devices for placing one or more oocytes/embryos in almost no liquid medium [3, 8–20]. Numerous carrier systems have been created using large or minimal volumes to allow direct contact with liquid nitrogen (open system), or conversely to avoid it (closed system), the latter to reduce the theoretical risk of contamination [21, 22].

In contrast to the application of vitrification for embryo cryopreservation, efforts to cryopreserve oocytes were largely unsuccessful for two decades. This was primarily a result of the oocytes' size, the formation of ice leading to organelle damage [23], and cryosensitivity of the oocyte membrane [24]. In 1999, Kuwayama successfully developed oocyte vitrification for a bone marrow transplant cancer patient, with the first baby born in February 2012. Consequently, oocyte vitrification became more widespread, facilitated in 2013 by the practice committees of the American Society for Reproductive Medicine (ASRM) and of the Society for Assisted Reproductive Technology removing its classification as an experimental procedure [25].

While many commercial vitrification systems are available, survival rates can vary significantly. However, vitrification systems that fail to support more than 96% survival will not preserve oocytes that are sensitive, weak, or fragile, such as those retrieved from cancer patients, or from women over 38 years old; in such cases, the use of an optimised system is critical. The minimal volume method, Cryotec Ready to Use (RtU) system, is the most efficient in terms of survival (99%–100%), embryo quality, implantation rates, and clinical pregnancy (59.8%–65.5%), with results comparable to those obtained using fresh oocytes or embryos [7, 8]. Post-warming survival rates of 90%–93% have been reported using Cryotop [4]. Further, data from 1000 ART Centres in 60 countries indicate successful vitrification of half a million embryos, with 140,000 vitrified oocytes waiting for IVF with no evidence of contamination using the low-cost and easy-to-perform Cryotec vitrification method (unpublished data).

The method is suitable for all ART patients, including those who wish to delay motherhood, oocyte donors, and patients with cancer.

Cryopreservation has become a mainstay of the ART laboratory while also supporting fertility preservation programs for patients with cancer and other indications. Vitrification has completely replaced slow freezing protocols due to its improved survival rates and clinical outcomes. Collectively, studies have demonstrated vitrification to be safe, effective, repeatable, and easily applicable in varied clinical circumstances, therefore justifying its application as a routine technique in the treatment of infertility and preservation of fertility. Herein, we describe a method for the vitrification and warming of oocytes and embryos for clinical use with the Cryotec RtU system.

2 Materials

There is no requirement for special machines or special conditions for introducing vitrification as a routine protocol to an ordinary IVF laboratory. Only standard laboratory devices, storage tanks, and proper training of embryologists are necessary for this purpose. Therefore, any laboratory can convert to routine vitrification within a few days without any specific setup [26].

2.1 Vitrification

1. Cryotec Ready to Vitri kit, consisting of three vitri plates preloaded with one well of Equilibration solution (ES) and two wells of Vitrification solution (VS), and four cryotecs (Fig. 1, *see* **Note 1**).
2. Laminar flow hood.
3. Stereo microscope (*see* **Note 2**).
4. Benchtop incubator.

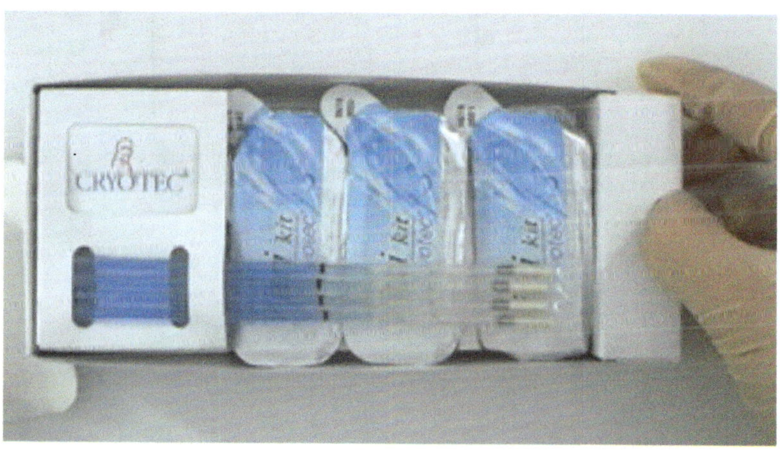

Fig. 1 Cryotech ready to Vitri kit

5. Stripper pipettes with tips 150 μL for oocytes, 200 μL for embryos.
6. Cooling rack filled with liquid nitrogen (*see* **Note 3**).
7. Timer (with count-up function).
8. Forceps.
9. Scissors to open the cryotec device.

2.2 Warming

1. Cryotec Ready to Warm kit, containing five warming plates (Fig. 2) preloaded with one square well containing 1 mL of Thawing solution (TS), one well of Diluent Solution (DS), and two wells of Washing Solution (WS) (*see* **Note 4**). Kits should be stored at 2–8°C until the expiration date. Do not use an expired kit.
2. Stereo microscope with heated warming stage turned on.
3. Cooling rack filled with liquid nitrogen (*see* **Note 3**).
4. Forceps.
5. Culture plates (for insemination or transfer).

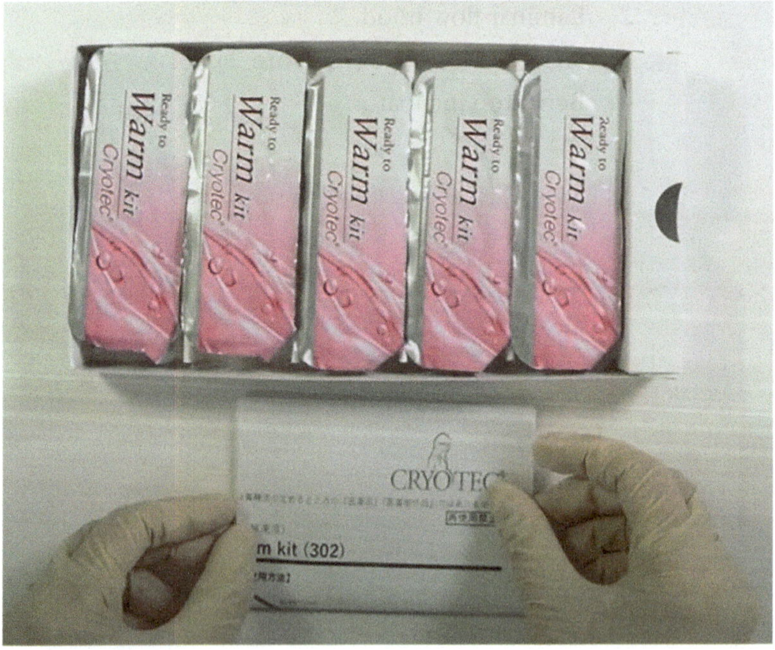

Fig. 2 Cryotech ready to Warm kit

3 Methods

The following methods are applicable to both oocytes and embryos. Oocytes and embryos should remain in culture until the time of vitrification. Care should be taken to avoid over-manipulation during vitrification and to minimise solution mixing.

3.1 Oocyte and Embryo Vitrification

The vitrification process should be performed at room temperature. Optimal lab temperature is 25–27°C.

3.1.1 Oocyte and Embryo Equilibration

1. Place the Ready to Vitri kit on a clean surface and allow it to equilibrate to 25–27°C for at least for 1 h.
2. In a laminar flow hood, carefully peel off the film by pressing the plate against the surface firmly with one hand.
3. Observe the oocytes/embryos prior to vitrification and note the size of the perivitelline space relative to the thickness of the zona pellucida (Fig. 3A, *see* **Notes 5 and 6**).
4. Aspirate one to three oocytes/embryos followed by a very small amount of culture medium (~1 mm) and pipette it at the surface of ES in the center of the well, expelling just the oocyte(s)/embryo(s) (*see* **Note 7**). Immediately start the timer (at 0:00). The oocyte/embryo will sink slowly to the bottom of the ES well.

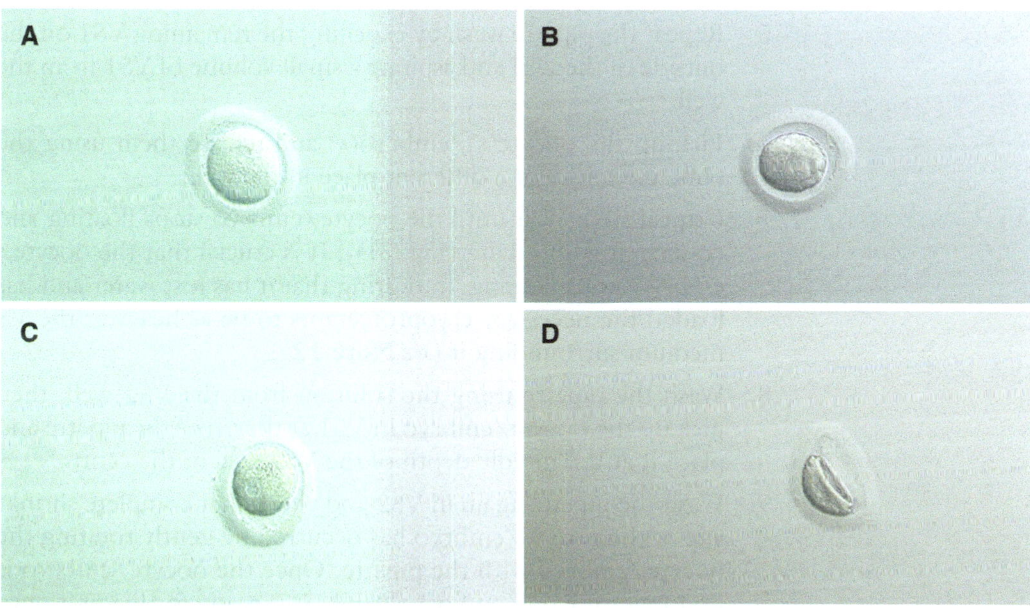

Fig. 3 Oocyte morphology through vitrification. (**A**) Oocyte morphology prior to vitrification, where size of the perivitelline space and thickness of the zona pellucida should be noted. (**B**) Morphology of the oocyte following incubation in equilibration solution. (**C**) Recovery morphology of oocytes in equilibration solution. (**D**) Shrinkage of the oocyte with incubation in vitrification solution 1

5. Confirm that shrinkage of the oocyte(s)/embryo(s) has occurred after 60–90 s (Fig. 3B, *see* **Note 8**), then turn off the light of the microscope.
6. Incubate oocytes and embryos in ES solution until full recovery is confirmed or the maximum equilibration time is reached, whichever is earlier (Fig. 3C, *see* **Note 9**).
7. While oocytes/embryos are equilibrating, prepare the required number of cryotec devices by cutting the wrapper and labelling each device with the necessary information according to in-house procedures.
8. Place the cryotec in the groove of the plate with the logo side up (*see* **Note 10**).

3.1.2 Vitrification

1. Aspirate the oocyte(s)/embryo(s) followed by a very small amount (~1 mm) of ES.
2. Place the pipette tip containing the oocyte(s)/embryo(s) at the bottom of the VS1 well, then expel just the oocyte/embryo (*see* **Note 11**).
3. Quickly wash the pipette by expelling all remaining ES on the outside of the well, then aspirate a small volume of VS1 and expel it from the pipette on the outside of the well again.
4. Pick up the oocyte(s)/embryo(s) at the tip of the pipette and release them in the centre of the well by moving the pipette horizontally.
5. Repeat the pipette wash by expelling the remaining VS1 on the outside of the well and aspirate a small volume of VS1 from the well.
6. Pick up the oocyte(s)/embryo(s) and release them using the same technique at a different place in the well.
7. Critical step: Wait until the oocyte/embryo stops floating and confirm it is shrinking (Fig. 3D). It is crucial that the oocyte/embryo stops floating, indicating that it has lost water and has loaded the necessary cryoprotectants to be as heavy as the VS medium surrounding it (*see* **Note 12**).
8. Wash the pipette using the solution from the VS2 well, then pick up the oocyte/embryo in VS1 at the tip of the pipette and place it at the middle depth of the VS2 well in the centre.
9. Wash the pipette again in VS2 and check that complete shrinkage of the oocyte/embryo has occurred by gently rotating the oocyte/embryo with the pipette. Once the oocyte/embryo is no longer floating and has shrunk, it is ready for loading onto the cryotec device (*see* **Note 13**).

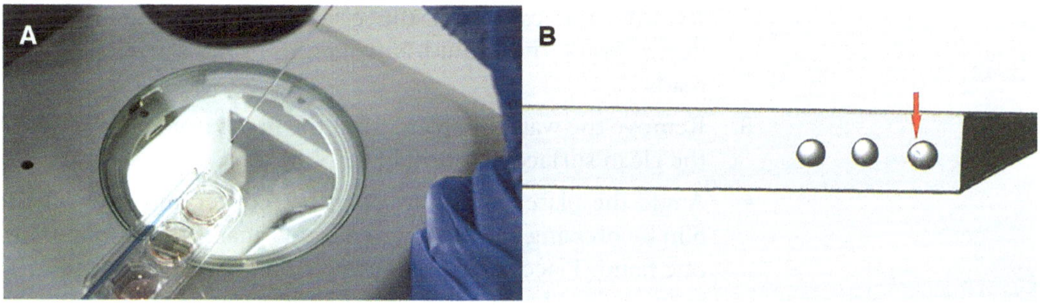

Fig. 4 Loading of oocytes/embryos onto the cryotec device. (**A**) Pipetting an oocyte onto a cryotec loaded into the groove of the vitrification plate with the logo side up. (**B**) Cryotec device with three droplets of VS2 solution. The arrow indicates the presence of an oocyte in the right-most drop

3.1.3 Loading Oocytes/Embryos onto Cryotec Devices

1. Aspirate one to three oocytes or one embryo into the tip of a pipette and make small droplets (~300–500 µm in diameter) near the triangle marker of the cryotec device (Fig. 4, *see* **Note 14**).
2. Check each droplet for the presence of the oocyte(s)/embryo(s) and, with the presence of the oocyte(s)/embryo(s) confirmed, submerge the cryotec into the Liquid Nitrogen immediately. Ensure the entire clear surface of the cryotec is submerged.
3. Shake the cryotec gently until bubbles are no longer appearing within the liquid nitrogen.
4. If you have just placed three oocytes in the cryotec and you are working with six or nine oocytes in the ES, repeat the vitrification procedure in the VS, taking special care to handle small volumes. Pipette the oocytes with a minimum volume of ES so as not to contaminate the VS with ES.
5. For capping, submerge the cap in liquid nitrogen using forceps and wait for the liquid nitrogen bubbles to stop.
6. Insert the cryotec into the cap and twist the cap using your fingers to tighten (*see* **Note 15**).
7. Full videos of the technique can be viewed at https://repro-life.jp/product/cryotec-rtu/

3.2 Oocyte and Embryo Warming

To warm oocytes/embryos, start preparation at least 2 h before warming is to be performed.

1. Place the warming plate in an incubator at 37 °C for at least 2 h or overnight. The wrapper should not be removed at this stage.

2. Remove the cap from the cryotec while maintaining the device inside the Liquid Nitrogen until the warming plate is ready.
3. Remove the warming plate from the incubator and place it on the clean surface of a hood (*see* **Note 16**).
4. Angle the plate slightly to open the corner and peel off the film by pressing the plate down firmly against the surface with one hand. Place the plate on the heated stage (*see* **Note 17**).
5. Critical step: Set the focus of the microscope on the TS well. Quickly remove the cryotec from liquid nitrogen and immediately insert the tip into TS (within 1 s). Keep the cryotec still in TS for 1 min (Fig. 5A). Start the timer as the cryotec is inserted into TS. The oocyte/embryo will detach slowly, and due to the high viscosity of the medium, it will remain in view and in focus (Fig. 5B).
6. After 1 min, while watching the oocyte(s)/embryo(s), gently remove the cryotec by sliding it so as not to agitate the medium.
7. Aspirate the oocyte/embryo followed by 3 mm of TS (Fig. 6, *see* **Note 18**).
8. Critical step: Slowly expel the TS into the center of the DS well at the bottom and ensure the oocyte/embryo remains inside this small amount of TS. This is very important to protect the oocyte/embryo from the osmotic shock (Fig. 7A).
9. Turn off the microscope light and note the time on the timer.
10. Critical step: After 2 min, aspirate the oocyte(s)/embryo(s) (Fig. 7B), followed by 3 mm of DS, then slowly release at the bottom of WS1 in the center of the well. Remove the pipette and ensure the oocyte/embryo remains inside this small amount of DS, again to avoid osmotic shock (Fig. 7C). Note the time on the timer.

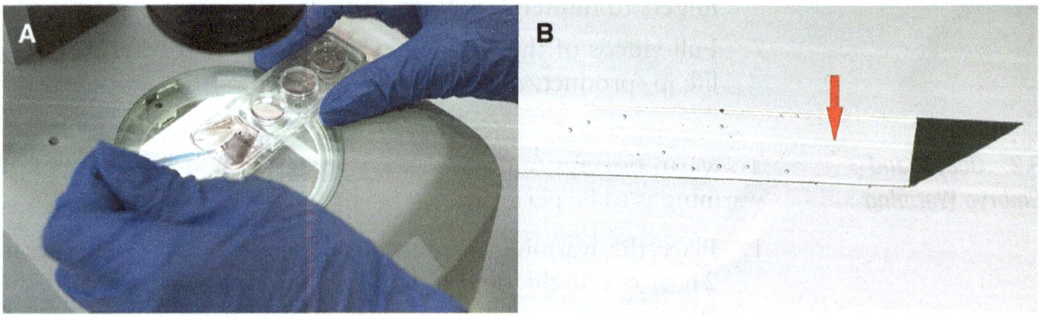

Fig. 5 Warming the cryotec device and oocyte observation. (**A**) The tip of the cryotec device is inserted into the thawing solution and observed under a microscope. (**B**) Oocyte (highlighted by the arrow) on the cryotec device after submersion in thawing solution

Fig. 6 Aspiration of the oocyte from the thawing solution into a pipette

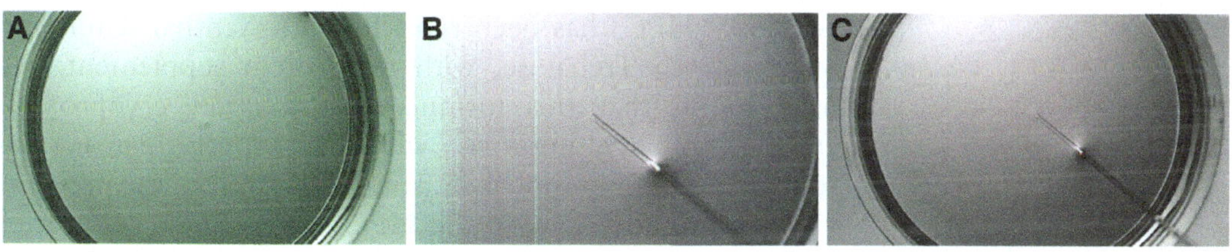

Fig. 7 Dilution of warming solution. (**A**) The oocyte remains within a small amount of thaw solution after transfer to the diluent solution. (**B**) Aspiration of the oocyte along with a small amount of diluent solution. (**C**) The oocyte remains within a small amount of diluent solution after transfer to the wash solution

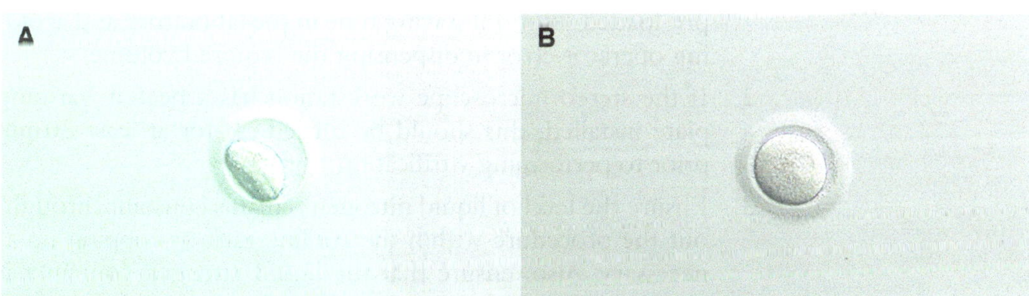

Fig. 8 Assessment of the oocyte during and after warming. (**A**) Appearance of the vitrified-warmed oocyte after transfer to the wash solution. (**B**) Recovery of the vitrified-warmed oocyte after 3 min in wash solution

11. Quickly assess the volume of the oocyte/embryo and turn off the microscope light (Fig. 8A).
12. After 3 min, check the recovery of the oocyte/embryo (Fig. 8B, *see* **Note 19**).

13. Aspirate the oocyte/embryo in the tip of the pipette and release it on the surface of the WS2 well to the left of centre. It will sink to the bottom, washing gently as it falls.
14. Aspirate the oocyte/embryo again with the tip of the pipette and release it to the right of centre at the surface of the WS2 well, where it will fall back to the bottom, producing another gentle wash.
15. Transfer the oocyte/embryo to a prepared and equilibrated culture dish.
16. Full videos of the technique can be viewed at https://repro-life.jp/product/cryotec-rtu/

4 Notes

1. Ready to Use (RtU) media contains ethylene glycol (EG) and dimethyl sulfoxide (DMSO) in MEM. Hydroxypropyl cellulose (HPC) has been replaced with Xanthan Gum and Sucrose with Trehalose, with no protein supplement. We have found that this formulation promotes cryoprotectant permeability, improves vitrification performance, and reduces processing time by 5.5 min compared to the previous Cryotec method. Due to the HPC-free formula, no precipitation under incubator temperature occurs, eliminating the need for separate incubation of warming solutions and plates. Further, the higher viscosity facilitates easier oocyte/embryo manipulation and lowers the risk of osmotic shock. Each of the Ready to Use plates is sealed in a blister pack, with sterile, pre-loaded solutions, saving time in the laboratory and avoiding operator error in dispensing the required volume.
2. If the stereo microscope workstation has a heated warming plate installed, this should be turned off for at least 30 min prior to performing vitrification.
3. Ensure the level of liquid nitrogen remains constant throughout the procedure within the cooling rack by topping up as necessary. Also, ensure that the liquid nitrogen container is kept away from culture dishes.
4. Each blister pack contains a warming plate preloaded with media composed of MEM, trehalose, and Xanthan Gum.
5. During the process of dehydration, the oocyte embryo will undergo shrinkage followed by re-expansion. The degree of oocyte/embryo shrinkage will depend on the oocyte/embryo but will be between 50% and 60%. By the end of the equilibration procedure, the oocyte/embryo should have

re-expanded to its original size noted prior to dehydration to ensure viability is maintained.

6. Much has been done and published regarding artificial shrinkage, including the suggestion that having a reduced volume of liquid inside the blastocoel improves survival in cryopreservation. However, the vitrification procedure quickly and completely dehydrates a vitrified blastocoel. Laser assistance is not required to improve either blastocyst dehydration or cryopreservation. These practices only provide improvements to methods for which survival is inherently low. The method described here reliably supports viability, therefore inducing damage to blastocysts does not achieve better results. Instead, we suggest gently dehydrating the large blastocyst (more than 300 μm) to the desired size (280 μm) and then performing vitrification with the normal procedure. Dehydration of hatched blastocysts can also be performed, although it is not desirable, as results are suboptimal. Ideally, before leaving the laboratory each day, check that there are no blastocysts larger than 200 μm so as not to find them hatched the next day. If you have blastocysts larger than 200 μm, vitrify them before leaving the lab.
7. We recommend handling no more than one to three oocytes in total in each Vitri plate, depending on the experience of the embryologist. If there is a large number of oocytes/embryos to be cryopreserved, do not overlap steps in an attempt to reduce the total time, as this will impact survival. An embryologist with extensive experience and proven results of 99.99% may be able to handle up to nine oocytes in a Vitri plate, placing the nine oocytes in ES, waiting for the first three to equilibrate as evidenced by recovery of the original volume before transfer to ES, and carrying out the procedure through submerging the cryotec in Liquid Nitrogen. They must then return to ES to take another three equilibrated oocytes and carry out the procedure before returning to ES for the third time and checking that the remaining oocytes are equilibrated. If more than three oocytes/embryos are processed it is critical to minimise the transfer of ES into VS.
8. Shrinkage time for each oocyte/embryo varies but is usually achieved within 60–90 s. Shrinkage is an indication that the oocyte/embryo is alive and its membrane permeability is normal.
9. For oocytes and blastocysts, the maximum recovery time is 13 min, for other cleavage stages, the maximum recovery time is 10 min. The operator should check recovery approximately every 30 s from minute 7. Full recovery will be evident once the oocyte/blastocyst re-expands to its pre-equilibration size.

Blastocysts can be more difficult to determine due to the loss of the blastocoel. Carefully observe the contact of the blastocyst with the inside of the zona pellucida; when it touches the zona at three points, we consider that equilibration is complete.

10. Placing the logo side up will enable identification of the side of the cryotec on which the oocyte/embryo has been vitrified when warming for transfer.
11. Due to the increased viscosity of the medium RtU the oocyte/embryo will not float rapidly to the surface but will remain in place floating very gently and slowly.
12. To confirm this, the oocyte/embryo is placed in focus at maximum magnification and if the focus is lost, it means that it is still ascending. If the focus remains stable, it has stopped floating and it is time to move on to VS2.
13. A specific incubation time is not noted as timing depends on the response of the oocyte as indicated.
14. When loading the cryotec with the oocyte/embryo in one drop, place your pipette as vertically as possible and just touch the cryotec sheet to form drops. Droplet size is important to achieve the necessary speed of cooling. The droplet should not be too big, between ⅓ and ½ of the width of the plastic sheet of the cryotec. Never place two oocytes/embryos in the same drop because the drop will inevitably be excessively large, thus preventing rapid cooling in the Liquid Nitrogen. Repeat making drops without paying attention to how many empty drops you place on the cryotec, merely confirm that the oocyte/embryo is in one of the drops. Do not try to remove any empty drops.
15. The cap of the device can also be fitted before plunging into Liquid Nitrogen, creating a closed system method, but great care must be taken to avoid touching the cryotec surface holding the oocyte(s)/embryo.
16. Before removing the warming plate from the incubator, make sure you have everything ready. The cryotec should be uncapped and in Liquid Nitrogen as close as possible to the working microscope, your pipette, and the work area cleared and ready to perform the procedure.
17. The warming solution (TS) must be at 37°C when the cryotec is submerged so this step should be completed quickly upon removing the warming plate from the incubator.
18. The 3 mm of TS medium (and then DS) that must be aspirated after collecting the oocyte/embryo from the TS are necessary to create a barrier within the DS medium (and then WS) to avoid osmotic shock due to the osmolar difference

between the media. This barrier must be large enough to fit the oocyte/embryo inside and small enough so that it can slowly mix with the surrounding DS. A marker on the warming plate can be used to check the amount (3 mm) of TS solution in the pipette.

19. Some studies have recently been published indicating that ultra-rapid warming may obtain "better results" [27–29]. It is simply impossible to obtain better results when working with optimal viability and indicates that ultra-rapid warming represents an improvement over the technique against which it is being compared. Avoiding osmotic shock and allowing the cell to adapt to the environment upon exiting vitrification is paramount. We conclude that this gradualism is absolutely necessary, and that only high-quality oocytes or blastocysts can survive ultrarapid warming while weak, sensitive or low-quality ones cannot. The best method that hundreds of ART Centers in the world apply is capable of preserving the viability and competence of all oocytes and embryos regardless of the quality and condition of those oocytes and embryos.

References

1. Rienzi L, Gracia C, Maggiulli R, LaBarbera AR, Kaser DJ, Ubaldi FM, Vanderpoel S, Racowsky C (2017) Oocyte, embryo and blastocyst cryopreservation in ART: systematic review and meta-analysis comparing slow-freezing versus vitrification to produce evidence for the development of global guidance. Hum Reprod Update 23(2):139–155. https://doi.org/10.1093/humupd/dmw038
2. Rall WF, Fahy GM (1985) Ice-free cryopreservation of mouse embryos at -196 degrees C by vitrification. Nature 313(6003):573–575. https://doi.org/10.1038/313573a0
3. Kuwayama M, Vajta G, Kato O, Leibo SP (2005) Highly efficient vitrification method for cryopreservation of human oocytes. Reprod Biomed Online 11(3):300–308. https://doi.org/10.1016/s1472-6483(10)60837-1
4. Tomas C, Alsbjerg B, Martikainen H, Humaidan P (2012) Pregnancy loss after frozen-embryo transfer—a comparison of three protocols. Fertil Steril 98(5):1165–1169. https://doi.org/10.1016/j.fertnstert.2012.07.1058
5. McLernon DJ, Bhattacharya S (2023) Quality of clinical prediction models in in vitro fertilisation: which covariates are really important to predict cumulative live birth and which models are best?. Best Pract Res Clin Obstet Gynaecol 86:102309. https://doi.org/10.1016/j.bpobgyn.2022.102309
6. Sugawara S, Shibasaki S, Sasaki C, Takahashi M, Hattori H, Toya M, Igarashi H, Kyono K (2024) Toreharōsu, kisantangamu tenka garasu-ka tōketsu hozon-eki no rinshō-teki yūyō-sei no kentō [A study on the clinical utility of vitrification solutions containing trehalose and xanthan gum]. J Mammalian Ova Res 41(1):S65
7. Sawabe Y (2024) Seishi oyobi hai-ban saibō ni okeru Ready to use tōketsu yūkai kitto no yūyō-sei [The effectiveness of ready to use vitrification and thawing kits for oocytes and blastocysts]. In: The 42nd Annual Meeting of the Japan Society of Fertilization and Implantation, Osaka, Japan
8. Martino A, Songsasen N, Leibo SP (1996) Development into blastocysts of bovine oocytes cryopreserved by ultra-rapid cooling. Biol Reprod 54(5):1059–1069. https://doi.org/10.1095/biolreprod54.5.1059
9. Vajta G, Holm P, Kuwayama M, Booth PJ, Jacobsen H, Greve T, Callesen H (1998) Open pulled straw (OPS) vitrification: a new way to reduce cryoinjuries of bovine ova and embryos. Mol Reprod Dev 51(1):53–58. https://doi.org/10.1002/(SICI)1098-2795(199809)51:1<53::AID-MRD6>3.0.CO;2-V

10. Kong IK, Lee SI, Cho SG, Cho SK, Park CS (2000) Comparison of open pulled straw (OPS) vs glass micropipette (GMP) vitrification in mouse blastocysts. Theriogenology 53(9):1817–1826. https://doi.org/10.1016/s0093-691x(00)00317-4
11. Tominaga K, Hamada Y (2001) Gel loading tips as container for vitrification in vitro-produced bovine embryos. J Reprod Dev 47(5):267–273
12. Chen SU, Lien YR, Cheng YY, Chen HF, Ho HN, Yang YS (2001) Vitrification of mouse oocytes using closed pulled straws (CPS) achieves a high survival and preserves good patterns of meiotic spindles, compared with conventional straws, open pulled straws (OPS) and grids. Hum Reprod 16(11):2350–2356. https://doi.org/10.1093/humrep/16.11.2350
13. Matsumoto H, Jiang JY, Tanaka T, Sasada H, Sato E (2001) Vitrification of large quantities of immature bovine oocytes using nylon mesh. Cryobiology 42(2):139–144. https://doi.org/10.1006/cryo.2001.2309
14. Kuleshova LL, Lopata A (2002) Vitrification can be more favorable than slow cooling. Fertil Steril 78(3):449–454. https://doi.org/10.1016/s0015-0282(02)03305-8
15. Mukaida T, Takahashi K, Kasai M (2003) Blastocyst cryopreservation: ultrarapid vitrification using cryoloop technique. Reprod Biomed Online 6(2):221–225. https://doi.org/10.1016/s1472-6483(10)61713-0
16. Kuwayama M, Vajta G, Ieda S, Kato O (2005) Comparison of open and closed methods for vitrification of human embryos and the elimination of potential contamination. Reprod Biomed Online 11(5):608–614. https://doi.org/10.1016/s1472-6483(10)61169-8
17. Huang CC, Lee TH, Chen SU, Chen HH, Cheng TC, Liu CH, Yang YS, Lee MS (2005) Successful pregnancy following blastocyst cryopreservation using super-cooling ultra-rapid vitrification. Hum Reprod 20(1):122–128. https://doi.org/10.1093/humrep/deh540
18. Gutnisky C, Alvarez GM, Cetica PD, Dalvit GC (2013) Evaluation of the cryotech vitrification kit for bovine embryos. Cryobiology 67(3):391–393. https://doi.org/10.1016/j.cryobiol.2013.08.006
19. Weng Y-W, Lin M-H, Kuo-Kuang Lee R, Li S-H, Jou Q-B, Chen YJ (2022) Is there an optimal timing interval between hCG trigger and oocyte vitrification?. Taiwan J Obstet Gynecol 61(3):485–488
20. Larman MG (2017) Appendix E: Rapid-i(TM): closed vitrification device by vitrolife. Methods Mol Biol 1568:335–342. https://doi.org/10.1007/978-1-4939-6828-2_25
21. Castellon LAR, Loya AGC (2015) Vitrification solutions: historical development. In: Allahbadia G, Kuwayama M, Gandhi G (Eds.) Vitrification in assisted reproduction. Springer, New Delhi, 9–15
22. Vajta G, Rienzi L, Ubaldi FM (2015) Open versus closed systems for vitrification of human oocytes and embryos. Reprod Biomed Online 30(4):325–333. https://doi.org/10.1016/j.rbmo.2014.12.012
23. Smith GD, Silva ESCA (2004) Developmental consequences of cryopreservation of mammalian oocytes and embryos. Reprod Biomed Online 9(2):171–178. https://doi.org/10.1016/s1472-6483(10)62126-8
24. Iussig B, Maggiulli R, Fabozzi G, Bertelle S, Vaiarelli A, Cimadomo D, Ubaldi FM, Rienzi L (2019) A brief history of oocyte cryopreservation: arguments and facts. Acta Obstet Gynecol Scand 98(5):550–558. https://doi.org/10.1111/aogs.13569
25. Practice Committees of the American Society for Reproductive Medicine and the Society for Assisted Reproductive Technology (2013) Mature oocyte cryopreservation: a guideline. Fertil Steril 99(1):37–43. https://doi.org/10.1016/j.fertnstert.2012.09.028
26. Elnahas A, Alcolak E, Marar EA, Elnahas T, Elnahas K, Palapelas V, Diedrich K, Al-Hasani S (2010) Vitrification of human oocytes and different development stages of embryos: an overview. Middle East Fertil Soc J 15(1):2–9
27. Gunst J, Vynck M, Hostens K, Standaert V, Roggeman S, van de Vijver A (2023) Comparative assessment of survival and clinical outcome between two commercial vitrification kits with different warming protocols after blastocyst culture: potential perspectives toward simplified warming procedures. Reprod Sci 30(11):3212–3221. https://doi.org/10.1007/s43032-023-01281-1
28. Liebermann J, Brohammer R, Wagner Y, Smith R, Even K, Hirshfeld-Cytron J, Uhler ML (2024) Fast and furious: successful survival and resumption of meiosis in immature human oocytes vitrified and warmed using a short protocol. Reprod Biomed Online 49(1):103976. https://doi.org/10.1016/j.rbmo.2024.103976
29. Liebermann J, Hrvojevic K, Hirshfeld-Cytron J, Brohammer R, Wagner Y, Susralski A, Jasulaitis S, Chan S, Takhsh E, Uhler M (2024) Fast and furious: pregnancy outcome with one-step rehydration in the warming protocol for human blastocysts. Reprod Biomed Online 48(4):103731. https://doi.org/10.1016/j.rbmo.2023.103731

Check for updates

Chapter 9

Microwave-Assisted Dehydration, Non-Cryogenic Storage, and Rehydration of Spermatozoa and Germinal Vesicles

Pei-Chih Lee and Pierre Comizzoli

Abstract

The microwave-assisted dehydration technique enables preservation of both male and female genomes in a dried state at non-cryogenic temperatures, eliminating the reliance on liquid nitrogen for long-term storage. The dehydration strategy outlined below utilizes trehalose, a natural disaccharide known to ensure survival of certain primitive species under drought conditions, to increase the dehydration tolerance of the cells. In this chapter, we describe step-by-step procedures for trehalose loading, microwave-assisted dehydration, sample storage, and rehydration of both spermatozoa and germinal vesicle stage oocytes. Tips for modification are provided to apply the protocols to different species of interest.

Keywords Spermatozoa, Germinal vesicles, Oocyte, Fertility preservation, Dehydration, Dry-preservation, Microwave, Trehalose

1 Introduction

Fertility preservation is an invaluable tool that has been implemented in human fertility clinics, the livestock industry, and wildlife conservation [1–3]. Cryopreservation is the current standard procedure for biobanking of spermatozoa and oocytes. However, this strategy relies on cryogenic temperatures for long-term storage that involves specialized infrastructure, a continuous liquid nitrogen supply, high maintenance costs, and risks of contamination. Dry preservation technology has been envisioned as a solution to overcome challenges associated with cryopreservation. Advantages of preserving gametes in a dried state span from the safe and simple handling of the samples to the flexibility of storing and transporting samples at non cryogenic temperatures [4, 5].

In this chapter, we describe the microwave-assisted dehydration technique, one of the few dehydration methods that have successfully been employed to preserve spermatozoa and germinal vesicles (GVs; Fig. 1) [6–8]. It has been reported that despite being immotile,

Alexandra J. Harvey and David K. Gardner (eds.), *The Preimplantation Embryo*, Methods in Molecular Biology, vol. 3038, https://doi.org/10.1007/978-1-0716-5292-3_9,

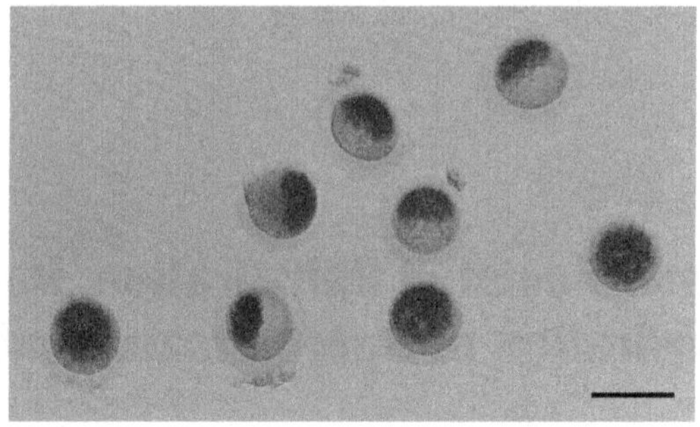

Fig. 1 Example of GV oocytes (domestic cat) with polarized lipid droplets used as starting biomaterials. Scale bar: 100 μm

dried-rehydrated spermatozoa retained the capacity to fertilize eggs and produce embryos after intracytoplasmic sperm injection [6, 9]. Structural and DNA integrity were maintained in dried-rehydrated GVs that may allow reconstitution of oocytes via GV transfer technique [8, 10]. This method utilizes microwave radiation to induce vibrations of dipolar water molecules. The vibrations result in an accelerated evaporation while remaining within a physiological range of temperature (below 40°C), enabling rapid and homogenous drying of the samples. The current dehydration strategy involves the use of trehalose—a natural disaccharide—as a protective agent (a.k.a. xero-protectant) to prevent dehydration injuries [11–13]. Trehalose can replace water molecules in the cells and help maintain the three-dimensional structure of macromolecules after water removal [14, 15]. Microwave drying leads to isothermal vitrification of the disaccharide sugar and generates a stable glass state for storage at non-cryogenic temperatures (above −20°C) [7, 16]. Compared to other drying methods (such as freeze-drying and convective drying), microwave-assisted dehydration offers rapid and homogenous water removal without preliminary freezing of samples [17].

We outline here protocols for the dry preservation of spermatozoa or GVs. Both descriptions contain the major steps of the procedure, including membrane permeabilization to allow loading of trehalose, microwave dehydration to induce trehalose vitrification, sample storage, and sample reanimation by rehydration. This technique was initially developed in the domestic cat model [9, 18, 19] and has been adapted for a few other mammalian species with similar outcomes (unpublished data). It is important to emphasize that the following protocols are provided to serve as a starting point for further optimizations in other species and/or for other research objectives.

2 Materials

2.1 Reagents and Solutions

1. Sperm handling medium, formulated for the specific species of interest, if available. For example, Ham's F10 medium with 5% fetal bovine serum, 25 mM HEPES, 1 mM pyruvate, 2 mM L-glutamine, 100 IU/mL penicillin, 100 μg/mL streptomycin (*see* **Note 1**).
2. Oocyte handling medium, formulated for the specific species of interest, if available. For example, Minimum essential medium with 4 mg/mL bovine serum albumin (BSA), 25 mM HEPES, 1 mM pyruvate, 2 mM L-glutamine, 100 IU/mL penicillin, 100 μg/mL streptomycin (*see* **Note 1**).
3. 10× Hemolysin solution: 100 μg/mL hemolysin in Dulbecco's phosphate-buffered saline (DPBS).
4. 2× Sperm hemolysin solution: dilute one-part 10× hemolysin solution with four-parts of sperm handling medium.
5. 1× GV hemolysin solution: dilute one part of 10× hemolysin solution in nine parts of oocyte handling medium to prepare a 1× working solution.
6. Sperm Trehalose solution: 0.3 M Trehalose in TE buffer (10 mM Tris, 1 mM EDTA) for sperm drying.
7. GV Trehalose solution: 1.1 M Trehalose in TE buffer (10 mM Tris, 1 mM EDTA) for GV drying.
8. Oocyte handling medium lacking BSA.

2.2 Microwave Drying System and Accessories

1. CEM SAM 255 microwave drying system (*see* **Note 2**).
2. 13-mm Swinnex filter holders (*see* **Note 3**).
3. 22 × 22 mm glass coverslips.
4. Press-To-Seal Silicone Isolators with 1 mm depth.
5. Dri-shield 2000 Moisture barrier bags. The bags can be cut into the desired size for storage.
6. Sample pads: Whatman Standard 14 filters cut into 10 mm rounds with a hole punch.

2.3 Other Equipment and Materials

1. Dissecting microscope (suggested magnification: 7–30×).
2. Slide warmer.
3. Dry bath or incubator.
4. Microcentrifuge and microcentrifuge tubes.

5. Heat sealer.
6. Stripper micropipetter and stripper tips with appropriate size for oocytes from the species of interest.
7. Pipettors (P20, P200, P1000).
8. Pipette tips.
9. Low retention pipette tips.
10. Flat-tip forceps.
11. Hole punch.
12. Microscope slides.
13. Sperm counting device, such as a hemocytometer.

3 Methods

This method outlines the process of dry-preservation of spermatozoa and germinal vesicles. The general principle is the same for both protocols, but details differ due to the different handling techniques for each sample type. The protocol for rehydration is also provided.

All procedures should be conducted at ambient temperature unless otherwise specified.

3.1 Biomaterial Collection

1. For sperm drying, semen or epididymal spermatozoa should be collected according to standard procedures established for the specific species. It is recommended to select only motile spermatozoa as starting materials. Suggested range of sperm concentration is 5–100 million per mL (*see* **Note 4**).
2. For GV drying, GV oocytes should be collected and denuded according to standard procedures established for the specific species (*see* **Note 5**).

3.2 Dry-Preservation of Spermatozoa

3.2.1 Trehalose Loading

1. To permeabilize the membrane, add 200 µL of 2× hemolysin solution to 200 µL of the sperm suspension (*see* **Note 6**) in a microcentrifuge tube. Incubate with lid closed in a dry bath or incubator at 37°C for 15 min (*see* **Note 7**).
2. Centrifuge at 300 × *g* for 8 min. Carefully remove as much supernatant as possible without disturbing the sperm pellet (*see* **Note 8**).
3. To rinse off the hemolysin solution, resuspend the sperm pellet in 500 µL of sperm handling medium (*see* **Note 9**).
4. Centrifuge at 300 × *g* for 8 min. Carefully remove the supernatant without disturbing the sperm pellet.

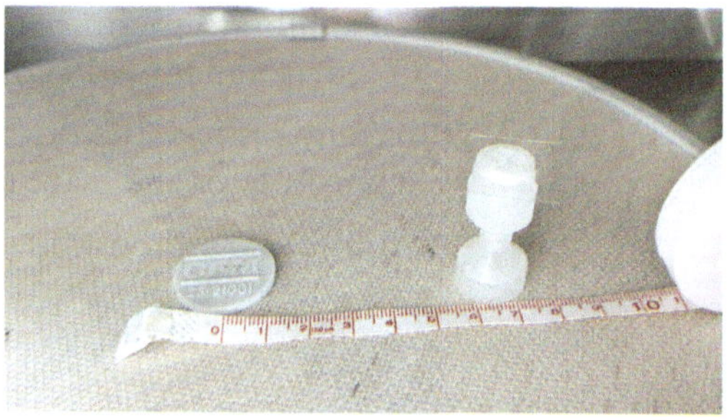

Fig. 2 Assembly of the adaptor and coverslip, and their placement on the microwave turntable

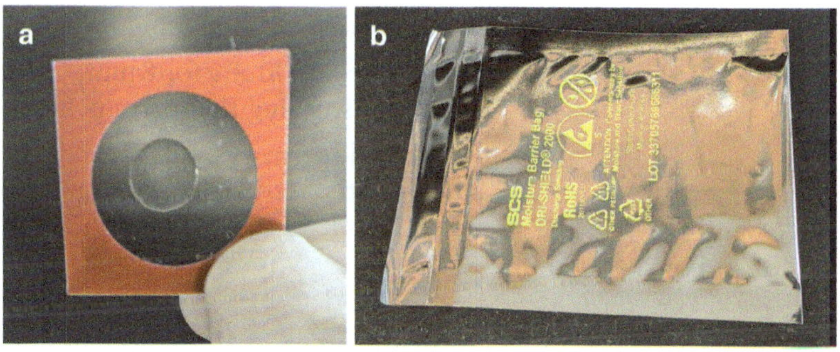

Fig. 3 Sample storage device. (**a**) A dried sperm sample covered with a silicone isolator. (**b**) A 4″ × 6″ (approximately 10 cm x 15 cm) moisture barrier bag used to store dried samples

5. Use low-retention tips to resuspend sperm in 200 μL of Sperm Trehalose solution (*see* **Note 10**). Incubate for 30 min.

3.2.2 Microwave Dehydration

1. From the microwave menu, set up dehydration parameters as follows: 20% power, 30 min dwell time, and 40°C temperature.
2. Remove the turntable from the microwave. Place clean adaptors on the turntable, then place a coverslip on each adaptor (Fig. 2, *see* **Note 11**).
3. Use low-retention tips to deposit 40 μL of sperm suspension onto the center of each coverslip.
4. Carefully return the turntable to the microwave.
5. Press “START” to begin 30 min of microwave dehydration.

3.2.3 Sample Storage

1. Cover the dried sample with press-to-seal silicone isolators to create a closed chamber (Fig. 3a).
2. Place each covered coverslip in a labelled moisture barrier bag. Seal the bag with a heat sealer (Fig. 3b).
3. Store the bags at the desired temperature (*see* **Note 12**).

3.2.4 Rehydration

1. Pre-equilibrate sperm handling medium at 37°C under a 5% CO_2 environment (*see* **Note 7**).
2. Carefully retrieve the dried sample from the bag. Remove the silicone isolator and place the coverslip into a 35 mm Petri dish.
3. Add 500 μL of prewarmed sperm handling medium onto the dried sample. Incubate at 37°C (*see* **Note 7**) on a slide warmer or incubator for 1 h. Gently shake the dish every 10 min (*see* **Note 13**).
4. Using a low retention pipette tip, carefully aspirate the sperm suspension then transfer it to a 1.5 mL microcentrifuge tube.
5. Rinse the coverslip and dish with another 500 μL of sperm handling medium. Combine with the previously collected portion in the microcentrifuge tube.
6. Centrifuge at 300 × *g* for 8 min. Carefully remove supernatant without disturbing the sperm pellet.
7. Resuspend sperm in 40 μL of medium. Assess sperm count and adjust to the desired concentration for downstream applications (Fig. 4, *see* **Note 14**).

3.3 Dry-Preservation of Germinal Vesicles

3.3.1 Trehalose loading

1. Prepare a drop of 200 μL of 1× hemolysin on a Petri dish.
2. Transfer GV oocytes to the hemolysin solution and incubate at 37°C for 15 min on a slide warmer to permeabilize the membrane (*see* **Note 7**).
3. Wash GV oocytes by transferring them to a fresh drop of 500 μL oocyte handling medium on a clean petri dish. Repeat washes three times (*see* **Note 15**).

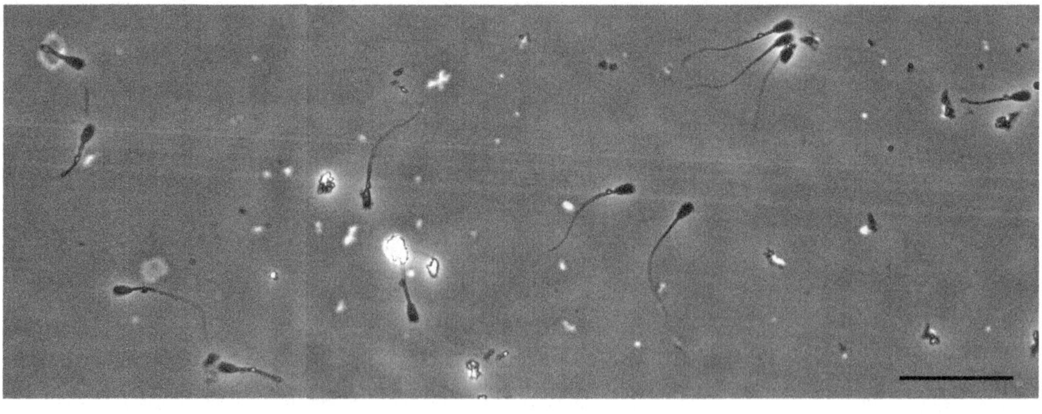

Fig. 4 Representative images of spermatozoa (white-tailed deer) after rehydration. Scale bar: 50 μm

4. Prepare 100 µL droplets of GV trehalose solution on clean petri dishes. Determine the number of droplets based on the number of oocytes to be dried. Each droplet will hold up to ten oocytes.
5. Transfer up to ten oocytes to each GV trehalose solution droplet. Incubate at room temperature for 10 min (Fig. 5a, *see* **Note 16**).

3.3.2 Microwave Dehydration

1. From the microwave menu, set the following dehydration parameters: 20% power, 30 min dwell time, and 40°C temperature.
2. Remove the turntable from the microwave. Place clean adaptors on the turntable. Using flat-tip forceps, place the sample pads on the adaptors (*see* **Note 11**).
3. Use a low-retention pipette tip to transfer up to ten oocytes in 40 µL of trehalose solution onto the center of each sample pad.
4. Carefully return the turntable to the microwave.
5. Press "START" to begin 30 min of microwave dehydration.

3.3.3 Sample Storage

1. Using flat-tip forceps, remove each sample pad from the microwave to an individually labelled moisture barrier bag (*see* **Note 17**). Mark the bag to indicate the side of the sample pad on which the oocyte has been deposited. Seal the bag with a heat sealer.
2. Store the bags at the desired temperature (*see* **Note 18**).

3.3.4 Rehydration

1. Prepare a Petri dish filled with prewarmed oocyte handling medium at 37°C (*see* **Note 7**).
2. Carefully cut open the storage bags without damaging the sample pads (*see* **Note 19**).
3. Retrieve the sample pad from the bag with flat-tip forceps and place it on a Petri dish with the oocyte-deposited side up.
4. To rehydrate GV oocytes, add 1 mL of rehydration medium on top of the sample pad (*see* **Note 20**).

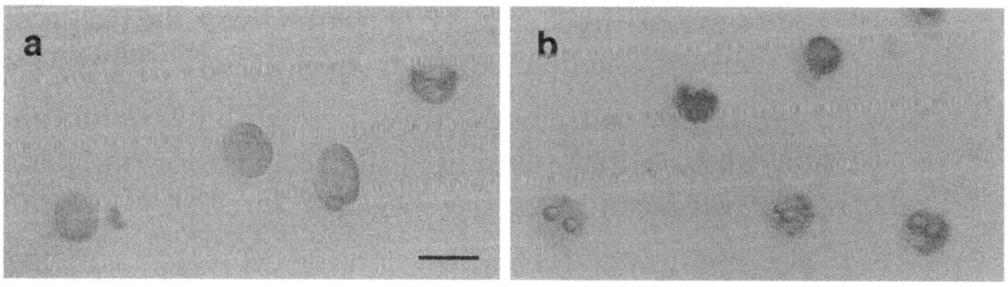

Fig. 5 Representative images of GV oocytes (domestic cat) (**a**) after trehalose exposure and (**b**) after rehydration. Scale bar: 100 µm

5. Use the forceps to flip the sample pad upside-down in the medium to release oocytes (*see* **Note 21**). Collect and transfer oocytes to the medium-filled dish (Fig. 5b, *see* **Note 22**). Repeat the step until all oocytes are recovered from the sample pad.
6. Incubate oocytes at 37°C (*see* **Note 7**) for 30 min to recover. GVs are then ready to be retrieved for downstream applications.

4 Notes

1. The provided medium recipe has been used for handling samples from domestic cats and other carnivorous species. It may be used for other clades if no conventional recipe is available from closely-related model species.
2. These protocols are developed exclusively with the use of the SAM 255 microwave system and have not been tested in other regular microwaves. This instrument uses an infrared sensor to monitor temperature, enabling temperature control based on a programmable threshold. This unique feature prevents overheating of thermally sensitive samples, such as sperm and oocytes. The unit is also equipped with a fabric turntable.
3. These filter holders are used as adaptors to hold sample pads during microwave drying. They are reusable for the drying procedure. To clean, the adaptors are washed three times with MilliQ H_2O, soaked in 70% ethanol overnight, and air-dried.
4. Recovery rate may vary for different species. Optimal starting concentration should be determined empirically.
5. While the preservation target is the GV, it is kept enclosed by the zona pellucida during the procedure for ease of manipulation. Oocytes are denuded to avoid the potential hindrance of trehalose loading by surrounding cumulus cells. If the oocytes have a high lipid content, polarization of lipid droplets can be achieved by centrifugation at 10,000 × *g* for 5 min (Fig. 1). This optional step will allow better GV visualization for retrieval for downstream applications.
6. Adjust the volume as necessary, depending on the initial volume of the sperm suspension. Keep the 1:1 volume ratio to ensure the final concentration of hemolysin is 1×.
7. Temperature and atmosphere should be modified based on the normal physiological temperature and handling media composition for the species of interest.
8. Hemolysin forms pores on the cytoplasmic membrane to enable trehalose loading. However, over-exposure of hemolysin may lead to cell rupture. While it is essential to remove as

much hemolysin solution as possible, disturbing the sperm pellet may result in loss of spermatozoa. Aspirate the solution in a slow, controlled manner and aim to leave no more than 50 μL of the solution behind.

9. Increase the wash volume if the sample contains a large number of spermatozoa.
10. Adjust the volume as necessary to reach the desired sperm concentration.
11. For sperm, it is ok to omit adaptors or put the coverslips on an alternative elevated surface (e.g., Petri dish lid) to allow easier retrieval. For GV oocytes, the adaptors are essential for proper drying. To achieve a consistent drying level, it is recommended to put the samples within 10 cm from the center of the turntable.
12. Storage bags can be cut into the desired size. The recommended storage temperature for spermatozoa dried with this protocol is −20°C. Keep the samples stored in an environment with minimum fluctuation of relative humidity and temperature. Proper storage conditions are essential to ensure that the samples remain in a stable glassy state and prevent crystallization that may damage the cells.
13. If possible, prevent the solution from spilling over the coverslip while gently shaking the sample. This may reduce sample loss after recovery.
14. Some spermatozoa may be lost during the procedure; therefore, it is better to re-assess the sperm concentration after rehydration. The average recovery rate for domestic cat sperm is 60%. Recovery rate may vary depending on the species.
15. As a general principle, minimize liquid carried over during oocyte transfer.
16. It is recommended to transfer GV oocytes with a stripper pipette to avoid diluting the trehalose solution with carried-over medium. Due to a change in light refraction, oocytes appear lighter in color when incubated in trehalose. Some oocytes may have a concave shape after incubating in the trehalose droplet due to the high osmolarity of the solution. The change in shape does not affect the structure of GVs inside the oocytes.
17. The sample pad will become rigid after drying. This indicates that trehalose vitrification has occurred. Place the sample pad closer to the bottom of the bag to prevent accidentally damaging it when opening the bag. Handle the storage bag with care to avoid damaging the glassy sample, and preferentially store it with the oocyte-deposited side up.
18. Recommended storage temperature for germinal vesicles dried with this protocol is 4°C. Keep the samples stored in an

environment with minimum fluctuation of relative humidity and temperature. Proper storage conditions are essential to ensure that the samples remain in a stable glassy state and prevent crystallization that may damage the cells. Previous studies indicated that DNA integrity within dried cat GVs can be sustained for at least 8 weeks using this protocol [7, 18].

19. Use fingers to gently locate the sample pad from outside of the bag before cutting it open to avoid damaging the pad.
20. It is normal that oocytes may not be easily visible on top of the sample pad due to the opaqueness of the fiberglass.
21. It is common to see vacuoles in the cytoplasm after drying/rehydration (Fig. 5b). This does not appear to affect GV structure [8]. In general, over 70% of the oocytes can be recovered following the standard rehydration procedure.
22. Oocytes sometimes float on the top of the liquid surface or at the edge of the liquid droplet due to surface tension. Adjusting the focus of the dissection microscope and carefully scanning the edge of the droplet will help locate these oocytes. Avoid tearing apart the sample pad because the scattered glass fibers may obscure oocytes.

References

1. Comizzoli P, Holt WV (2014) Recent advances and prospects in germplasm preservation of rare and endangered species. Adv. Exp. Med. Biol 753:331–356. https://doi.org/10.1007/978-1-4939-0820-2_14
2. Oldenhof H, Wolkers WF, Sieme H (2021) Cryopreservation of semen from domestic livestock: bovine, equine, and porcine sperm. Methods Mol Biol 2180:365–377. https://doi.org/10.1007/978-1-0716-0783-1_15
3. Antonouli S, Di Nisio V, Messini C, Daponte A, Rajender S, Anifandis G (2023) A comprehensive review and update on human fertility cryopreservation methods and tools. Front Vet Sci 10:1151254. https://doi.org/10.3389/fvets.2023.1151254
4. Comizzoli P, He X, Lee PC (2022) Long-term preservation of germ cells and gonadal tissues at ambient temperatures. Reprod Fertil 3(2):R42–R50. https://doi.org/10.1530/RAF-22-0008
5. Loi P, Anzalone DA, Palazzese L, Dinnyés A, Saragusty J, Czernik M (2021) Dry storage of mammalian spermatozoa and cells: state-of-the-art and possible future directions. Reprod Fertil Dev 33(2):82–90. https://doi.org/10.1071/rd20264
6. Lee PC, Zahmel J, Jewgenow K, Comizzoli P (2021) Desiccated cat spermatozoa retain DNA integrity and developmental potential after prolonged storage and shipping at non-cryogenic temperatures. J Assist Reprod Genet 39(1):141–151. https://doi.org/10.1007/s10815-021-02337-4
7. Wang S, Lee P-C, Elsayed A, Zhang F, Zhang Y, Comizzoli P, Elliott GD (2020) Preserving the female genome in trehalose glass at supra-zero temperatures: the relationship between moisture content and DNA damage in feline germinal vesicles. Cell Mol Bioeng 14(1):101–112. https://doi.org/10.1007/s12195-020-00635-y
8. Lee PC, Comizzoli P (2019) Desiccation and supra-zero temperature storage of cat germinal vesicles lead to less structural damage and similar epigenetic alterations compared to cryopreservation. Mol Reprod Dev 86(12):1822–1831. https://doi.org/10.1002/mrd.23276
9. Patrick JL, Elliott GD, Comizzoli P (2017) Structural integrity and developmental potential of spermatozoa following microwave-assisted drying in the domestic cat model. Theriogenology 103:36–43. https://doi.org/10.1016/j.theriogenology.2017.07.037
10. Graves-Herring JE, Wildt DE, Comizzoli P (2013) Retention of structure and function of the cat germinal vesicle after air-drying

and storage at suprazero temperature. Biol Reprod 88(6):139. https://doi.org/10.1095/biolreprod.113.108472
11. Tapia H, Koshland DE (2014) Trehalose is a versatile and long-lived chaperone for desiccation tolerance. Curr Biol 24(23):2758–2766. https://doi.org/10.1016/j.cub.2014.10.005
12. Arguelles JC (2014) Why can't vertebrates synthesize trehalose?. J Mol Evol 79(3-4):111–116. https://doi.org/10.1007/s00239-014-9645-9
13. Julca I, Alaminos M, Gonzalez-Lopez J, Manzanera M (2012) Xeroprotectants for the stabilization of biomaterials. Biotechnol Adv 30(6):1641–1654. https://doi.org/10.1016/j.biotechadv.2012.07.002
14. Jain NK, Roy I (2009) Effect of trehalose on protein structure. Protein Sci 18(1):24–36. https://doi.org/10.1002/pro.3
15. Golovina EA, Golovin A, Hoekstra FA, Faller R (2010) Water replacement hypothesis in atomic details: effect of trehalose on the structure of single dehydrated POPC bilayers. Langmuir 26(13):11118–11126. https://doi.org/10.1021/la100891x
16. Weng L, Elliott GD (2014) Determination of the relaxation characteristics of sugar glasses embedded in microfiber substrates. Mater Sci Eng C Mater Biol Appl 44:422–429. https://doi.org/10.1016/j.msec.2014.08.048
17. Weng L (2021) Technologies and applications toward preservation of cells in a dry state for therapies. Biopreserv Biobank 19(4):332–341. https://doi.org/10.1089/bio.2020.0130
18. Elliott GD, Lee PC, Paramore E, Van Vorst M, Comizzoli P (2015) Resilience of oocyte germinal vesicles to microwave-assisted drying in the domestic cat model. Biopreserv Biobank 13(3):164–171. https://doi.org/10.1089/bio.2014.0078
19. Lee PC, Comizzoli P (2024) Microwave-assisted dehydration, long-term storage at non-freezing temperatures, and rehydration of cat germinal vesicles. Biol Reprod 111(2):312–321. https://doi.org/10.1093/biolre/ioae060

Chapter 10

Non-Invasive Assessment of Embryo Physiology

Emma G. Whatley and David K. Gardner

Abstract

Through miniaturisation of enzymatic reactions, microfluorescence analyses enable the non-invasive assessment of embryo metabolism. Nutrient concentrations in embryo culture medium are quantified by ultramicrofluorescence (UMF) techniques, which utilise specialised constriction pipettes and a fluorescent microscope coupled with a photomultiplier tube and photometer, to quantitate nutrient levels in nanolitre or picolitre volumes of spent media. Here, the non-invasive method for analysis of embryo metabolism utilising spent culture media and UMF is described.

Keywords Constriction pipette, Enzymatic analysis, Fluorescent, Metabolism, Microfluorescence, Nanolitre, Picolitre, Single cell, Viability

1 Introduction

Embryonic metabolism is intimately linked with developmental viability. For example, a high rate of blastocyst glucose consumption is positively correlated with pregnancy outcome in the cow, mouse, and human [1–4], whilst excessive conversion of glucose to lactate (glycolytic flux) is indicative of a loss of metabolic control and diminished embryo viability [5]. Such conclusions regarding the metabolic "rate and fate" and the exact developmental outcomes of the embryo in question have been made possible by non-invasive methods to assess embryonic metabolism through the analysis of spent culture medium. Although limited in the ability to determine exact metabolic pathway activity, non-invasive metabolic assessments avoid the use of radiolabelled substrates or the destruction of cells, their integrity, or function. They are therefore a feasible method to assess metabolism as a developmental biomarker, with the potential for translation to clinical application.

Ultramicrofluorescence (UMF) is a highly reproducible method of non-invasive metabolic analysis [6]. UMF utilises coupled enzymatic reactions involving the generation or utilisation of pyridine nucleotides, $NAD(P)^+$ and NAD(P)H. Whilst NAD(P)H fluoresces at

Alexandra J. Harvey and David K. Gardner (eds.), *The Preimplantation Embryo*, Methods in Molecular Biology, vol. 3038, https://doi.org/10.1007/978-1-0716-5292-3_10,

495 nm when excited with light in the UV range (340 nm), $NAD(P)^+$ does not. This can be exploited to determine metabolite concentrations within spent medium samples. For example, by using the enzyme lactate dehydrogenase (LDH) to oxidise lactate to pyruvate concomitant with the reduction of NAD^+ to NADH, the increasing amount of NADH fluorescence is correlated with the concentration of lactate present in the sample when compared against a standard curve. The reverse reaction can be favoured by adjusting the reaction conditions to shift the equilibrium, and hence the decreasing amount of NADH fluorescence correlates with pyruvate concentration in the sample. Whilst the theory of this technique does not differ from other methods of fluorometric analysis, such as the use of a microplate reader or a NanoDrop spectrophotometer (technologies which utilise the same reactions listed here but require higher volumes), UMF is unique in that it enables the analysis of sub-microlitre (nanolitre and picolitre) volumes of sample with high accuracy, therefore facilitating single embryo or even single-cell metabolite analysis. This can be coupled with repeated measurements to determine metabolic rate in real time.

The analysis of such low-volume samples requires specialised equipment to produce and calibrate the pipettes used to deliver the sample, and a means of incubating the sub-microlitre reactions that avoids evaporation. Mroz and Lechene [7] described the production and use of constriction micropipettes utilising a microforge, and pipette calibration using tritiated water. They quantified urea produced from ammonium in the femtomolar range. Constriction pipettes were used to add samples to a reaction cocktail on a siliconized microscope slide submersed in paraffin oil. After incubation, reaction solutions were collected and housed in a capillary tube "fluorescence chamber" that was imaged on a fluorescence microscope to determine concentration. This method was further developed by Leese and colleagues [8] who housed reactions in nanolitre-sized drops on a siliconized microscope slide under mineral oil and directly measured the fluorescence of the reactions in the drops to determine extracted ATP, ADP, and AMP from within single mouse embryos and oocytes. They additionally identified the need to increase enzyme concentration in the reaction cocktail due to denaturation at the aqueous/oil interface using this method. Leese and Barton [9] then applied this method to the non-invasive analysis of embryo metabolism by measuring pyruvate and glucose uptake of mouse oocytes and embryos. The UMF method has since been developed further to encompass a range of metabolite assays exploiting the fluorescence of pyridine nucleotides. The analysis of glucose, lactate, pyruvate [9, 10], glutamine [11], aspartate [12], and β-hydroxybutyrate [13] has now been described, and Whatley [14] has additionally developed a fluorometric method to calibrate constriction pipette volume without the requirement for radioactive substrates or tritiated water.

This chapter outlines the method for the construction and non-radioactive calibration of glass constriction micropipettes, siliconization of microscope slides used for housing fluorescence reactions, UMF metabolic assays for glucose, lactate, pyruvate, aspartate, glutamine, and β-hydroxybutyrate, and the setup requirements for culturing embryos to produce spent medium samples for metabolic analysis in both clinical and research settings.

2 Materials

2.1 Embryo Culture for Metabolic Analysis: Research Setting

1. 60 mm Petri dish, embryo-tested.
2. Volumetric analytical syringe.
3. Sterile watchmaker's forceps.
4. Borosilicate glass pipettes.
5. Incubator.
6. Embryo culture medium
7. Paraffin oil.
8. Laminar flow hood.

2.2 Embryo Culture for Metabolic Analysis: Clinical Setting

1. Individual well embryo culture dish (such as the Vitrolife EmbryoSlide+ ic8 dish).
2. Embryo culture medium.
3. Paraffin oil.
4. Laminar flow hood fitted with stereomicroscope.
5. Incubator.

2.3 Siliconizing Microscope Slides

1. Fume hood.
2. Glass microscope slides.
3. Siliconizing agent (such as Sigmacote).
4. Reverse osmosis ultrapure water.
5. Slide holder.
6. Slide bath.

2.4 Construction, Calibration, and Siliconization of Glass Constriction Micropipettes

1. Dental burner.
2. Borosilicate glass capillaries (o.d. 1 mm, i.d. 0.8 mm).
3. Microforge.
4. Paperclip.
5. Watchmaker forceps.

6. Wax.
7. 16-gauge hollow metal rods.
8. Qubit fluorometer.
9. Qubit 1× dsDNA HS Assay Kit containing 1× dsDNA HS Working Solution, 1× dsDNA HS standard #1, and 1× dsDNS HS Standard #2.
10. Eppendorf PCR tubes (0.5 mL).
11. Nuclease-free pipettors and tips.
12. Siliconizing agent (such as Sigmacote).
13. Acetone.
14. Hydrochloric acid (1.0 M HCl).
15. Reverse osmosis ultrapure water.
16. 20 mL syringe with tubing (i.d. 1 mm) attached.
17. Vortex.

2.5 Metabolic Assay Reagents

Prepare all solutions with ultrapure reverse osmosis water and analytical grade reagents. Prepare reagents on ice and store at 4°C unless otherwise specified. Buffers should be prepared at room temperature, then stored at 4°C after pH adjustments. Reconstituted reagents can be stored and used for 2–3 months. Ensure reagents are weighed on a calibrated analytical balance, especially for standard curve generation as incorrectly weighed standards can affect quantitation. Gloves must be worn while preparing and using metabolic assay stocks to prevent the transfer of enzymes and lactate present on the skin that can interfere with the reactions.

2.5.1 Glucose Assay

$$\text{Glucose} + \text{ATP} \xrightarrow{\text{Hexokinase}} \text{Glucose} - 6 - \text{phosphate} + \text{ADP}$$

$$\text{Glucose} - 6 - \text{phosphate} + \text{NADP}^{+} \xrightarrow[\textit{dehydrogenase}\,(G6PDH)]{\textit{Glucose}-6-\textit{phosphate}} 6 - \text{phosphogluconate} + \text{NADPH} + H^{+}$$

1. 5 mM dithiothreitol (DTT) stock solution: Add 0.0077 g DTT to 10 mL water in a tube.
2. 37 mM magnesium sulphate ($MgSO_4$) stock solution: Add 0.0912 g $MgSO_4$ to 10 mL water in a tube.
3. 10 mM adenosine diphosphate (ADP) stock solution: Add 0.0423 g ADP to 10 mL water in a tube.

4. 10 mM $NADP^+$ stock solution: Add 0.394 g $NADP^+$ to 5 mL water in a tube.
5. 2 M NaOH stock solution: Add 8 g NaOH pearls to 100 mL water in a bottle.
6. EPPS buffer: 50 mM EPPS, pH 8.0. Weigh 2.52 g EPPS and transfer to a bottle containing 150 mL water. Add 10 mg of penicillin. Mix and adjust pH to 8.0 with 2 M NaOH. Make up to a final volume of 200 mL with water.
7. Glucose cocktail: 0.42 mM DTT, 3.09 mM $MgSO_4$, 0.42 mM ATP, 1.25 mM $NADP^+$, 14.11 U/mL hexokinase (EC 2.7.1.1), 7.06 U/mL glucose-6-phosphate dehydrogenase (EC 1.1.1.49) in EPPS buffer, pH 8.0. To prepare, combine 1250 μL EPPS buffer, 167 μL DTT stock solution, 167 μL $MgSO_4$ stock solution, 83 μL ATP stock solution, 250 μL $NADP^+$ stock solution, and 83 μL hexokinase (340 U/mL)/ glucose-6-phosphate (170 U/mL) in a tube (*see* **Note 1**). Store in 40 μL aliquots in microcentrifuge tubes at −20°C.
8. Glucose standards: 0.5 mM glucose in water. Add 0.045 g glucose to 5 mL water (50 mM stock) (*see* **Note 2**), then dilute 1:100 by adding 20 μL of this solution to 1980 μL water (0.5 mM). Use serial dilutions in water to produce 0.25, 0.125, 0.0625, 0.03125, and 0 mM glucose standards (*see* **Note 3**). Store in 40 μL aliquots at −20°C.

2.5.2 Lactate Assay

$$\text{Lactate} + \text{NAD}^+ \xrightarrow{\textit{Lactate dehydrogenase (LDH)}} \text{Pyruvate} + \text{NADH} + \text{H}^+$$

1. 50 mM NAD^+ stock solution: Add 0.0663 g NAD^+ to 2 mL water in a tube. Mix then aliquot into microcentrifuge tubes. Store at −20°C.
2. 2 M NaOH stock solution: Add 8 g NaOH pearls to 100 mL water in a bottle.
3. Glycine-hydrazine buffer: 1 M glycine, 1.6 M hydrazine, pH 9.4. Add 7.5 g glycine, 5.2 g hydrazine, and 0.2 g EDTA to 10 mL water in a bottle. Adjust pH with 2 M NaOH, mix well then dilute to a final volume of 100 mL with water.
4. Lactate cocktail: 5.27 mM NAD^+, 189 U/mL lactate dehydrogenase (EC 1.1.1.27) in glycine-hydrazine buffer, pH 9.4. To prepare, combine 36 μL glycine/hydrazine buffer (pH 9.4), 8 μL NAD^+ stock solution, 2 μL lactate dehydrogenase (7.202 kU/mL), and 30 μL water in a tube (*see* **Note 1**). This solution must be made fresh due to the instability of NAD^+ once in the buffer at pH 9.4.

5. Lactate standards: 1 mM lactate in water. Add 0.0560 g sodium-lactate to 5 mL water (100 mM stock) (*see* **Note 2**), then dilute 1:100 by adding 20 μL of this solution to 1980 μL water (1 mM). Use serial dilutions in water to produce 0.5, 0.25, 0.125, 0.0625, and 0 mM lactate standards (*see* **Note 3**). Store in 40 μL aliquots at −20°C.

2.5.3 Pyruvate Assay

$$\text{Pyruvate} + \text{NADH} + \text{H}^+ \xrightarrow{\textit{Lactate dehydrogenase (LDH)}} \text{Lactate} + \text{NAD}^+$$

1. 5 mM NADH stock solution: Add 0.0177 g NADH to 5 mL water. Store in a sterile tube wrapped in aluminium foil.
2. 2 M NaOH stock solution: Add 8 g NaOH pearls to 100 mL water in a bottle.
3. EPPS buffer: 50 mM EPPS, pH 8.0. Weigh 2.52 g EPPS and transfer to a bottle containing 150 mL water. Add 10 mg of penicillin. Mix and adjust pH to 8.0 with 2 M NaOH. Make up to a final volume of 200 mL with water.
4. Pyruvate cocktail: 0.12 mM NADH, 196 U/mL lactate dehydrogenase (EC 1.1.1.27) in EPPS buffer, pH 8.0. To prepare, combine 70 μL EPPS buffer (pH 8.0), 1.5 μL NADH stock solution, and 2 μL lactate dehydrogenase (7.202 kU/mL) in a tube (*see* **Note 1**). Store in 40 μL aliquots at −20°C.
5. Pyruvate standards: 0.5 mM pyruvate in water. Add 0.011 g sodium-pyruvate to 100 mL water (1 mM), then dilute 1:2 by adding 1 mL of this solution to 1 mL water (0.5 mM). Use serial dilutions in water to produce 0.25, 0.125, 0.0625, 0.03125, and 0 mM pyruvate standards (*see* **Note 3**). Store in 40 μL aliquots at −80°C (*see* **Note 4**).

2.5.4 Glutamine Assay

$$\text{Step 1:}\quad \text{Glutamine} + \text{H}_2\text{O} \xrightarrow{\textit{Glutaminase}} \text{Glutamate} + \text{NH}_3$$

$$\text{Step 2:}\quad \text{Glutamate} + \text{H}_2\text{O} + \text{NAD}^+ \xrightarrow{\textit{Glutamate dehydrogenase}} \text{oxoglutarate} + \text{NH}_4{}^+ + \text{NADH}$$

1. 50 mM NAD^+ stock solution: Add 0.0663 g NAD^+ to 2 mL water in a tube. Mix then aliquot into microcentrifuge tubes. Store at −20°C.

2. 10 mM adenosine diphosphate (ADP) stock solution: Add 0.0423 g ADP to 10 mL water in a tube.
3. 2 M NaOH stock solution: Add 8 g NaOH pearls to 100 mL water in a bottle.
4. Acetate buffer: 564 mM sodium acetate, 0.35% v/v acetic acid, pH 5.0. Resuspend 6.8 g of sodium acetate in 100 mL water in a bottle. In a separate bottle, aliquot out 2.9 mL of acetic acid and add to 97.1 mL water. To create the final acetate buffer, mix 68 mL of sodium acetate solution with 32 mL of acetic acid solution in a bottle.
5. Glycine-hydrazine buffer: 1 M glycine, 1.6 M hydrazine, pH 9.0. Add 7.5 g glycine, 5.2 g hydrazine, and 0.2 g EDTA to 10 mL water in a tube. Adjust pH with 2 M NaOH, mix well then dilute to a final volume of 100 mL with water in a bottle.
6. Glutamine cocktail (Step 1—conversion of glutamine to glutamate): 50 U/mL glutaminase (EC 3.5.1.2) in acetate buffer, pH 5.0. To prepare, combine 200 μL water, 200 μL acetate buffer, and 100 μL glutaminase (250 U/mL) in a tube. Prepare fresh.
7. Glutamate cocktail (Step 2—conversion of glutamate to malate): 1.6 mM NAD^+, 1.0 mM ADP, 100 U/mL glutamate dehydrogenase (EC 1.4.1.2) in glycine-hydrazine buffer, pH 9.0. To prepare, combine 84.8 μL glycine-hydrazine buffer, 3.2 μL NAD^+ stock solution, 10 μL ADP stock solution, and 2 μL glutamate dehydrogenase (5 kU/mL) in a tube (*see* **Note 1**). Must be made fresh.
8. Glutamine standards: 0.5 mM glutamine. Make a 50 mM glutamine stock by adding 0.0731 g glutamine to 10 mL water in a tube, mix well. Dilute to 0.5 mM by adding 50 μL of 50 mM glutamine to 4.95 mL water in a tube. Use serial dilutions to produce 0.25, 0.125, 0.0625, 0.03125, and 0 mM glutamine standards (*see* **Note 3**).

2.5.5 Aspartate Assay

$$\text{Aspartate} + \alpha - \text{ketoglutarate} \xrightarrow{\textit{Glutamic oxaloacetic transaminase}} \text{Oxaloacetate} + \text{Glutamate}$$

$$\text{Oxaloacetate} + \text{NADH} + H^+ \xrightarrow{\textit{Malate dehydrogenase}} \text{Malate} + NAD^+$$

1. 1.2 mM α-ketoglutarate stock solution: Add 0.0201 g α-ketoglutarate to 10 mL water in a tube, then dilute 1:10 by adding 1 mL of this solution to 9 mL water in a separate tube.

2. 5 mM NADH stock solution: Add 0.0177 g NADH to 5 mL water. Store in a sterile tube wrapped in aluminium foil.
3. Imidazole buffer: 30 mM imidazole, 20 mM imidazole-HCl, pH 7.0. In a bottle, resuspend 0.2402 g imidazole and 0.4192 g imidazole-HCl in 10 mL water. Adjust pH to 7.0 and dilute to a final volume of 50 mL with water.
4. Aspartate cocktail: 0.12 mM α-ketoglutarate, 0.15 mM NADH, 3.5 U/mL malate dehydrogenase (EC 1.1.1.37), 20 U/mL glutamate oxaloacetate transaminase (EC 2.6.1.1) in imidazole buffer, pH 7.0. To prepare, combine 83 μL imidazole buffer, 10 μL α-ketoglutarate stock solution, 3 μL NADH stock solution, 2 μL malate dehydrogenase (175 U/mL), and 2 μL glutamate oxaloacetate transaminase (1 kU/mL) in a tube (*see* **Note 1**). Store in 40 μL aliquots at −20°C.
5. Aspartate standards: 1 mM aspartate. Add 0.0133 g aspartate to 100 mL water (1 mM) in a bottle. Use serial dilutions in water to produce 0.1, 0.05, 0.025, 0.0125, 0.00625, and 0 mM aspartate standards (*see* **Note 3**). Store in 40 μL aliquots in microcentrifuge tubes at −20°C.

2.5.6 *β-Hydroxybutyrate Assay*

$$\beta - \text{hydroxybutyrate} + \text{NAD}^{+} \xrightarrow{\beta-hydroxybutrate\ dehydrogenase\,(3-BDH)} \text{Acetoacetate} + \text{NADH} + \text{H}^{+}$$

1. 50 mM NAD⁺ stock solution: Add 0.0663 g NAD⁺ to 2 mL water in a tube. Mix then aliquot into microcentrifuge tubes. Store at −20°C.
2. Tris-HCl buffer: 100 mM Tris, pH 9.5. Add 0.6057 g Tris to 40 mL water in a bottle. Mix and adjust pH to 9.5 with 1 M HCl, then dilute to a final volume of 50 mL with water.
3. Reconstituted β-hydroxybutyrate dehydrogenase (3-BDH): 5 U/mL 3-BDH (EC 1.1.1.30). Make a 25 U/mL stock by adding 1 mL cold (4°C) Tris–HCl buffer to a tube containing 25 U 3-BDH. Store 10 μL aliquots in microcentrifuge tubes at −80°C. To make 5 U/mL 3-BDH, thaw a 10 μL aliquot of 25 U/mL 3-BDH and add 40 μL cold Tris-HCl buffer. Make fresh from frozen aliquots.
4. B-hydroxybutyrate cocktail: 14 mM NAD⁺, 1.5 U/mL 3-BDH in 60 mM Tris-HCl buffer, pH 9.5. To prepare, combine 30 μL Tris-HCl buffer, 14 μL NAD⁺ stock solution, and 15 μL 3-BDH (5 U/mL) in a tube (*see* **Note 1**). Must be made fresh.
5. β-hydroxybutyrate standards: 1 mM β-hydroxybutyrate. Make a 40 mM β-hydroxybutyrate stock by adding 0.0504 g β-hydroxybutyrate to 10 mL water in a tube, mix well. Dilute to

1 mM by adding 100 µL of 40 mM βOHB to 3.9 mL water in a separate tube. Use serial dilutions to produce 0.5, 0.25, 0.125, 0.0625, and 0 mM βOHB standards (*see* **Note 3**). Store in 10 µL aliquots at −20°C.

2.6 Fluorometric Quantitation of NADH Fluorescence

1. Mercury lamp.
2. Inverted fluorescence microscope.
3. Shutter system/optical switch.
4. Photometer attachment.
5. Computer software to numericize light emission (such as MicroLab).
6. Stereomicroscope with stand-alone micromanipulator.

3 Methods

3.1 Embryo Culture for Analysis of Substrate Consumption or Production in a Research Setting

In a research setting, embryo metabolism can be analysed in either a linear or an endpoint manner. For the analysis of linear rates of substrate consumption/production, embryos should be cultured in sub-microlitre droplets (20–50 nL for mouse, or 200 nL to 1 µL for domestic/human embryos), with spent medium samples analysed hourly for at least 3 h. Incubation can therefore occur on a 37°C heated stage, in a HEPES or MOPS buffered culture medium. Alternatively, for end-point analyses for research, embryos can be cultured in larger volumes for longer durations (*see* **Note 5**), and incubation should therefore occur in a CO_2-gassed incubator in a bicarbonate-buffered culture medium.

1. Prepare research metabolic embryo culture dishes in a laminar flow hood (Fig. 1). In the base of an embryo-tested 60 mm Petri dish, create three 20 µL wash drops by first pipetting 10 µL drops of culture medium (*see* **Note 6**). Cover with 7 mL paraffin oil and top-up wash drops with a further 10 µL of medium. Use an eVol (volumetric analytical syringe) (for volumes ≥0.5 µL) or a calibrated constriction pipette (for volumes <0.5 µL) (*see* **Note 7**) to create metabolic culture drops under the oil (*see* **Note 8**). Make one culture drop per embryo, plus several others in which no embryo is placed so as to act as unspent controls (*see* **Note 9**). Equilibrate culture dishes for at least 4 h in the appropriate temperature and gas environment. Suggested culture drop volumes and culture durations for specific substrates and developmental stages are outlined in Table 1.
2. Wash embryos thoroughly in the three wash drops to ensure no carryover of substrates from non-metabolic culture medium. Move embryos to small individual incubation drops by using a

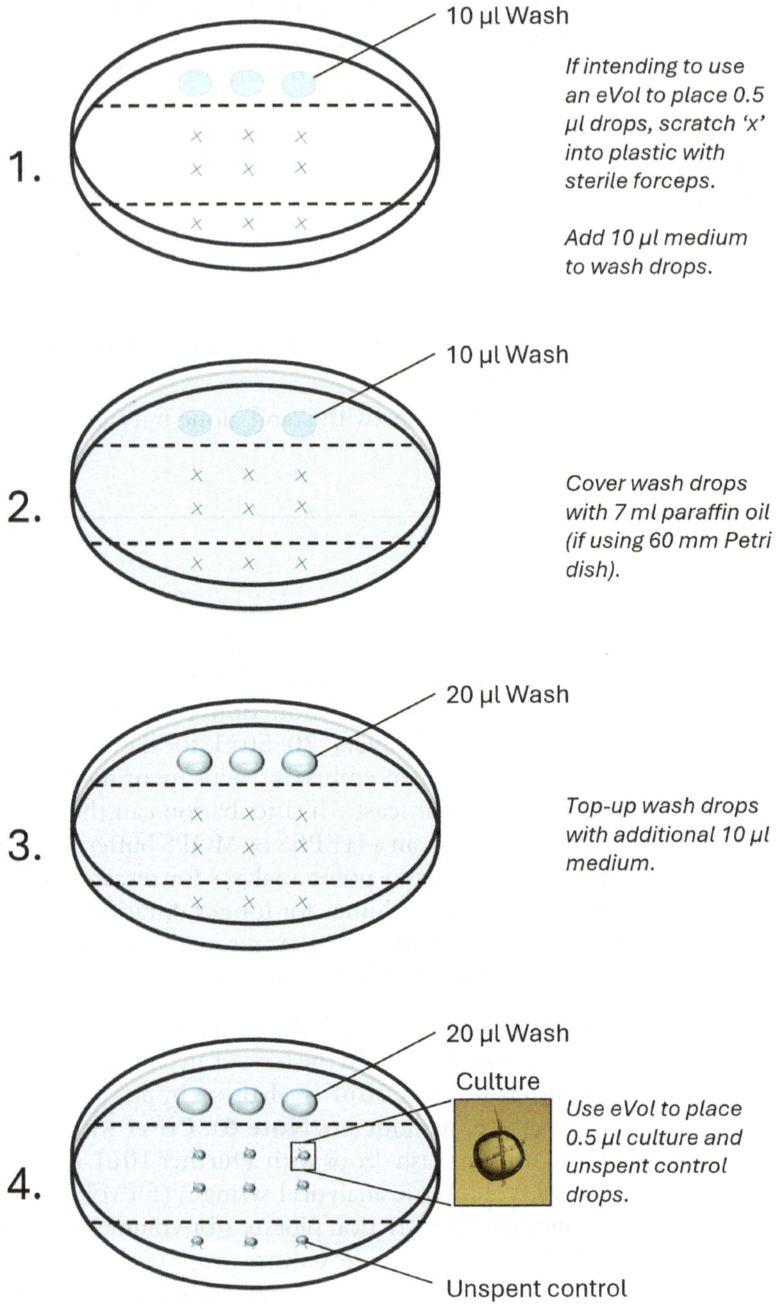

Fig. 1 Preparation of 60 mm embryo culture dish for metabolic analysis in a research setting. **(1)** If intending to culture embryos in 0.5 µL droplets, first use sterile watchmaker forceps to scratch an "x" where each drop will be placed. Next, place 10 µL metabolic culture medium for each wash drop. **(2)** Cover wash drops with 7 mL paraffin oil. **(3)** Top up wash drops with an additional 10 µL metabolic culture medium. **(4)** Place individual culture drops and unspent control drops directly under oil. The scratched 'x' is visible under the drop

pulled pasture pipette slightly larger than the embryo's diameter, with minimal medium uptake (*see* **Note 10**). Incubate embryos in the appropriate temperature and gas environment.

3. For linear analyses, take samples from the spent and unspent metabolic culture medium hourly throughout the experimental incubation and measure immediately via UMF. Alternatively, for end-point analyses, remove embryos from incubation drops (*see* **Note 11**) and store whole metabolic culture dishes at −80°C until time of analysis.

3.2 Embryo Culture for Analysis of Substrate Consumption or Production in a Clinical Setting

In a clinical setting, human embryo metabolism can be assessed by endpoint analysis. Incubation should occur in a CO_2-gassed incubator in a bicarbonate-buffered culture medium, and embryos must be cultured in single-well microwell dishes. One such dish is the Vitrolife EmbryoSlide+ ic8 dish, which will be used as an example in this chapter (Fig. 2a). Well of the well (WOW) culture dishes (such as the Vitrolife EmbryoSlide+ culture dish) (Fig. 2b) must not be used because it is a requirement that embryos do not have a shared medium source. The duration and timing of metabolic culture will depend on the metabolite of interest. For example, to assess glucose consumption, a 24 h metabolic culture period at the blastocyst stage is recommended, whereas for assessment of pyruvate consumption, metabolic culture is best performed during the cleavage stage (*see* **Note 12**). Refer to Table 1 for suggested metabolic culture and assay conditions.

1. Prepare a clinical single-well microwell embryo culture dish using a stereomicroscope within a laminar flow hood. To prepare the Vitrolife EmbryoSlide+ ic8 dish (Fig. 2a), use a micropipette to add 35 μL medium to each of the four wash wells. Next, using a stereomicroscope and calibrated micropipette, add 15 μL medium to each individual embryo culture well. Swiftly fill the dish with 1.6 mL paraffin oil to minimise evaporation. Using the stereomicroscope, assess whether any microbubbles are present within culture wells, and gently tap the dish to dislodge bubbles. Do not use a pipette to remove bubbles as this may cause volumetric changes within wells. Refrigerate culture dishes at 4°C for at least 3 h to further assist with the removal of microbubbles. Equilibrate the metabolic culture dish for at least 4 h in the appropriate gas and temperature environment.
2. Move embryos to metabolic culture microwells. Wash embryos thoroughly in the four wash drops to ensure no carry-over of substrates from non-metabolic culture medium. Subsequently, move individual embryos to microwells, ensuring minimal medium transfer. Leave at least one well unoccupied in the metabolic culture dish (or if the culture dish is full, co-incubate

Table 1

Suggested metabolic culture conditions and ultramicrofluorescence (UMF) metabolic assay conditions for the non-invasive analysis of human and mouse embryo metabolism of glucose, lactate, pyruvate, glutamine, aspartate and beta-hydroxybutyrate

			Metabolic culture conditions						UMF metabolic assay conditions				
Metabolic assay	Species	Developmental stage	Assumed metabolic rate (pmol/embryo/h) (consumption [−] or production [+])	Initial metabolite concentration in culture medium (mmol/L)	Culture volume (μL)	Initial metabolite amount in culture drop (pmol)	Culture duration (h)	Expected total amount consumed (−) or produced (+) (pmol/embryo)	Standard curve top concentration (mmol/L)	Sample:-Cocktail volume ratio	Reaction incubation time at 37°C (min)	Sample dilution factor required prior to analysis	Expected detected metabolite concentration in sample after dilution (mmol/L)
Glucose	Human	Blastocyst	−80	3.15[a]	15	47,250	24	−1920	0.5	1:3	2	10	0.3022
	Mouse	Blastocyst	−8	0.5[b]	0.5	250	8	−64	0.5	1:3	2	1	0.372
Lactate	Human	Blastocyst	+95	5.87[a]	15	88,050	24	+2280	1	1:3	2	10	0.6022
	Mouse	Blastocyst	+6	5.87[a]	0.5	2935	8	+48	1	1:3	2	10	0.5966
Pyruvate	Human	Oocyte or cleavage-stage	−36	0.32[c]	15	4800	24	−864	0.5	1:5	10	1	0.2624
	Mouse	Oocyte or cleavage-stage	−4	0.32[c]	0.5	160	8	−32	0.5	1:5	10	1	0.256
Glutamine	Human	Blastocyst	−4	0.5[a]	0.5	250	24	−96	0.5	1:4	5	1	0.308
Aspartate	Mouse	Blastocyst	−3	0.1[a]	0.144	14.4	3	−9	0.1	1:4	5	1	0.0375
B-hydroxy-butyrate	Mouse	Blastocyst	−2	2[b]	0.5	1000	8	−16	1	1:4	10	4	0.508

[a]Based on commercial G2 Plus medium for blastocyst stage

[b]Based on published studies with modified metabolic culture medium [13]

[c]Based on commercial G1 Plus medium for oocyte/cleavage stage Bolded values indicate parameters that may be modified to optimise culture and assay analyses

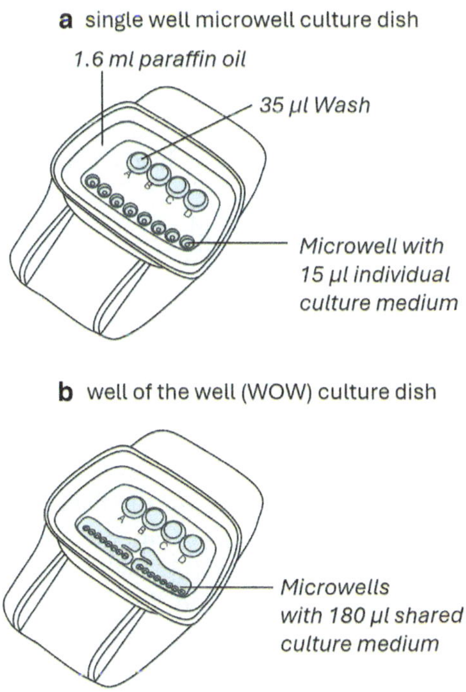

Fig. 2 A single well microwell culture, e.g., ic8 dish (**a**) should always be used for clinical assessment of embryo metabolism. Well of the well (WOW) culture dishes (**b**) contain a shared medium source and are inappropriate for metabolic analyses

another unoccupied dish); this will act as the unspent medium control sample. Once all embryos are in place, incubate the culture dish for 24 h in the appropriate gas and temperature environment.

3. Carefully remove embryos from the culture dish. Embryos should be moved to a new culture dish, vitrified, transferred, or discarded, as per relevant clinical procedures. Ensure minimal medium is removed from microwells during embryo extraction and avoid inter-sample contamination by using a clean pipette to move each embryo.
4. The remaining sample within microwells should be homogenised by pipetting, using a clean pipette for each sample. Store whole metabolic dishes at −80°C until the time of analysis.

3.3 Siliconizing Microscope Slides

Due to the minute volume of the enzymatic reactions analysed via the UMF method, these must occur submerged in paraffin oil to prevent evaporation. This is best conducted on siliconized microscope slides to prevent oil spillage and promote adherence of the reaction drops to the glass surface.

1. In a fume hood, load glass slides into a microscope slide holder and plunge into a microscope slide bath filled with undiluted

siliconizing agent for 1 min. Rinse slides in reverse osmosis ultrapure water for 1 min.

2. Lay siliconized microscope slides on an absorbent towel to dry overnight in a fume hood. Cover the upper surface with paper towel to prevent any dust accumulation.

3.4 Construction of Glass Micropipettes

1. Pull a borosilicate glass capillary (Fig. 3a) over the flame of a dental burner, producing an inner diameter of approximately 50–100 μm. Break the capillary into two halves (Fig. 3b), then create a hook on the thinnest end of each half by quickly passing the capillary through the flame of the dental burner; this will cause the capillary to fold back onto itself (Fig. 3c).
2. Mount the capillary on a microforge with the hooked glass pointing downwards. Midway along the tapered section create a constriction by using the filament of the microforge to heat and expand the glass (Fig. 3d). To form the tip of the pipette, place a light weight (i.e., half of a paperclip) on the hook of the capillary (Fig. 3e) and slowly heat the glass in the tapered area below the constriction (Fig. 3f). The weighted heated glass will stretch downwards forming a finer taper. Once stretched to the required diameter, quickly stop heating the glass. Use forceps to pull downwards on the weight, breaking off the remainder of the tip (*see* **Note 13**). A photograph of a glass micropipette with constriction is shown (Fig. 3g).
3. Mount the glass pipette into a 16-gauge stainless steel tube. Adjust the length of the pipette by pulling the wide end through the tube, then break off any excess overhang (*see* **Note 14**). Melt sealing wax in the flame of a dental burner and apply a droplet of wax to the top of the metal rod where the glass capillary extrudes. While the wax is still pliable, seal the gap between the glass and metal tube (*see* **Note 15**).
4. Micropipettes should be siliconized to prevent fluid droplets from sticking to the small internal diameter, thus increasing the accuracy of volume delivery. In a fume hood, with the pipette connected to a syringe via plastic tubing, siliconize the lumen of the pipette by repeatedly aspirating and expelling undiluted siliconizing agent for 30 s. Rinse the pipette by aspirating/expelling dilute (0.1 M) HCl, then water for 30 s each (*see* **Note 16**). Leave pipettes to dry overnight in an upright position (*see* **Note 17**).

3.5 Calibration of Glass Constriction Pipettes using a Qubit™ Fluorometer

A Qubit™ fluorometer and 1× dsDNA HS Assay kit are used to determine the volume of micropipettes (Fig. 4) without the requirement for radioactive substrates/tritiated water. This technique has been validated against the use of tritiated water.

1. Prepare a two-point standard (0 ng/μL and 10 ng/μL dsDNA) by adding 10 μL of each standard (provided in the kit) to 190 μL

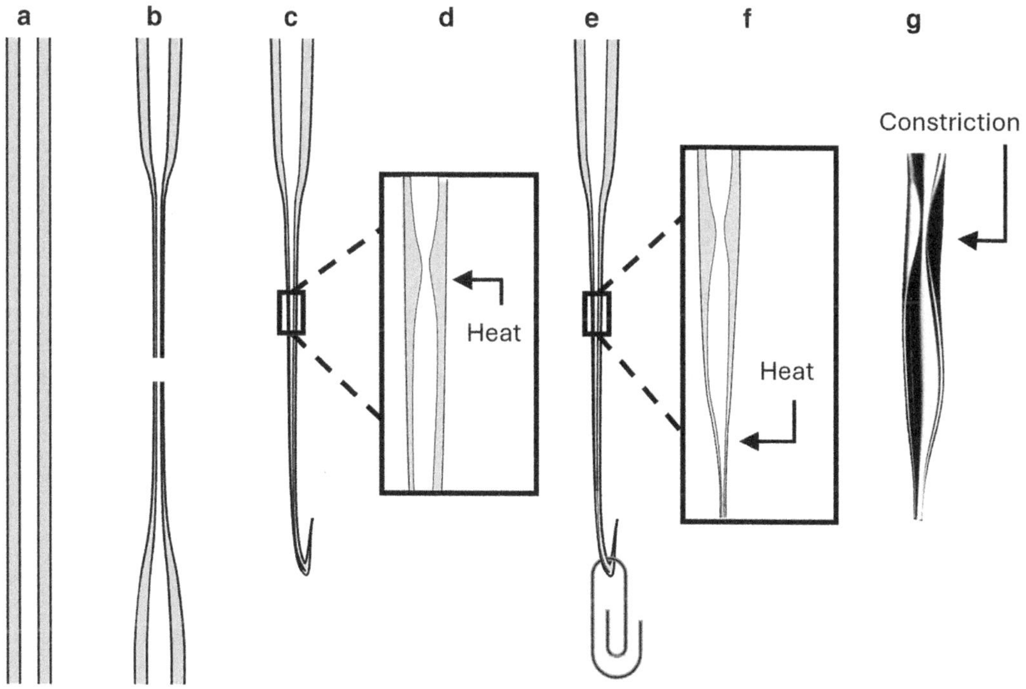

Fig. 3 Production of constriction pipettes for the delivery of volumes in the nano- and picolitre range. (**a**) Borosilicate capillary (length 150 mm), with an outer diameter of 1.0 mm and an inner diameter of 0.8 mm. The capillary is heated in the middle using a small yellow flame and pulled. (**b**) The capillary is then broken into two halves. Two pipettes can be prepared from one capillary. The size of the pipette is determined in part by how thin the capillary is pulled. **(c)** A hook is made by melting the end of the pulled capillary using a small flame. The capillary is then mounted in a microforge with the hook end down. **(d)** Using a heated platinum filament, a constriction is made in the pulled area of the capillary. The heated filament is placed as close to the capillary without touching, and the heat is increased until the glass implodes within the capillary, forming a constriction. **(e)** A small weight is placed on the hook, such as half a small paper clip. **(f)** The heating element is then moved down the capillary, and then heat is reapplied to the capillary. The distance between the initial constriction and where the capillary is melted for the second time will in part determine the volume of the pipette (the other factor being the inner diameter of the pulled capillary). As the glass melts, rather than imploding, the glass capillary is pulled to a tip as the weight on the hook pulls on the capillary. Do not let the sides of the capillary touch. Rather, stop applying heat once the capillary has been drawn to a small constriction. Use less heat at this time than when the original constriction was made, as the glass is much thinner once it starts to be pulled and will quickly seal if the filament is too hot. Once the capillary has been drawn out, the glass can be broken using a pair of watchmaker's forceps, by pulling on the capillary below the position of the tip. The capillary will break square, leaving an intact constriction pipette. Take the newly constructed pipette and mount it in a 16-gauge stainless-steel tube, and fix the pipette in place using sealing wax. The mounted pipette can then be siliconized and calibrated. Such a pipette can be re-used indefinitely, provided it is cleaned between uses and re-siliconized when required. (**g**) Photograph of a nanolitre constriction pipette

"working solution" (provided in the kit) in a clear 0.5 mL PCR Eppendorf tube. Vortex for 3 s then incubate at room temperature in the dark for 2–20 min. Measure the fluorescence of each standard using a Qubit™ fluorometer.

2. Place a droplet of 10 ng/μL dsDNA under paraffin oil on a siliconized microscope slide. With the constriction pipette that is to be volumetrically calibrated held in a micromanipulator, aspirate paraffin oil to the constriction of the micropipette (*x*μL; Figs. 3g and 4), followed by an equal volume (*x*μL) of 10 ng/μL dsDNA. Carefully dispense the total volume from the micropipette into a clear 0.5 mL PCR Eppendorf tube containing 200 μL "working solution." Vortex the PCR tube for 3 s and incubate in the dark at room temperature for 2–20 min prior to measuring fluorescence. Assess the volume of each pipette in triplicate. Use Eq. 1 to determine the volume of each micropipette, where QF is the value given by the Qubit™ fluorometer, and *x* is the volume (in μL) from the tip to the constriction of the pipette.

$$x = \frac{200(QF)}{10ng/ul - 2(QF)} \tag{1}$$

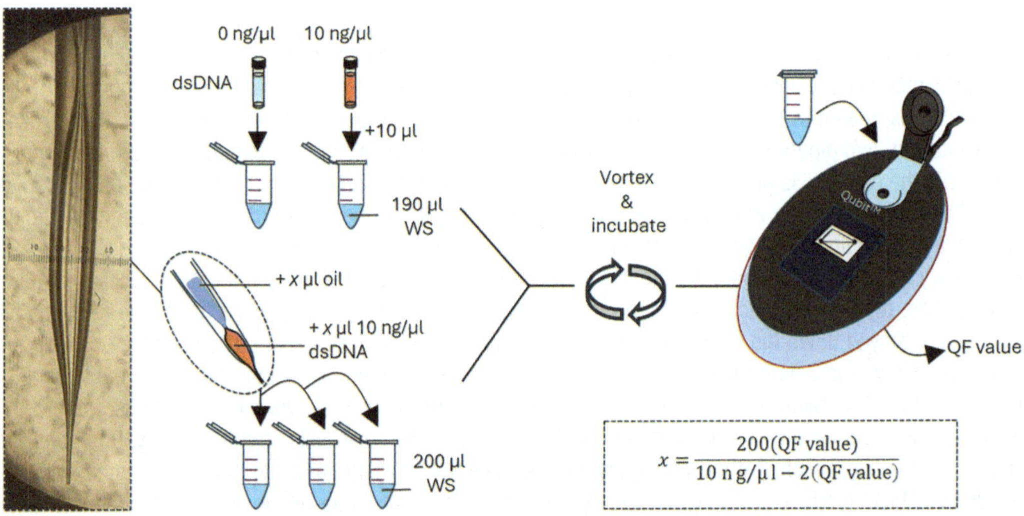

Fig. 4 Use of the Qubit™ 1× dsDNA HS Assay Kit to calibrate glass constriction micropipettes (photograph shown on left). This kit utilizes a Molecular Probes dye that binds to dsDNA, with subsequent fluorescence measured at 485/530 nm. It is designed to determine the dsDNA concentration when 10 μL of a sample is added to 190 μL working solution (WS). However, to calibrate micropipettes, the assay can be modified to determine the unknown volume of 10 ng/μL dsDNA that was added to 200 μL working solution, using the $C_1V_1 = C_2V_2$ principle. This is a more accessible and safe method of calibration compared with the traditional method of using tritiated water. To calibrate constriction pipettes, first run a two-point standard by adding 10 μL of 0 ng/μL and 10 ng/μL dsDNA to 190 μL WS. Vortex and incubate for 2–20 min in the dark, then measure on the Qubit™ fluorometer. Next, aspirate oil and then 10 ng/μL dsDNA to the constriction of the micropipette to be tested (*x* μL each) and expel into 200 μL WS in triplicate. Vortex then incubate reactions for 2–20 minin the dark, then measure using the Qubit™ fluorometer, with output value "QF". Use the equation shown to solve for *x* to determine pipette volume, taking the average of triplicates as the final volume

3.6 Ultramicrofluorescence (UMF)

Siliconized microscope slides should always be handled with gloves and should be wiped prior to use with a microscope lens wipe to ensure a clean surface, as dust particles and fibres may interfere with fluorescence readings. Constriction micropipettes can be washed by aspirating and expelling acetone then ultrapure water. Two constriction pipettes are required for each metabolic assay: a larger pipette to aliquot the reagents of the cocktail, and a smaller pipette to aspirate the standards and spent medium samples. The ratio of cocktail and sample pipette sizes is important, with this ratio requiring optimisation according to the pipette sizes that you have available (*see* **Note 18**).

Constriction micropipettes are controlled pneumatically by use of a micromanipulator with an air syringe and visualised under a stereomicroscope. The following steps should be performed in a dark room (with red light), at 37°C using heated stages, unless specified otherwise (all reagents must be kept on ice prior to use).

Spent medium samples may require dilution prior to analysis. This is dependent on the expected target metabolite concentration remaining in spent medium samples, and the range of the standard curve for analysis (*see* **Note 5**). Table 1 includes suggested dilution factors for samples prior to UMF analysis.

1. For the relevant metabolic assay, prepare a six-point standard curve. On one siliconized microscope slide (Slide A) (Fig. 5), use a P1000 pipette to pipette six medium-sized drops (approximately 150 μL) of paraffin oil. Using a smaller P10 pipette, place a smaller droplet (approximately 15 μL) of each standard solution under an oil drop to prevent evaporation, and carefully set this microscope slide aside.
2. Prepare a second microscope slide, which will be used to incubate and measure the standard curve reactions (Slide B) (Fig. 5), by first placing two adjacent large drops (approximately 300 μL) of oil on Slide B. On the stage of a stereomicroscope, use a P100 pipette to place a drop (approximately 30 μL) of the relevant reagent cocktail under one oil drop. Then, using the cocktail constriction pipette to aspirate the reagent cocktail solution to the level of the constriction, make small droplets under the second oil drop. Make at least six reagent cocktail drops (one for each standard).
3. Place Slide B on the unheated stage of an inverted fluorescent microscope connected to a shutter system, photometer, and computer software to detect the signal, and allow to cool to room temperature for 2 min. Measure the initial fluorescence of the oil and each cocktail drop by visualising through the ×20 or ×40 objective and exposing to UV light, typically for 0.5–4 s (keep exposure time standard for each drop), during which time photooxidation of NAD(P)H is negligible [9] (*see* **Note 19**).

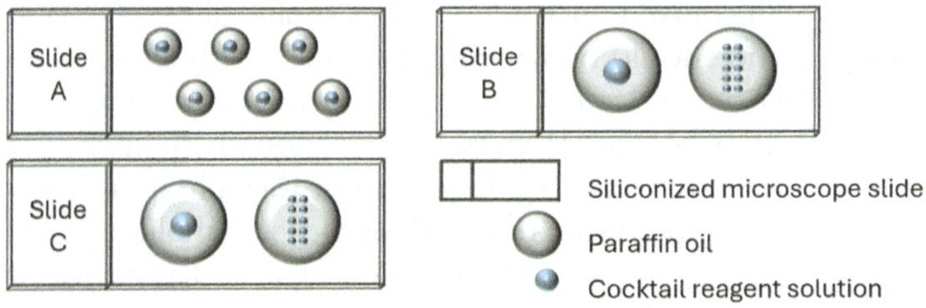

Fig. 5 Enzymatic reactions for ultramicrofluorescence are housed on siliconized microscope slides under paraffin oil to prevent evaporation. Slide **A** houses a small drop of each standard under paraffin oil, Slides **B** and **C** house the reaction cocktail drops, created with the larger "cocktail" constriction micropipette, where the incubation will take place. The addition of each standard from Slide **A** to a small cocktail drop on Slide **B** using the smaller "sample" constriction micropipette, or the addition of each spent medium sample to the cocktail drops on Slide **C**, initiates the enzymatic reaction. Before and after the addition of standards/samples to Slides **B/C**, the fluorescence of oil and small cocktail drops is quantified following UV light exposure for 0.5–4 s. The difference between initial and final oil fluorescence should be negligible, and the difference in cocktail drop readings with standards added should be linear ($R^2 \geq 0.99$). Substrate concentration within samples can be determined from a linear standard curve

4. Next, return Slide B to the stage of a stereomicroscope, use the sample constriction pipette to successively add each standard from Slide A to the cocktail drops on Slide B; this initiates the enzymatic reaction (*see* **Note 20**). Incubate Slide B in the dark until the reaction has gone to completion (*see* **Note 21** and Table 1).
5. Place Slide B back onto the unheated stage of the fluorescent microscope and cool to room temperature for 2 min. Re-measure the fluorescence of the oil and each cocktail drop after the addition of the sample. The difference between initial and final fluorescent measurements of the oil should be negligible, and the difference in cocktail drop readings with standards added should be linear ($R^2 \geq 0.99$). Create a new Slide B and repeat if an initial standard curve with $R^2 < 0.99$ is obtained. A new standard must be run on each experimental day prior to measuring samples (*see* **Note 22**).
6. Once a linear standard curve has been established, the concentration of substrates within samples can be measured against this standard curve. Prepare another slide (Slide C) (Fig. 5) with a cold cocktail as per the steps above for Slide B and measure the initial and final fluorescence of oil and cocktail drops following the addition of spent and unspent (control) embryo culture medium samples (*see* **Note 23**).

4 Notes

1. If the volumetric ratio of reagents is maintained, the final volume can be modified as required. Note the high concentration of enzyme added to UMF cocktail reagents, which is necessary to compensate for the degradation of the enzyme at the aqueous-oil interface. This issue is not present in other fluorometric assays where lower concentrations of enzyme can be utilised.
2. The osmotic concentration of stock solutions can be tested with an osmometer to ensure initial accuracy before proceeding with further dilutions. For example, a 50 mM glucose solution will have an osmolality of 50 mOsm/kg, whereas a 100 mM sodium-lactate solution will read 200 mOsm/kg.
3. Standards can alternatively be made up in a modified culture medium, which is devoid of the substrate to be analysed, in place of water. Doing so may increase the accuracy of results by including other substrates that will be present in spent culture medium samples that may contribute to fluorescence, such as aromatic amino acids. However, this is not essential as water standards yield robust results.
4. Pyruvate will degrade at −20°C, therefore standards and samples should be stored at −80°C.
5. When designing experimental culture conditions, it is important to consider:
   a. rates of metabolite consumption/production previously reported for specific species and developmental stages,
   b. the duration of the incubation period,
   c. the volume and frequency of sampling,
   d. the concentration of nutrients available within the metabolic culture drop, and
   e. the volume of the culture drop.

   Ideally, embryos should be incubated in a culture medium drop with an appropriate volume, substrate concentration, and for a duration whereby approximately 50% of the substrate is expected to be consumed by the embryo by the end of the incubation period. This will ensure detectability of substrate consumption and the ability to distinguish metabolic differences between experimental treatment groups if applicable. When nutrient concentrations cannot be modified (such as for clinical metabolic assessment of human embryos cultured in commercially produced medium), the culture medium volume and duration can instead be adjusted to optimise conditions. A final option is to dilute the spent medium sample to within a measurable range prior to analysis. Care must be taken to ensure accuracy of dilutions, to minimise introducing a source of error. For linear metabolic analyses, be mindful to use a culture drop with sufficient volume to facilitate multiple samples being removed, with enough volume remaining for continued culture. For example, mouse blastocysts are reported to consume glucose at a rate of ~20 pmol/embryo/h [15]. If a single mouse blastocyst was cultured in 0.5 μL medium with 0.5 mM glucose (250 pmol total glucose available) for 7 h, it would be expected to consume 140 pmol, equivalent to 56% of the available glucose being consumed. However, if the culture drop were larger, for example 2 μL (1000 pmol total glucose available), consumption of 140 pmol would equate to only 14% of the available glucose being consumed, consequently reducing the sensitivity of the assay.

In contrast, human blastocyst glucose consumption is reported to be ~90 pmol/embryo/h [16]. Therefore, with a total of 630 pmol glucose consumption expected over a 7 h incubation by a human blastocyst, a 0.5 μL culture drop would not have sufficient glucose available (only 250 pmol glucose available), whilst a 2 μL drop would be a more appropriate volume (1000 pmol glucose available) in this scenario. In the clinical example provided in Table 1, a human blastocyst cultured for 24 h in 15 μL medium containing 3.15 mmol/L glucose would be expected to produce a spent medium sample with 3.022 mmol/L glucose remaining. As this could not be measured against a 0.5 mmol/L glucose standard curve, the sample would first require a 1:10 dilution to bring it within a measurable range (0.3022 mmol/L).

One must also be mindful of transporter kinetics and ensure that in an attempt to increase detectability of substrate consumption (by reducing initial substrate concentration), the concentration within the culture drop is not reduced (either by experimental design or via consumption by the embryo) to a level below the EC50 of the relevant transporter, as this will significantly reduce uptake.

6. The composition of the culture medium should be appropriate to the substrate consumption/production to be analysed. Typically, the upper limit of UMF concentration detection is 1 mM, therefore metabolic culture media may require modification to reduce substrate concentration within this limit, or spent medium can be diluted to within a measurable range. For analysis of glycolytic flux (the conversion of 1 mol glucose to 2 mol lactate), metabolic culture medium should be prepared devoid of both lactate and pyruvate. This increases the likelihood that the lactate produced is derived from glycolysis

and not from the reduction of pyruvate via endogenous lactate dehydrogenase. However, lactate production from endogenous pyruvate derived from amino acid metabolism, or from glycogen, cannot be entirely excluded. Metabolic culture medium with substrate concentrations suitable for the experimental conditions can be made in-house using sterile and mouse embryo assay-safe reagents.

7. Precision of culture drop volumes is highly important, especially at lower volumes, whereby small deviations will result in a larger percentage difference in substrate availability. An eVol pipette (volumetric analytical syringe) can be used to make ≥0.5 μL drops, whereas smaller drops should be made with a calibrated constriction pipette. Test the accuracy and consistency of the eVol pipette by weighing drops of water expelled onto an analytical balance.

8. Culture drops should be made after oil coverage is placed, not before. This is to prevent evaporation and therefore volumetric, osmotic, and substrate concentration changes. For larger drop volumes (~1–5 ul) the eVol pipette (volumetric analytical syringe) can be used to place drops directly onto the culture dish surface under the oil. For smaller drop volumes (~0.5 μL), this is more difficult, and the culture medium preferentially sticks to the eVol syringe rather than transferring to the dish surface. To circumvent this, a small "x" can be scratched into the surface of the plastic dish using sterile watchmaker forceps prior to placing oil. A 0.5 μL culture drop can then be placed over each "x," with static attraction helping to retain the medium on the culture dish surface. If using the scratched "x" method, non-toxicity must be confirmed.

9. It is also important to measure samples of culture medium which have been co-incubated in the absence of an embryo ("unspent control" or "no embryo control"). This is used to determine the exact substrate concentration within the medium, and accounts for any potential breakdown that occurred during the incubation period. This control should be used as the baseline against which to determine rates of substrate consumption/production by embryos. If testing the effect of different culture media or incubation conditions on embryo metabolism, unspent controls for each type of medium/experimental group must be tested.

10. To move embryos into incubation drops, collect one embryo in the very tip of the pulled glass pasture pipette, then slightly touch the tip of the pipette to the very top of the incubation drop. This will cause the meniscus of the drop to connect with the pipette, and withdrawing the pipette will cause the

embryo to "snap" into the incubation drop with minimal medium carry-over.

11. A new clean pipette must be used to remove each embryo from its respective incubation drop to prevent cross-contamination of spent medium samples. Non-human/research embryos can be stained to determine cell number, in which case embryonic metabolism may be calculated on a per-cell basis.
12. A 24 h culture duration is suggested for human embryos at the blastocyst stage; however, the duration and timing can be modified depending on the substrate to be tested and the appropriate developmental stage. For metabolic analyses to be incorporated into clinical practice, it may be most appropriate to combine the transition into a metabolic culture dish with routine medium changeovers that occur for all patient embryos, such as a day 5 medium refresh. It should also be noted that specific consent may or may not be required to conduct additional medium changeovers, or for the utilisation of spent culture medium, depending on the requirements of local regulatory bodies.
13. A paperclip cut into two halves works well as a light weight that can be hung over the hook of the capillary. When forming the tip of the pipette, do not let the glass stretch so far that it breaks, as this will seal the capillary. After pulling the weight to remove excess glass, check through the eyepiece of the microforge that the tip of the glass pipette has broken off square. If it is jagged, you can delicately touch the end of the watchmaker forceps to the tip of the pipette to square it off.
14. It can be useful to make all your pipettes a consistent length. A convenient way to do this is to use the distance from the top of your forefinger to the middle knuckle as an approximate measure.
15. It can be useful to label pipettes using some masking tape wrapped around the metal tube and folded back onto itself, much like a flag. This also provides an area to grip the pipettes. It is not recommended to handle the pipettes by the waxed area, as the wax is delicate and can snap. If this does occur and the pipette is otherwise intact, wax can simply be reapplied to re-seal the glass and metal tube.
16. Always rinse pipettes in HCl then in water, and never the other way around to prevent violent boiling and splashing of the concentrated acid.
17. A convenient way to store the glass pipettes is standing upright in a piece of Perspex with small holes drilled partially into the surface. Store pipettes in a protected space, such as a

shelf in a closed cupboard, to protect from breakage and dust accumulation, which can clog pipettes.

18. Some trial and error is required to find the correct combination of sample and cocktail pipettes, and therefore the sample: cocktail ratio for each assay. A good starting point is a 1:4 ratio and adjust as required. For example, if the reaction is not going to completion (indicated by flattening of the curve at higher test reagent concentrations), this may be troubleshooted by decreasing the sample: cocktail ratio (i.e., increasing the cocktail pipette size or reducing the sample pipette size), by increasing the reaction incubation time, or by increasing the concentration of reactants (NAD[P]$^+$, but not NAD[P]H) and/or enzyme in the cocktail.
19. The initial readings of cocktail drops should all be relatively similar. If a cocktail drop produces a particularly high reading relative to all other drops, avoid using this drop for standards/samples as it will produce unreliable results. Make "spare" drops for this reason. High variability between initial cocktail drop readings is often an indicator that the mercury bulb within the fluorescent microscope needs replacing. A mercury bulb typically gives reliable excitation emission for up to 150–200 h.
20. When adding samples to cocktail droplets (either for the standard curve or for analytical samples), it is not necessary to wash micropipettes between samples so long as the entire sample is expelled from the pipette. As this process is conducted under a microscope, it is visually very clear when a partial sample has or has not been retained in the pipette. If bubbles begin to form/sample is retained on the lumen of the pipette, then the pipettes must be washed to prevent this from occurring. If washing does not improve sample retention, pipettes may need to be re-siliconized.
21. The incubation time for the reaction to reach completion will differ for each assay, depending on the substrate being analysed, and will require optimisation (take fluorescence readings of the reaction over time to determine when completion is reached). Suggested incubation times are listed in Table 1. To test if a reaction has gone to completion during optimisation, image the fluorescence at time intervals (e.g., at 1-min intervals) after initiating the reaction. If the reaction is one that involves the conversion of NAD(P)$^+$ to NAD(P)H, the fluorescence should initially increase in intensity. It will then plateau when the reaction has reached completion and may begin to decrease in intensity if imaging is continued as NAD(P)H becomes photooxidised. In reactions involving NAD(P)H to NAD(P)$^+$ conversion, the opposite will be true.

22. A new standard curve must be run if a sample or cocktail pipette becomes irreversibly blocked or broken and requires replacing. Do not measure samples using a new sample/cocktail pipette combination against a standard curve made with different pipettes, as the results will be incorrect. It is recommended to run a standard curve as close to the time of sample measurement as possible to keep environmental conditions consistent. Even small changes in room temperature can affect fluorescence readings. If you run multiple batches of samples against a single standard curve, pay attention that unspent controls are measuring at an expected concentration; changes in their levels may indicate a shift in environmental conditions and a new standard curve should be run.
23. It is recommended to measure a maximum of 10–12 samples at one time. If there are more samples than this to measure, create a new Slide C for each batch of measurements.

References

1. Renard J-P, Philippon A, Menezo Y (1980) In-vitro uptake of glucose by bovine blastocysts. Reproduction 58(1):161–164
2. Gardner DK, Leese HJ (1987) Assessment of embryo viability prior to transfer by the non-invasive measurement of glucose uptake. J Exp Zool 242(1):103–105
3. Gardner DK, Wale PL, Collins R, Lane M (2011) Glucose consumption of single post-compaction human embryos is predictive of embryo sex and live birth outcome. Hum Reprod 26(8):1981–1986
4. Ferrick L, Lee YSL, Gardner DK (2020) Metabolic activity of human blastocysts correlates with their morphokinetics, morphological grade, KIDScore and artificial intelligence ranking. Hum Reprod 35(9):2004–2016
5. Lane M, Gardner DK (1996) Selection of viable mouse blastocysts prior to transfer using a metabolic criterion. Hum Reprod 11(9):1975–1978
6. Gardner DK (2007) Non-invasive metabolic assessment of single cells. Methods Mol Med 132:1–9
7. Mroz EA, Lechene C (1980) Fluorescence analysis of picoliter samples. Anal Biochem 102(1):90–96
8. Leese HJ, Biggers JD, Mroz EA, Lechene C (1984) Nucleotides in a single mammalian ovum or preimplantation embryo. Anal Biochem 140(2):443–448
9. Leese H, Barton AM (1984) Pyruvate and glucose uptake by mouse ova and preimplantation embryos. J Reprod Fertil 72(1):9–13
10. Gardner DK, Leese HJ (1990) Concentrations of nutrients in mouse oviduct fluid and their effects on embryo development and metabolism in vitro. J Reprod Fertil 88(1):361–368
11. Gardner DK, Clarke RN, Lechene CP, Biggers JD (1989) Development of a non-invasive ultramicrofluorometric method for measuring net uptake of glutamine by single preimplantation mouse embryos. Gamete Res 24(4):427–438
12. Lee YSL, Gardner DK (2021) Early cleaving embryos result in blastocysts with increased aspartate and glucose consumption, which exhibit different metabolic gene expression that persists in placental and fetal tissues. J Assist Reprod Genet 38(12):3099–3111
13. Whatley EG, Truong TT, Wilhelm D, Harvey AJ, Gardner DK (2022) β-hydroxybutyrate reduces blastocyst viability via trophectoderm-mediated metabolic aberrations in mice. Hum Reprod 37(9):1994–2011
14. Whatley EG (2023) The metaboloepigenetic regulation of preimplantation embryo development and fetal programming by ketone bodies. PhD thesis, University of Melbourne
15. Lee YS, Thouas GA, Gardner DK (2015) Developmental kinetics of cleavage stage mouse embryos are related to their subsequent carbohydrate and amino acid utilization at the blastocyst stage. Hum Reprod 30(3):543–552
16. Ferrick L, Lee Y, Gardner DK (2019) Metabolic activity of human preimplantation embryos relates to their morphokinetics, morphological grade, KIDScore and artificial intelligence selection. In: ESHRE annual meeting, Vienna

Chapter 11

Evaluating Preimplantation Embryo Metabolism Through Imaging of Cellular Autofluorescence

Darren J.X. Chow, Tiffany C.Y. Tan, Kishan Dholakia and Kylie R. Dunning

Abstract

Optical imaging offers a non-invasive approach for assessing embryo viability. This may be achieved through capturing fluorescence from endogenous fluorophores that are associated with cellular metabolism. By recording and analysing natural autofluorescence emitted by molecules involved in metabolism, such as NADH and FAD, this technique offers crucial insights into the metabolic state of embryos, in the absence of exogenous labels. This approach is particularly valuable for preimplantation embryos developed in vitro, where the ability to identify those with high developmental potential may lead to improved outcomes in a clinical setting. Light sheet microscopy has recently emerged as a powerful imaging modality suitable for recording autofluorescence from the developing embryo. In contrast to standard point-scanning (confocal) approaches, it only illuminates the plane of interest, minimising light exposure and photodamage while enabling rapid 3D image acquisition. These aspects make light sheet microscopy particularly well-suited for metabolic imaging of live preimplantation embryos. In this chapter, we describe a general procedure to perform metabolic imaging of live mammalian embryos using light sheet microscopy. This protocol enables direct observation of metabolic changes within embryos in a spatial manner and provides guidelines for optimising imaging parameters to ensure accuracy and reproducibility.

Keywords Optical imaging, Murine preimplantation embryo, Confocal microscopy, Light sheet microscopy, Autofluorescence, Label-free, Metabolism

1 Introduction

Optical imaging has gained recent attention for its ability to leverage endogenous fluorophores for label-free investigation of embryo metabolic health [1–6]. This is achieved by exploiting the autofluorescent properties of key metabolic cofactors, reduced nicotinamide adenine dinucleotide (NADH) and flavin adenine dinucleotide (FAD) (Fig. 1), which play important roles in energy production during preimplantation embryo development [7]. As embryo metabolism is closely linked to viability, monitoring these metabolic pathways provides a valuable window into embryo quality without the need for external fluorescent labels or stains [5, 8].

Alexandra J. Harvey and David K. Gardner (eds.), *The Preimplantation Embryo*, Methods in Molecular Biology, vol. 3038,
https://doi.org/10.1007/978-1-0716-5292-3_11,

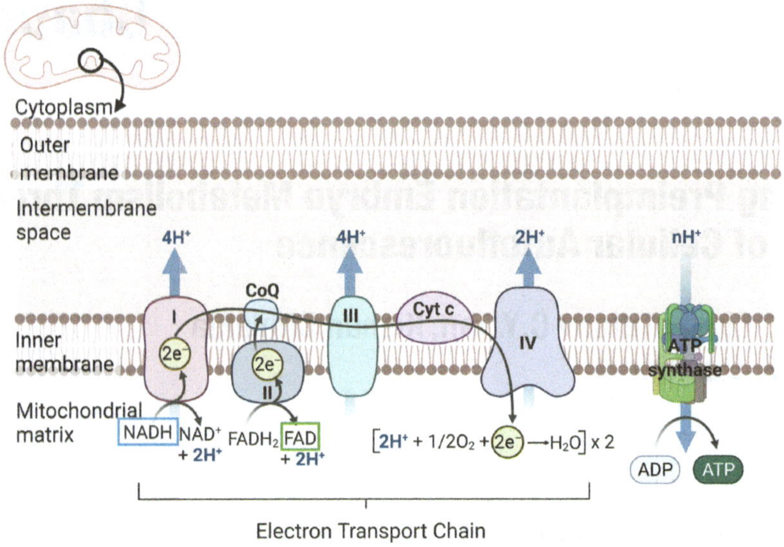

Fig. 1 Cellular metabolism can be assessed non-invasively through optical imaging of autofluorescent metabolic cofactors, NADH and FAD, which fuel the electron transport chain (ETC). Electrons from NADH and $FADH_2$ pass through ETC complexes, creating a proton gradient that drives adenosine triphosphate (ATP) production. NADH nicotinamide adenine dinucleotide, FAD flavin adenine dinucleotide. (Adapted from [5])

Recent advancements in light sheet microscopy have demonstrated the power of this imaging modality for label-free, 3D imaging of embryo autofluorescence [1]. By confining light illumination to a thin, sheet-like plane, light sheet microscopy reduces light exposure, thereby minimising photodamage and photobleaching [1]. This contrasts with confocal microscopy, a widely utilised "point-scanning" optical approach associated with increased levels of photodamage due to the use of a high-power laser [9, 10]. Furthermore, light sheet microscopy offers rapid image acquisition and improved signal-to-noise ratio when performing 3D, volumetric imaging, making it well-suited for metabolic imaging of live embryos where autofluorescent signals are weak [1].

Here, we outline procedures that serve as a starting point for acquiring volumetric images of preimplantation embryos using light sheet microscopy. Additionally, we emphasise the key considerations for sample preparation, optimisation of imaging parameters, and data analysis of metabolic activity to ensure accuracy and reproducibility.

2 Materials

All media, plasticware, and consumables used in this protocol should be pre-tested and certified for embryo culture in accordance with mouse embryo assay (MEA) standards [11].

2.1 Consumables

1. Glass bottom dish 35 mm.
2. Syringe filter (0.22 μm).
3. Syringe with plunger (1 mL, 10 mL, 20 mL).
4. 29-gauge insulin syringe (1 mL).
5. Non-treated culture dish (35 mm).
6. Non-treated culture dish (60 mm).
7. Eppendorf tube (0.6 mL).
8. Hand pipette, e.g., *Cook® Flexipet® Adjustable Handle* (0.25–3 μL).
9. Cook® Flexipet® Pipette tip (100 μm).
10. Hank's Balanced Salt Solution (HBSS).
11. Sodium chloride (NaCl).
12. Polyvinylpyrrolidone (PVP) or polyvinyl alcohol (PVA).
13. Paraffin oil (*Puriss.* grade).
14. Bovine serum albumin (Low fatty acid; BSA).
15. Hyaluronidase.
16. Round bottom tubes (14 mL)
17. Phosphate-buffered saline (PBS).
18. Single-channel manual micropipette (0.5–10 μL, 2–20 μL, 20–200 μL, 200–1000 μL) and compatible sterile pipette tips.
19. Milli-Q ultrapure water.

2.2 Media

1. HEPES-buffered wash medium (E.g., Research Wash media, ARTLab Solutions, Australia) supplemented with 4 mg/mL BSA (*see* **Note 1**). Filter-sterilise the solution using a 10 mL sterile syringe with plunger and a 0.22 μm syringe filter, dispensing into a sterile 15 mL round-bottom tube.
2. Bicarbonate-buffered embryo cleavage medium (E.g., Research Cleave media, ARTLab Solutions, Australia) supplemented with 4 mg/mL BSA (*see* **Note 1**). Filter-sterilise the solution using a 10 mL sterile syringe with plunger and a 0.22 μ m syringe filter, dispensing into a sterile 15 mL round-bottom tube.
3. Fertilisation medium (E.g., Research Fertilisation media, ARTLab Solutions, Australia) supplemented with 4 mg/mL BSA (*see* **Note 1**). Filter-sterilise the solution using a 10 mL sterile syringe with plunger and a 0.22 μm syringe filter, dispensing into a sterile 15 mL round-bottom tube.

2.3 Hormones

1. Equine chorionic gonadotropin (eCG) is reconstituted in 0.9% NaCl solution to make a stock solution of 50 IU/mL. Aliquot

into 1 mL syringes with 29 G needles. Store in −20°C until required.

2. Human chorionic gonadotropin (hCG) is reconstituted in 0.9% NaCl solution to make a stock solution of 50 IU/mL and loaded into 1 mL syringes with 29 G needles. Store in −20°C until required.

2.4 Equipment

1. Fine forceps (Dumont tweezers style 5).
2. Micro-scissors.
3. Gassed, humidified, temperature-controlled incubator (37°C, 5% O_2, 6% CO_2, balanced with N_2).
4. Dissecting microscope with heated stage (calibrated to 37°C at the sample).
5. Dry block heater with insert for 14 mL tubes (calibrated to 37°C).
6. Imaging software: e.g., FIJI/ImageJ: https://imagej.net/software/fiji/downloads or Imaris.
7. Light sheet microscope (LSM; Custom-built, open-top or commercial grade; system capable of maintaining sample at 37°C during imaging).
8. Custom 3D printed imaging chambers ([9]).

2.5 Animals

All experiments using animals require approval from the appropriate animal ethics committee. Food and water are to be provided *ad libitum* to ensure optimal health for ovarian hyperstimulation.

1. Prepubertal female mice (CBA × C57Bl/6 first filial generation (F1); 21–23 days old) to be obtained from a laboratory animal supplier (*see* **Note 2**).
2. Mature male mice (for mating) from a laboratory animal supplier.

3 Methods

This protocol outlines the workflow for imaging and analysing autofluorescence signals, specifically NAD(P)H and FAD, to assess metabolic activity in preimplantation mouse embryos. It covers sample preparation, imaging configuration, and quantitative analysis using FIJI/ImageJ and associated plugins. While the methods described here are optimised for mouse (CBA × C57Bl/6 F1 strain) embryos, they can be adapted to other mouse strains or mammalian species with appropriate adjustments to developmental timing, imaging parameters, and handling protocols.

3.1 Hormonal Stimulation and Media Preparation

The collection process depends on the required mode of fertilisation and the developmental stage of the embryo to be assessed.

1. Induce ovarian hyperstimulation by administering female mice with 5 IU of eCG (0.1 mL of 50 IU/mL stock per mice) intraperitoneally (i.p.) at 1800 h followed by 5 IU of hCG 48 h later at 1800 h. This promotes the development of numerous follicles and oocyte maturation/ovulation, respectively.
2. On the day of hCG administration, prepare the following media:
    1. 20 mL of Research Wash, and 15 mL each of Research Fertilisation and Research Cleave medium containing 4 mg/mL BSA in tubes (*see* **Note 1**).
    2. Aliquot a 1 mL drop of Research Fertilisation medium in a 60 mm culture dish and overlay this with 12 mL of paraffin oil (Fig. 2, *sperm capacitation dish*). Equilibrate at 37°C, 5% O_2, 6% CO_2, balanced with N_2 for 4–16 h prior to use (*see* **Note 3**).
    3. Aliquot two 180 μL and eight 90 μL drops of Research Fertilisation medium onto a 60 mm culture dish and overlay with 12 mL of paraffin oil (Fig. 2, in vitro *fertilisation dish*). Equilibrate at 37°C, 5% O_2, 6% CO_2, balanced with N_2 for 4–16 h prior to use.
    4. For procedures requiring handling in atmospheric conditions (on the bench), warm Research Wash medium in a tube on a heating block with inserts for 14 mL round bottom falcon tube calibrated to 37°C, for 4–16 h prior to use (*see* **Note 4**).
    5. Prepare two 40 μL and eight 20 μL drops of Research Cleave medium on a 35 mm culture dish and overlay with 12 mL of paraffin oil (Fig. 2, *embryo culture dish*). Equilibrate at 37°C, 5% O_2, 6% CO_2, balanced with N_2 for 4–16 h prior to use.

3.2 Sperm Collection for In Vitro Fertilisation

1. At 13 h post-hCG, euthanise a male mouse (of proven fertility), dissect the vas deferens, cauda, and corpus epididymis (see Fig. 3), and place into a round-bottom tube containing 3 mL of pre-warmed Research Wash medium (see Fig. 2a). Handle gently to avoid compressing or puncturing the cauda epididymis, as it contains mature spermatozoa.
2. Transfer the vas deferens, cauda, and corpus epididymis into 2 mL of Research Wash medium in a 35 mm dish (*see* Fig. 2b, *dissection dish*). Using fine forceps and micro-scissors, dissect and trim all fat or associated membranes attached to the vas deferens, cauda, and corpus epididymis to prevent contamination during IVF (Fig. 3c).

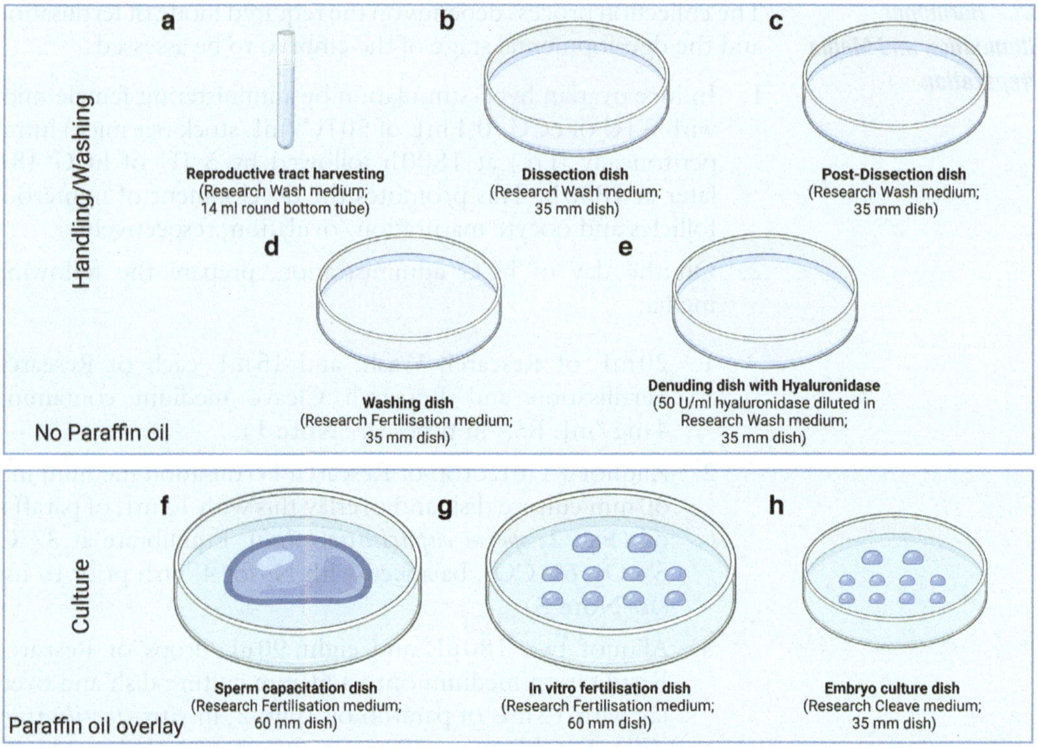

Fig. 2 Schematic representation of culture media for in vitro handling and culture. *Top row*: Depicted here are the different handling/culture media used at various stages of in vitro handling. Research Wash medium is used for harvesting male and female reproductive tracts (**a**); fat or tissue dissection (**b**); and post-dissection washing (**c**). Also shown here are Research Fertilisation medium for washing prior to culture (**d**) and a denuding dish containing hyaluronidase diluted in Research Wash medium (**e**) for removal of cumulus cells. *Bottom row*: All culture dishes are overlaid with paraffin oil and equilibrated at 37°C, 5% O_2, 6% CO_2, balanced with N_2 for 4–16 h prior to use. This row shows dishes containing Research Fertilisation medium for sperm capacitation (**f**) and in vitro fertilisation (**g**). Research Cleave medium is used for in vitro embryo culture (**h**). The schematic was generated using BioRender.com. All dish dimensions refer to diameter (e.g., 35 mm, 60 mm)

3. Next, wash the trimmed vas deferens, cauda, and corpus epididymis by transferring these into a new 35 mm dish containing 2 mL of Research Wash medium (see Fig. 2c, *post-dissection dish*) to remove any residual tissue, cells, lipid droplets, or blood prior to culture in vitro (see Fig. 3d).
4. Next, wash the trimmed vas deferens, cauda, and corpus epididymis in 2 mL of Research Fertilisation medium (see Fig. 2d) before placing them into the 1 mL drop for sperm capacitation dish (see Fig. 2f).
5. Proceed to bluntly dissect the vas deferens, cauda, and corpus epididymis in the sperm capacitation dish (see Fig. 2f) to release spermatozoa. To do this, hold the tissue gently with a

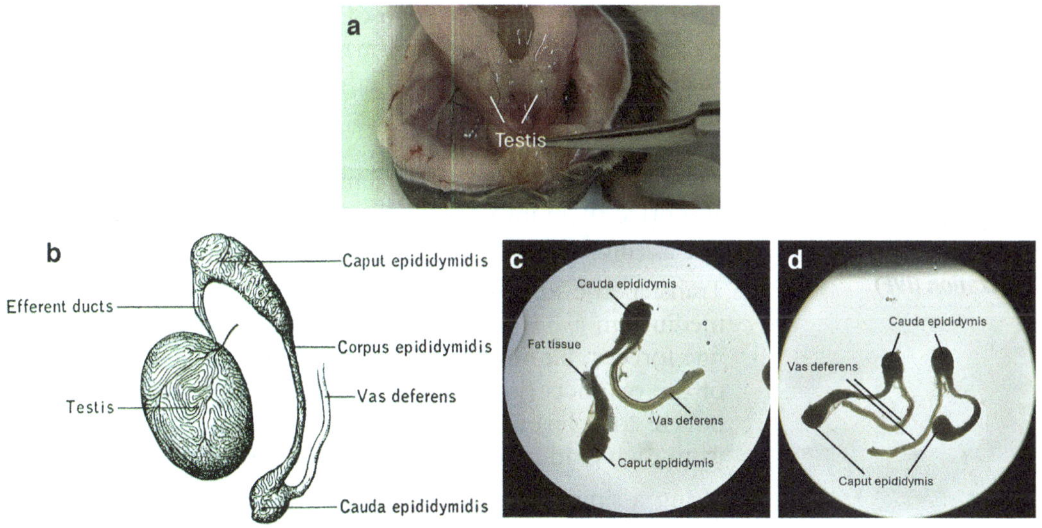

Fig. 3 Illustration of the testis, vas deferens, cauda, and corpus epididymis. (**a**) Mouse testes were exposed after removal of the skin and abdominal muscle, with the peritoneum carefully peeled back using sterile fine forceps. (**b**) Schematic diagram of the testis, vas deferens, cauda, and corpus epididymis (Adapted from [12]). (**c**) Top-down microscopic view of the vas deferens, cauda, and corpus epididymis prior to blunt dissection (*see* **Note 6**). (**d**) Top-down microscopic view of the vas deferens, cauda, and corpus epididymis after trimming off fat tissue

pair of fine forceps at one end to stabilise it in the dish. Using another pair of forceps or a blunt probe, carefully tease apart the epididymal tissue, applying gentle pressure to open the tubules (*see* **Note 5**). The cauda epididymis, in particular, contains mature sperm and is the main region targeted during this process (see Fig. 3d).

6. Following dissection, remove all tissue from the drop using fine forceps, and allow spermatozoa to capacitate at 37°C, 5% O_2, 6% CO_2, balanced with N_2 in a humified incubator for 45–60 min.

7. After capacitation, the sperm should be dispersed and appear hyperactive. When taking sperm from the dish for IVF, take only dispersed, motile sperm from the edge of the drop in the sperm capacitation dish (see Fig. 2f).

8. To count sperm for IVF, place 95 μL of water into a 0.6 mL Eppendorf tube. Using a fresh tip (as water will immobilize sperm), remove 5 μL of sperm from the edge of the drop in the sperm capacitation dish and mix well with the water in the tube. Load 10 μL of this immobilized sperm into the top and bottom of the hemocytometer chamber. Count five squares (usually four corners and middle) of both the top and bottom. Average the counts—there should be around 30–40 sperm. This will result in an insemination volume of around 1×10^4

sperm/mL in a final volume of 100 μL (10 μL sperm + 90 μL Research Fertilisation medium). If this is not the case, the insemination volume can be adjusted to compensate for low sperm numbers.

3.3 Collection of In Vivo Matured Oocytes for In Vitro Fertilization (IVF)

1. At 14 h post-hCG, euthanise female mice, dissect the oviducts, and place them into a round-bottom tube containing 3 mL of pre-warmed Research Wash medium (see Fig. 2a).
2. Transfer the oviducts into fresh 2 mL of Research Wash medium in a 35 mm dish (see Fig. 2b, *dissection dish*). Using fine forceps, bluntly dissect (i.e., tear open) the middle section of the oviducts (i.e., ampulla) to release the mature cumulus oocyte complexes (see Fig. 4a). These will typically be released from each oviduct in a cloud-like mass (see Fig. 4b, c).
3. Using a single-channel manual micropipette (2–20 μL), wash the cumulus mass in a new 35 mm dish containing 2 mL of Research Wash medium (see Fig. 2c) to remove any residual tissue, cells, lipid droplets, or blood prior to culture in vitro.
4. Next, transfer the cumulus mass into the in vitro fertilisation dish (see Fig. 2g) and wash twice in the 180 μL drops of Research Fertilisation medium, before transferring into the 90 μL Research Fertilisation drops at a density of one mass of cumulus oocyte complexes per 90 μL drop (see Fig. 2g).

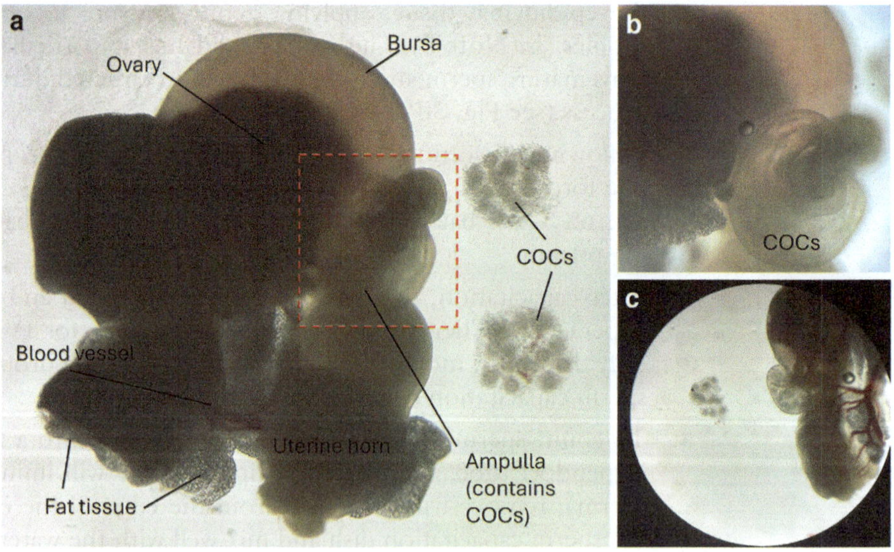

Fig. 4 Representative images of in vivo-derived cumulus oocyte complexes. (**a**) Dissected female mouse reproductive tract, including the ovary, bursa, surrounding fat tissue, ampulla, and uterine horn. Two masses of cumulus–oocyte complexes (COCs) obtained from a separate tract are positioned adjacent for reference. (**b**) Magnified view of COCs located within the ampulla, corresponding to the red dashed box in panel A. (**c**) Top-down microscope image showing all major structures as described in panel A

5. For IVF, aspirate 10 µL of spermatozoa from the edge of the drop in the sperm capacitation dish (see Fig. 2f) and add to the 90 µL drops containing mature cumulus oocyte complexes (see Fig. 2g; final insemination concentration: 1×10^4 sperm/mL). These cumulus oocyte complexes are ready to undergo fertilisation in vitro for 4 h in a humidified incubator at 37°C, 5% O_2, 6% CO_2, balanced with N_2.
6. After 4 h incubation, load a Flexipet tip into the Flexipet adjustable handle, following the manufacturer's guidelines (*see* **Note 7**). Verify that the tip is securely fastened to ensure precise manipulation and prevent dislodgement during use.
7. Following fertilisation, use the Flexipet or a single-channel manual micropipette (0.5–10 µL) to collect the resultant presumptive zygotes (1-cell stage embryos) and wash twice in the 40 µL drops of Research Cleave, before transferring into the 20 µL Research Cleave drops at a density of ten embryos per 20 µL drop for in vitro culture (see Fig. 2h).
8. For in vivo matured oocytes that undergo IVF, resultant embryos can be captured at specific developmental stages as follows: 2-cell (24 h post-IVF), 4-cell (48 h post-IVF), 8-cell (56 h post-IVF), morula (72 h post-IVF), and blastocyst stage (96 h post-IVF) embryos.

3.4 Collection of Embryos Derived from Natural Mating [11]

1. Administer female mice with 5 IU of eCG intraperitoneally (i.p.) at 1400 h followed by 5 IU of hCG i.p. 46 h later at 1200 h.
2. At 1600 h on the same day as hCG injection, house female and male mice (one female: one male per cage) together for overnight mating.
3. At 22 h post hCG, prepare a 35 mm dish containing a 500 µL drop of hyaluronidase diluted 1:1 with Research Wash (final concentration: 750 U/mL; recommended concentration should not exceed 1000 U/mL) and overlay with paraffin oil (see Fig. 2e). Warm the dish on a heating stage at 37°C before use (1 h).
4. At 23 h post-hCG, separate the male and female mice and confirm mating by the presence of a copulation plug. Next, euthanise female mice in accordance with ethical guidelines. Dissect the oviducts and place them into a round-bottom tube containing 3 mL of pre-warmed Research Wash medium (see Fig. 2a).
5. Transfer the oviducts into 2 mL of Research Wash medium in a 35 mm dish (see Fig. 2b). Using fine forceps, bluntly dissect the oviducts to release presumptive zygotes.

6. Using a single-channel manual micropipette (2–20 μL), transfer the presumptive zygotes into the denuding dish containing 750 U/mL of hyaluronidase in Research Wash medium in a 35 mm dish (see Fig. 2e).
7. Using a Flexipet pipette, gently denude the presumptive zygotes by repeated aspiration and expelling, ensuring they remain in hyaluronidase for no more than 2 min. This process effectively removes cumulus cells while minimizing exposure time.
8. Transfer the cleaned presumptive zygotes to a 35 mm dish containing 3 mL of fresh pre-warmed Research Wash medium (see Fig. 2c) to thoroughly rinse off the hyaluronidase solution and any remaining cumulus cells. The rinsing step eliminates any residual tissue, cells, lipid droplets, or blood before in vitro culture.
9. Using a Flexipet pipette or a single-channel manual micropipette (0.5–10 μL), wash all presumptive zygotes in the 40 μL Research Cleave drops twice before transferring into the 20 μ L Research Cleave drops at a density of ten embryos per 20 μL drop (see Fig. 2h) and culture to the stage desired in Research Cleave medium (*see* **Note 8**). Embryos are maintained under optimal culture conditions (humidified incubator, 37°C, 5% O_2, 6% CO_2, balanced with N_2).
10. To obtain stage-specific in vivo-derived embryos following natural mating—based on our experience with the CBA x C57BL/6 first filial generation (F1) mouse strain—embryos can be flushed at defined intervals post-hCG injection to isolate stage specific embryos: 2-cell (40–48 h), 4-cell (56–60 h), 8-cell (68–70 h), morula (72–76 h) and blastocyst stage (90–96 h) (*see* **Notes 8 and 9**).
11. If wanting to obtain embryos at later stages of development as described above, flush embryos from the reproductive tract. Harvest the oviducts and uteri and place them into a round-bottom tube containing 3 mL of pre-warmed Research Wash medium (see Fig. 2a).
12. Transfer the tissue into 2 mL of Research Wash medium in a 35 mm dish (see Fig. 2b). Using fine forceps and micro-scissors, carefully trim away excess fat and connective tissue. To flush embryos, hold the uterine horn and direct the flow depending on developmental stage: flush from the infundibulum towards the oviduct for early-stage embryos (2-, 4-, 8-cell, morula stages), and from the uterine lumen towards the uterus for late-stage embryos (blastocyst stage). To do this, hold the oviduct or uterine horn with forceps, and using a 29-gauge needle attached to a 1 mL syringe filled with pre-warmed Research Wash medium, gently insert the needle into the infundibulum (for oviduct flushes) or into the uterine

lumen (for later-stage embryos). Slowly flush the contents into the dish to recover the embryos.

13. Collect the embryos using a Flexipet pipette or a single-channel manual micropipette (0.5–10 μL) and transfer them into 2 mL of fresh pre-warmed Research Wash medium in a 35 mm dish. These embryos are now ready to use for imaging.

3.5 Metabolic Imaging with Light Sheet Microscopy

Given the variety of light sheet imaging systems available in commercial and research settings, this chapter provides a general overview for consideration when using light sheet microscopy for metabolic imaging. Here, we outline the procedures used for light sheet imaging of murine preimplantation embryos using our custom-built system. We do not cover the specific optical design parameters of a light sheet microscope here. For comprehensive details regarding both the design of a custom imaging chamber and the light sheet microscope, we refer the reader elsewhere [12–13].

Traditional sample preparation for light sheet microscopy typically relies on two common approaches: embedding in agarose or mounting free-hanging in media [14]. While these methods provide stability, they also present challenges. Specifically, agarose embedding can make sample retrieval difficult and may introduce mechanical stress negatively impacting cell viability. Free-hanging samples often experience drift and require precise positioning to maintain imaging consistency. To overcome these limitations, we use a specialised 3D-printed, V-shaped sample chamber (Fig. 5) [9, 15] designed for seamless sample retrieval without compromising imaging stability. This chamber design enables controlled positioning while avoiding the mechanical constraints of agarose embedding. Unlike conventional methods, our approach allows for efficient handling of delicate samples, facilitating sequential imaging, downstream processing, and improved experimental reproducibility. It also enables the use of microdrops of medium overlaid with oil, in the same manner as standard embryo culture techniques. Here, we present our protocol for imaging with light sheet microscopy using the 3D-printed sample chamber as previously described [9, 15].

1. Prepare the 3D-printed sample chamber by pipetting a 2 μL drop of Research Wash medium or any HEPES-buffered medium at the centre of the chamber (*see* **Notes 10 and 11**). Overlay with paraffin oil and warm the chamber on a heating stage for 4 h before use.

2. One hour before imaging, switch on the heating elements, which function to warm the immersion water in the light sheet system. This heated immersion water (i.e., Milli-Q ultrapure water) ensures that embryos are maintained at 37°C throughout the imaging process and is free of contaminants that could compromise image quality.

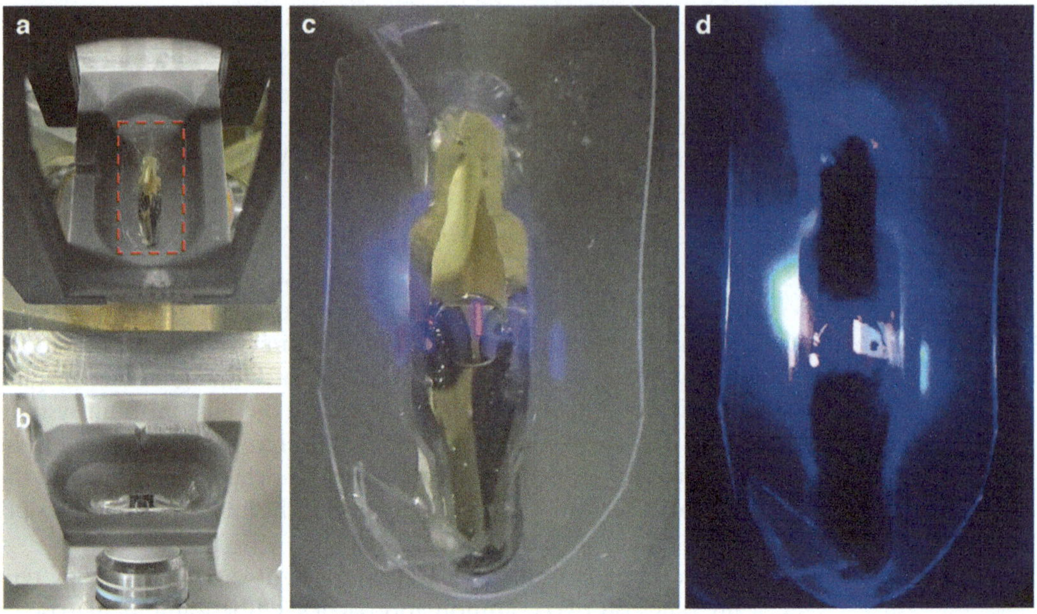

Fig. 5 Representative images of the 3D-printed sample chamber for light sheet imaging. (**a**) Top-down view of the custom 3D-printed imaging chamber mounted in a light-sheet (LS) microscope stage holder. (**b**) Side view of the same chamber positioned between two objectives. (**c**) Close-up of the red-boxed region highlighted in (a), with ambient lights on. The purple line indicates the light sheet (405 nm laser) in operation. (**d**) Same view as (c) but with lights off, enhancing visibility of the light sheet path

3. Prior to imaging, use a single-channel manual micropipette (0.5–10 μL) or Flexipet to transfer embryos from culture dish into the 2 μL drop located within the pre-equilibrated sample chamber on a dissecting microscope with heated stage.
4. Place the sample chamber containing embryos onto the microscope stage. Ensure the chamber is securely positioned, levelled, and centred. Gently lower the stage into the immersion water, taking care to avoid any contact with the water-immersion imaging objective.
5. Locate the embryo using the control software (e.g., Micro-Manager) on the microscope or manual actuation by the XYZ translation stage and adjust the focus until the embryo can be clearly visualised.
6. Identify the specific region of the embryo for imaging (*see* **Note 12**). For example, in a blastocyst-stage embryo, decide whether to focus on the inner cell mass, trophectoderm separately (applicable only if performing 2D imaging), or the whole embryo (applicable to 3D imaging).
7. Locate both ends of the embryo using low laser power (e.g., 0.5–1 μW) to prevent photobleaching and phototoxicity. Input these positions into the imaging software to define the z-stack range.

8. Configure imaging settings and optimise parameters based on factors including, but not limited to, wavelength, sample volume, z-slice spacing (e.g., how many micron steps), and scanning methods for 3D volumetric imaging as follows (*see* **Note 13**):
   1. Required Wavelength(s) (see Fig. 6):
      (a) NAD(P)H: Typically excited at 405 nm (*see* **Note 14**);
      (b) FAD: Typically excited at 488 nm (see recent review [5])
      (c) Alternative approach: 375 nm can be used for simultaneous excitation of both NAD(P)H and FAD [12].
   2. Sample volume: Entire volume of a mouse embryo (~80–100 μm)
   3. Z-slice spacing: 1 μm between z-planes
   4. Exposure time and resolution: 200 ms and 2×2 binning, respectively
   5. Laser power at the objective used for imaging: 3 mW for 405 nm and 8 mW for 488 nm, respectively. These power levels were independently measured at the objective using an optical power meter and were optimised to provide sufficient autofluorescence signals of NAD(P)H and FAD while minimising photobleaching and photodamage (*see* **Notes 15–17**).

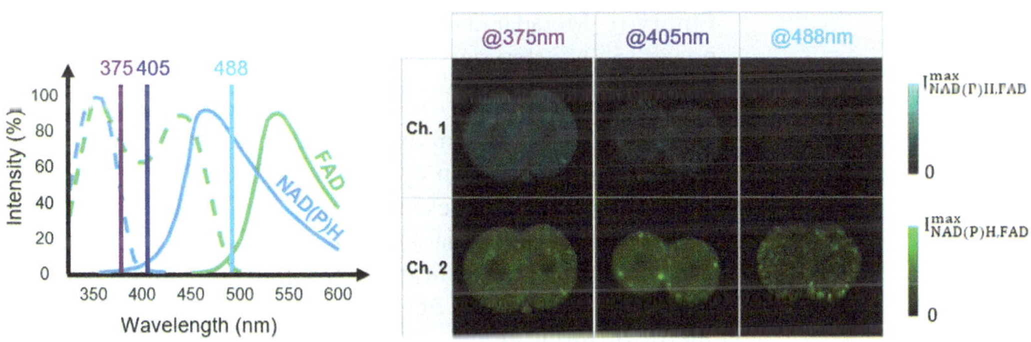

Fig. 6 Spectral profile of NAD(P)H and FAD. Left: Excitation (dashed lines) and emission (solid lines) spectra of NAD(P)H (blue) and FAD (green), with excitation wavelengths at 375 nm, 405 nm, and 488 nm marked (vertical solid line). The overlap of these lines (dashed line vs. vertical solid line) indicates the relative contributions of NAD(P)H and FAD to the detected signals. Right: Representative images of NAD(P)H (Channel 1; Ch1; emission detection range: 417–477 nm) and FAD (Channel 2; Ch2; emission detection range: 513–607 nm) fluorescence in a 2-cell stage murine embryo using 375 nm, 405 nm, and 488 nm excitation wavelengths. At 375 nm, NAD(P)H and FAD signals appear balanced or equal, while 405 nm and 488 nm excitation produce variable fluorescence intensities. Notably, although 405 nm is typically used for NAD(P)H excitation [3], the detected signal primarily originates from FAD. This suggests that 375 nm excitation alone can effectively co-excite both NAD(P)H and FAD, a finding we have recently validated in preimplantation murine embryos [35]. (Adapted from [35])

9. Initiate the imaging process and monitor it to ensure images are captured correctly. Example: Exclude or re-image if image quality is poor, or embryos drift out of the field of view during acquisition (*see* **Notes 18 and 19**).
10. Save the acquired images or z-stacks to the intended folder using unique identifiers [16] (e.g., LabName/UserName (first and last)/InstrumentName/YYYYMMDD/Group_Data_details.czi).
11. Proceed to image analysis using ImageJ/FIJI software.

3.6 Image Storage: Metadata Preservation

Perform regular backups of all images to an external hard drive and cloud-based storage to prevent data loss. Ensure that critical image metadata (e.g., z-step size, laser power, exposure time) is retained during saving or exporting to facilitate accurate replication and interpretation of results. This can be achieved through the following methods:

1. **Embedded metadata:** Save files in formats that automatically retain metadata, such as .tif, .czi, or .lsm, which are supported by most microscopy systems. These formats embed all relevant imaging settings within the image file itself, ensuring seamless access during analysis.
2. **External metadata files:** For proprietary microscope systems or formats lacking embedded metadata, export the metadata as a separate text or CSV file. Include key imaging parameters such as acquisition date and experimental details for future reference. Example: Create a metadata file named "2025_01_07_Embryo1_Metadata.txt" containing information such as laser power (5%), z-step size (2 μm), and exposure time (50 ms).

3.7 Image Analysis

A range of software platforms is available for analysing fluorescence microscopy data. Common options include FIJI/ImageJ, which offers a flexible, open-source environment for 2D and basic 3D analysis; and Imaris, a commercial software with advanced capabilities for 3D visualisation and quantification. The following method is based on FIJI for 2D analysis and 3D ROI Manager plugin in FIJI for 3D volumetric quantification.

3.7.1 2D Image Analysis

1. Open the saved 2D image files in analysis software (FIJI/ImageJ)
2. Identify and select the region of interest (ROI) corresponding to the embryo. This can be done by using the "Circular," "Polygon," or "Freehand" selection tools in FIJI to outline the entire embryo, or specific regions, such as the inner cell mass or trophectoderm layer, depending on the focus of the analysis.
3. Measure the average fluorescence intensity within the ROI to assess NAD(P)H and FAD fluorescence, i.e., use the "Analyse

> Measure" function in FIJI to calculate fluorescence intensity values. To improve the accuracy of intensity measurements, apply automated background subtraction (e.g., using "Process > Subtract Background" in FIJI), or manually subtract the intensity measured in an area where no embryos are present (Blank media) using the same ROI dimensions to control for background fluorescence.

4. Export intensity measurements to a CSV file for analysis by selecting "Results > File > Save As."

3.7.2 3D Image Analysis

1. Load the saved 3D image stack into FIJI for analysis. Ensure FIJI is updated with plugins for volumetric rendering and z-stack visualisation.
2. Split the channels (if required) using "Image→Colour→Split Channels" and label the resulting stacks appropriately, e.g., NAD(P)H and FAD.
3. Define the ROI for the embryo in the 3D volume by selecting a specific region using ROI tools or plugins (e.g., Launch 3D ROI manager by "Plugins→3D→3D ROI Manager").
4. Measure fluorescence intensities across the entire volume or specific planes to capture spatial variations in metabolic activity, e.g., use the "3D ROI manager" to calculate total and mean fluorescence intensity for NAD(P)H and FAD. You can do this either manually or automatically using threshold-based segmentation.
    1. To do this manually, use "Polygon, Freehand, or Brush" tools to outline a region on one plane, then use "Plugins→3D→3D Manager→Interpolate ROIs" to create a 3D ROI across slices, then click "Add" in the 3D ROI Manager to save it.
    2. To do this using automatic threshold-based segmentation, click "Image→Adjust→Threshold" and adjust for the signal (e.g., FAD or NAD(P)H) and apply, then use "Plugins→3D→3D Manager→Create ROI from mask."
    3. To measure signal intensity: In the 3D ROI Manager, select the ROI and click "Measure" to get the total intensity, mean intensity, volume, etc.
5. Generate a maximum intensity projection of the 3D stack if detailed volumetric analysis, such as topological variation in metabolic activity, is unnecessary. Example: Use "Image > Stacks > Z-project" in FIJI to generate a 2D projection from the 3D stack for quantifying fluorescence intensity.
6. If required, use the "Colocalization analysis" plugin to assess NAD(P)H and FAD co-localization for metabolic mapping.

7. Export volumetric fluorescence intensity data using the "Results > File > Save As" function.

3.8 Quantification and Analysis of Metabolic Activity

The interpretation of mitochondrial metabolism has evolved since its introduction by Britton Chance in 1979 [17]. Currently, there are two commonly used equations for assessing mitochondrial metabolism, which account for changes in the redox state of metabolic cofactors: FAD and NAD(P)H.

1. Chance's redox ratio: FAD/(FAD + NAD(P)H) [17–19] - Equation 1 (Eq. 1)
2. The FAD/NADH ratio: FAD/NADH [20, 21] - Equation 2 (Eq. 2)

While both metabolic cofactors can exist in either oxidized (NAD+, FAD) or reduced (NADH, $FADH_2$) forms, only NADH and FAD exhibit significant autofluorescence. Importantly, as NADH and NADPH have overlapping fluorescence spectra, optical imaging techniques fail to reliably distinguish between them and are collectively referred as NAD(P)H [22] (*see* **Note 14**). We have utilised Chance's method of calculating redox ratio, which normalizes FAD fluorescence against the total available redox pool (FAD+ NAD(P)H). This approach accounts for not only NADH oxidation, but also for NADPH-dependent antioxidant defence and biosynthetic activity, providing a more comprehensive measure of redox homeostasis. Further, this normalised redox ratio is also the preferred method for calculating redox ratio as it provides an upper and lower bound for ratio values, which enable accurate statistical comparisons that assume a normal distribution [23]. Although Eq. 2 can still provide insights into metabolic shifts by detecting complex I and II activity (via NADH and FAD fluorescence, respectively) in the electron transport chain, it fails to account for variations in the total redox pool, the role of NADPH in the protection against oxidative stress, and cytosolic redox dynamics. Consequently, the absolute fluorescence intensities may vary due to experimental conditions rather than true metabolic shifts, increasing the risk of misinterpretation. In contrast, Eq. 1 captures both oxidative phosphorylation activity and redox homeostasis, making it a more robust indicator of metabolic state (*see* **Note 13—*Data analysis*** for alternative approaches to interpret/calculate metabolic activity).

1. Apply the following formula to determine overall metabolic activity [17]:

$$Overall\ metabolic\ activity = \frac{FAD}{FAD + NAD(P)H} \quad (1)$$

Higher ratios typically indicate increased oxidative phosphorylation, while lower ratios suggest an increased reliance on glycolytic pathways [24].

2. Use metadata parameters such as exposure time and laser power to normalise fluorescence intensities. This ensures calculated redox ratios are consistent and comparable across experiments. For example:

   1. Note or record the metadata for each image generated, e.g., exposure time: 500 ms; laser power: 10 mW
   2. Measure the mean fluorescence intensity for NAD(P)H and FAD from the raw image (e.g., Image J/FIJI: 2700 arbitrary units for NAD(P)H)
   3. Normalise the intensity for NAD(P)H using the following formula:

      (a)

$$Normalized\ Intensity: \frac{Raw\ intensity}{Exposure\ time\ (s) \times Laser\ Power\ (mW)} \tag{2}$$

      (b)

$$\text{Example}: \frac{2700}{0.5 \times 10} = 540 \tag{3}$$

   4. Calculate redox ratio using the normalised intensity for NAD(P)H and FAD

(a)

$$Redox\ ratio = \frac{FAD_{norm}}{FAD_{norm} + NAD(P)H_{norm}} \tag{4}$$

4 Notes

1. The addition of BSA to culture media acts as a protein source (for cellular processes), to maintain osmotic balance, and to prevent adhesion of embryos to the surface of the culture dish. Following the addition of BSA to media, filter-sterilise the solution using 0.22 μm syringe filter before use. Alternatives, including synthetic polymers such as polyvinylpyrrolidone (PVP) at 3%–5%, or polyvinyl alcohol (PVA) at 0.3 mg/mL, can also serve as substitutes to prevent adhesion when mounting embryos for imaging. These should be filter-sterilised as described above for BSA.

2. Mice should be housed in groups (no more than six per cage at this age) and kept on a 12-h light/12-h dark cycle (lights on at 07:00 and off at 19:00) to maintain a consistent circadian rhythm. This helps to reduce stress, support appropriate social interactions, and promote physiological and reproductive cycles. These factors contribute to the collection of high numbers of oocytes/embryos (approximately 40 per mouse). Prepubertal female mice (younger than 1 month of age) are recommended for ovarian stimulation protocols, as they typically yield the highest number of oocytes or embryos in response to hormonal stimulation. However, if the experiment aims to investigate age-related differences, embryos from both young and old mice could be used. To contextualise age-related comparisons between mice and human, it is useful to note that the average lifespan of a mouse is approximately 24 months [25], whereas that of a human is around 80 years [26]. This allows for approximate age correlation between species when designing experiments with translational relevance. More nuanced comparisons require careful consideration, particularly in relation to reproductive lifespan and the selection of age-appropriate models, as the onset, duration, and decline of reproductive capacity may differ significantly between species.
3. Prior to use, equilibrate culture media overlaid with paraffin oil in the culture conditions in a humidified incubator at 37°C, 5% O_2, 6% CO_2, balanced with N_2, for 4–16 h. This ensures that the media, into which gametes and embryos will be placed, reach an optimal temperature and culture conditions to maintain embryo viability. The use of a paraffin oil overlay minimises media evaporation and fluctuations in temperature, pH, and osmolality.
4. HEPES buffering in Research Wash medium is required to maintain the pH between 7.2 and 7.4 under atmospheric conditions (e.g., embryo handling at the bench or live imaging). Embryos are highly susceptible to pH fluctuations, and maintaining a stable environment is critical to preserving their viability and ensuring that any observed metabolic signatures are not artifacts of pH instability, for example. For longitudinal imaging studies, the use of an imaging system that supports bicarbonate-buffered medium (e.g., Research Cleave medium) within a temperature- and gas-controlled chamber is recommended to maintain physiological conditions over extended periods.
5. Do not cut the epididymis. Instead, press or separate the tissue slowly so that sperm can be released into the surrounding medium while minimising tissue debris.

6. Fat tissue must be carefully removed before dissection.
7. For mouse embryos, we generally use a pipette tip with a 100 µm inner diameter. Murine preimplantation embryos, from the 1-cell stage until the blastocyst-stage (excluding expanded and hatching stages) are approximately 80 µm in diameter. This size ensures that the pipette is large enough to pick up embryos without compressing or damaging them, but small enough to maintain gentle control during handling and in a minimal volume of medium. If the pipette is too large, it may draw in excess fluid or contaminating cellular debris, or carry over a large volume of medium into the final treatment/imaging dish. This can result in altered concentration of any potential treatments in the final dish.
8. Based on our experience with mice from the CBA × C57Bl/6 first filial (F1) cross, the following timepoints can be used as a reference to collect particular stages of embryo development: 2-cell, 4-cell, 8-cell, morula, and blastocyst stage embryos. For in vivo matured oocytes that undergo IVF, resultant embryos can be captured at specific developmental stages as follows: 2-cell (24 h post-IVF), 4-cell (48 h post-IVF), 8-cell (56 h post-IVF), morula (72 h post-IVF), and blastocyst stage (96 h post-IVF) embryos. For embryos derived from natural mating but cultured in vitro from the 1-cell stage (presumptive zygote) onwards, specific developmental stages can be captured as follows: 2-cell (47 h post-hCG), 4-cell (60 h post-hCG), 8-cell (68 h post-hCG), morula (84 h post-hCG), and blastocyst stage (96 h post-hCG) embryos.
9. For time-point imaging studies, embryos can be cultured to specific stages, such as the 2-cell, 8-cell, morula, or blastocyst stage, depending on experimental requirements (*see* **Note 8** for the specific timings to capture embryos at various developmental stages). We recommend using separate cohorts of embryos to investigate specific stages of development as opposed to repeated imaging of the same cohort across development. This is to minimise the impact of frequent removal of the dish from optimal culture conditions, which can negatively alter embryo metabolism and affect viability. Alternatively, for longitudinal imaging studies–continuous observation of metabolic changes over time–we recommend using a benchtop incubator or culture chamber equipped with an integrated imaging system. This allows for real-time, time-lapse monitoring of metabolic activity throughout development without repeatedly removing the embryos from the controlled culture environment. Continuous imaging within a consistent incubator environment helps preserve optimal culture conditions while enabling repeated recording of metabolic dynamics across development.

10. Components within culture media may contain nutrients or additives that are autofluorescent, affecting the signal-to-noise ratio of images. Autofluorescent compounds, such as riboflavin [27] or phenol red [28], may emit fluorescence at wavelengths overlapping with endogenous cell fluorophores such as NAD(P)H and FAD. This overlap can affect the signal-to-noise ratio and complicate the interpretation of metabolic imaging data detected within embryos. This may lead to misinterpretation of cellular metabolic activity, particularly in low-signal conditions where the autofluorescence signal from the embryo may be similar in intensity to that of autofluorescent compounds in the media. Consequently, where possible, consider using culture media without phenol red or other highly autofluorescent components (e.g., HBSS) or apply appropriate background subtraction during image processing.
11. To minimise autofluorescence signal contamination, an amino-acid-free medium, such as Hank's Balanced Salt Solution (HBSS) by itself, or in combination with polyvinylpyrrolidone (PVP) or polyvinyl alcohol (PVA) (to prevent embryos adhering to the culture dish; *see* **Note 1**), can be used for imaging. The use of HBSS is beneficial as it emits minimal to no fluorescence when excited during fluorescence imaging. This is due to the absence of components with known autofluorescence (e.g., amino acids, phenol red). This minimizes the background autofluorescence, allowing for more accurate autofluorescence signal detection from the embryo. Alternatively, this could be overcome in post-processing by background subtraction of the signal obtained in blank media or areas where no cells are present.
12. When deciding between 2D and 3D imaging, consider the specific goals of your study and the resources available. 2D imaging involves capturing a single optical plane, typically the middle plane (i.e., the widest point of an embryo), which provides a quick and resource-efficient snapshot of the embryo. This approach is useful for initial observations to compare metabolic changes. However, it does not provide information on spatial variations in metabolic activity in different cell types and regions across the entire embryo. In contrast, 3D imaging captures the full volume of the embryo (from top to bottom). This can be achieved by scanning the light sheet or moving the sample stage in z-plane. This approach is generally recommended for comprehensive metabolic analyses, as it enables the study of topological variations in metabolic activity, cellular heterogeneity, and spatial dynamics over the whole 3D volume. This is particularly valuable for correlating developmental stage or differences in cell lineages with

metabolic function. However, this requires more time spent on imaging and resources (e.g., data size) when compared to 2D imaging.

13. There are a number of general considerations to take into account for 3D imaging. *Embryo size*: The size of murine embryos increases from approximately 80 µm at earlier stages to approximately 100 µm or more at the expanded blastocyst-stage before hatching. Ensure the z-stack settings are adjusted accordingly to capture the entire embryo volume at each stage of development. *Experimental Goals*: Use 2D imaging for rapid assessment or when targeting specific regions of interest with high magnification. Choose 3D imaging when the project objectives are to study spatial variations, developmental-stage differences, or lineage-specific variations. *Z-step size and resolution*: The step size between z-planes affects the level of detail captured. Smaller step sizes are ideal for detailed metabolic analysis, but it increases image acquisition time and risk of photobleaching. *Time Considerations*: Image acquisition time increases proportionally with the number of z-planes needed to cover the entire volume of the embryo. This can pose challenges when imaging multiple live embryos or time-lapse studies, where maintaining viability and limiting phototoxicity are critical. *Data Storage*: 3D imaging generates significantly larger datasets compared to 2D imaging. It is crucial to ensure adequate data storage for large 3D datasets, particularly for time-lapse experiments, to avoid the risk of halting the imaging process mid-acquisition. *Data analysis*: Apart from the equation for calculating the optical redox ratio for interpreting metabolic activity, others have also implemented alternative approaches, such as fluorescence lifetime microscopy (FLIM)-based approaches in confocal [29] or light sheet microscopy [30] configurations for metabolic imaging of human embryos and other cell types. Additionally, we and others have also demonstrated the use of phasor-based approaches, which circumvent the need for intensive post-processing computing power [12, 31]. We recently demonstrated the first use of light sheet microscopy at a single wavelength to detect dynamic metabolic shifts during preimplantation development in mouse embryos [12]. This approach enabled rapid volumetric imaging of cellular autofluorescence using a single 375 nm excitation wavelength (*see* Fig. 6; Note the overlapping excitation wavelength range of NAD(P)H and FAD with 375 nm). Further, a hardware-based separation method eliminated the need for computationally intensive post-processing.

14. It is essential to recognise that NADH and NADPH (reduced nicotinamide adenine dinucleotide phosphate) exhibit nearly

identical spectral properties, which is why they are often collectively referred to as NAD(P)H [22, 32]. NADPH plays a crucial role in the biosynthesis of fatty acids and nucleotides, and it also helps maintain the balance between reduced and oxidized glutathione, which helps protect against reactive oxygen species (ROS) [33].

15. Appropriate optimisation of imaging settings and acquisition techniques is essential for obtaining high-quality, interpretable data, especially when working with sensitive biological samples like live mammalian embryos. *Light intensity* Always begin by using the lowest light intensity or laser power that acquires clear and visible images. Ensure you follow laser safety principles for the laser source used. Example: Use low laser power (e.g., 1%; 0.5 μW) to locate and set the upper and lower boundaries of a z-stack for an embryo. Avoid high laser power as it can lead to photobleaching and phototoxicity, compromising embryo viability and altering metabolic measurements. *Detector Sensitivity*: Adjust the detector (camera) sensitivity (gain) or consider evaluating different camera models designed for low-light imaging (*see* recent review [34]) to optimise image quality while minimizing noise. In general: Start with a low gain setting and gradually increase it until the signal is clear without over-amplifying background noise. Avoid using excessively high gain, as it may introduce artifacts and compromise quantitative accuracy. Example: Ensure that the background signal appears black relative to the sample signal. Alternatively, check for signal intensity for the background by quantitatively measuring using software like ImageJ (FIJI) to avoid over-saturating the signals collected. *Exposure time*: Perform test exposures using non-experimental samples (e.g., fluorescent beads) to determine the optimal settings to obtain sufficient signal intensity for clear imaging. Prolonged exposure increases the risk of photobleaching and phototoxicity, which can negatively affect embryo development and distort metabolic measurements. *Resolution*: Ensure embryos are in focus before acquiring images. Begin by adjusting the objective using the phase contrast or brightfield mode to focus on the embryo. Once focused, switch to the desired excitation wavelength to collect for NAD(P)H or FAD and fine-tune the z-plane focus if required.
16. Troubleshooting—Low signal intensity: optimise laser power and exposure time: Incrementally increase laser power by 1%–3% if the signal remains weak. Adjust exposure time to achieve a balance between signal clarity and photobleaching.
17. Troubleshooting—Photobleaching: (a) reduce laser power: lower the laser power by 5%–10% and check whether the

fluorescence signal remains sufficient for imaging; (b) shorten exposure time: adjust the exposure time or dwell time to minimise photobleaching while maintaining clear fluorescence signals; (c) larger step size: increase the step size for z-stack imaging to maintain high-quality images while minimising light exposure per plane (e.g., switching from 1 μm to 2–4 μm between z-planes); and (d) use minimal laser power during pre-imaging setup: for instance, in 2D imaging, use the lowest laser power (e.g., 1%) to locate the embryo and define the region of interest. Focus the embryo and set the desired laser power (e.g., 4%) just before image acquisition. This minimises photobleaching and photodamage during alignment while ensuring high-resolution imaging. For 3D imaging: set the upper and lower boundaries of the z-stack using minimal laser power (e.g., 1%). Once boundaries are determined and the embryo is properly aligned, increase the laser power to the appropriate level for imaging acquisition to ensure clear and consistent volumetric data. This minimises the risk of photobleaching, especially at increased depths, preventing signal loss in half of the embryo.

18. Troubleshooting—Poor image quality: (a) clean optical components: ensure objectives and optical mirrors are free from dust, smudges, or debris that can degrade image clarity. Use a lens cleaning solution and lint-free wipes to gently clean the objective lens and other optical surfaces and (b) optimize focus and sample positioning: Adjust the focus and check that the embryo is properly mounted and located within the imaging field. Adjust the objective focus carefully to bring the embryo into focus. Ensure that the imaging chamber is stable and positioned centrally in the light path. Re-mount the sample chamber if tilted or misaligned.
19. Troubleshooting—Embryos drift out of focus during imaging: (a) minimise imaging drop volume: reduce the volume of the imaging drop to 1–2 μL of pH-buffered media. A smaller drop restricts embryo movement and improves sample stability during imaging and (b) check microscope objective z-position: Ensure the microscope objective is not exerting upward pressure against the base of the imaging chamber, as this can lift the chamber and cause focus drift. Lower the stage to prevent contact.

Acknowledgments

K.R.D. acknowledges the support of a Future Making Fellowship from the University of Adelaide and a Future Fellowship (FT240100291) from the Australian Research Council (ARC). The work mentioned in this study was funded by the National Health and Medical Research Council (APP2003786). K.D. thanks the support from an ARC Laureate Fellowship (Grant FL210100099).

References

1. Chow DJX, Tan TCY, Upadhya A, Lim M, Dholakia K, Dunning KR (2024) Viewing early life without labels: optical approaches for imaging the early embryodagger. Biol Reprod 110(6):1157–1174. https://doi.org/10.1093/biolre/ioae062
2. Sanchez T, Venturas M, Aghvami SA, Yang X, Fraden S, Sakkas D, Needleman DJ (2019) Combined noninvasive metabolic and spindle imaging as potential tools for embryo and oocyte assessment. Hum Reprod 34(12):2349–2361. https://doi.org/10.1093/humrep/dez210
3. Santos Monteiro CA, Chow DJX, Leal GR, Tan TC, Reis Ferreira AM, Thompson JG, Dunning KR (2021) Optical imaging of cleavage stage bovine embryos using hyperspectral and confocal approaches reveals metabolic differences between on-time and fast-developing embryos. Theriogenology 159:60–68. https://doi.org/10.1016/j.theriogenology.2020.10.012
4. Seidler EA, Sanchez T, Venturas M, Sakkas D, Needleman DJ (2020) Non-invasive imaging of mouse embryo metabolism in response to induced hypoxia. J Assist Reprod Genet 37(8):1797–1805. https://doi.org/10.1007/s10815-020-01872-w
5. Tan TCY, Dunning KR (2022) Non-invasive assessment of oocyte developmental competence. Reprod Fertil Dev 35(2):39–50. https://doi.org/10.1071/RD22217
6. Tan TCY, Mahbub SB, Campbell JM, Habibalahi A, Campugan CA, Rose RD, Chow DJX, Mustafa S, Goldys EM, Dunning KR (2021) Non-invasive, label-free optical analysis to detect aneuploidy within the inner cell mass of the preimplantation embryo. Hum Reprod 37(1):14–29. https://doi.org/10.1093/humrep/deab233
7. Sanchez T, Zhang M, Needleman D, Seli E (2019) Metabolic imaging via fluorescence lifetime imaging microscopy for egg and embryo assessment. Fertil Steril 111(2):212–218. https://doi.org/10.1016/j.fertnstert.2018.12.014
8. Thompson JG, Brown HM, Sutton-McDowall ML (2016) Measuring embryo metabolism to predict embryo quality. Reprod Fertil Dev 28(1-2):41–50. https://doi.org/10.1071/RD15340
9. Chow DJX, Schartner EP, Corsetti S, Upadhya A, Morizet J, Gunn-Moore FJ, Dunning KR, Dholakia K (2024) Quantifying DNA damage following light sheet and confocal imaging of the mammalian embryo. Sci Rep 14(1):20760. https://doi.org/10.1038/s41598-024-71443-x
10. Icha J, Weber M, Waters JC, Norden C (2017) Phototoxicity in live fluorescence microscopy, and how to avoid it. Bioessays 39(8):1700003. https://doi.org/10.1002/bies.201700003
11. Gardner DK, Lane M, Watson AJ (2004) A laboratory guide to the mammalian embryo, online edn. Oxford Academic. New York, NY https://doi.org/10.1093/oso/9780195142266.001.000
12. Hummel, KP, Richardson, FL, Fekete, E, 1966. Anatomy. Biology of the laboratory mouse, 2nd edn. Dover Publications Inc, New York
13. Kramer SN, Antarasen J, Reinholt CR, Kisley L (2024) A practical guide to light-sheet microscopy for nanoscale imaging: looking beyond the cell. J Appl Phys 136(9):091101. https://doi.org/10.1063/5.0218262
14. Weber M, Mickoleit M, Huisken J (2014) Light sheet microscopy. Methods Cell Biol 123:193–215. https://doi.org/10.1016/B978-0-12-420138-5.00011-2
15. Strnad P, Gunther S, Reichmann J, Krzic U, Balazs B, de Medeiros G, Norlin N, Hiiragi T, Hufnagel L, Ellenberg J (2016) Inverted light-sheet microscope for imaging mouse pre-implantation development. Nat Methods 13(2):139–142. https://doi.org/10.1038/nmeth.3690
16. Briney KA (2020) File naming convention worksheet. California Institute of Technology. https://authors.library.caltech.edu/records/mmnpf-cez11
17. Chance B, Schoener B, Oshino R, Itshak F, Nakase Y (1979) Oxidation-reduction ratio studies of mitochondria in freeze-trapped samples. NADH and flavoprotein fluorescence signals. J Biol Chem 254(11):4764–4771. https://doi.org/10.1016/S0021-9258(17)30079-0
18. Alhallak K, Rebello LG, Muldoon TJ, Quinn KP, Rajaram N (2016) Optical redox ratio identifies metastatic potential-dependent changes in breast cancer cell metabolism. Biomed Opt Express 7(11):4364–4374. https://doi.org/10.1364/BOE.7.004364
19. Quinn KP, Sridharan GV, Hayden RS, Kaplan DL, Lee K, Georgakoudi I (2013) Quantitative metabolic imaging using endogenous

fluorescence to detect stem cell differentiation. Sci Rep 3:3432. https://doi.org/10.1038/srep03432
20. Ostrander JH, McMahon CM, Lem S, Millon SR, Brown JQ, Seewaldt VL, Ramanujam N (2010) Optical redox ratio differentiates breast cancer cell lines based on estrogen receptor status. Cancer Res 70(11):4759–4766. https://doi.org/10.1158/0008-5472.CAN-09-2572
21. Skala MC, Riching KM, Gendron-Fitzpatrick A, Eickhoff J, Eliceiri KW, White JG, Ramanujam N (2007) In vivo multiphoton microscopy of NADH and FAD redox states, fluorescence lifetimes, and cellular morphology in precancerous epithelia. Proc Natl Acad Sci U S A 104(49):19494–19499. https://doi.org/10.1073/pnas.0708425104
22. Ghukasyan VV, Heikal AA (2014) Natural biomarkers for cellular metabolism: biology, techniques, and applications, 1st edn. CRC Press, Boca Raton. https://doi.org/10.1201/b17427
23. Georgakoudi I, Quinn KP (2012) Optical imaging using endogenous contrast to assess metabolic state. Annu Rev Biomed Eng 14:351–367. https://doi.org/10.1146/annurev-bioeng-071811-150108
24. Kolenc OI, Quinn KP (2019) Evaluating cell metabolism through autofluorescence imaging of NAD(P)H and FAD. Antioxid Redox Signal 30(6):875–889. https://doi.org/10.1089/ars.2017.7451
25. The Jackson Laboratory (n.d.) Aged C57BL/6J mice: an emerging translational model of human aging (White paper). The Jackson Laboratory. https://resources.jax.org/white-papers/whitepaper-aged-b6
26. Dutta S, Sengupta P (2016) Men and mice: Relating their ages. Life Sci 152:244–248. https://doi.org/10.1016/j.lfs.2015.10.025
27. AAT BioquestInc.(2025)Spectrum[Riboflavin]. https://www.aatbio.com/fluorescence-excitation-emission-spectrum-graph-viewer/riboflavin
28. Ettinger A, Wittmann T (2014) Fluorescence live cell imaging. Methods Cell Biol 123:77–94. https://doi.org/10.1016/B978-0-12-420138-5.00005-7
29. Shah JS, Venturas M, Sanchez TH, Penzias AS, Needleman DJ, Sakkas D (2022) Fluorescence lifetime imaging microscopy (FLIM) detects differences in metabolic signatures between euploid and aneuploid human blastocysts. Hum Reprod 37(3):400–410. https://doi.org/10.1093/humrep/deac016
30. Samimi K, Desa DE, Lin W, Weiss K, Li J, Huisken J, Miskolci V, Huttenlocher A, Chacko JV, Velten A, Rogers JD, Eliceiri KW, Skala MC (2023) Light sheet autofluorescence lifetime imaging with a single photon avalanche diode array. bioRxiv. https://doi.org/10.1101/2023.02.01.526695
31. Hedde PN, Cinco R, Malacrida L, Kamaid A, Gratton E (2021) Phasor-based hyperspectral snapshot microscopy allows fast imaging of live, three-dimensional tissues for biomedical applications. Commun Biol 4(1):721. https://doi.org/10.1038/s42003-021-02266-z
32. Galeotti T, van Rossum GD, Mayer DH, Chance B (1970) On the fluorescence of NAD(P)H in whole-cell preparations of tumours and normal tissues. Eur J Biochem 17(3):485–496. https://doi.org/10.1111/j.1432-1033.1970.tb01191.x
33. Koju N, Qin ZH, Sheng R (2022) Reduced nicotinamide adenine dinucleotide phosphate in redox balance and diseases: a friend or foe?. Acta Pharmacol Sin 43(8):1889–1904. https://doi.org/10.1038/s41401-021-00838-7
34. Peterkovic Z, Upadhya A, Perella C, Bajraktarevic A, Bautista Gonzalez RE, Lim M, Dunning KR, Dholakia K (2025) Optimizing image capture for low-light widefield quantitative fluorescence microscopy. APL Photonics 10(3):1–28. https://doi.org/10.1063/5.0245239
35. Morizet J, Chow D, Wijesinghe P, Schartner E, Dwapanyin G, Dubost N, Bruce GD, Anckaert E, Dunning K, Dholakia K (2023) UVA hyperspectral light-sheet microscopy for volumetric metabolic imaging: application to preimplantation embryo development. ACS Photonics 10(12):4177–4187. https://doi.org/10.1021/acsphotonics.3c00900

Check for updates

Chapter 12

Imaging Spatiotemporal RNA and Protein Dynamics in Mouse Embryos

Hongbin Jin and Jennifer Zenker

Abstract

To study subcellular real-time dynamics in preimplantation embryos of mammals such as mice, the gold standard is currently to microinject fluorescently tagged constructs to visualize their protein products, which has significantly advanced our knowledge of early mammalian embryogenesis. However, during early embryonic development, it is recognized that mRNAs are precisely localized in space and time inside single cells to ensure their gene products, such as proteins, are synthesized and available when and where they are needed. Here, we describe a method that enables the simultaneous real-time imaging of RNAs and proteins of interest at the subcellular level in living mammalian embryos. Additionally, we present imaging settings specifically tailored for live mouse embryos, using a confocal microscope to visualize the real-time interactions between proteins and RNAs, such as mRNA transport along microtubule filaments.

Keywords RNA, Fluorescence, Live imaging, Microtubule cytoskeleton, Mouse preimplantation embryo

1 Introduction

The transport of RNAs to specific subcellular regions is crucial for specifying cell fates and supporting different subcellular functions, especially during early embryogenesis of mammalian and non-mammalian species [1–5]. For example, in *Drosophila melanogaster* embryos, *bicoid* mRNA localizes to the anterior region to define the anterior–posterior axis [6]. In zebrafish embryos, *huluwa* mRNA is transported to the dorsal region during the first two cell cycles to establish the dorsal–ventral axis [7]. Furthermore, in mouse embryos, *Caudal type homeobox 2 (Cdx2)* mRNA is transported to the apical region at the 8-cell stage to specify trophectoderm fate [8]. During these processes, the microtubule cytoskeleton serves as an intracellular network of scaffolds, akin to roads that guide traffic to specific destinations and sites within a defined space, to enable mRNAs to be delivered to their specific subcellular locations that, in turn, facilitate site-specific, local protein translation [9–11].

Alexandra J. Harvey and David K. Gardner (eds.), *The Preimplantation Embryo*, Methods in Molecular Biology, vol. 3038,
https://doi.org/10.1007/978-1-0716-5292-3_12,

This conserved process is more efficient than transporting individual proteins before their delivery to distal sites within cells, from flies to fish to mammals [12–15]. Given that these processes are time- and location-dependent, the simultaneous visualization of the real-time dynamics of proteins and RNAs within living embryos has been a technical challenge keenly adopted by developmental biologists [16–20] that can help us better understand how RNAs and proteins orchestrate preimplantation development.

The process of fluorescently tagging proteins in live mouse embryos through microinjection of *in vitro* transcribed RNAs is a well-established technique that is widely used to visualize selected proteins of interest (Fig. 1A) [11, 21–25]. Fluorescently tagged proteins can also be used to detect RNAs through their interaction with specific RNA secondary structures, such as RNA stem loops [17]. RNA stem loops form when a single-stranded RNA molecule folds back on itself to create a double-stranded stem with a loop at the end, such as in the case of PP7 stem loops [26]. RNAs of interest can be tagged with PP7 stem loops at their 5′ or 3′ untranslated region (UTR), which can subsequently be detected by visualizing the PP7 coat protein (PCP) as a fusion protein tethered to a fluorescent tag of choice (Fig. 1B) [26, 27]. Similarly, MS2 stem loops can also be used to label RNAs and be detected by the fluorescently tagged MS2

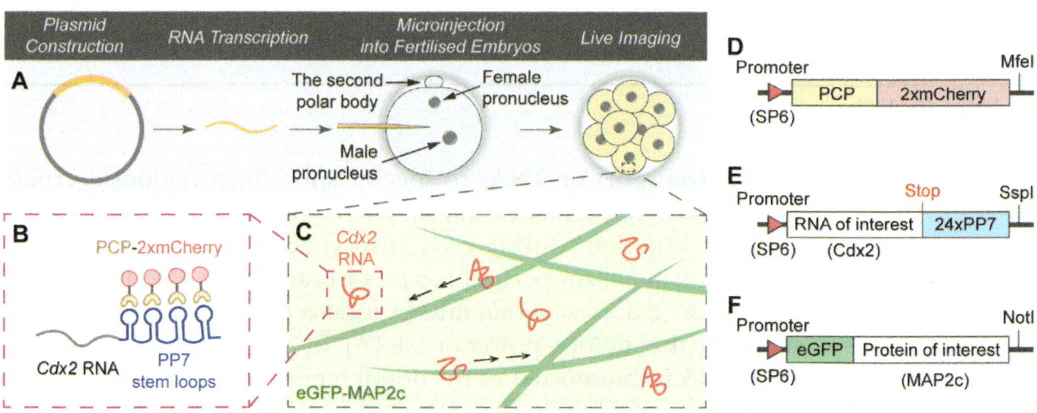

Fig. 1 Strategy to simultaneously visualize RNA and protein in live mouse embryos. (**A**) To express fluorescently tagged proteins in mouse embryos, RNA encoding the protein of interest is transcribed *in vitro* and microinjected into fertilized 1-cell stage embryos. The fertilized 1-cell stage embryos can be identified by the presence of the second polar body and the male pronucleus. (**B**) Fluorescently tagged PCPs (e.g., PCP-2xmCherry) detect PP7 stem loops tagged on the RNA of interest, allowing for visualization of the RNA within living embryos. (**C**) We exemplify this method to visualize *Cdx2* RNA transport along microtubule filaments labeled with eGFP-MAP2c in live mouse embryos. (**D**) PCP-2xmCherry was cloned into a plasmid with an *in vitro* transcription promoter (SP6) for RNA production. (**E**) An *in vitro* transcription promoter (SP6) was inserted upstream of the cDNA sequence of the RNA of interest and the PP7 tag for *in vitro* transcription of the RNA. (**F**) To visualize microtubule filaments in mouse embryos, MAP2c was cloned downstream of eGFP, driven by an SP6 promoter for *in vitro* transcription

coat protein (MCP) [28, 29]. This technique has seen widespread use across animal models, including human cells [30], as well as *Caenorhabditis elegans* [31], *D. melanogaster* [32], zebrafish [33], and mouse embryos [12]. Notably, this approach allows us to track the dynamic behavior of not only proteins, but also RNA granules in live embryos across various spatial and temporal scales, which are crucial for orchestrating developmental pathways and ensuring proper cellular functions. In this chapter, we exemplify this method by visualizing the dynamics of *Cdx2* RNA and microtubules in live mouse embryos (Fig. 1C).

In this chapter, we also describe imaging settings to trace the transport of RNA granules with second-range temporal resolution, while ensuring the continued development of embryos under examination.

2 Materials

2.1 Cloning and RNA In Vitro Transcription

1. PCR mix containing high-fidelity DNA polymerase (e.g., NEB Q5® High-Fidelity 2× Master Mix).
2. Colony PCR reagents: MangoTaq™ DNA Polymerase, 5× MangoTaq™ Colored Reaction Buffer, 50 mM $MgCl_2$, 100 mM dNTP, DMSO.
3. PCR machine.
4. 0.2 mL PCR tubes and 1.5 mL nuclease-free tubes.
5. Primers (*see* Table 1).
6. Mouse cDNA library.
7. Agarose gel electrophoresis setup.
8. Agarose.
9. Tris-Acetate-EDTA (TAE) buffer: 40 mM Tris, 20 mM acetic acid, 1 mM EDTA, pH 8.0.
10. 1% agarose gel: Add 0.5 g agarose to 50 mL TAE buffer (adjust the volume based on the size of the gel tray). Heat the mixture until the agarose is fully dissolved. Let it cool slightly before pouring it into the gel tray.
11. 1 kb DNA ladder.
12. Nuclease-free gel loading buffer.
13. UV spectrophotometer (e.g., NanoDrop).
14. DNA ligase or DNA assembly mix (e.g., NEBuilder® HiFi DNA Assembly Master Mix).
15. Competent cells.
16. Luria–Bertani (LB) medium.

Table 1
Primer sequences

Name	Sequence
pCS2+-F	AGGTTCTGGTATAGTGAGTCGTATTACGTAGATCCA
pCS2+-R	TTTTGGACATGGTAGTTCTAGAGGCTCGAGAGG
PCP-2xmCherry-F	CTCTAGAACTACCATGTCCAAAACCATCGTTCTT
PCP-2xmCherry-R	GACTCACTATACCAGAACCTCCACCCATATGC
pCS2+-cPCR-F	CCCACTTGGCAGTACATCAA
pCS2+-cPCR-R	TGTGGAATTGTGAGCGGATA
pcDNA4TO-F	TAAATCAAGCTAACGCGTATATCTGGCCC
pcDNA4TO-R	CTAGAACTGCCACCATGGGCG
SP6-F	TATACGCGTTAGCTTGATTTAGGTGACACTATAGAATACAAGC
SP6-R	CATGGTGGCAGTTCTAGAGGCTCGAGAGGC
GCN4-PP7-F	TGAAAGCTTCACGATTCGAAGATCGCATCGC
GCN4-PP7-R	CTCACGTATGACCGGTGCGGCC
Cdx2-F	ACCGGTCATACGTGAGCTACCTTCTGGACAAG
Cdx2-R	CGAATCGTGAAGCTTTCACTGGGTGACAGTG
pcDNA4TO-cPCR-F	GCGATGTACGGGCCAGATATAC
pcDNA4TO-cPCR-R	AAGGGCGAAAAACCGTCTAT

17. Antibiotics (e.g., Ampicillin).
18. LB agar plates, prepared with LB medium and 100 μg/mL ampicillin.
19. Plasmid mini-prep kit, including resuspension buffer, lysis buffer, neutralization buffer, wash buffer, and spin columns (e.g., Sangon Biotech).
20. Restriction enzymes: SspI-HF, MfeI-HF, NotI-HF, ClaI, and rCutSmart™ buffer.
21. DNA extraction and purification kit, including binding buffer, wash buffer, elution buffer, and spin columns (e.g., Sangon Biotech).
22. RNA *in vitro* transcription kit, including transcription buffer, nucleotide mix (ATP, GTP, CTP, UTP) with cap analog, SP6 RNA polymerase (e.g., mMESSAGE mMACHINE™ SP6 Transcription Kit) (*see* **Note 1**).
23. Milli-Q water.
24. Ribonuclease-free (RNase-free) water.
25. 70% Ethanol, diluted with RNase-free water.

Table 2
Plasmids

Plasmid name	Source
pCS2+ vector	NovoPro V011726
pHR-PCP-2xmCherry-CAAX	Addgene #74925
pcDNA4TO-24xGCN4_v4-kif18b-24xPP7	Addgene #74928
pCW57.1-Cdx2-3xT7	Addgene #200891
pCS2+eGFP-MAP2c	Zenker et al. [11]

26. Dry bath or incubator capable of maintaining temperature at 37°C (for plasmid linearization and RNA transcription) and 42°C (for bacterial transformation).
27. Centrifuge set to 4°C and operating at speeds over 12,000 × *g*.
28. −20°C freezer.
29. Shaking incubator set to 37°C.
30. Autoclaved glass beads.
31. Pipettes.
32. Nuclease-free filtered pipette tips.
33. Plasmids (see Table 2).

2.2 Mouse Embryo Collection and Culture

1. FVB/N and C57BL/6 mice are commonly used strains for embryo studies due to their high fertility rates. Here, we use 8- to 12-week-old wild-type FVB/N female mice and 8-week to 8-month-old wild-type FVB/N male mice to generate fertilized embryos for collection.
2. Superovulation hormones: pregnant mare serum gonadotrophin (PMSG), human chorionic growth hormone (HCG).
3. EZ-Squeeze pipette for mouse embryo manipulation.
4. Embryo handling media, such as M2 medium.
5. Embryo culture medium: Potassium Simplex Optimized Medium (KSOM).
6. Mineral oil.
7. Hyaluronidase: diluted in M2 media to a concentration of 10 mg/mL.
8. 35 mm culture dishes.
9. CO_2 incubator set to 37°C and 5% CO_2.

2.3 Mouse Embryo Microinjection

1. Injection buffer: 5 mM Tris, 5 mM NaCl, 0.1 mM EDTA, pH 7.4.
2. 60 mm culture dishes.
3. Inverted microscope with warming stage.
4. Micromanipulators (e.g., Narishige MMO-4, MM-94, and MM-94B).
5. Embryo holding microinjector (e.g., Narishige IM-11-2).
6. Holding capillary (e.g., Eppendorf VacuTip I).
7. Microinjector (e.g., Eppendorf FemtoJet 4i).
8. Microinjection capillary (e.g., Eppendorf Femtotip II).
9. Microloader pipette tips.
10. Embryo manipulation media: M2 or advanced KSOM, which can maintain pH stability when kept on the benchtop or under the microscope during microinjection.
11. Embryo culture media: KSOM.
12. Mineral oil.

2.4 Microscopy and Image Data Processing

1. Confocal microscope equipped with a light-protected incubation chamber, warming stage set to 37°C, and humified CO_2 supply set to 5% CO_2, and high-sensitivity detectors. Here, we used a Zeiss LSM780 confocal microscope configuration equipped with an incubation chamber and APD detectors (Fig. 2A).
2. Immersion media for water-immersion objectives (e.g., Immersol™ W).
3. An imaging dish that can be mounted onto the confocal microscope stage. We use an 8-well Nunc™ Lab-Tek™ chambered coverglass to accommodate the warming stage of the Zeiss LSM780 confocal microscope.
4. Image processing software: Zen (Zeiss) and Imaris (Bitplane).

3 Methods

3.1 Plasmid Construction for Visualizing RNA

The plasmid expression constructs pHR-PCP-2xmCherry-CAAX plasmid (Addgene ID 74925) and pcDNA4TO-24xGCN4_v4-kif18b-24xPP7 (Addgene ID 74928) [30] are used for exogenous expression of PCP and PP7 components to detect RNA in mouse embryos. As both plasmids do not encode promoter sequences upstream of the coding sequence for *in vitro* transcription, an additional cloning step is required: (1) either cloning the entire coding sequence into a separate plasmid backbone that already contains

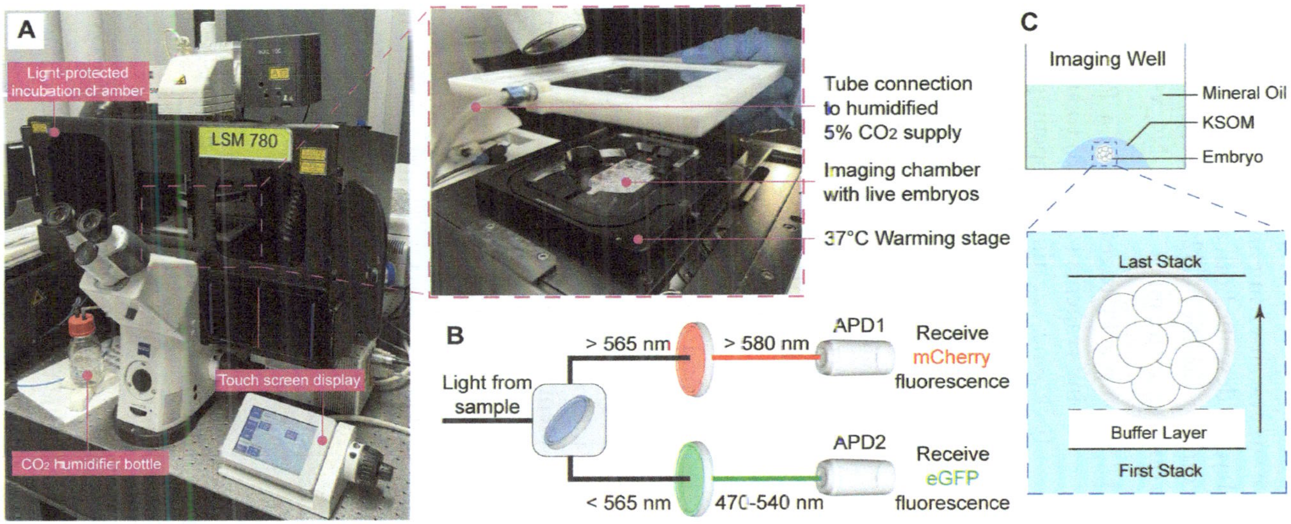

Fig. 2 Microscope setup, filter configuration, and imaging chamber for live imaging. (**A**) A ZEISS LSM780 confocal microscope equipped with a light-protected incubation chamber, humidified CO_2 supply, and warming stage. (**B**) Recommended light path, filter set, and detectors to visualize eGFP and mCherry. (**C**) Embryos are placed in pre-incubated KSOM media within the imaging well, and a Z-stack is configured to capture the entire 3D structure of the embryos

such a promoter, such as the pCS2$^+$ backbone with an SP6 promoter (ATTTAGGTGACACTATAG) or (2) inserting an *in vitro* transcription promoter (SP6, T7, or T3) upstream of the coding sequence. We generally recommend the first approach. However, for difficult PCR sequences, such as those containing GC-rich repeats, we suggest inserting an *in vitro* transcription promoter upstream. Here we describe both approaches, by cloning the PCP-2xmCherry sequence into the pCS2$^+$ backbone, and by clonal insertion of an SP6 promoter upstream of 24xGCN4_v4-kif18b-24xPP7.

3.1.1 Construction of pCS2$^+$ PCP-2xmCherry

1. Set up a PCR reaction in a PCR tube on ice with the following reagents (Table 3) to amplify and linearize the pCS2$^+$ backbone, creating an opening downstream of the SP6 promoter where the PCP-2xmCherry can be inserted.
2. Transfer the PCR tube to a PCR machine programmed as follows: initial denaturation at 98°C for 3 min, followed by 35 cycles of 98°C for 30 s, 56°C for 30 s and 72°C for 2 min and 10 s, then a final extension at 72°C for 5 min. At completion, the reaction can be held at 12°C.
3. Run 5 μL of the pCS2$^+$ backbone PCR product mixed with 1 μL of gel loading buffer on a 1% agarose gel to verify the correct size (4112 bp), then measure the concentration using a Nanodrop. A concentration higher than 200 ng/μL is preferred for further assembly.
4. Set up a PCR reaction in a PCR tube on ice with the following reagents (Table 4) to amplify the PCP-2xmCherry fragment from the pHR-PCP-2xmCherry-CAAX plasmid, which includes overlapping sequences with the linearized pCS2$^+$ backbone at both ends, allowing for their assembly (*see* **Note 3**).
5. Transfer the PCR tube to a PCR machine programmed as follows: initial denaturation at 98°C for 3 min, followed by 35 cycles of 98°C for 30 s, 60°C for 30 s and 72°C for 1 min and

Table 3
pCS2$^+$ PCR reaction mix (1×)

NEB Q5® High-Fidelity 2× Master Mix	12.5 μL
pCS2$^+$-F (10 μM)	1.25 μL
pCS2$^+$-R (10 μM)	1.25 μL
pCS2$^+$ vector (2 ng/μL) (*see* **Note 2**)	1 μL
Milli-Q water	9 μL
Total volume	25 μL

Table 4
PCP-2xmCherry PCR reaction mix (1×)

NEB Q5® High-Fidelity 2× Master Mix	12.5 μL
PCP-2xmCherry-F (10 μM)	1.25 μL
PCP-2xmCherry-R (10 μM)	1.25 μL
pHR-PCP-2xmCherry-CAAX (2 ng/μL)	1 μL
Milli-Q water	9 μL
Total volume	25 μL

15 s, then a final extension at 72°C for 5 min. At completion, the reaction can be held at 12°C.

6. Mix 5 μL of the PCP-2xmCherry PCR product with 1 μL of gel loading buffer and run on a 1% agarose gel at 140 V for 30 min, alongside a 1 kb ladder, to verify the correct size (2177 bp). Then measure the concentration using a Nanodrop. A concentration higher than 200 ng/μL is preferred for further assembly.
7. Assemble 200 ng of the linearized pCS2+ backbone with 300 ng of the PCP-2xmCherry fragment using 2.5 μL of NEBuilder® HiFi DNA Assembly Master Mix in a total reaction volume of 5 μL. Incubate in a PCR machine at 50°C for 1 h (lid heat at 105°C).
8. Perform bacterial transformation by thawing 50 μL of competent *E. coli* cells on ice, then adding 5 μL of the assembly reaction. Incubate on ice for 30 min, followed by heat shock in a dry bath at 42°C for 45 s. Return to ice for 2 min, then add 400 μL of LB medium and incubate in a shaking incubator at 37°C for 1 h. Plate 200 μL of the recovery mixture onto LB agar plates with ampicillin, using autoclaved glass beads to evenly spread the cells. Incubate the plate overnight in a 37°C incubator. Store the remaining 255 μL of the recovery mixture at 4°C for potential future use.
9. Prepare colony PCR reaction mixture (Table 5) for 10 colonies to select the correct colony with insertion of the fragment into the pCS2+ backbone. Use the colony testing primers pCS2+-cPCR-F and pCS2+ cPCR-R, which target sequences encoded within the pCS2+ backbone and which span the SP6 promoter, the coding sequence, as well as the SV40 poly(A) signal sequence (*see* **Note 4**). Mix thoroughly and distribute 10 μL into each PCR tube.
10. Pick a single colony from the overnight LB agar plate using a 10 μL filtered pipette tip. Use the tip to streak a line on a new

Table 5
Colony PCR reaction mix

Reagent	Reaction for 1 colony	Reactions for 10 colonies
MangoTaq DNA Polymerase	0.2 μL	2 μL
5× MangoTaq Colored Reaction Buffer	2 μL	20 μL
$MgCl_2$ (50 mM)	1 μL	10 μL
DMSO	0.375 μL	3.75 μL
dNTPs (100 mM)	0.1 μL	1 μL
$pCS2^+$-cPCR-F (10 μM)	0.5 μL	5 μL
$pCS2^+$-cPCR-R (10 μM)	0.5 μL	5 μL
Milli-Q water	5.325 μL	53.25 μL
Total volume	10 μL	100 μL

LB agar plate with ampicillin. Mark the colony number for future reference. Then, place the same tip into the PCR reaction tube corresponding to each colony for the PCR assay.

11. Transfer the PCR tube to a PCR machine programmed as follows: initial denaturation at 95°C for 3 min, followed by 30 cycles of 95°C for 30 s, 55°C for 30 s and 72°C for 1 min and 30 s, then a final extension at 72°C for 5 min. At completion, the reaction can be held at 12°C.
12. Run the colony testing PCR products on a 1% agarose gel at 140 V for 30 min, alongside a 1 kb ladder. Detection of a 2885 bp band is indicative of a positive colony with successful insertion of PCP-2xmCherry into the $pCS2^+$ backbone, to give rise to the final plasmid $pCS2^+$PCP-2xmCherry (Fig. 1D). Detection of a 728 bp band is indicative of a negative colony with no insertion (*see* **Note 5**).
13. Pick the positive colony in 5 mL LB media with ampicillin and culture it overnight in a 37°C shaking incubator (*see* **Note 6**).
14. Purify the $pCS2^+$PCP-2xmCherry plasmid from the overnight culture using a mini-prep kit, according to the manufacturer's instructions (*see* **Note 7**).

3.1.2 Construction of pcDNA4TO-SP6-24xGCN4_v4-kif18b-24xPP7

1. Set up a PCR reaction in a PCR tube on ice with the following reagents to amplify and linearize the pcDNA4TO-24xGCN4_v4-kif18b-24xPP7 plasmid (Table 6), creating an opening upstream of the 24xGCN4_v4-kif18b-24xPP7 sequence where the SP6 promoter can be inserted.

Table 6
pcDNA4TO-24xGCN4_v4-kif18b-24xPP7 PCR reaction mix (1×)

NEB Q5® High-Fidelity 2× Master Mix	12.5 μL
pcDNA4TO-F (10 μM)	1.25 μL
pcDNA4TO-R (10 μM)	1.25 μL
pcDNA4TO-24xGCN4_v4-kif18b-24xPP7 (2 ng/μL)	1 μL
Milli-Q water	9 μL
Total volume	25 μL

Table 7
SP6 promotor PCR reaction mix (1×)

NEB Q5® High-Fidelity 2× Master Mix	12.5 μL
SP6-F (10 μM)	1.25 μL
SP6-R (10 μM)	1.25 μL
pCS2+ vector (2 ng/μL)	1 μL
Milli-Q water	9 μL
Total volume	25 μL

2. Transfer the PCR tube to a PCR machine programmed as follows: initial denaturation at 98°C for 3 min, followed by 35 cycles of 98°C for 30 s, 56°C for 30 s and 72°C for 5 min and 10 s, then a final extension at 72°C for 5 min. At completion, the reaction can be held at 12°C.
3. Mix 5 μL of the pcDNA4TO-24xGCN4_v4-kif18b-24xPP7 PCR product with 1 μL of gel loading buffer and run on a 1% agarose gel at 140 V for 30 min, alongside a 1 kb ladder, to verify the correct size of DNA product (10,230 bp). Measure the concentration of the PCR product using Nanodrop. A concentration higher than 250 ng/μL is preferred for further assembly, as the pcDNA4TO-24xGCN4_v4-kif18b-24xPP7 backbone is much longer than the pCS2+ backbone (4112 bp).
4. Set up a PCR reaction in a PCR tube on ice with the following reagents (Table 7) to amplify a short fragment comprising the SP6 promoter from the pCS2+ empty vector plasmid, which includes overlapping sequences with the linearized

pcDNA4TO-24xGCN4_v4-kif18b-24xPP7 plasmid at both ends, allowing for their assembly.

5. Transfer the PCR tube to a PCR machine programmed as follows: initial denaturation at 98°C for 3 min, followed by 35 cycles of 98°C for 30 s, 60°C for 30 s and 72°C for 10 s, then a final extension at 72°C for 5 min. At completion, the reaction can be held at 12°C.
6. Mix 5 µL sample of the SP6 promoter PCR product with 1 µ L of gel loading buffer and run on a 1% agarose gel at 140 V for 30 min, alongside a 1 kb ladder, to verify the correct size of DNA product (114 bp). Measure the concentration of the PCR product using a Nanodrop. A concentration higher than 200 ng/µL is preferred for further assembly.
7. Combine 500 ng of the linearized pcDNA4TO-24xGCN4_ v4-kif18b-24xPP7 with 100 ng of the fragment containing the SP6 promoter using 2.5 µL of NEBuilder® HiFi DNA Assembly Master Mix in a total reaction volume of 5 µ L. Incubate in a PCR machine at 50°C for 1 h (lid heat at 105°C).
8. Perform bacterial transformation by thawing 50 µL of competent *E. coli* cells on ice, then adding 5 µL of the assembly product. Incubate on ice for 30 min, followed by heat shock in a dry bath at 42°C for 45 s. Return to ice for 2 min, then add 400 µL of LB medium and incubate in a shaking incubator at 37°C for 1 h. Plate 200 µL of the recovery mixture onto LB agar plates with ampicillin, using autoclaved glass beads to evenly spread the cells. Incubate the plate overnight in a 37°C incubator. Store the remaining 255 µL of the recovery mixture at 4°C for potential future use.
9. Pick ten individual colonies from the LB agar plate and transfer each colony into 5 mL of LB media with ampicillin, and culture overnight in a 37°C shaking incubator.
10. Purify the plasmids from the overnight culture using a miniprep kit, according to the manufacturer's instructions (*see* **Note 7**). Measure the concentration of the PCR product using a Nanodrop.
11. Digest 500 ng of plasmids with 0.5 µL of ClaI, 0.5 µL of MfeI-HF, 1 µL of rCutSmart buffer. Add Milli-Q water to reach a total volume of 10 µL and mix well. Incubate at 37°C for 2 h in a dry bath or incubator to verify which clones contain the correct insertion of the SP6 promoter (*see* **Note 8**).
12. Mix 5 µL of the digested plasmids with 1 µL of gel loading buffer, and run on a 1% agarose gel at 140 V for 30 min, alongside a 1 kb ladder. If a 134 bp band is detected, it indicates the successful insertion of the SP6 promoter. Store the remaining

undigested pcDNA4TO-SP6-24xGCN4_v4-kif18b-24xPP7 plasmid at −20°C for long-term storage.

3.1.3 Construction of Plasmid for RNA Visualization with PP7 Stem Loops

The *Kif18b* sequence in the pcDNA4TO-SP6-24xGCN4_v4-kif18b-24xPP7 plasmid can be replaced with the DNA sequence of RNAs of interest to be visualized (Fig. 1E). The replacement of *Kif18b* can be achieved through either a DNA ligation reaction of cloned fragments with AgeI and NheI restriction enzyme sites [12] or through a Gibson assembly reaction. Consequently, after *in vitro* transcription, the RNA of interest will be tagged with the 24× PP7 stem loops. Note that it is essential to include a stop codon before the PP7 stem loops. Without the stop codon, ribosomes may continue translating the cloned RNA sequence and potentially displace the PCP proteins that are bound to PP7 stem loops. Here, we introduce a method to replace *Kif18b* with a *Cdx2* cDNA sequence using Gibson assembly.

1. Set up a PCR reaction in a PCR tube on ice with the following reagents (Table 8) to amplify, linearize and remove the *Kif18b* fragment of the pcDNA4TO-SP6-24xGCN4_v4-kif18b-24xPP7 backbone to create an opening where the *Cdx2* cDNA fragment can be inserted.
2. Transfer the PCR tube to a PCR machine programmed as follows: initial denaturation at 98°C for 3 min, followed by 35 cycles of 98°C for 30 s, 60°C for 30 s and 72°C for 3 min and 55 s, then a final extension at 72°C for 5 min. At completion, the reaction can be held at 12°C.
3. Mix 5 μL sample of the pcDNA4TO-SP6-24xGCN4-24xPP7 PCR product with 1 μL of gel loading buffer and run on a 1% agarose gel at 140 V for 30 min, alongside a 1 kb ladder, to verify the correct size (PCR product length 7752 bp) Measure the concentration of the PCR product from Step 2 using Nanodrop.

Table 8
***Kif18b* removal PCR reaction mix (1×)**

NEB Q5® High-Fidelity 2X Master Mix	12.5 μL
GCN4-PP7-F (10 μM)	1.25 μL
GCN4-PP7-R (10 μM)	1.25 μL
pcDNA4TO-SP6-24xGCN4_v4-kif18b-24xPP7 (2 ng/μL)	1 μL
Milli Q water	9 μL
Total volume	25 μL

Table 9
Cdx2 amplification PCR reaction mix (1×)

NEB Q5® High-Fidelity 2× Master Mix	12.5 μL
Cdx2-F (10 μM)	1.25 μL
Cdx2-R (10 μM)	1.25 μL
pCW57.1-Cdx2-3xT7 (2 ng/μL)	1 μL
Milli-Q water	9 μL
Total volume	25 μL

4. Amplify the *Cdx2* cDNA sequence from a plasmid containing *Cdx2* coding sequence (e.g., pCW57.1-Cdx2-3xT7, Addgene ID 200891) or from a mouse cDNA library, including overlapping sequences at both ends that are complementary to the pcDNA4TO-SP6-24xGCN4-24xPP7 backbone, allowing for their assembly (Table 9; *see* **Note 3**).
5. Transfer the PCR tube to a PCR machine programmed as follows: initial denaturation at 98°C for 3 min, followed by 35 cycles of 98°C for 30 s, 60°C for 30 s and 72°C for 30 s, then a final extension at 72°C for 5 min. At completion, the reaction can be held at 12°C.
6. Mix 5 μL sample of the *Cdx2* PCR product with 1 μL of gel loading buffer and run on a 1% agarose gel at 140 V for 30 min, alongside a 1 kb ladder, to verify its correct size (product length 956 bp). Measure the concentration of the *Cdx2* PCR product using a Nanodrop.
7. Assemble 200 ng of the pcDNA4TO-SP6-24xGCN4-24xPP7 backbone with 150 ng of the cloned *Cdx2* fragment using 2.5 μL of NEBuilder® HiFi DNA Assembly Master Mix in a total reaction volume of 5 μL. Incubate in a PCR machine at 50°C for 1 h (lid heat at 105°C).
8. Perform bacterial transformation by thawing 50 μL of competent *E. coli* cells on ice, then adding 5 μL of the assembly product into the competent cells. Incubate on ice for 30 min, followed by heat shock in a dry bath at 42°C for 45 s. Return to ice for 2 min, then add 400 μL of LB medium and incubate in a shaking incubator at 37°C for 1 h. Plate 200 μL of the recovery mixture onto LB agar plates with ampicillin, using autoclaved glass beads to evenly spread the cells. Incubate the plate overnight in a 37°C incubator. Store the remaining 255 μL of the recovery mixture at 4°C for potential future use.
9. Prepare colony PCR reaction mixture (Table 10) for ten colonies to verify the insertion of the *Cdx2* fragment into the backbone. Use colony testing primers pcDNA4TO-cPCR-F

Table 10
Cdx2 insertion confirmation PCR reaction mix

Reagent	Reaction for 1 colony	Reactions for 10 colonies
MangoTaq DNA Polymerase	0.2 μL	2 μL
5× MangoTaq Colored Reaction Buffer	2 μL	20 μL
$MgCl_2$ (50 mM)	1 μL	10 μL
DMSO	0.375 μL	3.75 μL
dNTP (100 mM)	0.1 μL	1 μL
pcDNA4TO-cPCR-F (10 μM)	0.5 μL	5 μL
pcDNA4TO-cPCR-R (10 μM)	0.5 μL	5 μL
Milli-Q water	5.325 μL	53.25 μL
Total volume	10 μL	100 μL

and pcDNA4TO-cPCR-R that target the pcDNA4TO backbone, spanning the cloned SP6 promoter sequence, the desired coding sequence, as well as the SV40 poly(A) signal (*see* **Note 9**).

10. Pick a single colony from the overnight LB agar plate using a 10 μL filtered pipette tip. Use the tip to streak a line on a new LB agar plate with ampicillin and incubate at 37°C overnight. Mark the colony number for future reference. Then, place the same tip into the PCR reaction tube corresponding to each colony for the PCR assay.
11. Transfer the PCR tube to a PCR machine programmed as follows: initial denaturation at 95°C for 3 min, followed by 30 cycles of 95°C for 30 s, 57°C for 30 s and 72°C for 2 min and 35 s, then a final extension at 72°C for 5 min. At completion, the reaction can be held at 12°C.
12. Run the colony testing PCR product on a 1% agarose gel at 140 V for 30 min, alongside a 1 kb ladder. If a 5020 bp band is detected, the clone indicates a successful insertion event of the *Cdx2* cDNA into the plasmid construct, resulting in the final plasmid pcDNA4TO-SP6-24xGCN4_v4-Cdx2-24xPP7 (Fig. 1E). A colony without a successful insertion will exhibit a 6655 bp band in this colony PCR assay.
13. Pick the positive colonies and culture overnight in 5mL LB media with ampicillin in a 37°C shaking incubator (*see* **Note 6**).
14. Purify the pcDNA4TO-SP6-24xGCN4_v4-Cdx2-24xPP7 plasmid from the overnight culture of the clonal sample using a mini-prep kit, according to the manufacturer's instructions (*see* **Note 7**). Store the plasmid at −20°C.

3.2 RNA Preparation for Visualizing RNA

3.2.1 Plasmid Linearization and Purification

1. Select a specific restriction enzyme that cuts the plasmid after the poly(A) sequence to linearize it. This ensures that the *in vitro* transcribed RNA includes a poly(A) tail, allowing it to be properly translated into protein. Here we used SspI-HF to linearize pcDNA4TO-SP6-24xGCN4_v4-Cdx2-24xPP7, and MfeI-HF for pCS2$^+$ PCP-2xmCherry (*see* **Note 10**).
2. Mix 10 μg of pcDNA4TO-SP6-24xGCN4_v4-Cdx2-24xPP7 plasmid with 10 μL of rCutSmart buffer, and 2 μL of SspI-HF enzyme. Add Milli-Q water to reach a total volume of 100 μL. Mix well by repetitive pipetting.
3. For pCS2$^+$ PCP-2xmCherry plasmid, mix 10 μg of plasmid with 10 μL of rCutSmart buffer and 2 μL of MfeI-HF enzyme, then add Milli-Q water to reach 100 μL. Mix well by repetitive pipetting.
4. Incubate the enzyme digestion mixture in a dry bath or incubator at 37°C for 2–4 h.
5. Verify complete linearization by running a 5 μL sample of the enzyme-digested reaction product mixed with 1 μL of gel loading buffer on a 1% agarose gel at 140 V for 30 min, alongside a 1 kb ladder. Additionally, run 1 μL of undigested plasmid mixed with 1 μL of gel loading buffer in the next lane for comparison. The fully linearized plasmid should show a single band that appears larger than the corresponding faster-moving undigested circular plasmid band (*see* **Note 11**).
6. Purify the linearized plasmids using a DNA extraction and a purification kit by mixing the remaining 95 μL of the digestion reaction product with 285 μL of binding buffer. Mix well and load onto the column, spinning at 12,000 × *g* for 1 min. Wash the column with 500 μL of wash buffer and centrifuge at 12,000 × *g* for 1 min. Dissolve the linearized plasmid in 20 μL RNase-free water and collect in a clean tube (*see* **Note 12**).
7. Measure the concentration of the linearized pcDNA-4TO-SP6-24xGCN4_v4-Cdx2-24xPP7 and pCS2$^+$ PCP-2xmCherry plasmid using a Nanodrop. For optimal RNA *in vitro* transcription, the recommended concentration is higher than 100 ng/μL (*see* **Note 13**). Proceed with RNA transcription or store at −20°C for future use.

3.2.2 RNA In Vitro Transcription and Purification

1. Mix 0.5–1 μg of linearized plasmid with 2 μL of SP6 transcription buffer, 10 μL of NTP/CAP, and 2 μL of SP6 enzyme from the SP6 *in vitro* transcription kit. Add RNase-free water to reach a total volume of 20 μL. Mix well.

2. Incubate the mixture in a dry bath or incubator for 2–4 h at 37°C.
3. Mix 1 μL sample of the RNA with 1 μL of gel loading buffer and run on a 1% agarose gel at 150 V for 15 min, alongside a 1 kb DNA ladder, to confirm successful transcription (*see* **Note 14**).
4. Add 1 μL of DNase from the SP6 *in vitro* transcription kit to the reaction. Mix well and incubate at 37°C for 15 min to remove DNA template (*see* **Note 15**).
5. Add 40 μL of lithium chloride (LiCl) from the SP6 *in vitro* transcription kit to the transcribed RNA. Mix well and store in a −20°C freezer overnight to precipitate RNA (*see* **Note 16**).
6. Centrifuge at 4°C for 10 min at 12,000 × *g*. Carefully remove the supernatant.
7. Wash the RNA precipitate with 200 μL 70% ethanol.
8. Centrifuge at 4°C for 2 min at 12,000 × *g*. Carefully remove the supernatant and remove as much ethanol as possible from the tube without disturbing the RNA precipitate.
9. Allow the ethanol to evaporate at room temperature by leaving the lid open for at least 10 min, until it fully dries and turns into a white pellet.
10. Dissolve the RNA in 30 μL of RNase-free water.
11. Measure the concentration of purified RNA using a Nanodrop and store the RNAs in a −20°C freezer until ready for microinjection.

3.3 Preparation of RNA Samples for Visualizing Fluorescently Tagged Proteins

Proteins of interest can be tagged with a fluorescent marker of choice. The fluorescent marker can be cloned as a tether that is located on either the amino (N-) or carboxyl (C-) terminus of the protein, and this consideration is relative to whether the position of the fluorescent tag interferes with the function of the reference protein, or not. The DNA sequence of the protein and its fluorescent tag must be positioned downstream of an *in vitro* transcription promoter SP6, T7, or T3, such that the fluorescently tagged protein can be expressed in mouse embryos after microinjection of the *in vitro* transcribed RNAs. Here we used the pCS2$^+$eGFP-MAP2c plasmid [11], which tags microtubule-associated protein 2c (MAP2c) with enhanced GFP (eGFP) at the N-terminus to visualize microtubule filaments in microinjected mouse embryos (Fig. 1F).

1. Linearize 10 μg of the pCS2$^+$eGFP-MAP2c plasmid by adding 2 μL of NotI-HF enzyme and 10 μL of rCutSmart buffer. Add Milli-Q water to reach a total volume of 100 μL. Mix well by repetitive pipetting.

2. Incubate the enzyme digestion mixture in a dry bath or incubator at 37°C for 2–4 h.
3. Verify complete linearization by running a 5 μL sample of the enzyme-digested reaction product mixed with 1 μL of gel loading buffer on a 1% agarose gel at 140 V for 30 min, alongside a 1 kb ladder. Additionally, run 1 μL of undigested pCS2^{+}eGFP-MAP2c plasmid mixed with 1 μL of gel loading buffer in the next lane for comparison. The fully linearized plasmid should show a single band that appears larger than the undigested pCS2^{+}eGFP-MAP2c plasmid band (*see* **Note 11**).
4. Purify the linearized plasmids using a DNA purification kit by first mixing the remaining 95 μL of the digestion reaction product with 285 μL of binding buffer. Mix well and load onto a column, spinning at 12,000 × *g* for 1 min. Wash the column with 500 μL of wash buffer and centrifuge at 12,000 × *g* for 1 min. Dissolve the linearized plasmid in 20 μL RNase-free water and collect in a clean tube (*see* **Note 12**).
5. Measure the concentration of the linearized pCS2^{+}eGFP-MAP2c plasmid using a Nanodrop. For optimal RNA *in vitro* transcription, the recommended concentration is higher than 100 ng/μL (*see* **Note 13**). Proceed with RNA transcription or store at −20°C for future use.
6. Mix 0.5–1 μg of linearized pCS2^{+}eGFP-MAP2c plasmid with 2 μL of SP6 transcription buffer, 10 μL of NTP/CAP, and 2 μL of SP6 enzyme from the SP6 *in vitro* transcription kit. Add RNase-free water to reach a total volume of 20 μL. Mix well.
7. Incubate the mixture in a dry bath or incubator for 2–4 h at 37°C.
8. Mix 1 μL RNA product with 1 μL of gel loading buffer and run on a 1% agarose gel at 150 V for 15 min, alongside a 1 kb DNA ladder, to confirm successful transcription (*see* **Note 14**).
9. Add 1 μL of DNase from the SP6 *in vitro* transcription kit to the RNA transcription reaction. Mix well and incubate at 37°C for 15 min to remove DNA template (*see* **Note 15**).
10. Add 40 μL of lithium chloride (LiCl) from the SP6 *in vitro* transcription kit to the transcribed RNA. Mix well and store in a −20°C freezer overnight to precipitate RNA (*see* **Note 16**).
11. Centrifuge at 4°C for 10 min at 12,000 × *g*. Carefully remove the supernatant.
12. Wash the RNA precipitate with 200 μL 70% ethanol.

13. Centrifuge at 4°C for 2 min at 12000 × *g*. Carefully remove the supernatant and remove as much ethanol as possible from the tube without disturbing the RNA precipitate.
14. Allow the ethanol to evaporate at room temperature by leaving the lid open for at least 10 min, until it fully dries and turns into a white pellet.
15. Dissolve the RNA in 30 μL of RNase-free water.
16. Measure the concentration of purified RNA using a Nanodrop and store the RNAs in a −20°C freezer until ready for microinjection.

3.4 Embryo Microinjection and In Vitro Culture

1. Conduct all animal work in accordance with institutional animal ethics guidelines.
2. Superovulate fertile wildtype female mice by intraperitoneal injection of 5 IU of PMSG, followed by 5 IU of HCG 48 h later.
3. Mate superovulated female mice with male stud mice overnight. Check for successful mating the next morning by detection of a vaginal plug in female mice.
4. Collect all 1-cell stage mouse embryos from the ampulla of fallopian tubes of female mice into 37°C 200 μL prewarmed M2 Media overlaid with mineral oil (*see* **Note 17**).
5. Add 6 μL of hyaluronidase to M2 media to a final concentration of 300 μg/mL to digest cumulus cells surrounding embryos and culture in the incubator for 20 min.
6. Collect the embryos from hyaluronidase digestion and transfer to preequilibrated 20 μL KSOM medium overlaid with mineral oil. Culture embryos in an incubator set to 37°C and 5% CO_2 until ready for microinjection. The optimal time for microinjection is 24 h post-hCG injection.
7. Select fertilized embryos from unfertilized oocytes. Fertilized 1-cell stage mouse embryos will show the extrusion of the second polar body and male pronucleus shortly after fertilization (Fig. 1A). Unfertilized oocytes cannot develop but can potentially be used for other purposes such as oocyte immunostaining; if not utilized, they can be discarded.
8. Dilute and mix all the RNAs required (*eGFP-MAP2c*, *Cdx2-24xPP7*, *PCP-2xmCherry*) in injection buffer to a total sample volume of 10 μL. The final concentrations of RNAs in this experiment used were: 400 ng/μL for *eGFP-MAP2c*, 300 ng/μL *Cdx2-24xPP7*, and 100 ng/μL *PCP-2xmCherry* (*see* **Note 18**).

9. Centrifuge the RNA mixture at 4°C for 5 min at a speed of 12,000 × *g* and then keep it on ice until microinjection into mouse embryos (*see* **Note 19**).
10. Microinject the RNA mixture into the cytoplasm of 1-cell stage mouse embryos. Set the compensation pressure, injection pressure, and the injection time of the FemtoJet™ 4i microinjector based on the opening of the microinjection needle (*see* **Note 20**). Here we set compensation pressure to 80 hPa, injection pressure to 800 hPa, and adjusted the injection time between 0.8 and 1.2 s.
11. Culture microinjected embryos in KSOM media in an incubator set to 37°C and 5% CO_2 until they reach the predetermined stages required for the planned imaging experiments.

3.5 Visualizing Fluorescently Tagged RNA and Protein in Real-Time

3.5.1 Preparations to Perform Live Imaging on Preimplantation Mouse Embryos

1. Add 10 μL of KSOM medium to the center of one well of an imaging chamber, overlay the droplet with 300 μL of mineral oil, and preequilibrate it in an incubator set to 37°C and 5% CO_2 overnight.
2. Transfer embryos into the preincubated imaging chamber droplet through an EZ-Squeeze pipette or by mouth pipetting, and culture in the incubator until the imaging experiment (*see* **Note 21**).
3. Set the temperature of the incubation chamber and the warming stage (Fig. 2A) of the microscope to 37°C and the CO_2 concentration to 5% (*see* **Note 22**).
4. Turn on the lasers needed for fluorescence excitation at least 30 min before imaging to allow them to warm up and stabilize.
5. Select an appropriate lens to observe the mouse embryos. Here we used a 63× water immersion objective (numerical aperture = 1.2, working distance = 280 μm), and applied one drop of immersion fluid.
6. Mount the imaging chamber containing the embryos onto the warming stage (Fig. 2A inset). Close the lid of the incubation chamber to keep the embryo in a dark environment.
7. Adjust the position and focus to find embryos through the ocular lens.

3.5.2 Settings for Live Imaging

For effective live imaging of embryos under the microscope while maintaining their development, it is important to ensure that optimal environmental conditions are in place. Additionally, in terms of laser scanning microscopy, using low laser power and short scanning times are both crucial factors that can preserve embryos under observation. To achieve this, the use of highly sensitive detectors, such as avalanche photodiodes (APDs), as well as optimized microscopy settings including an enlarged pinhole, fast scanning speed, and restricted imaging region. These adjustments help minimize laser exposure and laser-scanning times, which, in turn, serves to minimize harm to the living embryos. Optimized imaging settings are also critical to obtain the best signal-to-noise ratio for high-quality images and movies for visualizing subcellular dynamics within single cells of living preimplantation mouse embryos.

The methods described here are based on the use of a ZEISS LSM780 confocal microscope. Similar acquisition parameters can be set on alternative confocal microscopes, but may require some optimization.

1. Choose appropriate excitation laser wavelengths, filters, and high-sensitivity detectors for fluorescence excitation and detection in the embryos. Here, we used a 561 nm laser to excite mCherry and configured the APD1 detector with a filter to capture wavelengths longer than 580 nm. For eGFP, we used a 488 nm laser and detected it with the APD2 detector using a filter set to the 470–540 nm range (*see* **Note 23**) (Fig. 2B).
2. Set the pinhole size to 3.0 Airy units (*see* **Note 24**).
3. Optimize the digital gain based on the expression levels of each fluorophore. If fluorescence expression is low and requires higher laser power for clear visualization, increase the digital gain up to 3× to compensate and reduce the need for higher laser power.
4. Adjust the laser power to the lowest possible setting that still allows for clear visualization, considering the expression levels of the proteins, pinhole size, and digital gain settings described above (*see* **Note 25**).
5. Set the image size per slice to 512 × 512 pixel dimensions. Larger image sizes can be used, but this will increase the laser scanning time on the embryos, which, in turn, affects embryo quality.
6. Optimize the imaging speed according to the experimental requirements. Lower imaging speeds can achieve higher quality images by increasing pixel dwell time, but this extends laser scanning time on embryos. Conversely, faster imaging can be employed to track rapid movements of granules in embryos, which can result in lower image quality.

3.5.3 3D Imaging of Whole Mouse Embryos

1. Zoom in on the area of interest such that a single embryo fills the frame. Here, we performed a 1.3× zoom when using a 63× objective to image one mouse embryo.

2. Use the "Z-Stack" function to set the first and last stack for imaging the entire 3D volume of the embryo. Set the first focal plane with a few additional buffer layers (about 10 μm) below the initial focus to account for embryo movement and set the last focal plane at the furthest depth of the embryo (Fig. 2C) (*see* **Note 26**).
3. Set the interval between slices to 1–1.5 μm, resulting in 50–70 slices per embryo. Some confocal microscopes may provide recommended intervals based on the pinhole size.
4. Start imaging to capture 3D images of embryos at different developmental stages (Fig. 3) (*see* **Note 27**).

3.5.4 Temporal Imaging of a Region of Interest

1. After capturing the 3D image of the whole embryo, deselect the "Z-Stack" function, and find a focal plane and region of interest where the RNA and protein of interest, in this case, *Cdx2* RNA granules and microtubule filaments, show favorable detection levels and patterns based on the captured 3D image.
2. Zoom in on the focal plane as needed based on the size, speed, and transport range of the RNA granules being tracked (Figs. 1C and 3).
3. To reduce laser scanning time and the interval between time-lapse images, use the "Regions" function to define a small region of interest (*see* **Note 28**).
4. Increase the imaging speed to shorten the pixel dwell time to meet experimental requirements.
5. Select "Acquisition" and verify that the scanning time per slice (interval of the time-lapse movies) and image quality meet the experimental needs after setting the region of interest and imaging speed (*see* **Note 29**). Here, for visualizing *Cdx2* RNA transport along microtubule filaments, we aimed for imaging every 2 s.
6. Select the "Time Series" function to set the interval of the time series to 0 if images are to be taken continuously. Adjust the number of time series based on experimental requirements. Here, we captured a 100 time series' for each experiment.
7. Start time-lapse imaging of the region of interest (Fig. 3 insets).

3.5.5 Computational Image Data Analysis

We recommend using Imaris to process the 3D images and 2D movies of the embryos, adjusting the color, contrast, and brightness for each channel (Fig. 3). For the 2D movie visualizing RNA transport, identify the time period during which RNA moves along the microtubule filament and capture images of the RNA movement at each time point (Fig. 3 insets). Imaris can also be used to segment cells or regions of interest, measure the transport speed and distance of

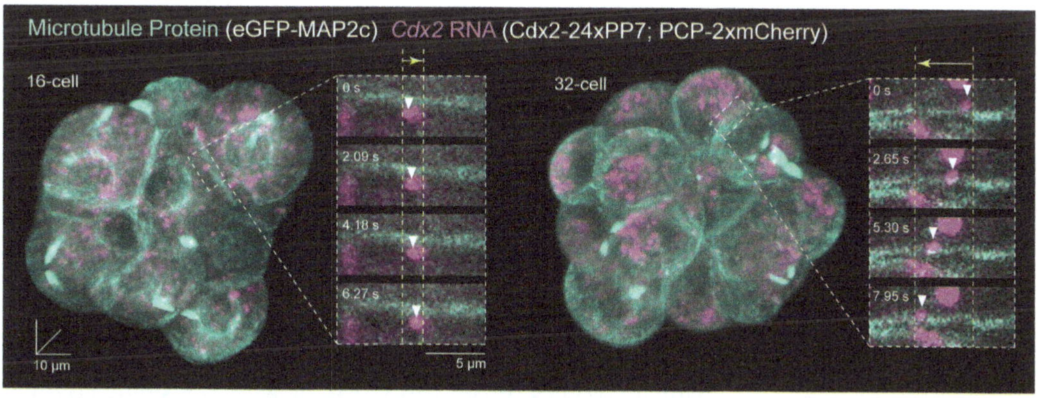

Fig. 3 3D image and 2D time-lapse images of a mouse preimplantation embryo, showing the microtubule cytoskeleton (cyan) and *Cdx2* RNA (pink) at the 16- and 32-cell stages (left and right, respectively). In the white dotted boxes, different transport speeds of *Cdx2* RNA along a microtubule filament at 16- versus 32-cell stages are shown (yellow arrows), as observed through short-interval live imaging

granules, quantify fluorescence intensity, and determine the volume of fluorescently-labeled particles.

4 Notes

1. The choice of RNA *in vitro* transcription kit depends on the promoter located in the 5′ UTR of the coding sequence, which could be SP6, T7, or T3 (e.g., Invitrogen™ mMESSAGE mMACHINE™ SP6/T7/T3 Transcription Kit). The kit should include a cap analog to enable the addition of a 5′ cap to the synthesized mRNA, enhancing its stability and ensuring proper translation in live mouse embryos.
2. When using a plasmid as a PCR template, the plasmid should be diluted to a low concentration, such as 2 ng/μL. Using a high plasmid concentration may cause PCR failure or excessive plasmid contamination in the PCR product, compromising downstream applications.
3. To assemble two or more DNA fragments (e.g., backbone and insert in this experiment), design 15–25 nucleotide complementary sequences at the fragment ends to enable seamless annealing during the Gibson assembly reaction. These overlaps, typically introduced via PCR primers, should be error-free, free of secondary structures, and nonself-complementary to ensure efficient assembly.
4. This primer pair can be used for most colony PCR assays using the pCS2$^+$ backbone to test for successful insertion of the coding sequence.

5. If no colonies with correct plasmids are identified after colony PCR, pick additional colonies from the agar plate and repeat the colony testing PCR. If there are insufficient colonies on the plate, plate the remaining transformation recovery mixture onto a new LB agar plate with ampicillin and incubate overnight at 37°C to grow more colonies. To improve the likelihood of obtaining the desired plasmid, consider increasing the ratio of the fragment to backbone during the cloning assembly reaction.
6. To ensure single colonies of clonal origin are isolated, optionally streak the colony onto a separate plate to obtain individual colonies. From this streak, isolate 3–5 single colonies, then verify each of these colonies by colony PCR using primer pair pCS2$^+$-cPCR-F and pCS2$^+$-cPCR-R for pCS2$^+$ backbone plasmids, or pcDNA4TO-cPCR-F and pcDNA4TO-cPCR-R for pcDNA4TO backbone plasmids. Once verified, grow the plasmids from the confirmed single colonies in LB media with ampicillin in a 37°C shaking incubator overnight.
7. Purify the plasmids using a mini-prep kit by first harvesting the overnight bacterial culture by centrifuging it at 12,000 × *g* for 1 min. Discard the supernatant and resuspend the pellet in 250 μL of resuspension buffer. Add 250 μL of lysis buffer and mix gently to lyse the cells. Neutralize the lysate with 350 μL of neutralization buffer and centrifuge at 12,000 × *g* for 10 min to pellet cellular debris. Transfer the supernatant to a clean column and wash with 500 μL of wash buffer to remove contaminants. Elute the plasmid DNA with 30 μL of Milli-Q water and collect the purified plasmid in a clean tube. Measure the DNA concentration using a Nanodrop and store the plasmid at −20°C for long-term storage.
8. To identify the correct plasmid comprising the inserted fragment by restriction enzyme digestion, two enzymes should be chosen: one that has a unique restriction site in the backbone and is not found within the inserted fragment, as well as one that targets a restriction site within the inserted fragment but does not cut the vector backbone.
9. pcDNA4TO-cPCR-F and pcDNA4TO-cPCR-R primer pair can be used for most colony PCR assays using the pcDNA4TO backbone to test for successful insertion of the coding sequence.
10. RNA *in vitro* transcription templates can also be prepared by PCR amplification of a fragment comprising the promoter, coding sequence, and poly(A) signal. The primer pair pCS2$^+$-cPCR-F and pCS2$^+$-cPCR-R can be used for the pCS2$^+$ backbone, and pcDNA4TO-cPCR-F and pcDNA4TO-cPCR-R can be used for the pcDNA4TO backbone.

After amplification, purify the PCR product with a DNA purification kit before using it as a template. When using a PCR product as a template for *in vitro* transcription, only 100–200 ng of the purified PCR product should be added to the reaction. Using higher amounts may result in failed transcription.

11. All plasmids must be linearized completely before they can be used as RNA *in vitro* transcription templates, with no circular plasmid band visible on the gel. If circular plasmid bands are still present, add an additional 1 μL of enzyme to the existing reaction and extend the incubation time by 2 h to ensure complete DNA digestion, and repeat gel electrophoresis to confirm.
12. Since the purified linearized plasmid will be used directly in *in vitro* transcription in the next step, it is recommended to dissolve the linearized plasmid in RNase-free water during purification to minimize RNA degradation.
13. If the final concentration after purification is low, it can be increased by either using a higher amount of plasmid for linearization, decreasing the volume of diluent, or washing the column with 60°C RNase-free water to elute any remaining product.
14. To minimize RNA degradation when running RNA on a gel, the gel should be freshly made, and the voltage should be increased to a level that shortens the electrophoresis time without causing excessive heat that could melt the gel. Here, we used 150 V for RNA electrophoresis in 1% gel. RNA is single-stranded compared to a double-stranded DNA ladder; thus, its migration as a detectable band within the gel may be slightly below the expected size (the length from the SP6 promoter to the enzyme cut site).
15. DNA contamination or carry-over residues from the preparation of samples for microinjection can affect the development of injected embryos. Therefore, degrading the DNA template after RNA *in vitro* transcription is essential when preparing RNA for embryo microinjections.
16. RNA can also be purified with a spin column-based purification kit (e.g., QIAGEN RNeasy kit).
17. Prepare all media required for mouse embryo isolation, culture, and microinjection before the experiment, and prewarm them in a 37°C incubator with 5% CO_2 for several hours or overnight.
18. To minimize the undesired background signal of unbound PCP-2xmCherry, it is recommended to use a 1:3 ratio of *PCP-2xmCherry* to *RNA-24xPP7* or lower when preparing the RNA mix for injection into mouse embryos.

19. Centrifuge the 10 μL RNA mixture right before performing microinjection and avoid freezing this mixture again. Centrifugation helps remove any possible impurities in the mixture that could block the microinjection needle. Repeated freezing and thawing can increase the accumulation of impurities in the sample.
20. The FemtoJet microinjector can be substituted with another electric microinjector, such as the Narishige IM-400. If no electric microinjector is available, manual microinjectors like Eppendorf's CellTram or Narishige's IM-21 can be used, though these require precise manual control.
21. Up to 20 embryos are recommended in a 10 μL KSOM droplet. Place the embryos in the KSOM media of the imaging well as early as possible, ideally one day before imaging. This allows the embryos to settle at the very bottom of the chamber, which will substantially eliminate the movement of embryos during live imaging.
22. It is best to set up the environmental conditions as early as possible, ideally several hours before imaging. This helps stabilize the environment and minimizes airflow within the chamber, which could otherwise cause the embryos to move.
23. It is recommended to image all fluorescence channels simultaneously in a single track to reduce scanning time, rather than a sequential imaging approach.
24. Smaller pinholes are possible but require higher laser power to achieve good image quality, which might negatively affect the development of the embryos. A wider pinhole allows more light to pass through, which can reduce the need for high laser power and minimize damage to embryos, though it may lead to increased image blurring due to more out-of-focus light applied. A pinhole size of 3.0 Airy units was chosen alongside lower laser power settings, which achieved good embryo imaging quality.
25. It is recommended that the combined laser power should not exceed 3% to maintain optimal embryo quality while imaging.
26. The sinking of embryos is less significant during short imaging sessions. However, for long-term overnight live imaging, embryos tend to sink significantly to the bottom of the imaging chamber. Therefore, we recommend placing the embryos in the imaging chamber as early as possible and adding a 10 μ m space layer below the initial focus. When setting the first and last focal planes for scanning, ensure that both fluorophores are encompassed within the upper and lower focal planes.

27. To capture images of embryos at different stages, we recommend keeping the embryos in the embryo incubator between imaging sessions rather than in the incubation chamber of the microscope to ensure normal development. At each desired time point, embryos should be retrieved from the incubator, placed into the microscope incubation chamber, and imaged at the specific developmental stage.
28. When drawing a region of interest, the length of the vertical edge is the main factor that determines the scanning time. The laser and detectors move across each scanning line to capture the fluorescence, and the process of changing lines is the most time-consuming part of scanning. Therefore, a shorter vertical edge reduces the number of line changes, significantly decreasing the scanning time and enabling short-interval, fast imaging. Alternatively, use the rotate function to set the region of interest with the shortest possible vertical edge. Bi-directional scanning mode can halve the scan time compared to mono-directional scanning as the bidirectional mode scans in a zig–zag pattern, but this will slightly reduce the signal-to-noise ratio. We recommend using bi-directional scanning mode when imaging fast-moving granules, such as RNAs.
29. The interval of the time-lapse movies depends on the expected speed of the structure of interest (RNA or protein) being tracked. If the structure of interest moves very fast, the interval has to be in the second- or subsecond range. If the structure of interest moves more slowly, the interval might be rather in the range of minutes. The interval time depends on the duration of each frame being imaged, which can be controlled by the pixel dwell time, the size of the region of interest, and pixel size.

Acknowledgments

This work was supported by the National Health and Medical Research Council (NHMRC) Ideas Grant APP2002507 and Investigator Grant APP2009409 to J.Z., and the Sylvia and Charles Viertel Senior Medical Fellowship to J.Z. The Australian Regenerative Medicine Institute is supported by grants from the State Government of Victoria and the Australian Government. We thank Julian Heng (ORCID: 0000-0002-0378-7078) for English language editing of this article, Lin Zhang (Tsinghua University) for a thorough review, and Azelle Hawdon (ORCID: 0000-0002-0378-7078) for plasmids for visualizing RNA.

References

1. Buxbaum AR, Haimovich G, Singer RH (2015) In the right place at the right time: visualizing and understanding mRNA localization. Nat Rev Mol Cell Biol 16(2):95–109. https://doi.org/10.1038/nrm3918
2. King ML, Messitt TJ, Mowry KL (2005) Putting RNAs in the right place at the right time: RNA localization in the frog oocyte. Biol Cell 97(1):19–33. https://doi.org/10.1042/bc20040067
3. Martin KC, Ephrussi A (2009) mRNA localization: gene expression in the spatial dimension. Cell 136(4):719–730. https://doi.org/10.1016/j.cell.2009.01.044
4. Weil TT, Parton RM, Davis I (2010) Making the message clear: visualizing mRNA localization. Trends Cell Biol 20(7):380–390. https://doi.org/10.1016/j.tcb.2010.03.006
5. Suter B (2018) RNA localization and transport. Biochim Biophys Acta 1861(10):938–951. https://doi.org/10.1016/j.bbagrm.2018.08.004
6. Spirov A, Fahmy K, Schneider M, Frei E, Noll M, Baumgartner S (2009) Formation of the bicoid morphogen gradient: an mRNA gradient dictates the protein gradient. Development 136(4):605–614. https://doi.org/10.1242/dev.031195
7. Yan L, Chen J, Zhu X, Sun J, Wu X, Shen W, Zhang W, Tao Q, Meng A (2018) Maternal Huluwa dictates the embryonic body axis through β-catenin in vertebrates. Science 362(6417). https://doi.org/10.1126/science.aat1045
8. Skamagki M, Wicher KB, Jedrusik A, Ganguly S, Zernicka-Goetz M (2013) Asymmetric localization of Cdx2 mRNA during the first cell-fate decision in early mouse development. Cell Rep 3(2):442–457. https://doi.org/10.1016/j.celrep.2013.01.006
9. Das S, Vera M, Gandin V, Singer RH, Tutucci E (2021) Intracellular mRNA transport and localized translation. Nat Rev Mol Cell Biol 22(7):483–504. https://doi.org/10.1038/s41580-021-00356-8
10. Hawdon A, Aberkane A, Zenker J (2021) Microtubule-dependent subcellular organisation of pluripotent cells. Development 148(20). https://doi.org/10.1242/dev.199909
11. Zenker J, White MD, Templin RM, Parton RG, Thorn-Seshold O, Bissiere S, Plachta N (2017) A microtubule-organizing center directing intracellular transport in the early mouse embryo. Science 357(6354):925–928. https://doi.org/10.1126/science.aam9335
12. Hawdon A, Geoghegan ND, Mohenska M, Elsenhans A, Ferguson C, Polo JM, Parton RG, Zenker J (2023) Apicobasal RNA asymmetries regulate cell fate in the early mouse embryo. Nat Commun 14(1):2909. https://doi.org/10.1038/s41467-023-38436-2
13. St Johnston D, Beuchle D, Nüsslein-Volhard C (1991) Staufen, a gene required to localize maternal RNAs in the Drosophila egg. Cell 66(1):51–63. https://doi.org/10.1016/0092-8674(91)90138-o
14. Zhou Y, King ML (1996) Localization of Xcat-2 RNA, a putative germ plasm component, to the mitochondrial cloud in Xenopus stage I oocytes. Development 122(9):2947–2953. https://doi.org/10.1242/dev.122.9.2947
15. Maegawa S, Yasuda K, Inoue K (1999) Maternal mRNA localization of zebrafish DAZ-like gene. Mech Dev 81(1-2):223–226. https://doi.org/10.1016/s0925-4773(98)00242-1
16. Zhang L, Chen L, Chen J, Shen W, Meng A (2020) Mini-III RNase-based dual-color system for in vivo mRNA tracking. Development 147(22). https://doi.org/10.1242/dev.190728
17. Braselmann E, Rathbun C, Richards EM, Palmer AE (2020) Illuminating RNA biology: tools for imaging RNA in live mammalian cells. Cell Chem Biol 27(8):891–903. https://doi.org/10.1016/j.chembiol.2020.06.010
18. Dufourt J, Bellec M, Trullo A, Dejean M, De Rossi S, Favard C, Lagha M (2021) Imaging translation dynamics in live embryos reveals spatial heterogeneities. Science 372(6544):840–844. https://doi.org/10.1126/science.abc3483
19. Le P, Ahmed N, Yeo GW (2022) Illuminating RNA biology through imaging. Nat Cell Biol 24(6):815–824. https://doi.org/10.1038/s41556-022-00933-9
20. Chen M, Sui T, Yang L, Qian Y, Liu Z, Liu Y, Wang G, Lai L, Li Z (2022) Live imaging of RNA and RNA splicing in mammalian cells via the dcas13a-SunTag-BiFC system. Biosens Bioelectron 204:114074. https://doi.org/10.1016/j.bios.2022.114074
21. Zenker J, White MD, Gasnier M, Alvarez YD, Lim HYG, Bissiere S, Biro M, Plachta N (2018) Expanding actin rings zipper the mouse embryo for blastocyst formation. Cell 173(3):776–791.e717. https://doi.org/10.1016/j.cell.2018.02.035
22. Gu B, Bradshaw B, Zhu M, Sun Y, Hopyan S, Rossant J (2022) Live imaging YAP signalling in mouse embryo development. Open Biol 12(1):210335. https://doi.org/10.1098/rsob.210335
23. Morris SA, Guo Y, Zernicka-Goetz M (2012) Developmental plasticity is bound by

pluripotency and the Fgf and Wnt signaling pathways. Cell Rep 2(4):756–765. https://doi.org/10.1016/j.celrep.2012.08.029
24. Cheng S, Altmeppen G, So C, Welp LM, Penir S, Ruhwedel T, Menelaou K, Harasimov K, Stützer A, Blayney M, Elder K, Möbius W, Urlaub H, Schuh M (2022) Mammalian oocytes store mRNAs in a mitochondria-associated membraneless compartment. Science 378(6617):eabq4835. https://doi.org/10.1126/science.abq4835
25. So C, Menelaou K, Uraji J, Harasimov K, Steyer AM, Seres KB, Bucevičius J, Lukinavičius G, Möbius W, Sibold C, Tandler-Schneider A, Eckel H, Moltrecht R, Blayney M, Elder K, Schuh M (2022) Mechanism of spindle pole organization and instability in human oocytes. Science 375(6581):eabj3944. https://doi.org/10.1126/science.abj3944
26. Lim F, Peabody DS (2002) RNA recognition site of PP7 coat protein. Nucleic Acids Res 30(19):4138–4144. https://doi.org/10.1093/nar/gkf552
27. Wu B, Chao JA, Singer RH (2012) Fluorescence fluctuation spectroscopy enables quantitative imaging of single mRNAs in living cells. Biophys J 102(12):2936–2944. https://doi.org/10.1016/j.bpj.2012.05.017
28. Bertrand E, Chartrand P, Schaefer M, Shenoy SM, Singer RH, Long RM (1998) Localization of ASH1 mRNA particles in living yeast. Mol Cell 2(4):437–445. https://doi.org/10.1016/s1097-2765(00)80143-4
29. Peabody DS (1993) The RNA binding site of bacteriophage MS2 coat protein. EMBO J 12(2):595–600. https://doi.org/10.1002/j.1460-2075.1993.tb05691.x
30. Yan X, Hoek TA, Vale RD, Tanenbaum ME (2016) Dynamics of translation of single mRNA molecules in vivo. Cell 165(4):976–989. https://doi.org/10.1016/j.cell.2016.04.034
31. Li Z, Zhang P, Zhang R, Wang X, Tse YC, Zhang H (2021) A collection of toolkit strains reveals distinct localization and dynamics of membrane-associated transcripts in epithelia. Cell Rep 35(5):109072. https://doi.org/10.1016/j.celrep.2021.109072
32. Halstead JM, Lionnet T, Wilbertz JH, Wippich F, Ephrussi A, Singer RH, Chao JA (2015) Translation. An RNA biosensor for imaging the first round of translation from single cells to living animals. Science 347(6228):1367–1671. https://doi.org/10.1126/science.aaa3380
33. Westerich KJ, Chandrasekaran KS, Gross-Thebing T, Kueck N, Raz E, Rentmeister A (2020) Bioorthogonal mRNA labeling at the poly(A) tail for imaging localization and dynamics in live zebrafish embryos. Chem Sci 11(11):3089–3095. https://doi.org/10.1039/c9sc05981d

Chapter 13

Embryo Biopsy

Maurizio Poli, Christian Ottolini and Antonio Capalbo

Abstract

Trophectoderm (TE) biopsy is the preferred method for obtaining embryo specimens for preimplantation genetic testing (PGT). Containing approximately 5–10 cells, TE biopsies offer a stronger DNA amplification template, enhancing diagnostic accuracy and representing the whole embryo's chromosomal makeup better than single-cell biopsies. Performing TE biopsy demands advanced micromanipulation expertize. The procedure involves using laser technology alongside micromanipulation to separate the specimen from the blastocyst. Different methods can be employed based on the blastocyst's features and laboratory conditions. Biopsy tubing requires exceptional manual dexterity and coordination. Achieving an efficient PGT program relies on optimized blastocyst culture, biopsy, tubing techniques, and reliable cryopreservation processes.

Keywords IVF, Preimplantation genetic testing, Blastocyst biopsy, Micromanipulation, Trophectoderm biopsy

1 Introduction

Embryo biopsy was first applied to human embryos in the early 1990s, initially at the cleavage stage. Since its introduction, trophectoderm (TE) biopsy has progressively replaced blastomere biopsy in IVF laboratories worldwide, made possible by advancing culture systems that support embryos until day 5, 6, or 7 postfertilization. TE cell sampling ensures more reliable preimplantation genetic testing (PGT) and does not impair embryo development [1, 2]. TE biopsy offers significant advantages over blastomere biopsy. Primarily, fewer cells are removed (2–10 out of 100 versus 1 out of 8), minimizing potential harm to the embryo [1]. Furthermore, at the blastocyst stage, the inner cell mass (ICM) and TE are distinct, reducing the risk of disrupting the embryo proper. Procedurally, TE biopsy focuses on embryos that have undergone genome activation and reached advanced stages, avoiding unnecessary work. Analytically, multicellular TE biopsies provide a more reliable DNA template, lowering the risk of failed amplification or inconclusive results in aneuploidy

Alexandra J. Harvey and David K. Gardner (eds.), *The Preimplantation Embryo*, Methods in Molecular Biology, vol. 3038, https://doi.org/10.1007/978-1-0716-5292-3_13,

and monogenic testing. Additionally, blastocysts demonstrate better resilience to cryopreservation than cleavage-stage embryos, improving IVF/PGT efficiency [3, 4].

While preimplantation genetic testing for monogenic conditions (PGT-M) enjoys widespread scientific and clinical agreement, the cost-effectiveness of PGT for aneuploidies (PGT-A) remains debated. However, decreasing costs and high-throughput next-generation sequencing (NGS), which simultaneously assesses chromosomal and monogenic embryo features, could enhance PGT-A's clinical value. As a highly complex procedure, embryo biopsy and subsequent handling demand exceptional technical skill and precision. Establishing and maintaining a successful PGT program depends on three core factors: standardized IVF methods, refined cryopreservation protocols, and proficient biopsy techniques. Effective embryo biopsy training and ongoing performance monitoring are essential to achieving optimal outcomes in PGT programs.

2 Materials

Consumables and solutions employed in embryo biopsy and tubing procedures must be produced by certified manufacturers. They must be sterile and used according to the instructions for use provided by the manufacturer. Sterile technique must be adopted at all times to avoid specimen contamination.

2.1 Biopsy Dish Preparation

1. Laminar airflow cabinet with stereomicroscope.
2. Micromanipulation dish (e.g., 60 mm Petri, ICSI dish).
3. HEPES or MOPS Buffered medium supplemented with serum albumin (5–10 mg/mL).
4. PVP for biopsy pipette priming (optional, *see* **Note 1**).
5. Mineral/paraffin oil (warm, non-gassed for biopsy).
6. Thermostat incubator at 37°C in air for buffered media.
7. Single channel pipette 10–20 μL.
8. Pipette tips, 10–20 μL.
9. Serological pipette controller.
10. 5–10 mL serological pipettes.
11. 300 μm stripper pipette.

2.2 Micromanipulation

1. Inverted microscope equipped with micromanipulators and microinjectors.
2. Laser.
3. Holding and biopsy micropipettes 30 μm ID, flat 35° angle (inner diameter and angle may vary according to operator's/laboratory's preference).

4. Incubator for embryoculture.

2.3 Biopsy Washing and Tubing

Biopsy washing and tubing should be carried out in a laminar flow cabinet.

1. Laminar airflow cabinet with stereomicroscope.
2. Micromanipulation dish (e.g., 60 mm Petri, ICSI dish).
3. Single channel pipette 100–200 μL.
4. Serological pipette controller.
5. 100–200 μL tips.
6. PCR tubes (200 μL).
7. Stripper pipette.
8. Stripper for capillaries of 100–140 μm diameters.
9. Cooling rack.
10. Biopsy washing and loading buffer (generally provided by the genetic laboratory) (*see* **Note 2**).
11. Minicentrifuge for PCR tubes.

3 Methods

TE biopsy should be conducted exclusively by trained professionals, as their expertize is crucial for achieving consistent and reproducible results [5] (*see* **Note 3**).

3.1 Biopsy Dish Preparation

1. On the day prior to the biopsy procedure, aliquot sufficient HEPES/MOPS buffered medium supplemented with human serum albumin and mineral oil to be used for manipulation the day after into round-bottom tubes and equilibrate overnight in a thermostat/incubator at 37°C in air.
2. On the day of the biopsy procedure, wear personal protective equipment and work in the laminar airflow cabinet with a heated stage turned off and at room temperature (*see* **Note 4**).
3. Prepare as many dishes as there are embryos to process (*see* **Note 5**).
4. Label the dish (not the lid) with date of setup, type of media, patient identifier (e.g., name, DOB, and unique ID number), RFID tag or barcode, and embryo ID number (*see* **Note 6**).
5. Using a single channel pipette with 10–20 μL tip, aliquot three drops of 10 μL HEPES/MOPS buffered medium supplemented with human serum albumin and a 5 μL drop of PVP (optional, *see* **Note 1**) and overlay with sufficient prewarmed mineral oil to cover the drops (e.g., ~4 mL in a square ICSI dish, or ~5 mL in a 60 mm Petri dish) (Fig. 1A).

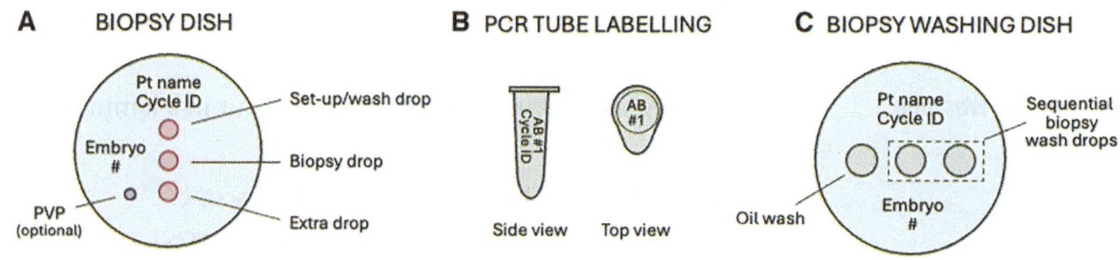

Fig. 1 Biopsy, washing dishes, and PCR tube setup. (**A**) An example of a biopsy dish setup displaying the patient's name, treatment ID number, and embryo unique number. HEPES-buffered media should be used for 10 μL drops. An optional 5 μL drop of PVP can be used to coat the inside of the biopsy pipette. (**B**) PCR tubes should be labeled both on the lid (patient's initials and embryo unique number) and the side (patient's initials, embryo unique number, and treatment ID number). (**C**) An example of biopsy wash dish where 50 μL of biopsy washing solution (generally provided by the genetic service) is plated on a dish labeled with patient's name, treatment ID number, and embryo unique number

6. Equilibrate the dish for at least 1 h in a thermostat incubator at 37°C in air.

3.2 Trophectoderm Biopsy

1. Equip the micromanipulator with holding and biopsy pipettes.
2. Assess and select the blastocyst(s) to be biopsied. Blastocysts should display a blastocoelic cavity and a recognizable ICM.
3. A second witness operator must ensure that the patient ID and embryo ID in the culture and biopsy dishes match [6] (*see* **Note 7**).
4. Collect the blastocyst from the culture dish using a 300 μm diameter stripper pipette, wash it in the first drop of the biopsy dish, then move it to the second drop.
5. Carefully, move the biopsy dish to the micromanipulator stage.
6. Assess and prime the pipettes in the first drop by aspirating and expelling medium (*see* **Note 1**).

3.2.1 Zona Pellucida Opening and TE Contact

1. Move the micromanipulators to the second drop of medium.
2. Orientate the ICM at 7–8 o'clock and firmly anchor the blastocyst with the holding pipette (*see* **Note 8**).
3. According to the specific laser settings, apply a few laser pulses to the zona pellucida around the 2 o'clock mark to obtain a hole of approximately 10 μm (*see* **Note 9**) (Fig. 2A). Avoid damaging TE cells (*see* **Note 10**).
4. If the operation involves non-herniating TE cells requiring intrazonal acquisition, position the biopsy pipette on the same focal plane as the zona opening, lean the biopsy pipette against the hole, and gently expel medium to detach the TE cells from the internal surface of the zona pellucida [7] (Fig. 2B, C).
5. Should the TE cells remain attached to the zona pellucida, retract the biopsy pipette from the embryo and gently push

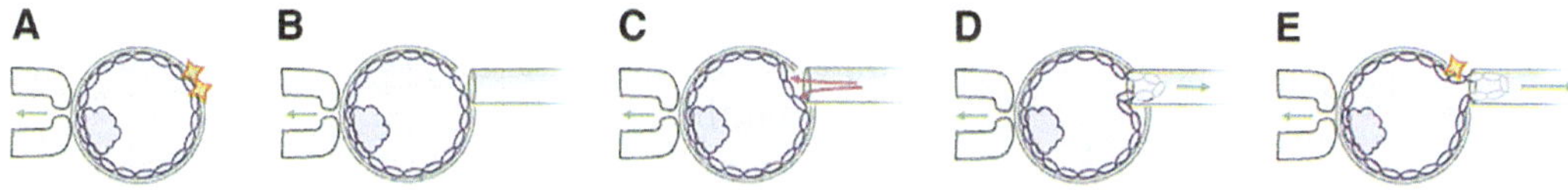

Fig. 2 Typical sequence of a zona pellucida opening procedure. (**A**) Anchor the blastocyst to the holding pipette and open the zona pellucida, firing a few laser pulses at the 2 o'clock position. (**B**) Contact the embryo with the biopsy pipette and ensure an open passage to the TE cells. (**C**) Expel some media onto the TE cells to favor their detachment from the zona pellucida. (**D**) Aspirate 5–10 cells with the biopsy pipette from within the zona pellucida. (**E**) Retrieve the biopsy pipette from within the zona pellucida and expose the TE cells. Fire a short laser pulse on a cell junction at 12 o'clock of the herniating portion to reduce pressure and facilitate TE cells' engagement into the biopsy pipette

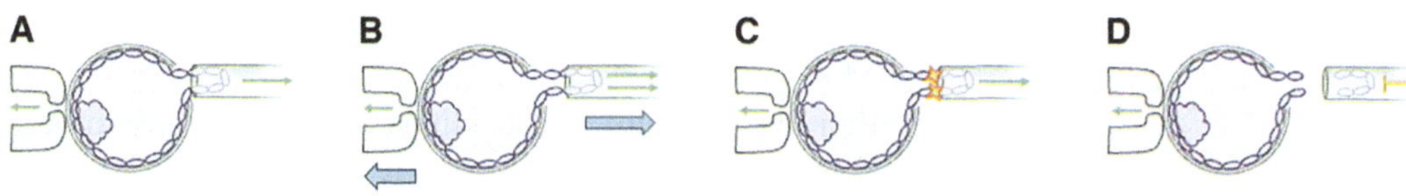

Fig. 3 Typical sequence of a TE biopsy procedure using the pulling method. (**A**) Maintain a firm hold on the embryo with both pipettes. (**B**) Gently stretch the extremities of the embryo by widening the holding and pulling pipette. (**C**) Fire a few laser pulses on the gap junctions of the exposed TE cells. (**D**) Promptly counterbalance suction in the biopsy pipette as soon as the biopsy fragment is released from the main embryo

from the outer side of the zona pellucida, then return to Step 4.

6. Once sufficient space has been created by expelling medium against the TE cells, approach the TE with the biopsy pipette and slowly aspirate 5–10 cells to expose them outside of the zona pellucida (Fig. 2D) (*see* **Note 11**).
7. A short laser pulse on a cell junction at 12 o'clock of the herniating portion may be applied to reduce pressure and facilitate TE cells' engagement into the biopsy pipette (Fig. 2E).

3.2.2 Biopsy Procedure

a) **Pulling Method**

The pulling method is particularly useful when the embryo is not fully hatched. It can be employed on herniating and non-herniating blastocysts.

1. Apply tension by maintaining negative pressure on both pipettes and widening the area between holding and biopsy pipettes, stretching the embryo, and exposing the junctions between TE cells (Fig. 3A, B) (*see* **Note 12**).
2. Apply laser pulses (4–7) on the thinner areas of the stretched TE, aiming at cell junctions (Fig. 3C).
3. Promptly counterbalance the aspiration pressure, as soon as the biopsy detaches from the blastocyst (Fig. 3D) by

turning clockwise the injector device connected to the biopsy pipette. Release the biopsied fragment into the drop away from the blastocyst.

4. In the case the biopsied fragment cannot be identified in the drop, expel all the remaining medium in the third drop to verify the sample has not been retained inside the biopsy pipette.

b) **Flicking Method (see Note 13)**

1. Maintaining the biopsy pipette in place, release the blastocyst from the holding pipette (Fig. 4A).
2. Position the lower edge of the holding pipette against the upper edge of the biopsy pipette, maintaining both pipettes in focus (Fig. 4B).
3. With a swift movement (flicking), bring the biopsy pipette up. Using the friction between the pipettes, excise the portion of TE cells from the main embryo (Fig. 4C, D).
4. In case the first movement does not detach the target cells from the embryo, invert the starting position of the pipettes by placing the holding pipette against the lower edge of the biopsy pipette and repeat.
5. Release the biopsied fragment away from the blastocyst and verify sample is not retained in the biopsy pipette.

3.3 Removal of Blastocyst from the Biopsy Dish

1. After completion of the biopsy procedure, the embryo can be transferred back to culture or to media in preparation for vitrification (*see* **Note 14**).
2. In either case, the transfer must be witnessed by a second operator to ensure embryo traceability (Patient ID and Embryo ID at a minimum) (*see* **Note 7**).

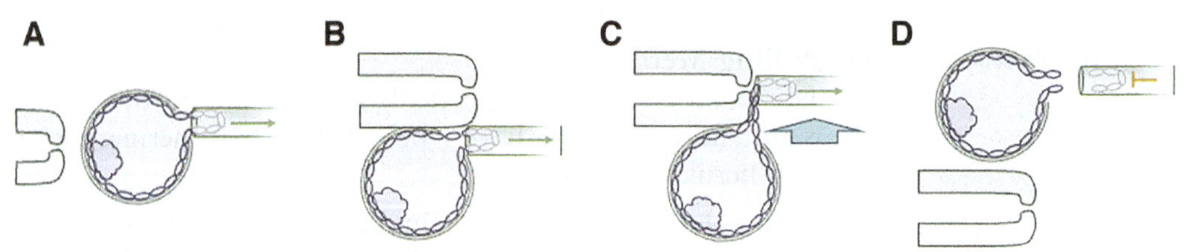

Fig. 4 Typical sequence of a TE biopsy procedure using the flicking method. (**A**) Maintain a firm hold on the TE cells with the biopsy pipette while releasing the embryo from the holding pipette. (**B**) Position the lower edge of the holding pipette against the upper edge of the biopsy pipette. (**C**) With a swift movement (flicking), bring the biopsy pipette up. Using the friction between the pipettes, excise the portion of TE cells from the main embryo. (**D**) Promptly counterbalance suction in the biopsy pipette as soon as the biopsy fragment is released from the main embryo

3.4 Biopsy Washing and Collection (Tubing)

Handling of the fragment must be performed while wearing personal protective equipment (*see* **Note 4**).

3.4.1 Preparation for Washing and Tubing Procedures

1. Working in the laminar airflow cabinet with the heated stage turned off and at room temperature (*see* **Note 15**), label a micromanipulation dish (e.g., 60 mm Petri, ICSI dish) with patient ID and embryo ID.
2. Label PCR tubes with the patient's initials and embryo ID on both the side and the lid of the PCR tube. Cycle ID should also be written on the side of the tube (Fig. 1B).
3. Prepare each biopsy wash dish immediately before use by aliquoting three drops of 50 µL of washing buffer (Fig. 1C).
4. Aliquot 1 µL of washing buffer into the prelabelled PCR tube (optional) (*see* **Note 16**).
5. Preset the stripper pipette to a volume of 1 µL (or 2 µL if the PCR tube is not primed with 1 µL of washing buffer). After tubing of the biopsy specimen, the final volume in the PCR tube must be no more than 2 µL (*see* **Notes 16 and 17**).

3.4.2 Biopsy Washing and Tubing

1. Prime a sterile stripper pipette capillary (100–135 µm diameter) with media from the first drop of the biopsy washing dish.
2. Expel washing buffer onto the TE biopsy specimen and collect it from the biopsy dish.
3. Slowly swipe the tip of the stripper capillary through the first biopsy washing drop multiple times without releasing the specimen. This action will rid of the oil on the surface of the capillary (Fig. 5A).
4. Transfer the specimen to the second drop in the biopsy washing dish and expel the remainder of the solution outside the drop.
5. Gently wash the biopsy specimen by repeating the following sequence several times: release the specimen into the washing drop, then expel any remaining fluid outside the drop to avoid contamination. Next, draw up fresh medium from a clean area of the drop, away from the specimen, and gently expel a small amount over the specimen to rinse it. Aspirate the specimen again and move it to a different area within the same drop. Repeat this process as needed to ensure thorough washing while minimizing mechanical stress (Fig. 5B).
6. Transfer the biopsy specimen to the third drop and gently wash in the same manner as performed for the second drop (Fig. 5C).

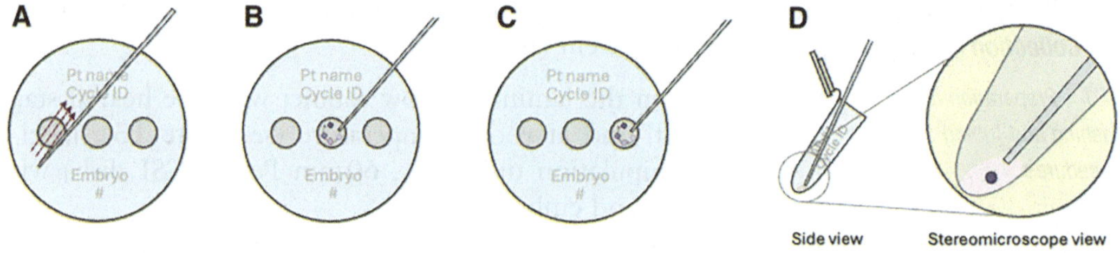

Fig. 5 Typical sequence of a biopsy washing and tubing procedure. (**A**) After collecting the biopsy fragment, gently swipe the capillary pipette across the first wash drop multiple times to rid it of the oil on its surface. (**B**) Release the biopsy fragment in the second wash drop and expel all the remaining fluid outside the drop. Rinse the fragment multiple times, moving it around the drop. (**C**) Repeat the sequence in the third drop. (**D**) Load the biopsy fragment in the PCR tube, visualizing it through the stereomicroscope. Release the whole volume contained in the capillary to reach the target loading volume

7. Collect the biopsy specimen with the capillary, releasing the plunger of the stripping pipette to reach the target loading volume.
8. Take hold of the PCR tube or secure it to a holding device and open the lid without touching the inner part to avoid contamination (*see* **Note 4**).
9. Place the PCR tube under the stereomicroscope and focus on the bottom of the tube.
10. Looking through the stereomicroscope, transfer the biopsy specimen into the PCR tube and visualize its release inside the container. Empty the rest of the pipette to reach the final target volume (Fig. 5D).
11. Immediately after releasing the cells in the tube, wash the pipette in the last drop of the biopsy washing dish to ensure that the cells are not retained inside the pipette.
12. In the case the biopsy specimen was not expelled in the PCR tube, prepare and label a new PCR tube and discard the old one.
13. Place the PCR tube containing the specimen in the cooling rack to minimize DNA degradation.
14. After completion of the procedure, keep the sample at −20°C until shipment (*see* **Note 17**).

3.5 Post-Biopsy Operations

Diagnostic reports from genetic analysis are typically provided through secure methods, such as encrypted emails, online portals, or certified links. Once received, the report should be securely stored, and its summary (e.g., embryo diagnosis: euploid, aneuploid, affected, carrier, or normal) transcribed onto the document tracking embryo development. This transcription process must be double-witnessed and signed by qualified personnel (*see* **Notes 7 and 18**).

4 Notes

1. According to the operator's preference, the biopsy pipette can be primed with PVP. PVP can reduce action within the biopsy pipette and create a non-adhesive surface, helping to prevent the specimen from sticking to the pipette walls.
2. Washing buffer is generally made of PBS supplemented with a compound that reduces stickiness of the cells (e.g., HSA, BSA, and PVA). These compounds are generally at a concentration between 1 and 5 mg/mL. Washing and loading buffers can have the same composition (*see* **Note 16**). Alternatively, loading buffers may have a composition tailored to improve downstream molecular processes (*see* **Note 17**).
3. IVF centers should maintain a team of at least two qualified operators to perform biopsy and tubing procedures, ensuring uninterrupted service. Operator performance should be regularly evaluated through the analysis of key indicators to detect any decline in standards.
   Three primary areas should be assessed:
   - Technical: The proportion of conclusive diagnoses (aneuploid/euploid) versus inconclusive results (e.g., amplification failures or non-concurrent outcomes).
   - Biological: The rates of embryo degeneration post-biopsy and embryo survival following warming.
   - Clinical: The pregnancy rate and live birth rate per transfer.
4. Protective personal equipment must be worn, including sterile gloves, a face mask, and a surgical cap.
5. Depending on the operator's competence and laboratory SOP, more embryos can be processed in the same biopsy dish—although each culture media drop is single use.
6. Traceability can be ensured by labeling the bottom part of the dish, as the lid can be removed and misplaced during biopsy operations.
7. Active double witnessing is essential to confirm that the patient ID matches the embryo's unique identifier across all devices and dishes during key stages: (i) transferring the embryo to the biopsy dish, (ii) tubing the biopsy sample, (iii) initiating blastocyst vitrification, (iv) recording PGT analysis results in the patient's worksheet, and (v) beginning the warming procedure.
8. If the blastocyst is already hatching, then there is no need to open the zona pellucida, and biopsy can start from step 4. However, if the portion hatching involves or is in proximity

to the ICM, then an additional opening in the zona pellucida is required, preferably on the opposite side of the blastocyst.

9. Laser power settings are typically fixed and differ by manufacturer. The hole size on the zona pellucida can be modified by adjusting the pulse length, which is also influenced by the microscope setup and laboratory conditions, such as dish types. Calibrate the pulse length to produce a hole approximately 10 μm in size.
10. When opening the zona pellucida, apply laser shots aiming from the outer area, moving progressively toward the embryo, avoiding directly lasering trophectoderm cells.
11. Aspiration of intrazonal TE cells can be attempted when the biopsy pipette can make full contact with detached cells that can provide a robust anchor point so that the portion to be excised can be exposed outside of the zona pellucida.
12. The distance between the pipette varies according to the characteristics of the embryo. Note that if too much stretching is applied, the embryo may get damaged by irreparably breaking its architecture or, in the case of suction loss. If too little stretching is applied, intercellular junctions may not be sufficiently exposed, causing subsequent laser pulses to damage TE cells.
13. The flicking method is particularly useful for fully hatched blastocysts, where the absence of the zona pellucida necessitates precise control of aspiration pressure on the holding pipette to avoid cellular damage. Limited suction makes it challenging to stretch TE cells for laser ablation. In such cases, the flicking technique can be applied alone or alongside laser use.
14. If the embryo is placed back in culture, vitrification should be carried out before its complete re-expansion.
15. Depending on the brand and model, heated stages may require up to several hours to cool down fully. To prevent potential loss of calibration or induction of thermal stress, switching the heated surface on and off frequently should be avoided. For steps that do not require a controlled temperature, such as washing and collecting the biopsy specimen, it is advisable to use a separate dissection microscope with a room temperature surface. This approach helps preserve the integrity of the heated stage and ensures proper handling conditions. All procedures and temperature settings should be locally validated and documented according to laboratory protocols.
16. Preloading the PCR tube with loading buffer can help minimize contamination carryover from the biopsy washing drop.

Since downstream cell lysis and DNA amplification reactions have fixed volumes, it is essential to precisely control the starting volume containing the biopsy specimen. Most protocols require a total input volume of 2 μL, which can be achieved either by transferring the entire 2 μL from the final biopsy washing drop or by combining 1 μL of preloaded loading buffer with 1 μL of the biopsy suspension. Accurate volume management at this step is critical to ensure reproducibility and integrity of the genetic analysis.

17. When partnering with an external PGT service laboratory, the IVF center and genetic laboratory must align their protocols, including those for labeling, tubing, and shipping, and validate the process. For example, tubing protocol and required loading volumes may slightly vary across PGT services and should be approved prior to clinical applications. A risk analysis plan, including preventive and corrective actions, should involve the IVF center, genetic center, and transport company. Specialized refrigerated kits from the PGT provider are used to prevent sample degradation during transit, with each shipment accompanied by supporting documentation.
18. If the diagnostic result is inconclusive (e.g., due to failed amplification), the cryopreserved blastocyst can be warmed and subjected to a repeat TE biopsy. Recent studies indicate that re-biopsied and re-vitrified/warmed euploid blastocysts achieve lower clinical outcomes comparable to those of euploid blastocysts biopsied only once. Nonetheless, the risks of re-biopsy need to be evaluated on a per-case basis and weighted against the risks associated with the lack of a genetic diagnosis [8].

References

1. Scott Jr RT, Upham KM, Forman EJ, Zhao T, Treff NR (2013) Cleavage-stage biopsy significantly impairs human embryonic implantation potential while blastocyst biopsy does not: a randomized and paired clinical trial. Fertil Steril 100(3):624–630. https://doi.org/10.1016/j.fertnstert.2013.04.039
2. Tiegs AW, Tao X, Zhan Y, Whitehead C, Kim J, Hanson B, Osman E, Kim TJ, Patounakis G, Gutmann J, Castelbaum A, Seli E, Jalas C, Scott Jr RT (2021) A multicenter, prospective, blinded, nonselection study evaluating the predictive value of an aneuploid diagnosis using a targeted next-generation sequencing-based preimplantation genetic testing for aneuploidy assay and impact of biopsy. Fertil Steril 115(3):627–637. https://doi.org/10.1016/j.fertnstert.2020.07.052
3. Kuwayama M (2007) Highly efficient vitrification for cryopreservation of human oocytes and embryos: the Cryotop method. Theriogenology 67(1):73–80. https://doi.org/10.1016/j.theriogenology.2006.09.014
4. Zaca C, Bazzocchi A, Pennetta F, Bonu MA, Coticchio G, Borini A (2018) Cumulative live birth rate in freeze-all cycles is comparable to that of a conventional embryo transfer policy at the cleavage stage but superior at the blastocyst stage. Fertil Steril 110(4):703–709. https://doi.org/10.1016/j.fertnstert.2018.05.012
5. Capalbo A, Ubaldi FM, Cimadomo D, Maggiulli R, Patassini C, Dusi L, Sanges F, Buffo L, Venturella R, Rienzi L (2016) Consistent and reproducible outcomes of blastocyst biopsy and aneuploidy screening across different biopsy practitioners: a multicentre study

involving 2586 embryo biopsies. Hum Reprod 31(1):199–208. https://doi.org/10.1093/humrep/dev294
6. Cimadomo D, Ubaldi FM, Capalbo A, Maggiulli R, Scarica C, Romano S, Poggiana C, Zuccarello D, Giancani A, Vaiarelli A, Rienzi L (2016) Failure mode and effects analysis of witnessing protocols for ensuring traceability during PGD/PGS cycles. Reprod Biomed Online 33(3):360–369. https://doi.org/10.1016/j.rbmo.2016.06.002
7. Capalbo A, Rienzi L, Cimadomo D, Maggiulli R, Elliott T, Wright G, Nagy ZP, Ubaldi FM (2014) Correlation between standard blastocyst morphology, euploidy and implantation: an observational study in two centers involving 956 screened blastocysts. Hum Reprod 29(6):1173–1181. https://doi.org/10.1093/humrep/deu033
8. Li Piani L, Petrone P, Brutto M, De Vos A, Van Der Kelen A, Vaiarelli A, Rienzi L, Conforti A, Cimadomo D, Verpoest W (2025) A systematic review and meta-analysis of double trophectoderm biopsy and/or cryopreservation in PGT: balancing the need for a diagnosis against the risk of harm. Hum Reprod Update 31(2):102–115. https://doi.org/10.1093/humupd/dmae031

Chapter 14

Non-Invasive Preimplantation Genetic Testing (niPGT)

Luis Navarro-Sánchez and Carmen Rubio Lluesa

Abstract

Selection of a normal euploid embryo for transfer is a crucial step in assisted reproduction treatment. While trophectoderm (TE) biopsy has been widely employed, the consequences of removing cells on embryo viability remain unknown. In contrast, non-invasive preimplantation genetic testing (niPGT) can determine an embryo's chromosomal content without the need to perform a biopsy. This strategy obtains valuable information by analyzing the genetic material released by the blastocyst into the culture medium, with high concordance with TE biopsies, inner cell mass biopsies, and the whole embryo. To achieve this, the culture medium sample must contain enough good-quality embryonic material, which can be influenced by sample collection timing. Further, minimizing the risk of contamination remains crucial, as contamination can result in misdiagnosis. For these reasons, performing niPGT in an IVF laboratory requires slight modifications to the standard embryo culture protocol. Here, we describe the protocol to maximise the probability of obtaining an informative and accurate niPGT result.

Keywords Culture medium, cfDNA, Blastocyst, Preimplantation, Fresh/frozen, Contamination, Validation, Prioritization, niPGT

1 Introduction

Selecting the best embryo for transfer remains crucial in assisted reproduction treatment. Embryos with aneuploidies and/or carrying pathogenic mutations can reach the blastocyst stage with optimal morphological scores and, thus, undergo transfer to the maternal uterus. Depending on the type of abnormality, particular embryos will not implant, others will implant but end in miscarriage, and some may result in an affected newborn [1, 2]; thus, normal euploid embryos must be selected. Various molecular techniques have been applied for preimplantation genetic testing (PGT) to determine the embryo's genetic content. The most commonly employed approach involves analyzing a trophectoderm (TE) biopsy from the blastocyst; however, this invasive procedure entails technical, economic, and even ethical/legal challenges. In addition, the possibility of harming the embryo remains a constant concern for doctors and patients, even

Alexandra J. Harvey and David K. Gardner (eds.), *The Preimplantation Embryo*, Methods in Molecular Biology, vol. 3038, https://doi.org/10.1007/978-1-0716-5292-3_14,

given the minimal amount of risk when in the hands of experienced biopsy operators.

Non-invasive approaches have been developed in recent years to overcome said limitations. Most non-invasive (ni)PGT efforts have focused on aneuploidy detection, with less research conducted on the non-invasive detection of monogenic diseases and carriers of translocations. Non-invasive approaches prioritize the transfer of blastocysts with a higher likelihood of resulting in a healthy baby without discarding any of the cohort.

During embryo development, distinct molecules (e.g., metabolites, RNA, or DNA) are released into the surrounding culture medium. The most successful results for niPGT have been obtained when analyzing cell-free DNA (cfDNA) released by the blastocyst. Studies from various research groups have demonstrated high cfDNA informativity (range between 55.6% and 100%) and high concordance when comparing, for the same embryo, cfDNA results to TE biopsies (range between 33.3% and 89.1%), inner cell mass biopsies, or even the whole blastocyst (range between 45.5% and 93.8%) [3].

Although the equipment and materials used do not influence the results obtained from the cfDNA present in the culture medium [4], the manner of embryo handling remains fundamental, and a specific protocol must be followed to obtain robust results. This chapter therefore outlines the collection of culture media samples for niPGT analysis, noting specific modifications to standard embryo culture practices to collect sufficient cfDNA for analysis.

2 Materials

2.1 Equipment

1. Laminar flow hood (preferably with an ultraviolet [UV] lamp).
2. Stereomicroscope.
3. Automatic pipette: volume range 1–10 μl or 2–20 μl, depending on the type of incubator used for embryo culture.
4. Handling device for denudation and embryo handling.
5. Incubator for embryo culture (either CO_2 or multi-gas).

2.2 Consumables

1. RNA/DNA-free sterile tips with filters.
2. Flexible polycarbonate/glass capillaries for denudation, embryo handling, and washes (diameters between 140 and 290 μm for distinct steps of the protocol).
3. 60 mm petri dishes or microdroplet culture dish.
4. Embryo culture dishes are suitable for use in the desired embryo incubator system.
5. 0.2 mL PCR tubes.

6. Cold-rack stored at −20°C and kept inside the freezer for at least 12 h before media collection.

2.3 Reagents

1. IVF culture medium.
2. Oil for embryo culture.
3. Surface decontaminant.

3 Methods

Contamination control remains an essential consideration before beginning any step of the protocol. Always use gloves when handling any materials/samples employed in this protocol. Embryologists performing this protocol must wear a disposable laboratory gown with long sleeves, a cap, and a mask. The laminar flow should be on during the entire protocol.

Clean the surface of the cabinet and the external surface of each piece of equipment and all materials implicated in the protocol (e.g., stereomicroscope, tip boxes, automatic pipettes, capillary boxes, and handling device) (*see* **Note 1**).

If available, switch on the UV lamp in the laminar flow hood for 15 min before handling materials and samples; furthermore, all materials except embryos and reagents must be exposed to UV light before starting any niPGT procedure.

3.1 Protocol for Fresh Cycles

Here, we describe the protocol to follow in the IVF laboratory when performing niPGT during a fresh cycle—i.e., from egg retrieval until day 6 (or 7) of embryo development without previous vitrification at the blastocyst stage (summarized in Fig. 1; [4, 5]). This protocol can also be applied to cycles with previously vitrified oocytes. niPGT supports both intracytoplasmic sperm injection (ICSI) and IVF cycles.

Controlled ovarian stimulation, induction of ovulation, egg retrieval, and denudation can be performed using the standard operating procedures of each laboratory.

3.1.1 Postdenudation Washing and Embryo Culture (Day 0/1–Day 4)

Following the completion of denudation for ICSI, or following IVF, extra washes must be performed in a washing dish to ensure that no carryover of maternal cellular material occurs (*see* **Note 2**).

1. Using the same medium as used for the mechanical denudation procedure, prepare a washing dish with a maximum of three rows of four droplets of 30 μL fresh culture medium covered with oil. Pre-equilibrate dishes at 37°C overnight, or for a minimum of 4 h.
2. Wash up to three oocytes individually through one row of four droplets using the same capillary (suggested diameter:

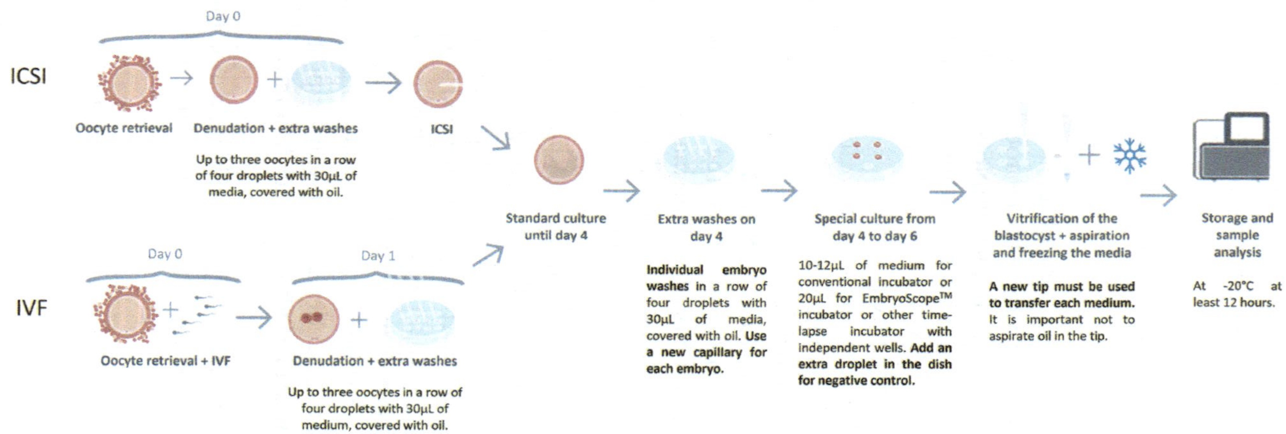

Fig. 1 Schematic diagram summarising the niPGT protocol for fresh cycles. Figure presents the stepwise protocol to perform niPGT in ICSI and IVF fresh cycles, encompassing the day-by-day embryo culture steps (day 0 to day 6) through to the analysis of samples

140–150 μm); the capillary can be used for the next set of oocytes if it remains clean after washing the first three oocytes.

3. Culture each oocyte according to the standard protocol used by the laboratory until day 4 of embryo development after the postdenudation extra washes.

3.1.2 Embryo Washes for niPGT Culture (Day 4)

Additional extra washes should be performed in a washing dish on day 4.

1. Prepare washing dishes with a maximum of three rows of four droplets of 30 μL fresh culture medium and covered with oil (*see* **Note 3**). Pre-equilibrate dishes at 37°C overnight, or for a minimum of 4 h.
2. Wash each embryo individually through one row of four droplets using a new capillary per embryo (suggested diameter: 175–190 μm); use only one row to wash one embryo (*see* **Note 4**).

3.1.3 Embryo Culture for niPGT (Day 4—Day 6/7)

1. Prepare an embryo culture dish according to the type of culture device used in the laboratory (*see* **Note 5**). One additional droplet/well should be added for each patient as a negative control; the negative control should be exposed to the same conditions as the clinical samples but without previous contact with an embryo. Cover the drops with oil then equilibrate the dish at 37°C for at least 4 h—overnight.
2. Transfer each embryo to an individual droplet/well of culture medium and culture to the blastocyst stage (day 6 or 7).

3.1.4 Media Collection (Day 6/7)

1. Remove blastocysts from each culture drop/well for vitrification using a new capillary for each embryo (suggested diameter ~290 μm) on day 6 (or 7) (*see* **Note 6**).
2. Use an automatic pipette adjusted to the drop/well volume to be collected for media aspiration (*see* **Note 7**). A new tip box should be opened, and a new filter tip must be used to aspirate media from each individual culture drop/well. Place a sterile filter tip vertically into the drop/well until it touches the bottom of the culture dish/well. Aspirate the culture medium being careful not to aspirate oil; wipe the tip on the edge of the culture dish to remove any excess oil around the tip before the next step.
3. Transfer each medium drop/well to an individual, empty, and correctly labelled PCR tube (*see* **Note 8**) by placing the tip with the media sample into the bottom of the PCR tube before gently dispensing the medium. Avoid touching the walls of the PCR tube and check that the entire volume is present at the bottom of the PCR tube.

4. Place the PCR tubes containing the aspirated medium in the frozen cold-rack and store them at −20°C for at least 12 h before in-house analysis or shipment to the reference genetics laboratory (*see* **Note 9**).

3.2 Protocol for Frozen Cycles

Here, we describe the protocol to follow by the IVF laboratory when performing niPGT during a frozen cycle (i.e., with previously vitrified day 5 or day 6 blastocysts; summarized in Fig. 2; [6]).

3.2.1 Embryo Washing and Culture (Day 5/6)

1. Thaw vitrified day 5/6 blastocyst(s) individually, as per the standard clinic protocol. Check blastocyst viability after thawing. niPGT can only be performed with viable blastocysts.
2. Check for the presence of maternal cumulus cells surrounding the blastocysts and, if necessary, remove any present by gentle pipetting using a capillary.
3. Extra washes should be performed in a washing dish. Prepare a washing dish with a maximum of three rows of four droplets of 30 μL fresh culture medium covered with an oil overlay. Equilibrate at 37°C overnight, or for a minimum of 4 h (*see* **Note 3**).
4. Wash each embryo individually through the four droplets of one of the rows using a new capillary per embryo (suggested diameter: 190–290 μm). Only wash one embryo per row (*see* **Note 4**).
5. Prepare an embryo culture dish according to the type of culture device used in the laboratory and pre-equilibrate at 37°C (*see* **Note 5**). One additional droplet/well should be added for each patient as a negative control; the negative control should be exposed to the same conditions as the clinical samples but without previous contact with an embryo.
6. Place each embryo in an individual drop/well of culture medium (*see* **Note 5**). The time in culture will depend on the day of vitrification. For day-5 vitrified blastocysts, thawed blastocysts should be cultured overnight (~24 h). For day-6 vitrified blastocysts, thawed blastocysts should be cultured for a minimum of 8 h.

3.2.2 Media Collection (Day 6)

1. Remove blastocysts from each culture drop/well for re-vitrification using a new capillary per blastocyst (suggested diameter ~290 μm).
2. Using an automatic pipette adjusted for the droplet/well volume to be collected for media aspiration (*see* **Note 7**), fitted with a new filter tip for each individual culture drop/well, place the sterile filter tip into the drop/well vertically until it touches the bottom of the culture dish/well. Aspirate the medium, avoiding aspirating any oil; wipe the tip on the edge of the culture dish to remove any excess oil around the tip before the next step.

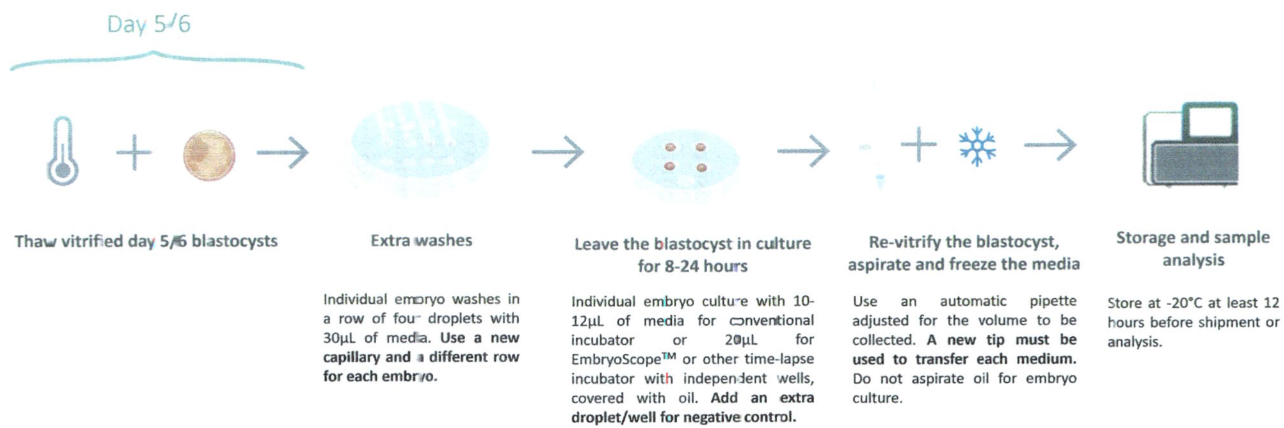

Fig. 2 Schematic diagram summarising the niPGT protocol for frozen cycles. Figure presents the stepwise protocol used for niPGT in frozen cycles, from blastocyst vitrification to sample analysis

3. Transfer media from each embryo and the negative control to individual, empty, and correctly identified PCR tubes (*see* **Note 8**) by placing the tip with the media sample into the bottom of the PCR tube before gently dispensing the medium. Avoid touching the walls of the PCR tube and check that the entire volume is present at the bottom of the PCR tube.
4. Place the PCR tubes containing the aspirated media in the frozen cold rack and store them at −20°C for at least 12 h before in-house analysis or shipment to the reference genetics laboratory (*see* **Note 9**).

3.3 Genetic Analysis

The analysis of media samples can be performed with different amplification methods and sequencing platforms depending on the information of interest. Due to the nature of cfDNA, already available PGT methods must be adapted for niPGT. The algorithms that interpret the results are of particular relevance; the sensitivity and specificity of the test might be deeply affected by the absence of specific actions to adapt previously available methods to niPGT, affecting the accuracy of the results [7, 8]. Mosaicism threshold is one critical aspect, for example, to be analyzed in order to develop a successful niPGT strategy.

The purpose of non-invasive approaches is also different: they are used as a prioritization system to choose the best embryo for transfer without discarding any blastocyst in the cohort. This approach requires a different reporting structure when sharing information with patients. This relevant aspect must be considered when clinically implementing these approaches.

3.4 Validation

Before performing the niPGT with clinical samples, the IVF laboratory must perform validation. The validation process is necessary to check the successful implementation of the modified niPGT culture conditions. A group of embryos must be cultured following the protocol outlined above, and then, for each blastocyst, the medium and other embryonic samples, such as a TE biopsy or the whole blastocyst (donated for research), must be collected, analyzed, and the results compared. The evaluation of parameters such as media informativity, concordance between the pair of embryonic samples, and presence of contamination (external/maternal) determines validation success.

4 Notes

1. We recommend dedicating an isolated area exclusively to the niPGT protocol; if possible, the materials should only be used for niPGT cases, especially for culture dishes, tips, media aliquots, and PCR tubes.

2. Removing all cumulus cells (or as many as possible) without compromising the sample's viability represents a crucial step, as the presence of maternal DNA can interfere with the results and result in misdiagnosis.
3. We recommend opening a fresh media aliquot to prepare the washing and culture droplets to reduce the potential for contamination.
4. Before transferring the embryo from one wash droplet to the next, a small volume from the new clean droplet should be aspirated with a capillary. This reduces the carryover of external/maternal contaminating DNA and cell debris from the old to the new drop.
5. The volume of the droplet/well depends on the type of incubator used: 10–12 μL if working with conventional incubators, 20 μL for EmbryoScope™ or other time-lapse incubators with independent culture wells. The protocol cannot be performed when sharing media between different droplets/wells. If the droplets merge or if the wells are connected, the medium, and thus, the released cfDNA, is shared between the different embryos, and, therefore, the results cannot be linked to a specific blastocyst.
6. Although working with day 5 blastocysts represents common practice in IVF laboratories, reaching day 6 (or 7) of embryo development is mandatory when performing niPGT for fresh cycles. Neither the amount nor the quality of cfDNA present in the culture medium droplet will suffice to perform the test if collected on day 5 of culture [9, 10].
7. Preset the pipette with a lower volume than the total volume of the droplet to avoid oil aspiration (e.g., 8–9 μl for a 10 μl droplet [conventional incubator], or 15–18 μl for a 20 μl well [EmbryoScope™ incubator, or other time-lapse incubators with independent wells]).
8. Label the PCR tubes prior to media collection (e.g., write patient identification information, followed by the number of the embryo, or a C, for the negative control).
9. Store the media in the freezer at −20°C for at least 12 h. Keeping the samples at different temperature (liquid nitrogen, −80°C, fridge, or at room temperature) or shorter time will affect the stability of the DNA and, therefore, increase the rate of non-informative results.

References

1. Rubio C, Pehlivan T, Rodrigo L, Simon C, Remohi J, Pellicer A (2005) Embryo aneuploidy screening for unexplained recurrent miscarriage: a minireview. Am J Reprod Immunol 53(4):159–165. https://doi.org/10.1111/j.1600-0897.2005.00260.x

2. Campos-Galindo I, Garcia-Herrero S, Martinez-Conejero JA, Ferro J, Simon C, Rubio C (2015) Molecular analysis of products of conception obtained by hysteroembryoscopy from infertile couples. J Assist Reprod Genet 32(5):839–848. https://doi.org/10.1007/s10815-015-0460-z
3. Navarro-Sanchez L, Garcia-Pascual C, Rubio C, Simon C (2022) Non-invasive preimplantation genetic testing for aneuploidies: an update. Reprod Biomed Online 44(5):817–828. https://doi.org/10.1016/j.rbmo.2022.01.012
4. Rubio C, Navarro-Sanchez L, Garcia-Pascual CM, Ocali O, Cimadomo D, Venier W, Barroso G, Kopcow L, Bahceci M, Kulmann MIR, Lopez L, De la Fuente E, Navarro R, Valbuena D, Sakkas D, Rienzi L, Simon C (2020) Multicenter prospective study of concordance between embryonic cell-free DNA and trophectoderm biopsies from 1301 human blastocysts. Am J Obstet Gynecol 223(5):751.e1–751.e13. https://doi.org/10.1016/j.ajog.2020.04.035
5. Sakkas D, Navarro-Sanchez L, Ardestani G, Barroso G, Bisioli C, Boynukalin K, Cimadomo D, Frantz N, Kopcow L, Andrade GM, Ozturk B, Rienzi L, Weiser A, Valbuena D, Simon C, Rubio C (2024) The impact of implementing a non-invasive preimplantation genetic testing for aneuploidies (niPGT-A) embryo culture protocol on embryo viability and clinical outcomes. Hum Reprod 39(9):1952–1959. https://doi.org/10.1093/humrep/deae156
6. Ardestani G, Banti M, Garcia-Pascual CM, Navarro-Sanchez L, Van Zyl E, Castellon JA, Simon C, Sakkas D, Rubio C (2024) Culture time to optimize embryo cell-free DNA analysis for frozen-thawed blastocysts undergoing non-invasive preimplantation genetic testing for aneuploidy. Fertil Steril 122(3):465–473. https://doi.org/10.1016/j.fertnstert.2024.04.037
7. Huang L, Bogale B, Tang Y, Lu S, Xie XS, Racowsky C (2019) Non-invasive preimplantation genetic testing for aneuploidy in spent medium may be more reliable than trophectoderm biopsy. Proc Natl Acad Sci U S A 116(28):14105–14112. https://doi.org/10.1073/pnas.1907472116
8. Li X, Hao Y, Chen D, Ji D, Zhu W, Zhu X, Wei Z, Cao Y, Zhang Z, Zhou P (2021) Non-invasive preimplantation genetic testing for putative mosaic blastocysts: a pilot study. Hum Reprod 36(7):2020–2034. https://doi.org/10.1093/humrep/deab080
9. Rubio C, Rienzi L, Navarro-Sanchez L, Cimadomo D, Garcia-Pascual CM, Albricci L, Soscia D, Valbuena D, Capalbo A, Ubaldi F, Simon C (2019) Embryonic cell-free DNA versus trophectoderm biopsy for aneuploidy testing: concordance rate and clinical implications. Fertil Steril 112(3):510–519. https://doi.org/10.1016/j.fertnstert.2019.04.038
10. Yeung QSY, Zhang YX, Chung JPW, Lui WT, Kwok YKY, Gui B, Kong GWS, Cao Y, Li TC, Choy KW (2019) A prospective study of non-invasive preimplantation genetic testing for aneuploidies (NiPGT-A) using next-generation sequencing (NGS) on spent culture media (SCM). J Assist Reprod Genet 36(8):1609–1621. https://doi.org/10.1007/s10815-019-01517-7

Chapter 15

Analysis of Relative Telomere Length and mtDNA Content in Individual Oocytes and Embryos by Quantitative Real-Time PCR (qPCR)

Yasmyn E. Winstanley, Darryl L. Russell and Rebecca L. Robker

Abstract

Oocytes and embryos are highly susceptible to intrinsic and extrinsic stressors, making assays for biomarkers of embryo quality of high importance. This has been technically challenging due to the very small amounts of material in embryos, leading to the use of assays requiring pooled embryos which mask inter-embryo variability. This protocol details a qPCR assay to analyze relative telomere lengths and mtDNA content in individual oocytes and embryos. Each is singly collected and lysed in a minimal volume, then qPCR is used to independently amplify two nuclear (telomere and Rn18S) and one mitochondrial DNA sequence for quantification of relative telomere length or mtDNA content that is normalized for cell number by the reference gene. This methodology enables robust and precise measurements of these molecular biomarkers in physiological contexts where tissue availability is limited. Importantly, this assay allows detection of the natural variability in mtDNA and telomeric DNA content between individual oocytes and embryos.

Keywords Telomere, Mitochondrial DNA, Mouse, Oocyte, Embryo, qPCR

1 Introduction

Oocyte developmental competence and embryo quality are highly susceptible to maternal and environmental stressors. Robust assays that can detect and quantify molecular biomarkers in very small amounts of cellular material, such as mammalian embryos, are needed as tools to analyze embryo quality. Two aspects of embryo quality that are of increasing interest are mitochondrial DNA (mtDNA) and telomere length. Existing assays for each of these biomarkers are limited by numerous technical challenges, mainly relating to sensitivity and precision. Here, we describe an assay with rigorous internal normalization and quality controls that accurately measures both biomarkers in an individual oocyte or embryo.

Mitochondrial function is essential for the normal progression of mammalian preimplantation development. mtDNA is a ~16 kb

Alexandra J. Harvey and David K. Gardner (eds.), *The Preimplantation Embryo*, Methods in Molecular Biology, vol. 3038, https://doi.org/10.1007/978-1-0716-5292-3_15,

circular DNA molecule that encodes 13 proteins of the electron transport chain, and thus mtDNA content is a critical component of overall mitochondrial function and cellular physiology. mtDNA replication occurs during oogenesis, with mtDNA levels highly abundant in the ovulated egg [1–6]. In some species, mtDNA replication is regulated during embryogenesis, but in most it remains stable until near the time of implantation when replication is dramatically induced to support the energy needs of the growing embryo [6, 7]. mtDNA is highly responsive to maternal and in vitro environmental signals [6]. As such, mtDNA has been used as a biomarker of embryo quality in agricultural species [2, 8, 9], mice [10], and in humans [5, 11, 12].

Telomere length is another molecular marker of embryo quality. Telomeres are repetitive DNA sequences that cap and protect the ends of linear chromosomes. Telomeres shorten with every cell division and also erode due to reactive oxygen (ROS) exposure; thus, telomere length is considered a marker of cellular lifespan and genomic stability [13, 14]. Telomere length shortens in oocytes with aging [15–17], however, preimplantation embryogenesis is one of the few contexts where telomere elongation occurs. Telomere length in embryos increases during the first few cell cycles [15, 18] and then a larger increase occurs around the morula and blastocyst stages [15, 19]. This telomere elongation is needed to support cell division throughout the life of the resultant offspring. Therefore, measuring telomere length in embryos may provide a surrogate marker of genome stability and cellular lifespan.

Assays for analyzing telomere length in preimplantation embryos are susceptible to artifacts due to poor normalization for input cell number or poor probe efficiency, and often have low sensitivity thus requiring pooled embryos, preventing quality assessments for individual embryos. We have developed a Quantitative Real-Time PCR (qPCR) assay with the capacity to measure these two important biomarkers (relative telomere length and mtDNA content) in parallel, in the same individual embryo, with sensitivity to detect minor perturbations in this biology [6, 15]. Importantly, the addition of a calibrator sample allows direct comparison of assay readouts across different PCR plates and experimental days. Here, we describe this qPCR assay for analysis of single mouse oocytes and embryos, but it could be adapted for other mammalian species through the use of species-specific reference primers (*see* **Note 1**). Samples are individually frozen and lysed in a minimal volume, after which DNA is isolated from every single oocyte or embryo and diluted accordingly for each assay. qPCR is then used to independently amplify two nuclear DNA sequences (telomere and Rn18S) and the mitochondrial genome using DNA from every single oocyte or embryo. The reference sequence (Rn18S) is reflective of input cell number and thus, when used to normalize either the mtDNA or telomere readouts, allows direct comparison of these measures in embryos of different developmental stages (*see* **Note 2**). Overall, this assay enables the investigation of both mtDNA and telomere kinetics in individual oocytes and embryos of all stages during maturation and preimplantation development, and the study of telomeres and mtDNA in cases of normal versus perturbed embryo development. Importantly, this method offers the opportunity to further our understanding of the underlying intrinsic oocyte and embryo deficiencies that are associated with lowered fertility, and how these may be responsible for observed offspring phenotypes.

2 Materials

2.1 Calibrator Sample Preparation

1. DNA isolation kit, e.g., a QIAamp® DNA Micro kit including QIAamp MinElute Columns, 2 mL collection tubes, Buffer ATL, Buffer AL, Buffer AW1, Buffer AW2, Buffer AE, Carrier RNA, and Proteinase K (*see* **Notes 3 and 4**).
2. 96%–100% molecular grade ethanol.
3. Sterile microcentrifuge tubes.
4. Micropipettes.
5. Minimal retention barrier tips.
6. Sterile PCR-grade water.
7. Thermomixer or heated orbital mixer.
8. Centrifuge suitable for microcentrifuge tubes.
9. Vortex.
10. DNA spectrometer.

2.2 Sample Collection

1. Nunc™ IVF Multi Dish, four well (non-treated).
2. PBS–PVP: 1×PBS (Ca^{2+}, Mg^{2+} free), 1 mg/mL polyvinylpyrrolidone (PVP).
3. Pipettors.
4. Minimal retention barrier tips.
5. Sterile tubes (suitable for thermocycler) for sample collection and lysis (*see* **Note 5**).
6. Liquid nitrogen.
7. Sterile PCR-grade water.
8. Centrifuge.
9. Thermal cycler.

2.3 Sample Lysis and Dilution

1. Micropipettes.
2. Minimal retention barrier tips.
3. Sterile tubes for mtDNA dilution (*see* **Note 6**).
4. Lysis buffer: 50 mM Tris-HCl, pH 8.0, 1 mM EDTA, 0.5% Tween-20, 200 μg/mL Proteinase K (*see* **Note 7**).
5. Sterile PCR-grade water.
6. Centrifuge.

2.4 Quantitative Real-Time PCR (qPCR)

The qPCR assay analyzes three sequences, telomeric DNA [15, 20], a 273 bp fragment in the 12 S rRNA region of mitochondrial genome (referred to as mtDNA), as previously described [21, 22], and a 91 bp fragment of the Rn18S gene, as previously described [6, 15, 23] (*see* **Note 8**).

1. 2× SYBR Green PCR Master Mix (e.g., Power SYBR Green PCR Master Mix, *see* **Note 9**).
2. Primer sequences (see Table 1).
3. Sterile PCR-grade water.
4. Sterile tubes (1.5 mL) for master mix.
5. Pipettors.
6. Minimal retention barrier tips.
7. Real-time PCR machine.
8. Real-time PCR reaction plates.
9. Optical real-time PCR adhesive film.
10. Film sealer.

Table 1
qPCR primer sequences

Primer target	Forward sequence	Reverse sequence
Telomere	5′ CGG TTT GTT TGG GTT TGG GTT TGG GTT TGG GTT TGG GTT 3′	5′ GGC TTG CCT TAC CCT TAC CCT TAC CCT TAC CCT TAC CCT 3′
mtDNA	5′ CGT TAG GTC AAG GTG TAG CC 3′	5′ CCA GAC ACA CTT TCC AGT ATG 3′
Rn18S (mouse-specific)	5′ AGA AAC GGC TAC CAC ATC CAA 3′	5′ CCT GTA TTG TTA TTT TTC GTC ACT ACC T 3′

3 Methods

3.1 Calibrator Sample

Calibrator DNA can be sourced from any mouse tissue; however, we isolate DNA from a whole ovary from a mouse of a similar genetic background to those from which oocytes or embryos to be analyzed are derived. DNA isolation for the calibrator sample can be done via any in-house method or commercial kit, such as phenol extraction, salting-out, or column-based methods. Here, the QIAGEN kit is used following the manufacturer's instructions for "Isolation of Genomic DNA from Tissues". It is important to wear gloves at all times when handling samples or kit components and to work in a clean area to avoid contamination. All buffers outlined are supplied with the kit; buffers AW1 and AW2 must first be reconstituted with ethanol, according to the manufacturer's instructions.

1. Place a tissue sample of less than 10 mg in weight into a 1.5 mL microcentrifuge tube (*see* **Notes 10 and 11**).
2. Add 180 μL Buffer ATL and equilibrate to room temperature (15–25°C).
3. Add 20 μL proteinase K (supplied) and mix for 15 s by pulse-vortexing. The sample is then incubated in a thermomixer or heated orbital incubator at 56°C overnight.
4. Add 200 μL of Buffer AL, close the lid, and mix for 15 s by pulse-vortexing (*see* **Note 12**).
5. Add 200 μL ethanol (96%–100%), close the lid, and thoroughly mix for 15 s by pulse-vortexing. Incubate the mixture is incubated for 5 min at room temperature (15–25°C).
6. Briefly centrifuge the tube to remove any drops from inside the lid, then carefully transfer the entire lysate to the center of a QIAamp MinElute column (in a 2 mL collection tube). Close the lid, and centrifuge at 6000 × *g* for 1 min. Place the QIAamp MinElute column in a clean 2 mL collection tube and discard the collection tube containing the flow-through.
7. Carefully open the QIAamp MinElute column and add 500 μ L Buffer AW1 to the center of the column. Close the lid, and centrifuge at 6000 × *g* for 1 min. Place the QIAamp MinElute column in a clean 2 mL collection tube. Discard the collection tube containing the flow through.
8. Carefully open the QIAamp MinElute column and add 500 μ L Buffer AW2 to the center of the column. Close the lid, and centrifuge at 6000 × *g* for 1 min. Place the QIAamp MinElute column in a clean 2 mL collection tube. Discard the collection tube containing the flow-through.

9. Centrifuge samples at full speed (20,000 × *g*) for 3 min to dry the membrane completely.
10. Place the QIAamp MinElute column in a clean 1.5 mL microcentrifuge tube and discard the collection tube containing the flow-through. Carefully open the lid of the QIAamp MinElute column and apply 50 μL sterile PCR-grade water to the center of the membrane (*see* **Notes 13 and 14**).
11. Close the lid and incubate at room temperature (15–25°C) for 5 min (*see* **Note 15**). Centrifuge at full speed (20,000 × *g*) for 1 min.
12. Quantify concentration and purity using a DNA spectrophotometer and dilute using sterile PCR-grade water to 2.5 ng/μL (*see* **Note 16**).

3.2 Sample Collection

This protocol has been used with mouse oocytes and embryos from a wide variety of developmental stages, including GV and MII oocytes, fertilized zygotes, 2-, 4-, and 8-cell embryos, morula, whole blastocyst, immunosurgically purified ICM, and manually dissected ICM and TE.

1. Add 500 μL of PBS-PVP to each well of a 4-well Nunc™ IVF Multi Dish.
2. Transfer and wash samples three times in PBS-PVP (*see* **Note 17**).
3. Transfer individual samples to collection tubes (*see* **Note 6**) in 1 μL of PBS–PVP (*see* **Note 18**).
4. Immediately freeze the sample in liquid nitrogen and store at −80°C until use.

3.3 Sample Lysis and Dilution

Due to the low amount of genomic DNA present in individual oocytes and early-stage preimplantation embryos, samples are lysed in a minimal volume (*see* **Note 19**). Due to the relative abundance of mtDNA compared to nuclear DNA, a portion of the lysed sample is further diluted for mtDNA analysis.

1. Briefly centrifuge the sample collection tube to remove drops from inside the lid and prevent sample loss.
2. Add 9 μL of lysis buffer.
3. Incubate samples in a thermal cycler (*see* **Note 20**) at 55°C for 2 h (lid temperature > 90°C), followed by 95°C for 10 min (to inactivate proteinase K).
4. Allow the sample to cool to 4°C, then briefly centrifuge the sample tube to remove drops from inside the lid and prevent sample loss.

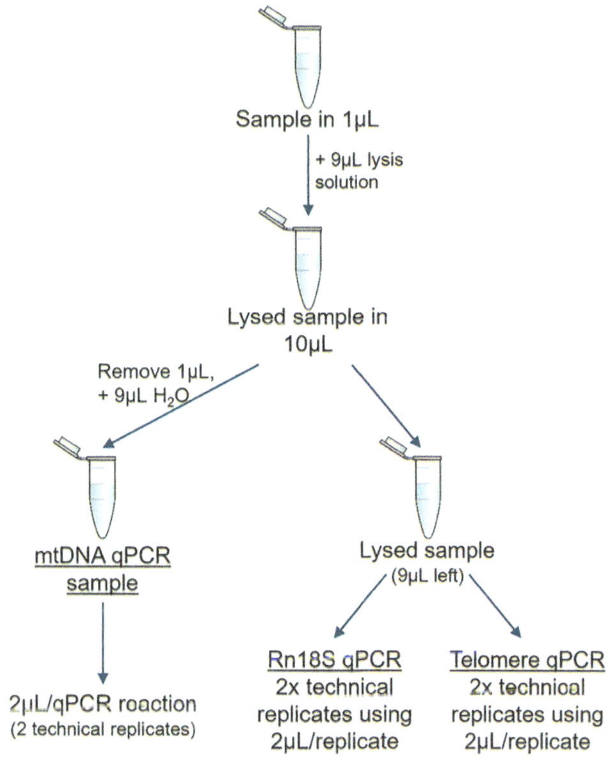

Fig. 1 Sample lysis and dilution workflow. Samples are collected in 1 µL of PBS–PVP, and 9 µL of lysis buffer is added. After lysis, 1 µL of the sample is aliquoted into a second tube containing 9 µL of sterile PCR-grade water to be used for mtDNA qPCR reactions (i.e., "mtDNA qPCR sample"), while the remaining 9 µL of the sample (i.e., "lysed sample") is used for qPCR reactions with the telomere and Rn18S primers

5. Remove 1 µL of sample and transfer to a new tube containing 9 µL of sterile PCR-grade water. This diluted lysate will be used for mtDNA analysis (*see* **Note 21**, Fig. 1).

3.4 Quantitative Real-Time PCR (qPCR)

This qPCR assay analyzes three sequences: the telomeres, a region of the mitochondrial genome (referred to as mtDNA), and the Rn18S (reference) gene. The polymerase cycling reactions can be performed in any real-time PCR thermal cycler, here we describe the use of a QuantStudio 12 K Flex Real-Time PCR System (ThermoFisher). Pipetting accuracy is critical for the integrity of this assay, particularly in the qPCR reaction setup. Further, the small sample volume is only enough for a single assay, and none will be remaining that can be re-assayed. All primers were previously characterized [20, 21, 23] and are compatible with the same cycling conditions, and thus, for each sample, all three primer pairs should be run in parallel in a single run (Fig. 2).

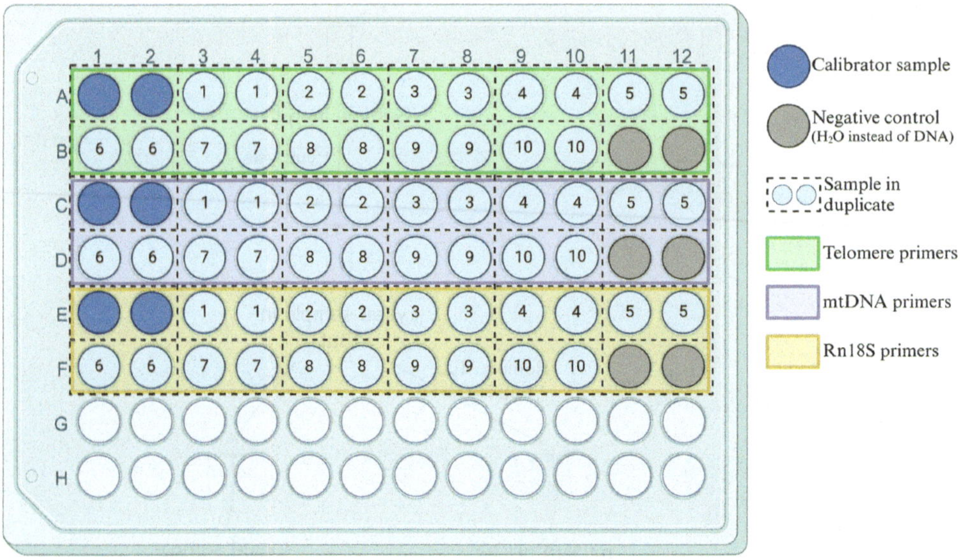

Fig. 2 Example of an assay set up on a 96-well qPCR plate. All three primer sets (i.e., telomere, mtDNA, Rn18S) are compatible with the same cycling conditions, and it is recommended to run all three primer sets for each sample on the same assay run

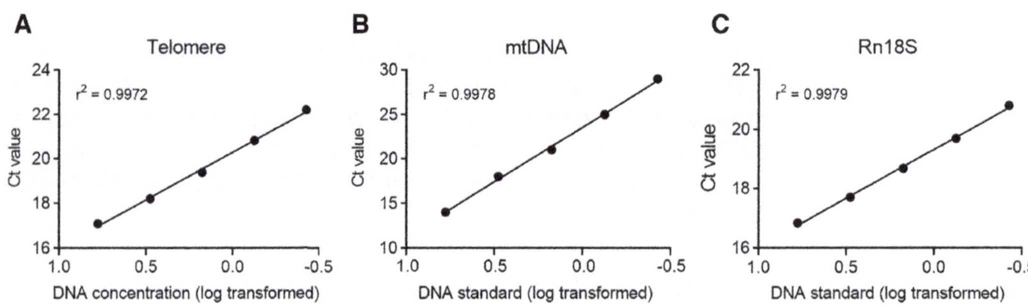

Fig. 3 Validation of primer efficiency. DNA from the calibrator sample (whole mouse ovary) was serially diluted to produce five concentrations of DNA (6, 3, 1.5,0.75, and 0.375 ng/μL) to evaluate the efficiency of the telomere primers (**A**), mtDNA primers (**B**), and Rn18S primers (**C**) under the same cycling conditions. To test for linearity, DNA concentration was log10 transformed for data plotting with Ct values, and r^2 values from linear regression analysis are shown. Each data point is derived from triplicate qPCR reactions

1. To validate the efficiency of the primers to detect low amounts of input DNA, five serial dilutions (6, 3, 1.5, 0.75, and 0.375 ng/μL) of ovary calibrator DNA are assayed for all 3 primer pairs. Resulting r^2 values above 0.99 have been obtained for the primers detailed in this protocol (Fig. 3). A similar r^2 value should be obtained if adapting this assay to different primers/species.

2. The telomere qPCR assays use 0.25 μM of each primer (forward and reverse, *see* **Note 1**), 10 μL SYBR Green PCR Master Mix, and 2 μL of lysis sample in a final 20 μL reaction volume (*see* **Note 22**).
3. The mtDNA qPCR assays use 0.1 μM of each primer (forward and reverse), 10 μL SYBR Green PCR Master Mix, and 2 μL of diluted lysis sample in a final 20 μL reaction volume.
4. The Rn18S qPCR assays use 0.25 μM of each primer (forward and reverse), 10 μL SYBR Green PCR Master Mix, and 2 μL of lysis sample in a final 20 μL reaction volume.
5. Thermal cycling conditions are 95°C for 10 min, 40 cycles of 95°C for 15 s, 60°C for 30 s, and 72°C for 30 s, followed by standard melt–curve analysis.

The calibrator sample is included in duplicates for each primer pair per run. Experimental samples should be assayed in duplicates for each primer pair per run. A negative control (replacing the 2 μL of sample with 2 μL of sterile PCR-grade water) should be included in duplicates for each primer pair per run (Fig. 2).

3.5 Relative Telomere Length and mtDNA Content Calculation

Once the raw qPCR data (Ct values) have been acquired, the accuracy of pipetting and the validity of sample values should be verified prior to any calculations being performed.

1. Verify that the Ct values of the negative control wells show no detectable product indicative of contamination of the reagent mixtures (*see* **Note 23**).
2. Technical accuracy between sample duplicates should be confirmed (i.e., Ct variation between duplicates should be less than 0.1 (10%)).
3. The Rn18S reference gene is used to normalize results for input cell (or chromosome) number, and unexpected variations could be indicative of aneuploidy or other corruptive variability in the input. Ct values across samples should also be scrutinized, particularly comparing samples with defined cell numbers (e.g., oocyte, 2-, 4-, and 8-cell). As Rn18S reflects changes in cell numbers, samples of the same oocyte or pre-compaction embryo type should have similar Rn18S Ct values (Fig. 4) (*see* **Notes 24 and 25**).
4. Monitor the Ct values of the calibrator sample for each primer pair across different runs, and particularly the ΔCt of the calibrator (i.e., telomere Ct–Rn18S Ct, and mtDNA Ct–Rn18S Ct) as this should remain consistent across runs even on different experimental days. Large variations in the ΔCt of the calibrator may indicate inaccuracy of preparation of qPCR assays, sample loading, or issues with DNA quality or purity of the calibrator.

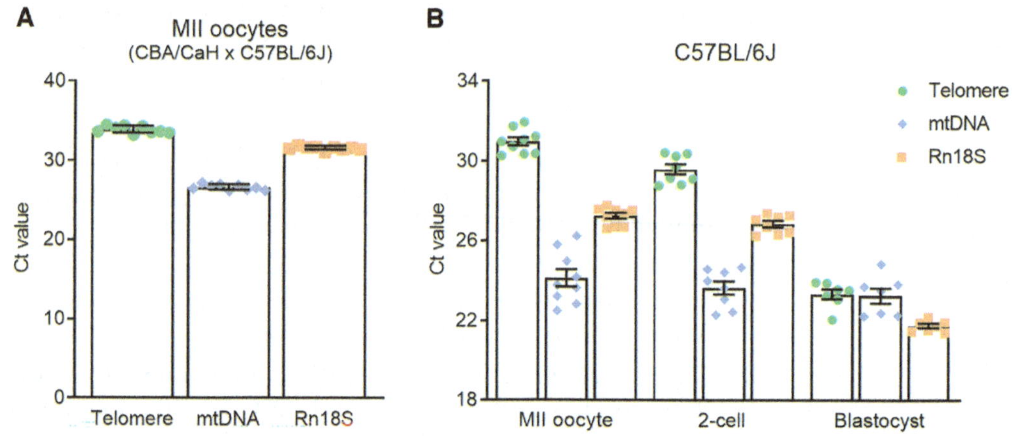

Fig. 4 Example Ct values from mouse oocyte and embryo samples. (**A**) Ct values resulting from PCR reactions for telomere, mtDNA, and Rn18S using DNA from $n=9$ MII oocytes derived from CBA F1 (CBA/CaH × C57BL/6 J) mice. (**B**) Ct values resulting from PCR reactions for telomere, mtDNA, and Rn18S using DNA from $n=9$ MII oocytes, $n=8$ 2-cell embryos, and $n=7$ whole blastocysts derived from C57BL/6 J mice

5. Once the raw data has been scrutinized, the relative telomere length or mtDNA content can be calculated. First, the average Ct of the duplicates is calculated, which is then used for the final calculations. Then, the $2^{-\Delta\Delta Ct}$ calculation is used, where the ΔCt of the calibrator (cal) is subtracted from the ΔCt of the sample (spl) to give the $\Delta\Delta$Ct (i.e., $2^{-(\Delta Ct\ spl)-(\Delta Ct\ cal)}$). Calculation formulae are below.

$$\text{Relative telomere length} = 2^{-(\text{Telo Ct spl} - \text{Rn18S Ct spl}) - (\text{Telo Ct cal} - \text{Rn18S Ct cal})}$$

$$\text{Relative mtDNA content} = 2^{-(\text{mtDNA Ct spl} - \text{Rn18S Ct spl}) - (\text{mtDNA Ct cal} - \text{Rn18S Ct cal})}$$

4 Notes

1. When adapting this protocol for a different species, select a reference gene that is multi-copy (so that it is detectable in single oocytes/embryos) and use bioinformatic software to design appropriate primers to the target sequences. Following primer design, the annealing temperature and primer concentration must be optimized. The annealing temperature can be calculated using online calculators (or may be supplied by the manufacturer), the temperatures may not be the same for all primers (usually between 55°C and 65°C). To optimize the annealing temperature, perform qPCR assays with

a range of annealing temperatures beginning approximately 3–5°C lower than the lowest recommended annealing temperature and increasing incrementally to 3–5°C higher than the recommended temperature. The optimal annealing temperature will (1) result in the lowest Ct value, (2) yield a melt curve indicating detection of a specific product and no primer dimers, and (3) be highly reproducible between replicate reactions. To optimize primer concentration, multiple qPCR assays with varying final concentrations of the forward and reverse primers (recommended between 100 and 800 nM) should be performed. Evaluate the Ct values for reproducibility between replicate reactions and the melt curve for any indication of primer dimers (i.e., more than one peak present on the melt curve, which can occur when primer concentration is too high). Once a concentration has been selected for the forward and reverse primer (they may not always be used at the same concentration) a standard curve should be run using at least five logs of input DNA concentration to calculate the efficiency, which should be within the 90%–110% range.

2. This assay allows comparison between embryos of different stages of development, however, we advise caution when comparing embryos from two different strains of mice as there can be variation between strains in Rn18S content in the genome (Fig. 4). For example, MII oocytes derived from CBA F1 mice (derived from CBA/CaH × C57BL/6J) returned an average Rn18S Ct value of 31.6 (Fig. 4A), while MII oocytes from C57BL/6J mice returned an average Rn18S Ct value of 26.65 (Fig. 4B), a Ct difference of approximately 4.9. Despite this, the MII oocytes from the two strains returned similar telomere Ct values (CBA F1 33.61 ± 0.24, C57BL/6J 31.98 ± 0.61, mean ± SD, Fig. 4).
3. Ensure adequate calibrator DNA is available to complete the intended number of experiments, as Ct values may change between different calibrator DNA samples.
4. QIAgen kits are suitable for the calibrator sample DNA isolation if there are abundant cells/tissues. These types of kits are not suitable for individual oocytes and embryos due to the need to elute in relatively large volumes (the manufacturer recommends no less than 20 μL) and the loss of at least some DNA on the column membrane.
5. Samples will undergo lysis and remain in the original collection tubes for the PCR. Therefore, the tubes that are used for sample collection should be suitable for use in a thermal cycler (used for lysis incubation). Thermal cyclers typically accommodate 0.5 or 0.2 mL tubes.

6. For mtDNA dilution, we recommend 0.5 or 0.2 mL tubes as the dilution volume is small (final volume 10 μL).
7. Lysis buffer should be stored at 4°C and used within 1 week.
8. Primer stocks should be aliquoted into small volumes to avoid freeze/thawing, as this practice leads to primer degradation and negatively impacts qPCR results. Once an aliquot has been thawed, it is recommended to discard any excess rather than refreeze.
9. SYBR green does not use a hot-start enzyme and is therefore sensitive to temperature changes; thus, all reagents should be kept cold (e.g., on ice) during qPCR preparation. Additionally, SYBR green is light sensitive; extended exposure to light should be avoided. The method described here uses a 2× SYBR Green PCR Master Mix. If using a different concentration SYBR Master Mix, the volume of SYBR Green used per reaction needs to be adjusted accordingly, as well as the volume of water, to result in a final dilution of 1×. For example, if using 5× SYBR Green PCR Master Mix, the amount of SYBR used per reaction will reduce to 4 μL, and the amount of water will increase by 6 μL to give a final SYBR concentration of 1× in a 20 μL reaction.
10. Tissue can be fresh or frozen. If the tissue is frozen, do not allow it to thaw prior to the addition of Buffer ATL (Step 2).
11. If using an ovary, or any tissue with adjacent adipose tissue, make sure the tissue is clean of fat before beginning the DNA isolation protocol.
12. If isolating from small amounts of tissue, such as an ovary, addition of Carrier RNA (included in kit) is recommended as per the manufacturer's protocol (1 μL dissolved carrier RNA per 200 μL of buffer AL added to the sample). Due to the small volume of carrier RNA used, it is recommended to prepare a master mix for the number of samples plus one extra reaction, to account for pipetting error, then pipette it into the sample reaction tubes.
13. Buffer AE (which contains EDTA) is supplied in the kit; however, this can interfere with applications sensitive to high pH. Therefore, we recommend using PCR-grade water instead.
14. For isolation of DNA from a mouse ovary, 50 μL is recommended. A larger piece of tissue should use a larger volume of water (up to 100 μL is recommended by the manufacturer).
15. The minimum recommended incubation time is 1 min, but incubation for 5 min generally increases DNA yield. A 5 min incubation is recommended for DNA isolated from a mouse ovary.

16. Diluted DNA should be stored at −20°C, but do not freeze-thaw the calibrator DNA as it can lead to DNA degradation. We recommend making small aliquots of an equivalent volume to that required for one experiment. Once an aliquot has been thawed, we recommend discarding any excess.
17. Components of oocyte and embryo culture media could impair lysis or qPCR efficiency, samples should be thoroughly washed. The addition of PVP to the PBS prevents cellular adherence to the plasticware.
18. Sample loss can occur at this step due to the cells adhering to the inside of the pipette tip. This is more likely with samples that are emerging from their zona pellucida (i.e., a hatching blastocyst) or those that no longer have a zona pellucida (e.g., immunosurgically purified ICM or manually dissected ICM or TE). To prevent this, we recommend priming the pipette tip by pre-wetting it with an entire volume (1 µL) of PBS-PVP (three times) prior to transferring a sample into a collection tube. Further, we recommend drawing up approximately half the volume (0.5 µL) of PBS–PVP into the pipette tip prior to picking up the sample so that the sample is not in contact with the meniscus. A new pipette tip is recommended for each sample in case a previous sample is retained within the tip.
19. Due to the small final volume of the sample, we recommend performing the lysis step no earlier than the day before qPCR to minimize the risk of sample loss/evaporation.
20. We recommend using a thermal cycler because the heated lid provides consistent heating and prevents the formation of condensation on the tube lid.
21. When measuring samples with a larger number of cells (> 20 cells), including blastocysts, rather than doing a separate dilution for mtDNA, the entire lysis product (in 10 µL) can be further diluted by adding 10 µL of sterile PCR-grade water (to a final volume of 20 µL), and this more dilute sample can be used for all for the assays (telomere, mtDNA, and Rn18S qPCR).
22. To ensure homogeneity of the reaction mix across all reactions, prepare sufficient reaction mix for the number of samples to be analyzed, plus a few extra reactions, to account for pipetting error, then pipette into sample reaction tubes.
23. If product is detected, this can indicate (1) contamination of the primer stocks or assay master mix, or (2) artifact generation that may arise when primer stocks undergo repeated thaw and refreezing (*see* **Note 8**).

24. When the assay is performed correctly, with precise pipetting, samples with equivalent input (i.e., technical replicates) will deviate by less than 0.25 cycles. Larger deviations may indicate an error that would corrupt the final result.
25. Be aware that a change in Ct of 1 indicates a halving or doubling (depending on the direction of change) of product, and such changes should be reasonably consistent with expected cell numbers. An example of Ct values changing consistently with cell number: MII oocytes return an Rn18S Ct of 27.26 ± 0.15, 2-cell embryos 26.86 ± 0.16 (a difference of 1 Ct), and whole blastocysts 21.77 ± 0.11 (mean ± standard deviation, Fig. 4B).

Acknowledgments

This work was supported by grants from the National Health and Medical Research Council to RLR and DLR. The authors thank Donna Holland for contributing constructive comments on the manuscript.

References

1. Mahrous E, Yang Q, Clarke HJ (2012) Regulation of mitochondrial DNA accumulation during oocyte growth and meiotic maturation in the mouse. Reproduction 144(2):177–185. https://doi.org/10.1530/REP-12-0113
2. Mao J, Whitworth KM, Spate LD, Walters EM, Zhao J, Prather RS (2012) Regulation of oocyte mitochondrial DNA copy number by follicular fluid, EGF, and neuregulin 1 during in vitro maturation affects embryo development in pigs. Theriogenology 78(4):887–897. https://doi.org/10.1016/j.theriogenology.2012.04.002
3. May-Panloup P, Vignon X, Chretien MF, Heyman Y, Tamassia M, Malthiery Y, Reynier P (2005) Increase of mitochondrial DNA content and transcripts in early bovine embryogenesis associated with upregulation of mtTFA and NRF1 transcription factors. Reprod Biol Endocrinol 3: 65. https://doi.org/10.1186/1477-7827-3-65
4. Monnot S, Samuels DC, Hesters L, Frydman N, Gigarel N, Burlet P, Kerbrat V, Lamazou F, Frydman R, Benachi A, Feingold J, Rotig A, Munnich A, Bonnefont JP, Steffann J (2013) Mutation dependance of the mitochondrial DNA copy number in the first stages of human embryogenesis. Hum Mol Genet 22(9):1867–1872. https://doi.org/10.1093/hmg/ddt040
5. Santos TA, El Shourbagy S, St John JC (2006) Mitochondrial content reflects oocyte variability and fertilization outcome. Fertil Steril 85(3):584–591. https://doi.org/10.1016/j.fertnstert.2005.09.017
6. Winstanley YE, Liu J, Adhikari D, Gonzalez MB, Russell DL, Carroll J, Robker RL (2024) Dynamics of mitochondrial DNA copy number and membrane potential in mouse pre-implantation embryos: responses to diverse types of oxidative stress. Genes (Basel) 15(3):367. https://doi.org/10.3390/genes15030367
7. St John JC (2019) Mitochondria and female germline stem cells—a mitochond-A Mitochondrial DNA perspective. Cells 8(8): 852. https://doi.org/10.3390/cells8080852
8. El Shourbagy SH, Spikings EC, Freitas M, St John JC (2006) Mitochondria directly influence fertilisation outcome in the pig. Reproduction 131(2):233–245. https://doi.org/10.1530/rep.1.00551
9. Spikings EC, Alderson J, St John JC (2007) Regulated mitochondrial DNA replication during oocyte maturation is essential for successful porcine embryonic development. Biol Reprod

76(2):327–335. https://doi.org/10.1095/biolreprod.106.054536
10. Wai T, Ao A, Zhang X, Cyr D, Dufort D, Shoubridge EA (2010) The role of mitochondrial DNA copy number in mammalian fertility. Biol Reprod 83(1):52–62. https://doi.org/10.1095/biolreprod.109.080887
11. Klimczak AM, Pacheco LE, Lewis KE, Massahi N, Richards JP, Kearns WG, Saad AF, Crochet JR (2018) Embryonal mitochondrial DNA: relationship to embryo quality and transfer outcomes. J Assist Reprod Genet 35(5):871–877. https://doi.org/10.1007/s10815-018-1147-z
12. Reynier P, May-Panloup P, Chretien MF, Morgan CJ, Jean M, Savagner F, Barriere P, Malthiery Y (2001) Mitochondrial DNA content affects the fertilizability of human oocytes. Mol Hum Reprod 7(5):425–429. https://doi.org/10.1093/molehr/7.5.425
13. Hayflick L, Moorhead PS (1961) The serial cultivation of human diploid cell strains. Exp Cell Res 25:585–621. https://doi.org/10.1016/0014-4827(61)90192-6
14. Rudolph KL, Chang S, Lee HW, Blasco M, Gottlieb GJ, Greider C, DePinho RA (1999) Longevity, stress response, and cancer in aging telomerase-deficient mice. Cell 96(5):701–712. https://doi.org/10.1016/s0092-8674(00)80580-2
15. Winstanley YE, Rose RD, Sobinoff AP, Wu LL, Adhikari D, Zhang QH, Wells JK, Wong LH, Szeto HH, Piltz SG, Thomas PQ, Febbraio MA, Carroll J, Pickett HA, Russell DL, Robker RL (2025) Telomere length in offspring is determined by mitochondrial-nuclear communication at fertilization. Nat Commun 16(1):2527. https://doi.org/10.1038/s41467-025-57794-7
16. Keefe DL, Marquard K, Liu L (2006) The telomere theory of reproductive senescence in women. Curr Opin Obstet Gynecol 18(3):280–285. https://doi.org/10.1097/01.gco.0000193019.05686.49
17. Yamada-Fukunaga T, Yamada M, Hamatani T, Chikazawa N, Ogawa S, Akutsu H, Miura T, Miyado K, Tarin JJ, Kuji N, Umezawa A, Yoshimura Y (2013) Age-associated telomere shortening in mouse oocytes. Reprod Biol Endocrinol 11: 108. https://doi.org/10.1186/1477-7827-11-108
18. Liu L, Bailey SM, Okuka M, Munoz P, Li C, Zhou L, Wu C, Czerwiec E, Sandler L, Seyfang A, Blasco MA, Keefe DL (2007) Telomere lengthening early in development. Nat Cell Biol 9(12):1436–1441. https://doi.org/10.1038/ncb1664
19. Schaetzlein S, Lucas-Hahn A, Lemme E, Kues WA, Dorsch M, Manns MP, Niemann H, Rudolph KL (2004) Telomere length is reset during early mammalian embryogenesis. Proc Natl Acad Sci USA 101(21):8034–8038. https://doi.org/10.1073/pnas.0402400101
20. Cawthon RM (2002) Telomere measurement by quantitative PCR. Nucleic Acids Res 30(10):e47. https://doi.org/10.1093/nar/30.10.e47
21. Kameyama Y, Ohnishi H, Shimoi G, Hashizume R, Ito M, Smith LC (2010) Asymmetrical allocation of mitochondrial DNA to blastomeres during the first two cleavages in mouse embryos. Reprod Fertil Dev 22(8):1247–1253. https://doi.org/10.1071/RD10076
22. Wu LL, Russell DL, Wong SL, Chen M, Tsai TS, St John JC, Norman RJ, Febbraio MA, Carroll J, Robker RL (2015) Mitochondrial dysfunction in oocytes of obese mothers: transmission to offspring and reversal by pharmacological endoplasmic reticulum stress inhibitors. Development 142(4):681–691. https://doi.org/10.1242/dev.114850
23. de Frutos C, Lopez-Cardona AP, Fonseca Balvis N, Laguna-Barraza R, Rizos D, Gutierrez-Adan A, Bermejo-Alvarez P (2016) Spermatozoa telomeres determine telomere length in early embryos and offspring. Reproduction 151(1):1–7. https://doi.org/10.1530/REP-15-0375

Chapter 16

Quantitative Analysis of Embryo Gene Expression by Real-Time PCR

Krishna Chaitanya Pavani, Juri Gnagnarelli, Ann Van Soom, Katrien Smits and Luc Peelman

Abstract

Quantitative real-time PCR (RT-qPCR) is a powerful and precise tool for analyzing gene expression, particularly suited for studying the embryonic transcriptome. Its high sensitivity enables the detection of low-abundance nucleic acids, which are essential for examining gene expression in early-stage embryos, where cellular diversity and rapid transcriptional changes pose unique challenges. This chapter provides a comprehensive guide to designing and conducting RT-qPCR experiments for preimplantation embryos, with a focus on accurate quantification of messenger RNA (mRNA) levels. The technique's extensive dynamic range, sequence specificity, and reliability make it invaluable for comparing mRNA levels across diverse biological samples and for validating data from other molecular approaches, such as microarrays or next-generation sequencing. This chapter outlines essential steps and considerations for obtaining accurate, reproducible gene expression data in embryonic research.

Keywords Quantitative real-time PCR (RT-qPCR), Gene expression analysis, mRNA quantification, Embryos

1 Introduction

Real-time quantitative reverse transcription PCR (RT-qPCR) is widely recognized as a leading technique often regarded as the gold standard for precise gene expression analysis, allowing researchers to quantify mRNA levels across various biological contexts [1, 2]. Despite its strengths in sensitivity and widespread adoption, RT-qPCR has limitations, particularly in the presence of inhibitors and with low-abundance targets, where it can yield variable results. Digital PCR (dPCR) offers an alternative, enabling absolute quantification without the need for standard curves, and is less affected by inhibitors. However, dPCR is generally more resource-intensive and less accessible, making RT-qPCR the preferred method in many research settings [3]. Its ability to quantify low-abundance transcripts,

Alexandra J. Harvey and David K. Gardner (eds.), *The Preimplantation Embryo*, Methods in Molecular Biology, vol. 3038,
https://doi.org/10.1007/978-1-0716-5292-3_16,

including low mRNA levels, makes it especially valuable for studying gene expression in early embryos, where limited sample sizes and high transcriptomic variability present significant challenges [4, 5]. Accurate quantification in embryo studies is essential to understanding developmental processes, as gene expression patterns at these stages are tightly regulated and play critical roles in cellular differentiation and embryo viability [6–8].

Recent advances in single-cell and low-input RNA analysis, combined with refined RNA amplification methods, have made it feasible to apply RT-qPCR to individual or small pools of preimplantation embryos [9–11]. However, these methods also introduce new sources of variation, including RNA, integrity, and PCR amplification efficiency, which can impact the reliability of expression data. The use of exogenous RNA spikes as reference standards has shown promise for mitigating these issues, particularly in studies where starting material is limited [12, 13]. While adding spikes during extraction can help account for sample variability, controlling for the ratio of spike RNA to endogenous RNA remains challenging, particularly with very low cell numbers [14–16].

In addition to ensuring accurate normalization, researchers must also consider factors specific to embryo development, such as rapid transcriptome changes and genetic variability among embryos. These characteristics require specialized protocols for comparing gene expression across different developmental stages and embryo types [17, 18].

This chapter provides a comprehensive overview of RT-qPCR tailored for bovine embryonic gene expression studies but suitable for other mammalian species [18]. It outlines key considerations for designing experiments that produce reliable and reproducible data, from RNA extraction and reverse transcription to quantification and normalization strategies, and describes the steps necessary to quantify mRNA expression levels of eight commonly used reference genes at various preimplantation stages. By discussing methods suited to preimplantation embryos, we aim to equip researchers with the tools to conduct precise gene expression analysis, ultimately contributing to a better understanding of developmental biology and refining techniques in assisted reproductive technologies.

2 Materials

2.1 Consumables for Bovine In Vitro Embryo Production

1. CO_2 incubator set to 38.5°C and 5% CO_2.
2. Trigas incubator set to 38.5°C with a humidified gas atmosphere of 5% CO_2, 5% O_2, and 90% N_2.
3. Thermos.
4. 18 G and 20 G needles.

5. Haemocytometer.
6. Sterile conical tubes (15 and 50 mL).
7. Sterile Eppendorf tubes (0.5 and 1.5 mL).
8. Sterile measuring cylinders.
9. Sterile serological pipettes.
10. Sterile syringes (1–10 mL).
11. Sterile 0.22 μm cellulose acetate filters.
12. Cryovials (1.5, 2, and 4 mL).
13. Coverslips (22 × 22 mm).
14. Parafilm.
15. Stereomicroscope.
16. Centrifuge.
17. Warming Plate.
18. Four-well culture dishes.
19. Petri dishes (35, 60, and 90 mm).
20. Water bath (38.5°C) for thawing semen samples.
21. Sterile pipettes and filter pipette tips (10, 100, and 1000 μL).
22. Laminar flow or biosafety cabinet.
23. Vortex.
24. Vacuum pump for media filtration.
25. 1000 mL 0.22 μm bottle-top vacuum filters.
26. Straws of frozen bull semen.
27. Sterile scissors.

2.2 Media and Reagents for In Vitro Embryo Production and Collection of Bovine Embryos

In vitro embryo production protocols can vary between laboratories; here, we provide details based on our lab's protocol using a serum-free culture system [19]. Alternatively, in vivo-derived embryos can be flushed for analysis (as detailed in [14]). Prepare all media in a laminar flow or biological safety cabinet.

1. Physiological Saline Solution: Normal saline supplemented with kanamycin (25 mg/mL)
2. B60 water (water for injections, 1000 mL bottle).
3. Bovine Serum Albumin (BSA): Fraction V, Fatty acid free.
4. TCM-199 Earle's Medium.
5. Epidermal growth factor (EGF): Lyophilized powder, molecular weight of ~6 kDa. Store at 4°C.
6. EGF stock solution (10 μg/mL): Take a sterile 15 mL conical tube and label it as TCM199 + BSA. Fill the tube with 12 mL of fresh sterile TCM199 using a sterile 10 mL serological

pipette, then add 0.0144 g of BSA. Mix the solution gently by inverting the tube until the BSA dissolves completely. Filter the solution through a 0.22 μm filter into a new sterile 15 mL conical tube and label this tube as TCM199 + BSA.2. Transfer 10 mL of the filtered solution into a new sterile 15 mL conical tube and label it as EGF Stock. Aspirate 2 mL of the sterile TCM199-BSA solution using a sterile 10 mL syringe and a sterile 18 G needle, inject the solution into the EGF vial, and mix gently until the powder dissolves. Recover the dissolved EGF solution with the syringe and transfer it into the EGF stock tube, repeating the aspiration and transfer two more times to rinse the EGF vial completely. Finally, aliquot the EGF stock solution into sterile 2 mL cryotubes in 1.1 mL volumes using micropipettes with filtered tips. Store the EGF stock at −80°C for a maximum of 6 months.

7. EGF work solution (1 μg/mL): Thaw one aliquot of EGF stock solution at room temperature. Fill a sterile 15 mL conical tube with 12 mL of fresh sterile TCM 199 Earle's medium using a 10 mL serological pipette and add 0.0144 g of BSA. Label the tube TCM199 + BSA and mix gently by rolling between the hands. Filter the solution through a 0.22 μm filter into a new sterile 15 mL conical tube and label TCM199 + BSA.2. Transfer 9 mL of this solution into another sterile 15 mL conical tube and add 1 mL of the thawed EGF stock using a micropipette with a sterile filter tip. Label the tube EGF Work, homogenize by gentle inversion, and aliquot the solution into sterile 1 mL cryovials in volumes of 250–500 μL. Store at −20°C for a maximum of 1 month.
8. In vitro maturation (IVM) medium: Bicarbonate-buffered TCM-199 supplemented with 50 μg/mL gentamicin and 20 ng/mL epidermal growth factor (EGF work solution). Sterilize with a 0.22 μm filter and store at 4°C.
9. 90% Percoll®
10. 45% and 90% Percoll® Gradient Solutions: For the preparation of the Percoll gradient, label two sterile 15 mL conical tubes as "Percoll 45%" and "Percoll gradient." Using a sterile 2 mL serological pipette, dispense 1.5 mL of Percoll 90% into the tube designated for the 45% solution. With the same pipette, transfer 2 mL of Percoll 90% into the tube labeled as Percoll gradient. To generate the 45% Percoll suspension, using a new sterile 2 mL serological pipette, add 1.5 mL of SP-TL (−) BSA medium to the tube labeled Percoll 45%. The suspension should be mixed carefully by aspirating and dispensing several times with the same pipette until homogeneity is achieved. Subsequently, overlay the 90% Percoll in the "Percoll gradient" tube with 2 mL 45% Percoll by pipetting

gently to prevent disruption of the interface and ensure the proper formation of the density gradient.

11. Heparin working solution (25 μg/mL): Dissolve 0.040 g of Heparin in 20 mL B60 water. Filter through a 0.22 μm filter and aliquot (160 μL) into 0.5 mL sterile Eppendorf tubes. Store at −20°C.
12. HEPES-TALP medium: Fill a 1000 mL measuring cylinder with 800 mL B60 water. Add the following components: 6.666 g of NaCl (final concentration 114 mM); 2 mL of Phenol Red; 3.1 mL of 1 M KCl (final concentration 3.1 mM); 3 mL of 0.1 M NaH_2PO_4 (final concentration 0.3 mM); 2.1 mL of 1 M $CaCl_2 \cdot 2H_2O$ (final concentration 2.1 mM); 4 mL of 0.1 M $MgCl_2 \cdot 6H_2O$ (final concentration 0.4 mM); 0.1680 g of $NaHCO_3$ (final concentration 2 mM); 0.1100 g of NA-pyruvate (final concentration 1 mM); 3.0576 mL of NA-lactate (final concentration 21.8 mM); and 5 mL of Gentamycin (final concentration 50 μg/mL). Add B60 water up to 1000 mL. Close the 1000 mL measuring cylinder with parafilm and mix to dissolve. Reserve 150 mL of the mixture in a 250 mL measuring cylinder and label "IVF-TALP." Dissolve 0.34 g of BSA (final concentration of 6 μM) in the remaining 850 mL of solution and add 8.5 mL of 1 M HEPES solution (final concentration of 9.9 mM). Filter the solution with a vacuum pump and a 0.22 μm bottle-top vacuum filter. Store at 4°C for a maximum of 14 days. Label the solution "HEPES-TALP" (*see* **Note 1**). Check the osmolarity; the final osmolarity of the solution should be adjusted to fall within the range of 275–285 mOsm/kg H_2O. Filter with a syringe and a 0.22 μm filter. Store the media at 4°C for a maximum of 14 days.
13. IVF-TALP medium: To the reserved 150 mL of "IVF-TALP," add 0.3 g $NaHCO_3$ (final concentration of 23.8 mM). Close the cylinder with parafilm and mix to dissolve the $NaHCO_3$. Check the osmolarity, which should be within the range of 305–315 mOsm/kg H_2O. Filter with a syringe and a 0.22 μm filter. Store the media at 4°C for a maximum of 14 days.
14. SP-TL (−) BSA medium: Pour 100 mL of IVF TALP medium into a fresh 100 mL measuring cylinder and label SP-TL (−) BSA medium. The final solution was prepared by adding 1 mL of 1 M HEPES stock solution to 100 mL of solvent, resulting in a final concentration of 9.9 mM. The osmolarity should fall between 315 and 325 mOsm/kg H_2O. Filter with a syringe and a 0.22 μm filter. Store the media at 4°C for a maximum of 14 days.
15. SP-TL (+) BSA medium: Pour 50 mL of SP-TL (−) BSA in a fresh 50 mL measuring cylinder. Add 0.3 g of BSA to the

50 mL solution to a final concentration of 0.902 mM. Close the measuring cylinder with parafilm. Label the solution SP–TL (+) BSA. The osmolarity should fall between 315 and 325 mOsm/kg H_2O. Filter with a syringe and a 0.22 μm filter. Store the media at 4°C for a maximum of 14 days.

16. Enriched IVF-TALP: IVF-TALP medium enriched with 6 mg/mL bovine serum albumin (BSA) and 25 μg/mL heparin.
17. SOF stock A (10× concentrated salt solution): Fill a sterile 50 mL tube with 30 mL B60 water then add the following components: 3.145 g NaCl (final concentration 1076 mM); 0.267 g of KCl (final concentration 71.6 mM); 0.081 g of K H_2PO_4 (final concentration 9.3 mM); 0.091 g of $MGSO_4$ (final concentration mM); and 300 μL of Na-Lactate (final concentration 41.85 mM). Dissolve by shaking the tube, then add B60 water up to 50 mL. Stable for 6 months at 4°C.
18. SOF stock B (10× bicarbonate solution): Fill a sterile 50 mL tube with 30 mL B60 water, then add 1.05 g of $NaHCO_3$ (final concentration 250 mM). Allow the $NaHCO_3$ to dissolve, then add B60 water to 50 mL. Stable for 6 months at 4°C.
19. SOF stock C (72.7 mM pyruvate solution): To a fresh sterile 15 mL tube, add 10 mL of B60 water using a sterile serological pipette, then add 0.80 g of sodium pyruvate.
20. SOF stock D (236 mM calcium chloride solution): To a sterile 15 mL tube, add 10 mL of B60 water, then add 0.262 g of $CaCl_2$. Stable for 6 months at 4°C.
21. Gentamycin stock solution (50 mg/mL): To a new sterile 15 mL conical tube, add 5 mL of sterile B10 physiological solution using a 5 mL sterile serological pipette. To this, add 0.250 g of gentamicin powder and allow to dissolve completely. The weighing container is rinsed with a small volume of the same solution, and this wash is returned to the conical tube to ensure full recovery of the compound. The resulting solution is aliquoted under sterile conditions. Aliquots of 110 μL are dispensed into 1.5 mL microcentrifuge tubes, and aliquots of 25 μL are dispensed into 0.5 mL tubes. Stable for 1 year at −20°C.
22. 0.5% w/v Phenol red. Stable for up to 1 year at room temperature and protected from light.
23. Synthetic Oviductal Fluid (SOFaa) stock: For a final volume of 75 mL, combine the following in a 100 mL measuring cylinder: 56 mL of B60 water; 0.0075 g of sodium citrate dihydrate (final concentration of 0.34 mM); 0.0375 g of Myoinositol (final concentration of 2.77 mM) Pipette 7.5 mL

of SOF stock A; 7.5 mL of SOF stock B; 750 μL of SOF stock C; and 750 μL of SOF stock D. Add 1.5 mL of 50× BME (final concentration 2% non-essential amino acids) and 150 μL of 100× MEM (final concentration 1% essential amino acids), 150 μL of 220 mM Glutamine (final concentration 0.44 mM), 75 μL of Gentamycin (final concentration 10 μg/mL), and 150 μL of Phenol red (final volume 5 μg/mL). The osmolarity range should fall between 275 and 285 mOsm/kg. Sterilize with a 0.22 μm filter and store at 4°C.

24. Insulin stock, 86.1 μM: In a fresh sterile 15 mL tube, fill with 10 mL of B60 water, add 0.005 g of Insulin and 40 μL of 1 N HCl. Mix by shaking, sterilize with a 0.22 μm filter, and aliquot (200 μL) into Eppendorf tubes. Store for up to 1 year when stored at −80°C.
25. Transferrin stock, 6.25 μM: Fill a fresh sterile 15 mL tube with 10 mL of B60 water, add 0.005 g of Transferrin. Mix by shaking, then sterilize using a 0.22 μm filter and aliquot (200 μL) into Eppendorf tubes. Store for up to 1 year when stored at −80°C.
26. Sodium selenite stock, 28.9 μM: In a fresh sterile 15 mL tube, fill with 10 mL of B60 water, add 0.005 of Sodium Selenite, and mix by shaking.
27. Sodium selenite working stock: Dilute the 100× Sodium Selenite solution in a fresh sterile 15 mL tube by pipetting 9.9 mL of B60 water, then add 100 μL of 100× Sodium Selenite solution. Sterilize with a 0.22 μm filter and prepare 200 μL aliquots in Eppendorf tubes. Store for up to 1 year at −80°C.
28. SOFaa culture medium, supplemented with 0.4% BSA and 5 μg ITS. In a fresh 15 mL tube, pipette 10 mL of SOFaa stock, then add 0.4 g of BSA. Sterilize using a 0.22 μm filter. Add 10 μL each of insulin stock, transferrin stock, and sodium selenite working stock per mL of SOF culture media.
29. Phosphate-buffered saline (PBS) Solution (no calcium, no magnesium), pH 7.4, sterilized using a 0.22 μm filter.
30. Paraffin Oil tested for embryo culture.

2.3 RNA Extraction and cDNA Synthesis

The choice of extraction method should be guided by the biological material type, the need for high RNA yield, and the available laboratory setup (*see* **Note 1**).

1. An RNA Isolation Kit suitable for high-quality total RNA recovery from pico-scale samples, used according to the manufacturer's protocol. Here, we outline a protocol using the PicoPure RNA Isolation Kit containing Conditioning Buffer (CB), Extraction Buffer (XB), 70% Ethanol (EtOH), Wash

Buffer 1 (W1), Wash Buffer 2 (W2), Elution Buffer (EB), RNA purification columns, and collection tubes and microcentrifuge tubes.

2. RNase-free DNase suitable for on-column use, such as RQ1 RNase-free DNase containing DNase I stock solution and RDD buffer.
3. cDNA Synthesis Kit, e.g., iScript Reverse Transcriptase Supermix, containing iScript Reverse Transcriptase (a modified MMLV-derived enzyme optimized for reliable and sensitive cDNA synthesis; *see* **Note 2**), 5× iScript Reaction Mix, and Nuclease-free water.
4. Tris-HCl; 10 mM Tris-HCl, pH 8.0 in nuclease-free water.
5. Housekeeping primers (e.g., GAPDH; Table 1) and primers to amplify genes of interest.
6. Thermocycler.
7. Microcentrifuge tubes (RNase Free tubes: 1.5 and 2 mL).
8. Ethanol (75%–100%). Molecular grade, diluted to 70% with nuclease-free water for RNA extraction.
9. Microcentrifuge.
10. Plate spinner (if using PCR plates for quantification).
11. Vortex.
12. Disposable gloves.
13. Spectrophotometer (e.g., Nanodrop or Bioanalyzer).

2.4 Real-Time PCR (qPCR) Assay

1. Real-Time PCR instrument (e.g., iCycler iQ Real-Time PCR Detection System).
2. Compatible RNase-free real-time PCR plates and seal, or tubes and caps.
3. 2× SYBR Green Mastermix (*see* **Note 2**), e.g., iQ™ SYBR® Green Supermix.
4. Specific primers (*see* Table 1), designed using primer design software (*see* **Notes 3–5**), were prepared at a final concentration of 10 μM per reaction.
5. PCR-grade water.
6. Pipettors.
7. Filter pipette tips.
8. Disposable gloves.

Table 1
Bovine reference gene primers commonly used for RT-qPCR

Gene	Genbank accession number	species	Sequence	Product size (bp)	Ta (°C)	% Homology
ACTB	AY141970	Cow	5′-CCTCACGGAACGTGGTTACA-3′ 5′-TCCTTGATGTCACGCACAATTT-3′	87	58	100%
GAPD	XM_618013	Cow	5′-TTCAACGGCACAGTCAAGG-3′ 5′-ACATACTCAGCACCAGCATCAC-3′	119	62	100%
Histone H2A	U62674	Mouse	5′-GTCGTGGCAAGCAAGGAG-3′ 5′-GATCTCGGCCGTTAGGTACTC-3′	182	60	82%
TBP	NM_003194	Human	5′-CCTAAAGACCATTGCACTTCG-3′ 5′-CTTCACTCTTGGCTCCTGTG-3′	146	57	94%
HPRT1	AF176419	Cow	5′-TGCTGAGGATTTGGAGAAGG-3′ 5′-CAACAGGTCGGCAAAGAACT-3′	154	58	100%
SDHA	NM_174178	Cow	5′-GCAGAACCTGATGCTTTGTG-3′ 5′-CGTAGGAGAGCGTGTGCTT-3′	185	60	100%
YWHAZ	BM446307	Cow	5′-GCATCCCACAGACTATTTCC-3′ 5′-GCAAAGACAATGACAGACCA-3′	120	60	97%
18S rRNA	AF176811	Cow	5′-AGAAACGGCTACCACATCCA-3′ 5′-CACCAGACTTGCCCTCCA-3′	169	62	100%

3 Methods

This protocol outlines the method for the analysis of gene expression in in vitro produced bovine embryos; however, it can be adapted for embryos from other species (*see* **Note 6**). Up to 20 genes of interest can be assessed from pools of 5–20 embryos, with RNA yield depending on both embryo number and developmental stage (Table 2, *see* **Note 7**). Earlier-stage embryos, such as two-cell embryos, contain significantly less RNA than later-stage embryos like blastocysts. Consequently, gene expression analysis from early-stage embryos may require pooling a larger number of embryos to obtain sufficient RNA for reliable quantification, whereas individual blastocysts may provide adequate RNA for single-embryo analysis. For in vivo-derived embryos, the protocol remains the same, except for slight modifications in the number of embryos pooled per sample (*see* **Note 1**).

3.1 In Vitro Embryo Production and Collection of Embryos

Culture embryos according to the preferred method of embryo culture. Routine bovine in vitro embryo production methods are described below [19].

3.1.1 Oocyte Collection and Maturation

1. At least 2 h prior to use, prepare IVM dishes by pipetting 500 μL of IVM medium into each well of a four-well plate. Add 1.5 mL TCM 199 Earle's medium in the center of the four-well plate to maintain humidity and place in a 5% CO_2 incubator for equilibration.

Table 2

Typical results after isolating RNA from eight blastocyst stage embryos using an Agilent Bioanalyzer system

Number of embryos	RNA concentration (pg/μL)	RIN value	rRNA ratio [28 S/18 S]	RNA area
8	736	10	2.7	259.5
8	812	10	2.3	286.2
8	996	10	2.6	351.2
8	404	9.9	2.6	142.4
8	781	10	2.2	275.4
10	1462	10	2.5	515.3
8	647	10	2.6	228.0

RNA concentration varies across samples depending on the embryo developmental stage and other factors. All samples were derived from blastocyst-stage embryos, with sample 6 (1462 pg/μL) showing nearly twice the RNA yield compared to other samples. While the higher embryo count (10 vs. 8) may contribute to this difference, additional factors such as developmental stage composition (e.g., the presence of hatching or expanded blastocystsEmbryoblastocyst with higher RNA content), embryo quality, or minor variations in RNA extractionBlastocystRNA extraction efficiency can also influence RNA yield. To reduce variability, standardizing embryo staging, pooling strategy, and processing conditions is recommended

2. Warm HEPES-TALP medium in a 38.5°C water bath overnight, or at least 2–3 h prior to start oocyte collection.
3. Warm Physiological Saline Solution in a 38.5°C water bath overnight, or at least 2–3 h prior to ovary collection.
4. Collect bovine ovaries from a local abattoir and transport them in a thermos containing warm Physiological Saline Solution within 2 h. Upon arrival in the lab, wash ovaries three times in warm Physiological Saline Solution.
5. Aspirate cumulus-oocyte complexes (COCs) from 4 to 8 mm diameter antral follicles using an 18-gauge needle attached to a 10 mL syringe. Transfer aspirate to 60 mm petri dishes and search for immature cumulus-oocyte complexes (COCs). Select only viable oocytes with uniform cytoplasm surrounded by three or more compact layers of cumulus cells.
6. Following two washes in HEPES-TALP medium, transfer the selected oocytes, in groups of 60 COCs, into prepared wells of IVM medium. Incubate oocytes in a 5% CO2 atmosphere at 38.5°C for 22 h to complete maturation.

3.1.2 Sperm Preparation and In Vitro Fertilization (IVF)

1. Prepare IVF dishes by pipetting 500 μL of enriched IVF-TALP medium into the left wells and 230 μL into the right wells of a four-well plate. Fill the middle compartment with 1.5 mL TCM 199 Earle's medium. Incubate and equilibrate dishes at 38.5°C in a 5% CO2 incubator, for at least 2 h before use.
2. On the day of fertilization, at 21 h of IVM, prepare Percoll gradients by pipetting 2 mL of 90% Percoll solution into a sterile 15 mL conical tube. Prepare a 45% Percoll solution by carefully mixing 1.5 mL of 90% Percoll solution with 1.5 mL SP-TL (−) BSA medium. Carefully layer this 45% Percoll solution over the 90% Percoll solution to create a gradient, ensuring a clear interface is visible between both layers (*see* **Notes 8 and 9**). Allow the Percoll gradients to equilibrate at room temperature prior to use.
3. Thaw the bull sperm straw(s), with sterile scissors, cut off the sealed end of the straws close to the edge and insert the open end into a sterile 1.5 mL tube. Then cut off the opposite end of the straws to release the contents into the tube. Using a sterile 1-mL syringe fitted with a 20 G needle, aspirate the suspension and carefully layer it on top of the 45% Percoll solution, taking care not to disrupt the gradient interface. Centrifuge the gradient at 700 × *g* for 30 min at 22°C.
4. Following centrifugation, carefully remove the supernatant above the sperm pellet without disturbing the pellet (*see* **Note 9**). Wash the sperm pellet by resuspending it gently in 5 mL of SP-TL (+) BSA medium. Centrifuge again at 100 × *g* for 10 min at 22°C.

5. Carefully remove the supernatant; the sperm pellet is retained in approximately 100 μL of 5 mL of SP-TL (+) BSA medium. To resuspend the pellet, the tube is gently tapped until the sperm are fully dispersed, forming a homogeneous suspension.
6. Remove 5 μL of the sperm suspension and dilute it in 45 μL of sterile water in a sterile 0.5 Eppendorf tube. Using a haemocytometer, determine the concentration of sperm.
7. Resuspend the final sperm pellet in the appropriate calculated volume of enriched IVF medium to achieve a concentration of 2 million sperm/mL. Adjust the sperm concentration to 2×10^6 spermatozoa/mL using enriched IVF-TALP medium.
8. Wash matured oocytes through the previously filled 500 μL IVF-TALP medium wells, then transfer 40–60 in 20 μL to the IVF wells on the right side of the dish containing 230 μL of IVF-TALP.
9. Add 250 μL of sperm suspension into the IVF wells containing 250 μL IVF-TALP medium and matured oocytes, resulting in a final sperm concentration of 1×10^6 spermatozoa/mL in a final volume of 500 μL enriched IVF-TALP medium. Co-incubate oocytes with sperm for 21–22 h.

3.1.3 Embryo Culture

1. Prepare four-well dishes, using one plate per 100 zygotes. Add 1.5 mL of SOFaa culture medium to the middle of each four-well dish. Prepare 5 μL SOF culture medium droplets in each well, taking the medium from the center of the dish. Carefully overlay each well with 900 μL of paraffin oil using a micropipette. Add a further 45 μL SOFaa culture medium droplets in each well and ensure the oil completely covers the 50 μL droplets; if not, add more oil until fully covered. Place the dishes in a CO_2 incubator and equilibrate overnight at 38.5°C.
2. Place a sterile 50 mL Falcon tube of HEPES-TALP in a medium warming incubator at 38.5°C overnight prior to starting embryo culture.
3. Following IVF, remove presumptive zygotes from fertilization wells and transfer them into a 15 mL conical tube containing 2.5 mL of pre-warmed HEPES-TALP medium. Vortex the presumptive zygotes for 3 min to remove excess cumulus cells and sperm.
4. Transfer the suspension into a 35 mm petri dish containing warmed HEPES-TALP medium. Identify and collect fully denuded embryos, ensuring only clean zygotes are selected for subsequent culture. Transfer individual embryos to prepared embryo culture dishes (*see* **Note 10**).

5. Incubate embryos at 38.5°C in an atmosphere containing 5% CO_2, 5% O_2, and 90% N_2 until they reach the desired stages, e.g., two-cell (24–36 h post insemination; hpi), eight-cell (48–64 hpi), blastocyst (168 hpi), and hatched blastocyst (up to 192 hpi).
6. Prior to collecting embryos for RNA extraction and analysis, wash all embryos three times in PBS, then transfer them in groups of 20 to 1.5 mL RNase-free microcentrifuge tubes (*see* **Notes 11 and 12**). Prepare at least three biological replicates of 20 pooled embryos for each developmental stage of interest (e.g., two-cell, eight-cell, blastocyst, and hatched blastocyst) (*see* **Note 13**).
7. Store at −80°C until RNA extraction.

3.2 RNA Extraction

When extracting RNA from delicate samples like oocytes or embryos, it is important to customize the protocol based on the sample size (*see* **Note 7**). A 10–30 mg sample of bovine tissue (heart, kidney, liver, muscle, lung, and placenta) should be processed in parallel for standard curve generation (*see* **Note 14**). Here, the PicoPure RNA Isolation Kit is used following the manufacturer's recommendations (*see* **Note 15**). Prepare all reactions on ice.

1. Place microcentrifuge tubes containing collected embryos on ice to thaw.
2. Mix the Extraction Buffer (XB) by pipetting thoroughly. If precipitate is present, warm the vial at 42°C to dissolve completely.
3. Add 50 μL of Extraction Buffer (XB) to each group of 20 pooled embryos from the two-cell, eight-cell, blastocyst, and hatched blastocyst stages. Incubate at 42°C for 30 min (*see* **Note 16**).
4. For bovine tissue, homogenize by pipetting each tissue sample (approximately five cuboidal pieces of 1.5 mm per side are required to reach a volume in the range of 10–30 mg) in 100 μL of Extraction Buffer (XB). Mix thoroughly.
5. Centrifuge the sample at 3000 × *g* for 2 min and collect the lysate, avoiding any pellet or cellular debris.
6. Pipette 250 μL of Conditioning Buffer (CB) onto the filter membrane of each RNA purification column. Incubate for 5 min at room temperature, then centrifuge at 16,000 × *g* for 1 min and discard the flow-through.
7. Add 50 μL of 70% ethanol to the embryo lysate. Mix gently by pipetting up and down without vortexing.

8. Transfer the lysate and ethanol mixture to the preconditioned RNA purification column. Centrifuge at 100 × *g* for 2 min, then at 16,000 × *g* for 30 s. Discard the eluate.
9. Add 100 µL of Wash Buffer 1 (W1) to the column. Centrifuge at 8000 × *g* for 1 min and discard the flow-through.
10. Prepare the DNase reaction mix on ice according to the manufacturer's instructions and apply it to the column. For RQ1 RNase-free DNase, add 5 µL of DNase I stock solution (2.72 Kunitz Units/µL in water) to 35 µL of RDD buffer for each sample, gently mixing by inverting or soft pipetting without vortexing, as DNase I is highly sensitive to physical denaturation. Briefly centrifuge to collect residual liquid from the sides of the tube.
11. For genomic DNA removal, treat the bound RNA with two units of RQ1 DNase directly on the column. Incubate at room temperature for 15 min, then continue with the next wash step (*see* **Note 17**).
12. Add 100 µL of Wash Buffer 2 (W2) to the column and centrifuge at 8000×g for 1 min. Discard the flowthrough.
13. Add another 100 µL of Wash Buffer 2 (W2) and centrifuge at 16,000 × *g* for 2 min. Ensure there is no remaining wash buffer on the membrane; if any remains, centrifuge the tube again for 1 min at 16,000 × *g* (*see* **Note 18**).
14. Transfer the RNA purification column to a fresh RNase-free 0.5 mL microcentrifuge tube. Add 11 µL of Elution Buffer (EB) directly onto the membrane, incubate at room temperature for 1 min, then centrifuge at 1000 × *g* for 1 min, followed by 16,000 × *g* for 1 min, to elute the RNA (*see* **Note 19**).
15. Determine the concentration of neat (undiluted) RNA using a spectrophotometer (e.g., Nanodrop; *see* **Note 20**).
16. Store isolated RNA at −80°C for future use (*see* **Note 21**) or proceed to cDNA synthesis.

3.3 cDNA Synthesis

cDNA synthesis can be performed using either total RNA or poly(A)+-selected RNA primed with oligo(dT), random primers, or a gene-specific primer. The method below uses the iScript™ cDNA Synthesis Kit.

1. Thaw all components of the iScript cDNA Synthesis Kit (Bio-Rad) at room temperature. Mix the 5× iScript Reaction Mix thoroughly in case any precipitate is present.
2. In a 0.2 mL PCR tube, prepare reverse transcription reaction mix by combining 4 µL of 5× iScript Reaction Mix, 1 µL of iScript Reverse Transcriptase, 1 µg of pooled total RNA

(embryo), and nuclease-free water for a 20 μL final volume per sample. The same reaction mix should be prepared in triplicate for tissue.

3. In parallel, prepare an identical reaction omitting the reverse transcriptase enzyme (–RT reaction). Combine 4 μL of 5× reaction mix, 1 μg of pooled total RNA, and nuclease-free water for a final reaction volume of 20 μL (*see* **Note 22**).
4. Place the reaction tubes in a thermocycler and incubate under the following conditions: Prime for 5 min at 25°C; reverse transcription for 20 min at 46°C; and then RT inactivation for 1 min at 95°C. As an optional step, reactions can be held at 4°C until further processing.
5. Use 2.0 μL of sample to determine the concentration of RNA in order to estimate the extraction yield (Table 2).
6. Dilute the remaining cDNA 2.5 times in 10 mM Tris-HCl, pH 8.0, for downstream use.
7. Pool the triplicate tissue cDNA reactions. Generate a standard curve by making a 10-point serial dilution in 10 mM Tris-HCl, pH 8.0. Aliquot the pooled stock to avoid repeat freeze–thaw cycles. Store synthesized cDNA at –20°C until use in qPCR (*see* **Note 23**).

3.4 Real-Time-qPCR

Perform all PCR reactions in a 20 μL total volume. The steps outlined below have been developed for use with an iCycler iQ Real-Time PCR Detection System but are applicable to other Real-Time PCR machines (*see* **Note 24**). Reactions can be performed in PCR tubes or PCR plates suitable for the Real-Time PCR machine available.

1. For each gene of interest, generate a 10-point standard curve. Prepare a master mix consisting of 10 μL iQ™ SYBR® Green Supermix, 0.5 μL of each sequence-specific primer (10 μ M stock, forward and reverse primers; final concentration 250 nM), and 7 μL nuclease-free water. Prepare sufficient master mix to analyze each tissue dilution in quadruplicate, as well as run a no template control (NTC) in duplicate, allowing extra of each component for pipetting error.
2. Pipette 18 μL of master mix into each tube/well of a PCR plate.
3. Add 2 μL of diluted cDNA template for a final reaction volume of 20 μL.
4. Set up NTC reactions by replacing the cDNA template with the same volume of nuclease-free water (*see* **Note 25**).
5. Seal the tubes/plate, then spin to ensure the reaction is at the bottom of the tube/well.

6. Program the real-time PCR machine with the following conditions: Initial denaturation at 95°C for 2–3 min to activate Taq DNA polymerase, followed by 45 cycles of: Denaturation at 95°C for 15–20 s, and combined annealing and extension at 60°C for 40 s, with fluorescence measured at the end of each cycle.
7. Calculate PCR efficiencies using the standard curve generated from the ten-point dilution of pooled tissue cDNA (*see* **Note 5**).
8. To analyze embryo samples, prepare sufficient master mix to analyze each sample in duplicate, as well as to run a no template control (NTC) in duplicate for each gene of interest, allowing extra of each component for pipetting error. Each reaction should consist of 10 μL iQ™ SYBR® Green Supermix, 0.5 μL of each sequence-specific primer (10 μM stock, forward and reverse primers; final concentration 250 nM), and 7 μL nuclease-free water.
9. Pipette 18 μL of master mix into each tube/well of a PCR plate.
10. Add 2 μL of diluted cDNA template (either from pooled embryos or the full cDNA equivalent of a single embryo; *see* **Note 26**) or nuclease-free water (NTC) for a final reaction volume of 20 μL.
11. Program the real-time PCR machine using the same conditions as for the standard curve: Initial denaturation at 95°C for 2–3 min to activate Taq DNA polymerase, followed by 45 cycles of: Denaturation at 95°C for 15–20 s, and combined annealing and extension at 60°C for 40 s, with fluorescence measured at the end of each cycle.
12. Generate a melt curve at the end of the amplification to confirm a single gene-specific peak and detect any primer-dimer formation by heating from 70°C to 95°C in 0.5°C increments, with a dwell time of 10 s at each increment, while continuously monitoring fluorescence (*see* **Note 27**).

3.5 Control Gene Optimization and Gene Expression Data Analysis

Accurate normalization is essential in RT-qPCR to minimize errors introduced by variations in sample preparation, RNA quality, and reverse transcription efficiency (*see* **Note 28**). In the context of embryo gene expression analysis, careful selection of normalization methods is crucial due to the limited starting material and the dynamic nature of gene expression across developmental stages.

The expression stability of several candidate housekeeping genes should then be evaluated using algorithms such as geNorm (Link: https://genorm.cmgg.be/), NormFinder (Link:http://moma.dk/normfinder-software), or BestKeeper (Link: https://

www.gene-quantification.de/bestkeeper.html) (*see* **Note 29**). These programs assess pairwise variation and rank genes according to their stability across all samples. Following the recommendations of Vandesompele [10] and subsequent embryo-specific studies, normalization should ideally be based on the geometric mean of the two to three most stable genes. This approach reduces normalization error compared with single-gene normalization, particularly important for embryo studies where expression levels are often low and variable.

By implementing this pilot validation testing several candidate housekeeping genes across representative samples, selecting the most stable set using statistical algorithms, and applying the geometric mean for normalization, researchers can ensure robust and reproducible gene expression data in embryo RT-qPCR studies.

3.5.1 Control Gene Selection for Normalization

The Internal Control Gene-Stability Measure (M), as outlined here, provides a quantitative assessment of the stability of candidate reference (housekeeping) genes by examining pairwise variation. This measure, introduced to enhance the robustness of data normalization, is especially valuable for experiments with dynamic samples, such as embryonic development stages [10] (*see* **Note 30**).

1. Pairwise Variation (V): For every pair of internal control genes j and k, an array A_{jk} is created, containing the log2-transformed expression ratios of gene j relative to gene k across all samples. Specifically, for each sample i, a_i represents the normalized expression level derived from Cq values, typically calculated using the formula $a_i = 2 - \mathrm{Cq}$

$$A_{j,k,i} = \log_2\left(\frac{a_{ij}}{a_{ik}}\right)$$

   where $Ai_{i,j}$ and $A_{i,k}$ represent the normalized expression levels of genes j and k in sample i, respectively. Thus, the array A_{jk} consists of all values $A_{j,k,i}$ for $i = 1, 2, \ldots, m$, where m is the total number of samples.

2. Calculation of Pairwise Variation (V_{jk}): The pairwise variation V_{jk} for each control gene pair is calculated as the standard deviation of the elements in A_{jk}.

$$V_{jk} = \mathrm{st.dev}(A_{jk})$$

3. Gene-Stability Measure (M): The stability measure M for each internal control gene j is then calculated as the arithmetic mean of all pairwise variations V_{jk}, where n represents the total number of internal control genes

$$M_j = \frac{1}{n-1} \sum_{k \neq j} V_{jk}$$

4. Lower M values indicate more stable genes, which are preferable for normalization. This approach allows researchers to select the most stable reference genes, reducing the likelihood of normalization errors.

3.5.2 Relative Gene Expression Analysis

When analyzing RT-qPCR data, it is recommended to use the mean Cq value from PCR duplicates, as outlined in this protocol. While triplicates are generally advised for increased reliability, duplicates are commonly used when sample availability is limited, as is the case with bovine preimplantation embryos. This approach balances data accuracy with the need to preserve precious biological material.

The $\Delta\Delta Cq$ method is commonly used to quantify relative gene expression, where data are expressed as fold-changes relative to a control sample or reference condition.

1. Calculate the ΔCq for each sample

$$\Delta C_q = C_{q\text{target}} - C_{q\text{reference}}$$

where Cq_{target} is the quantification cycle (Cq) value of the gene of interest and $Cq_{\text{reference}}$ is the Cq value of the reference gene (housekeeping gene), used to normalize variations in RNA input.

2. Then calculate the $\Delta\Delta Cq$:

$$\Delta\Delta C_q = \Delta C_{q\text{sample}} - \Delta C_{q\text{control}}$$

where $\Delta Cq_{\text{sample}}$ is the ΔCq value of the experimental sample, and$\Delta Cq_{\text{control}}$ is the ΔCq value of the control or reference sample.

3. Finally, calculate the fold-change in gene expression:

$$\text{Fold change} = 2^{-\Delta\Delta Cq}$$

4. Present results as raw fold-change data, or as a ratio of treatment vs. control.

4 Notes

1. The selection of an RNA extraction method depends on several factors, including the type of biological material, required RNA purity, yield, and downstream applications. For instance, TRIzol-based methods are highly effective for extracting RNA from complex biological samples but may require additional cleanup to remove contaminants. Silica column-based kits (e.g., RNeasy Mini Kit, PureLink RNA Mini Kit) provide a faster and more straightforward approach, reducing contamination risk while ensuring reproducibility. Laboratories with limited equipment may prefer spin-column-based kits, as they offer ease of use while still allowing for DNase treatment and RNA integrity assessment, both of which are standard in many commercially available kits. However, for applications requiring highly pure RNA, such as RNA sequencing, additional quality control steps, such as extended DNase treatment, RNA cleanup procedures, and rigorous integrity assessment using Bioanalyzer or TapeStation—are recommended to ensure optimal RNA quality.
2. As an optional internal control, the TaqMan® RNase P Control Reagents kit can be included in the qPCR reaction setup as an endogenous internal reference for normalization and verification purposes. Following RNA isolation using the PicoPure RNA Isolation Kit and cDNA synthesis using the iScript cDNA Synthesis Kit, incorporate the RNase P primers and probes by adding 10 μL into the qPCR reaction to enable multiplex detection, subtracting the volume added from the water. The inclusion of this control is particularly beneficial when working with low quantity RNA samples to confirm successful isolation, reverse transcription efficiency, and PCR amplification. Any commercially available 2× SYBR Green Mastermix containing an internal reference dye (e.g., ROX) may be substituted, provided it is used according to the manufacturer's instructions.
3. In RT-qPCR, carefully designed primers are essential for successful and accurate gene expression analysis. While primers can be sourced from published studies, an effective strategy for designing primers is to use online tools such as IDT's PrimerQuest (available at https://eu.idtdna.com/PrimerQuest/Home/Index) or NCBI's Primer-BLAST (https://www.ncbi.nlm.nih.gov/tools/primer-blast/). Primers should have specific characteristics to ensure efficient amplification and minimize non-specific binding. When designing primers for gene expression studies, it is beneficial to target exon–exon junctions to avoid amplification from

any contaminating genomic DNA. This approach ensures that the primers selectively amplify cDNA derived from mRNA, thus accurately reflecting gene expression levels. For optimal performance in SYBR Green-based assays, primers are recommended to have a GC content between 30% and 80%, a melting temperature (Tm) of approximately 58–60°C, and a length of at least 18 bases. Shorter amplicons, ideally between 50 and 150 base pairs, are preferable as they enhance amplification efficiency and yield. To minimize primer-dimer formation and non-specific amplification, the 3′ end of the primer should have no more than two consecutive G or C nucleotides within the last five bases. This prevents excessive primer self-complementarity and reduces the risk of off-target priming during PCR amplification.

4. Secondary structures within the target region can impede primer binding and reduce amplification efficiency. To predict these structures, tools like mFold (http://unafold.rna.albany.edu/?q=mfold/DNA-Folding-Form) can be used. Primers should not bind to regions with more than six consecutive base pairs, as these may form strong stem-loop structures. Additionally, ΔG values below -9 kcal/mol indicate highly stable structures that could interfere with primer annealing. Avoiding GC-rich regions (>60%) and selecting target sites with minimal hairpin formation (stems ≤6 bases, loops ≥3 bases) can further improve primer accessibility and ensure efficient amplification. By following these steps, researchers can design primers that are both specific and efficient for RT-qPCR applications in gene expression studies.
5. Primer validation is a crucial step in ensuring that primers amplify the target gene with high efficiency and consistency. For comparative RT-qPCR analyses, primer efficiency should be verified using non-critical samples (e.g., tissue samples) to ensure reliable quantification and normalization. Validation is typically performed by running RT-qPCR on a series of cDNA dilutions, using at least 1:1, 1:10, 1:100, and 1:1000 dilutions, but preferably a ten-point standard curve. Each dilution should be tested in triplicate to minimize experimental variability.

   Primer efficiency can be calculated for each primer set from a plot of the cycle threshold (Ct) values against dilutions. The slope of the curve is then used to determine efficiency (E) using the formula:

$$\text{Efficiency}\,(\%) = (10 - 1/\text{slope} - 1) \times 100$$

   Efficiencies between 90% and 110% (or a slope between -3.3 and -3.6) are generally considered ideal for RT-qPCR experiments. This range indicates that the primers amplify the target gene effectively and consistently across a range of concentrations, enabling accurate and reproducible quantification of gene expression. Ideally, all primer sets to be assessed within a given experiment should have similar efficiencies, indicating that the template is being equivalently amplified for each target. Reliable primer efficiency ensures that RT-qPCR data can be confidently used for comparisons between different biological samples.

Additionally, non-template controls (NTCs) should be included, replacing the cDNA template with nuclease-free water and RT-reaction to confirm the absence of contamination and to validate specificity. Likewise, analysis of a melt curve should be performed to ensure the amplification of a single PCR product.

6. This gene expression analysis method, optimized for in vitro-produced bovine embryos, can be effectively adapted to embryos from multiple species, including humans [20, 21], mice [22], and horses [14, 23]. RT-qPCR methods described in this protocol, encompassing critical steps like RNA extraction, cDNA synthesis, primer validation, and rigorous normalization strategies, are broadly applicable and reliable for comparative embryonic transcriptomic studies.
7. When working with delicate samples such as oocytes and embryos, protocol customization is necessary to maximize RNA yield while minimizing RNA degradation. Key modifications include reducing sample handling time, using low-volume reagents to accommodate small sample sizes (low cell numbers), and incorporating RNA stabilizers (e.g., RNAprotect Tissue Reagent) to preserve RNA integrity. Additionally, optimizing lysis conditions and selecting a kit designed for low-input samples (e.g., PicoPure RNA Isolation Kit) can improve RNA recovery. Special attention should be given to minimizing RNase contamination and avoiding unnecessary freeze-thaw cycles, as RNA from small biological samples is particularly susceptible to degradation.
8. The clear interface between the 45% and 90% Percoll layers is crucial because it ensures the proper separation of motile, viable sperm from immotile, dead sperm cells and seminal debris. The distinct interface allows effective isolation of high-quality sperm in the pellet after centrifugation, improving fertilization efficiency and embryo development outcomes.
9. To carefully remove the supernatant layers above the sperm pellet after centrifugation, use a sterile Pasteur pipette or syringe fitted with a fine needle. Starting from the very top of the tube, gently aspirate the Percoll layers without disturbing

or touching the sperm pellet at the bottom. Gradually lower the pipette tip as the upper layers are removed, maintaining a controlled and steady aspiration. It is critical to leave a small residual volume of liquid (~100 μL) above the pellet to prevent accidental loss or disruption of motile sperm.

10. In this protocol, we describe using synthetic oviduct fluid (SOF) supplemented with 0.4% bovine serum albumin (BSA) in combination with insulin, transferrin, and selenium (ITS) for embryo culture based on our previous paper [19]. In our earlier work [18], we used 5% fetal bovine serum (FBS) instead of BSA. However, using 5% FBS increases the risk of adverse effects associated with serum, such as large offspring syndrome and inconsistent developmental outcomes, which are often linked to unknown serum factors. For best results, BSA and ITS should be of high purity and free from RNase and DNase to maintain the culture's integrity and improve embryo viability.

11. For in vivo-derived bovine embryos, a few adjustments to the in vitro protocol are necessary. Typically, fewer embryos (average of 8–10 embryos from 8 to 10 single ovulating mares or cows, average of 5–6 embryos in superovulated cows) are collected in vivo, requiring modifications in reagent volumes and handling. As with in vitro samples, in vivo embryos should be washed in DPBS to remove contaminants and stored pooled at −80°C in RNase-free lysis buffer (10 mM dithiothreitol, 1 U/μL RNasin Plus RNase inhibitor, 0.5% Igepal CA-630 in nuclease-free water) to preserve RNA integrity. Due to potential environmental and biological variability, RNA yield and quality may differ, necessitating careful optimization during RNA extraction. These considerations ensure the protocol is appropriately adapted for in vivo-derived samples.

12. When pooling embryos into microcentrifuge tubes, it is essential to verify that all embryos have been evacuated by using a stereomicroscope, as embryos can adhere to the pipette during handling. To minimize embryo loss due to adhesion, use Low Protein Binding Microcentrifuge Tubes, which help prevent embryos from sticking to the tube walls, thereby ensuring accurate counts and preserving sample integrity. It is also advisable to pre-rinse the pipette tips with the same medium used for embryo collection to reduce static forces and prevent embryos from adhering to the tip. Gentle pipetting and visually confirming the transfer after each step can further improve accuracy and reduce the risk of sample loss.

13. In accordance with the methodology proposed in [24], pooling approximately 20 bovine embryos per sample ensures

both adequate RNA yield and reproducible transcript quantification across key stages of preimplantation development.

14. When extracting RNA from pooled embryos or tissue, it is important to adjust the RNA input to align with the optimal range specified by the RNA extraction kit. Researchers should refer to the manufacturer's guidelines for each kit, as some methods (e.g., TRIzol-based extraction) are more tolerant of variable RNA concentrations than silica column-based kits, which often have stricter loading limits. Overloading a column-based kit can lead to inefficient binding, reduced RNA recovery, or column clogging. To prevent overloading spin columns or saturating downstream reactions, it is important to obtain a rough estimate of RNA quantity before extraction. This can be done by adjusting the amount of starting material or diluting the sample before extraction to avoid exceeding the column's binding capacity. For kits that require specific RNA input amounts (e.g., for downstream cDNA synthesis or real-time qPCR), dilution should be performed carefully to maintain consistency across samples. Conversely, for samples with very low RNA concentrations (e.g., single oocytes or small numbers of embryos), it may be necessary to perform a concentration step using ethanol precipitation or a specialized RNA concentrator kit to ensure sufficient RNA input for analysis.

15. While this protocol outlines RNA extraction using the PicoPure RNA Isolation Kit, other commercially available kits (e.g., RNeasy Mini Kit, PureLink RNA Mini Kit, or TRIzol-based methods) can also be compatible with the downstream real-time PCR assays described. Users should confirm that their chosen kit reliably yields high-quality RNA with minimal inhibitors, particularly for small-scale embryo samples. In our experience, phenol–chloroform methods sometimes result in lower yields and require additional cleanup steps to remove carryover contaminants [25], making them less suited for sensitive RT-qPCR assays with limited input material.

16. The PicoPure™ RNA Isolation Kit is optimized for recovering high-quality total RNA from small or challenging samples, such as embryos. Before adding Extraction Buffer (XB), keep pooled embryos (e.g., groups of 20) on ice to minimize RNase activity. Once XB is added, move the sample immediately to the 42°C incubation step—XB stabilizes the RNA by denaturing nucleases, so there is no need to maintain low temperature afterwards. For best results, ensure all reagents are at recommended temperatures, and thoroughly dissolve any precipitates in XB by warming at 42°C if necessary. Avoid vortexing the sample to prevent RNA degradation; instead, use gentle pipetting to mix reagents. Use only the kit's

components or those specified in the protocol to avoid RNase contamination or reduced yields.

17. A thorough DNase treatment step is strongly advised to eliminate any genomic DNA when the extracted RNA is destined for sensitive applications (e.g., RT-qPCR). To maximize efficiency, prepare the DNase reaction mixture immediately before use, keep it on ice, and store DNase I in single-use aliquots at −20°C. This practice minimizes freeze-thaw cycles and preserves enzyme activity for consistent DNA digestion.
18. It is essential to eliminate all traces of Wash Buffer 2 before adding the Elution Buffer; residual salts or ethanol can reduce RNA yield or interfere with downstream analysis. If any wash buffer remains on the membrane after the initial centrifugation, spin again at 16,000 × g for 1 min. Visually confirm that the membrane is free of liquid droplets before proceeding to the elution step.
19. For precise elution, pipette the Elution Buffer directly onto the column membrane, touching the pipette tip to the membrane to maximize absorption. If RNA quantitation is challenging due to low yields, note that fluorescence-based methods may be more accurate than UV spectrophotometry for small-scale samples.
20. After extraction, RNA should be quantified using a spectrophotometer (e.g., Nanodrop) or fluorometer (e.g., Qubit) to determine the precise concentration needed for downstream applications. If the RNA concentration is too high, it can be diluted using RNase-free water or the kit's elution buffer to achieve the desired input for subsequent steps. For RT-qPCR, a final RNA concentration of approximately 400 pg/μL is generally suitable, although this may vary depending on the kit and target RNA. While smaller loading volumes can help prevent overloading, this is often impractical due to already limited sample volumes, making dilution the preferred approach for ensuring consistent RNA input.
21. Samples should be stored at −80°C to prevent RNA degradation and should not be thawed more than three times. For small sample volumes, adding RNase-free water or wash buffer can help reach the required final volume. For larger sample volumes, it is essential to adjust the RNA input to match the requirements of the chosen extraction kit. Many kits have an optimal RNA input range, and excessive RNA amounts may require scaling down to prevent column saturation or inefficient recovery. This is particularly relevant for high-yield samples such as pooled embryos or large tissue biopsies, where RNA extraction efficiency can be compromised if the RNA load exceeds the kit's processing capacity.

22. When using larger amounts of input RNA (>1 μg), the reaction should be scaled up (for example, 40 μL reaction for 2 μ g or 100 μL for 5 μg) to ensure optimum synthesis efficiency.
23. To verify genomic DNA removal, perform a Quantitative PCR reaction. Briefly, qPCR is a technique that amplifies a specific DNA sequence while simultaneously measuring the amount of DNA in real time using fluorescent dyes or probes. Each 20 μL qPCR reaction contains 10 μL of iQ™ SYBR® Green Supermix, 0.5 μL of each (forward and reverse) GAPDH primer (10 μM stock; final concentration 250 nM), 2 μL of cDNA or –RT sample, and 7 μL nuclease-free water. Alternative primer sets can be used, provided they span an exon junction. Amplification is performed in a real-time PCR instrument using the following cycling program: initial denaturation at 95°C for 3 min, followed by 40 cycles of 95°C for 15 s (denaturation) and 60°C for 30 s (combined annealing/extension). A melt curve analysis is included at the end of the run to verify the specificity of the amplified product. Negative controls (no template) and –RT samples are included in each experiment. If there is no amplification (or a high Cq value approaching "no signal") in the –RT sample, this indicates minimal genomic DNA contamination. If clear amplification is detected in the –RT sample, there is likely residual genomic DNA present, and you may need to review your DNase treatment or sample handling procedures.
24. When using a real-time PCR machine other than the iCycler iQ Real-Time PCR Detection System, optimal cycling conditions (such as temperature, time for each step, and ramp rates) may need adjustments. As specified in the SYBR Green Master Mix manual, different mixes, as well as different machines, may have unique thermal profiles and ramping characteristics. For instance, the recommended annealing temperatures for SYBR Green assays can range between 55°C and 65°C. Running a pilot test on your specific machine with your selected SYBR mix is recommended to optimize these parameters and to ensure consistent and reliable amplification results.
25. A valid NTC should not yield any amplification curve or signal. If the NTC shows amplification, investigate potential primer-dimer formation or reagent contamination before proceeding with more sensitive embryo samples.
26. For experiments requiring data from single embryos (e.g., examining inter-embryo variability), use the entire cDNA yield from one embryo's RNA extraction without pooling. In contrast, when multiple embryos (e.g., 20) are pooled for a more robust RNA yield, 2.5 μL of diluted cDNA no longer

represents exactly one embryo but provides a consistent fraction of the total pooled sample. This pooling strategy is often preferred when the primary goal is to enhance RNA yield and improve detection sensitivity, rather than to track individual-embryo differences. If single-embryo quantification (e.g., for embryo-to-embryo variation studies) is preferred, skip the pooling step and use that embryo's entire cDNA yield. Conversely, if multiple embryos are pooled for increased RNA yield, 2.5 μL of the diluted cDNA typically represents only a fraction of one embryo but ensures consistent template input across all qPCR reactions.

27. When assessing the melt curve, confirm that only a single, specific product peak is present. A single, high-temperature peak indicates the intended amplicon. Primer-dimers typically melt at lower temperatures, producing a separate (often broader) peak. In RT-qPCR, genomic DNA contamination sometimes appears as a peak at a higher temperature than the main product (especially if the gDNA amplicon is larger when primers are designed over intro–exon boundaries). Always include negative controls (no-template and minus-RT) and check for extra peaks or shoulders. If multiple peaks are visible, adjust primer concentrations, annealing conditions, or use DNase treatment to ensure the observed melt peak corresponds exclusively to the target amplicon.
28. According to the MIQE (Minimum Information for Publication of Quantitative Real-Time PCR Experiments) guidelines [9], rigorous normalization and validation procedures are essential in RT-qPCR studies to ensure the accuracy and reproducibility of gene expression measurements. In complex systems, such as embryonic development, normalization strategies must account for variations in RNA quantity and reverse transcription efficiency across samples, making careful selection of stable reference genes critical [10, 11]. Standard approaches involve normalizing target gene expression levels against reference genes, often called housekeeping genes, assumed to have stable expression across conditions. However, studies indicate that commonly used reference genes, such as GAPDH and ACTB, may not be stable across all experimental setups, particularly in early embryos or during comparisons of different embryo production methods like somatic cell nuclear transfer (SCNT) and in vitro fertilization (IVF) [12, 13]. As a result, each experimental design will require validation of reference genes within the context of the variables being assessed.
29. In practice, it is recommended to pre-validate multiple candidate housekeeping genes before analyzing genes of interest. A minimum of three to four candidate reference genes

should be selected, as reliance on a single control gene, such as GAPDH, can introduce substantial bias. Common candidates tested in bovine and equine embryos include ACTB, GAPDH, HPRT1, SDHA, YWHAZ, and 18S rRNA, but the optimal set may differ depending on developmental stage, embryo origin (in vivo, IVF, or SCNT), or treatment condition. Validation should be performed on representative samples that reflect the full experimental design, i.e., across the developmental stages and treatment groups that will later be analyzed (for example, pooled two-cell, eight-cell, morula, and blastocyst stage embryos from both control and treatment groups). This ensures that reference gene stability is assessed under the same biological variation that will be present in the experiment. Where embryo numbers are limiting, pooled samples (10–20 embryos per pool) provide sufficient RNA for testing stability. In contrast, when single-embryo variability is of interest, the entire RNA yield from individual embryos may be used.

30. Alternatively, the Single Control Normalization Error (*E*) method can be used to evaluate the stability and reliability of internal control genes by comparing their performance across multiple samples. This method is especially useful when multiple potential reference genes are being assessed to determine the most stable gene for the experimental conditions being examined.

To perform Single Control Normalization Error analysis, compute *E* for every combination of reference genes (*j, k)* and for every pair of samples (*p* and *q*), where *ai* represents the normalized expression level of gene in sample *i*, which is typically calculated by transforming Ct values into relative quantities, for example using the formula:

$$a_i = 2^{-\mathrm{cti}}$$

$$E_j, k = \frac{\left(\frac{a_{ij}}{a_{ik}}\right)_p}{\left(\frac{a_{ij}}{a_{ik}}\right)_q}$$

This equation represents the fold difference in gene expression between two samples when normalized to either reference gene *j* or *k*. Ideally, a low *E* value indicates minimal variation between reference genes, indicating their reliability for normalization.

References

1. Jozefczuk J, Adjaye J (2011) Chapter Six—Quantitative real-time PCR-based analysis of gene expression. In: Jameson D, Verma M, Westerhoff HV (Eds.) Methods in enzymology. Academic Press, United States, vol. 500:99–109. https://doi.org/10.1016/B978-0-12-385118-5.00006-2
2. VanGuilder HD, Vrana KE, Freeman WM (2008) Twenty-five years of quantitative PCR for gene expression analysis. Biotechniques 44(5):619–626. https://doi.org/10.2144/000112776
3. Taylor SC, Laperriere G, Germain H (2017) Droplet digital PCR versus qPCR for gene expression analysis with low abundant targets: from variable nonsense to publication quality data. Sci Rep 7(1):2409. https://doi.org/10.1038/s41598-017-02217-x
4. Wong ML, Medrano JF (2005) Real-time PCR for mRNA quantitation. Biotechniques 39(1):75–85. https://doi.org/10.2144/05391RV01
5. Ivan Bower N, Joachim Moser R, Robert Hill J, Arabella Lehnert S (2007) Universal reference method for real-time PCR gene expression analysis of preimplantation embryos. Biotechniques 42(2):199–206. https://doi.org/10.2144/000112314
6. Niemann H, Wrenzycki C (2000) Alterations of expression of developmentally important genes in preimplantation bovine embryos by in vitro culture conditions: implications for subsequent development. Theriogenology 53(1):21–34. https://doi.org/10.1016/S0093-691X(99)00237-X
7. Steuerwald N, Cohen J, Herrera RJ, Brenner CA (1999) Analysis of gene expression in single oocytes and embryos by real-time rapid cycle fluorescence monitored RT–PCR. Mol Hum Reprod 5(11):1034–1039. https://doi.org/10.1093/molehr/5.11.1034
8. Hasegawa Y, Taylor D, Ovchinnikov DA, Wolvetang EJ, de Torrenté L, Mar JC (2015) Variability of gene expression identifies transcriptional regulators of early human embryonic development. PLoS Genet 11(8):e1005428. https://doi.org/10.1371/journal.pgen.1005428
9. Eberwine J, Yeh H, Miyashiro K, Cao Y, Nair S, Finnell R, Zettel M, Coleman P (1992) Analysis of gene expression in single live neurons. Proc Natl Acad Sci U S A 89(7):3010–3014. https://doi.org/10.1073/pnas.89.7.3010
10. Vandesompele J, De Preter K, Pattyn F, Poppe B, Van Roy N, De Paepe A, Speleman F (2002) Accurate normalization of real-time quantitative RT-PCR data by geometric averaging of multiple internal control genes. Genome Biol 3(7):research0034.0031. https://doi.org/10.1186/gb-2002-3-7-research0034
11. Wang Y, Navin Nicholas E (2015) Advances and applications of single-cell sequencing technologies. Mol Cell 58(4):598–609. https://doi.org/10.1016/j.molcel.2015.05.005
12. Davis JW, De Sousa PA, Schultz RM (1996) Transient expression of translation initiation factor eIF-4C during the 2-cell stage of the preimplantation mouse embryo: identification by mRNA differential display and the role of DNA replication in zygotic gene activation. Dev Biol 174(2):190–201. https://doi.org/10.1006/dbio.1996.0065
13. Temeles GL, Ram PT, Rothstein JL, Schultz RM (1994) Expression patterns of novel genes during mouse preimplantation embryogenesis. Mol Reprod Dev 37(2):121–129. https://doi.org/10.1002/mrd.1080370202
14. Smits K, Goossens K, Van Soom A, Govaere J, Hoogewijs M, Vanhaesebrouck E, Galli C, Colleoni S, Vandesompele J, Peelman L (2009) Selection of reference genes for quantitative real-time PCR in equine in vivo and fresh and frozen-thawed in vitro blastocysts. BMC Res Notes 2(1):246. https://doi.org/10.1186/1756-0500-2-246
15. Lun ATL, Calero-Nieto FJ, Haim-Vilmovsky L, Göttgens B, Marioni JC (2017) Assessing the reliability of spike-in normalization for analyses of single-cell RNA sequencing data. Genome Res 27(11):1795–1806. https://doi.org/10.1101/gr.222877.117
16. Athanasiadou R, Neymotin B, Brandt N, Wang W, Christiaen L, Gresham D, Tranchina D (2019) A complete statistical model for calibration of RNA-seq counts using external spike-ins and maximum likelihood theory. PLoS Comput Biol 15(3):e1006794. https://doi.org/10.1371/journal.pcbi.1006794
17. Zhang X, Ding L, Sandford AJ (2005) Selection of reference genes for gene expression studies in human neutrophils by real-time PCR. BMC Mol Biol 6(1):4. https://doi.org/10.1186/1471-2199-6-4
18. Goossens K, Van Poucke M, Van Soom A, Vandesompele J, Van Zeveren A, Peelman LJ (2005) Selection of reference genes for quantitative real-time PCR in bovine preimplantation embryos. BMC Dev Biol 5(1):27. https://doi.org/10.1186/1471-213X-5-27
19. Wydooghe E, Heras S, Dewulf J, Piepers S, Van den Abbeel E, De Sutter P, Vandaele L, Van Soom A (2014) Replacing serum in culture medium with albumin and insulin, transferrin and selenium is the key to successful bovine

embryo development in individual culture. Reprod Fertil Dev 26(5):717–724

20. Dobson AT, Raja R, Abeyta MJ, Taylor T, Shen S, Haqq C, Pera RAR (2004) The unique transcriptome through day 3 of human preimplantation development. Hum Mol Genet 13(14):1461–1470. https://doi.org/10.1093/hmg/ddh157
21. Yan L, Yang M, Guo H, Yang L, Wu J, Li R, Liu P, Lian Y, Zheng X, Yan J, Huang J, Li M, Wu X, Wen L, Lao K, Li R, Qiao J, Tang F (2013) Single-cell RNA-Seq profiling of human preimplantation embryos and embryonic stem cells. Nat Struct Mol Biol 20(9):1131–1139. https://doi.org/10.1038/nsmb.2660
22. Hamatani T, Carter MG, Sharov AA, Ko MSH (2004) Dynamics of global gene expression changes during mouse preimplantation development. Dev Cell 6(1):117–131. https://doi.org/10.1016/S1534-5807(03)00373-3
23. Klein C, Rutllant J, Troedsson MHT (2011) Expression stability of putative reference genes in equine endometrial, testicular, and conceptus tissues. BMC Res Notes 4(1): 120. https://doi.org/10.1186/1756-0500-4-120
24. Robert C, McGraw S, Massicotte L, Pravetoni M, Gandolfi F, Sirard M-A (2002) Quantification of housekeeping transcript levels during the development of bovine preimplantation embryos. Biol Reprod 67(5):1465–1472. https://doi.org/10.1095/biolreprod.102.006320
25. Pavani KC, Baron EE, Faheem M, Chaveiro A, Da Silva FM (2015) Optimisation of total RNA extraction from bovine oocytes and embryos for gene expression studies and effects of cryoprotectants on total RNA extraction. Cytol Genet 49(4):232–239. https://doi.org/10.3103/S0095452715040076

Chapter 17

Analysis of mtDNA Methylation

Camila Bruna de Lima and Marc-André Sirard

Abstract

Whole-genome methyl sequencing facilitates the characterization of DNA methylation profiles within a reference genome. Recent identification of mitochondrial DNA (mtDNA) as an additional regulatory pathway within embryos paves the way for examining the impact of assisted reproductive technologies and other environmental factors on gamete and embryo developmental programming. Here, we describe the protocol for preparing libraries tailored to detect cytosine methylation in mtDNA extracted from minimal input, typical of experiments involving reproductive samples such as oocytes and embryos.

Keywords Mitochondria, mtDNA, Cytosine methylation, Epigenetics

1 Introduction

Mitochondria play important roles in metabolism and the regulation of cellular processes, including apoptosis [1]. Mitochondrial function has been shown to be critical for oocyte quality and preimplantation embryo development [2, 3], with mitochondria being inherited exclusively maternally [4]. Each mitochondrion possesses multiple copies of mitochondrial DNA (mtDNA), a 16 kB circular genome encoding 13 oxidative phosphorylation proteins (subunits of complexes I, III, IV, and V), mitochondria-specific small and large ribosomal RNA fragments and 22 transfer RNA molecules [5]. More recently, mitochondria, and metabolites more generally, have been implicated in the regulation of epigenetics [6]. Methylation of cytosines is one of the most important epigenetic modifications of nuclear DNA, regulating gene expression. However, unlike nuclear encoded mitochondrial genes, mtDNA lacks nucleosomes or histones, and rather self-organizes into structures called nucleoids [7]. Consequently, the existence of mtDNA methylation as an autonomous epigenetic mechanism has been the subject of controversial reports in the scientific literature [8]. Despite this, recent evidence suggests that mtDNA can be methylated at one cytosine residue [9] and that

Alexandra J. Harvey and David K. Gardner (eds.), *The Preimplantation Embryo*, Methods in Molecular Biology, vol. 3038, https://doi.org/10.1007/978-1-0716-5292-3_17,

different mtDNA methylation signatures may have significance in programming development [10, 11].

For many years, whole-genome bisulfite sequencing (WGBS) was the gold-standard approach to determine base-level CpG methylation in the nuclear genome (nDNA). However, with advances in technology, major technical issues raised questions about the reliability of the technique for evaluating mtDNA methylation [12, 13]. For instance, the secondary and even tertiary structure of mtDNA is known for causing incomplete conversion of 5-methylcytosine to uracil [13, 14], leading to incorrect estimations of the presence and the quantity of cytosine methylation. The treatment with bisulfite is also known for causing DNA loss and degradation, which can hinder downstream analysis [15]. Therefore, in order to correctly assess and characterize mtDNA methylation across different cell types, it was necessary to improve current techniques used for mtDNA extraction and library preparation, and consequently obtain better quality material to overcome the concerns related to mtDNA methylation detection.

Here, we describe a simple method for mtDNA recovery from bovine oocytes and blastocysts, coupled with an enzyme-based conversion protocol, i.e., without the use of bisulfite. This workflow has a number of advantages, including steps to ensure maximal conversion efficiency and to reduce DNA damage, thereby counteracting the bias introduced by conversion-resistant cytosines and enabling the detection of 5-methylcytosine and 5-hydroxymethylcytosine from fewer sequencing reads. This method has proven to be stable and replicable, allowing high-coverage sequencing of total DNA from oocytes and blastocysts (including mtDNA) with very small sample input (<10 ng) [11].

2 Materials

Prepare all solutions using ultrapure water and analytical grade reagents. Plastic supplies should be sterile and nuclease-free.

2.1 Consumables and Reagents

1. DNA Microprep Kit (e.g., Zymo Research Quick-DNA Microprep Kit containing DNA elution buffer, g-DNA Wash Buffer, Genomic Lysis Buffer, DNA Pre-wash Buffer, Collection Tubes, and Zymo-Spin IC Columns; *see* **Note 1**).
2. Enzymatic Methyl-seq Kit (e.g., New England Biolabs NEBNext Enzymatic Methyl-seq Kit containing Control DNA CpG methylated pUC19, Control DNA Unmethylated Lambda, NEBNext Ultra™ II End Prep Reaction Buffer, NEBNext Ultra II End Prep Enzyme Mix, NEBNext Ultra II Ligation Master Mix, NEBNext Ligation Enhancer NEBNext EM-seq™ Adapter, Elution Buffer, TET2 Reaction Buffer, TET2 Reaction Buffer Supplement, Oxidation Supplement, DTT, Oxidation Enhancer, TET2, Fe(II) Solution (*see* **Note 2**), Stop Reagent, APOBEC, APOBEC Reaction Buffer, BSA, NEBNext Q5U™ Master Mix).

3. 1% w/v PBS-PVP.
4. Ultrapure water.
5. Zymo Research M-Digestion Buffer.
6. Molecular grade ethanol; 100%, 80%, and 70%. Prepare dilutions with ultrapure water.
7. 0.1 N NaOH.
8. PCR strip tubes and caps.
9. Nuclease-free water.
10. 20 mg/mL Proteinase K.
11. Nuclease-free PCR-grade microcentrifuge tubes (1.5 mL and 0.6 mL).
12. 1× TE buffer: 10 mM Tris, 1 mM EDTA, pH 8.0.
13. Microtubes for ultrasonification (e.g., Covaris microTUBE-50).
14. Manual pipettors.
15. Pipette tips.

2.2 Equipment

1. Instrument for DNA shearing (e.g., Covaris Ultrasonicator).
2. Thermocycler.
3. Heated tube rack.
4. Magnetic stand.
5. TapeStation or BioAnalyzer for library visualization and quantification.
6. Vortex.
7. Centrifuge for microcentrifuge tubes.
8. Multichannel pipette.

3 Methods

This method has been optimized for bovine mature and immature oocytes (pools of $n = 30$) and Day 7 bovine blastocysts (pools of $n = 5$). All procedures should be carried out at room temperature, unless otherwise stated. The workflow for DNA extraction and library preparation is summarized in Fig. 1.

3.1 Sample Preparation and Storage

Oocytes and embryos should be devoid of cumulus cells either by mechanical (e.g., pipetting/vortex) or chemical methods (e.g., hyaluronidase treatment). Oocytes or embryos should also be washed at least three times in a buffer solution (e.g., PBS-PVP 1% w:v, *see* **Note 3**) to remove cellular debris and dilute any remaining culture medium prior to storage in a tube in a maximum volume of 3 μL.

Sample Digestion	Digestion of oocyte/embryo samples with proteinase K
Total DNA extraction	Extraction of Total DNA (gDNA + mtDNA) with Quick DNA ™ Microprep kit (Zymo Research)
DNA fragmentation	Input is below 10 ng of gDNA. Fragments have an average size of 300 bp.
End repair/ dA-Tailing	DNA is end-repaired and dA-tailed using NEBNext® Ultra II reagents.
Ligation of EM-Seq adapters	DNA is ligated to the EM-Seq adaptors using NEBNext® Ultra II reagents.
Oxidation of 5mC and 5hmC	The enzyme TET2 and the Oxidation Enhancer reagent protect 5mC/5hmC from deamination
Deamination of C to U	The enzyme APOBEC deaminates cytosines to uracils (oxidized 5mC or 5hmC are not deaminated)
PCR amplification	Library is amplified using NEBNext® index primers.
Sequencing	Sequencing is done using an Illumina platform.

Fig. 1 Workflow for DNA extraction and library preparation for the analysis of mtDNA methylation

3.2 Sample Digestion with Proteinase K

This step is necessary to remove the zona pellucida, as well as to break down the protein components of the cell membrane and allow access to the DNA for extraction.

1. Thaw samples on ice.
2. Briefly spin down the tubes containing the samples.
3. Add 10 μL M-Digestion Buffer on top of the 3 μL sample.
4. Add 1 μL Proteinase K (20 mg/mL) to the mix.
5. Complete to 20 μL with nuclease-free water.
6. Incubate for 20 min at 50°C in a thermal cycler with the heated lid off.

3.3 Total DNA Extraction

This step uses the Zymo Research Quick-DNA Microprep kit (*see* **Note 1**). Before starting, dilute the elution buffer (1:6) in nuclease-free water and warm to 70°C in a dry heated tube rack.

1. Add 80 μL of Genomic Lysis Buffer to the 20 μL of liquid sample (*see* **Note 4**).
2. Mix briefly by vortexing, then let stand at room temperature for 5 min.
3. Transfer the mixture to a Zymo-Spin™ IC Column in a Collection Tube (supplied with the kit).
4. Centrifuge at ≥10,000 × *g* for 1 min. Discard the Collection Tube with the flow-through.
5. Transfer the Zymo-Spin™ IC Column to a new Collection Tube. Add 200 μL of DNA Pre-Wash Buffer to the spin column. Centrifuge at ≥10,000 × *g* for 1 min.

6. Add 500 μL of g-DNA Wash Buffer to the spin column. Centrifuge at ≥10,000 × *g* for 1 min.
7. Transfer the spin column to a clean 1.5 mL tube (not supplied with the kit).
8. Add 18 μL of DNA Elution Buffer (diluted 1:6 in nuclease-free water and previously warmed to 70°C) to the spin column.
9. Incubate for 2–5 min at room temperature and then centrifuge at ≥10,000 × *g* for 30 s to elute total DNA. The eluted DNA can be used immediately or stored at ≤−20°C for future use.

3.4 Total DNA Fragmentation

1. Add positive (puc19) and/or negative (lambda DNA) controls to each microcentrifuge tube containing the samples (supplied with the kit, *see* **Note 5**).
2. Spike each sample with 1 μL of diluted pUC19 (0.001 ng) control DNA and 1 μL of diluted unmethylated lambda DNA (0.02 ng).
3. Add elution buffer (from the DNA microprep kit) to a total volume of 50 μL and transfer the total volume to the Covaris microTUBE-50 for DNA shearing.
4. Proceed with Total DNA shearing using ultrasonification. The following parameters have been optimised for use on a Covaris ultrasonicator:

   Time = 224 s
   Peak Power = 50 W
   Duty Factor = 20%
   Cycles/Burst = 200

3.5 Enzymatic Conversion and Library Preparation for Sequencing

Conversion and library preparation have been optimised using the New England Biolabs NEBNext Methyl-seq kit (*see* **Note 6**).

3.5.1 End Prep of Sheared DNA

On ice, for each sample, mix the kit components in a sterile nuclease free PCR tube as follows (*see* **Note 7**):

1. Add 50 μL of the Fragmented DNA from Step 3.3.
2. Add 7 μL of NEBNext Ultra II End Prep Reaction Buffer.
3. Add 3 μL of NEBNext Ultra II End Prep Enzyme Mix.
4. Then, using a pipette, mix the entire volume up and down at least ten times. Perform a quick spin to collect the liquid from the walls of the tubes (*see* **Note 8**).
5. Place the tubes in a thermal cycler with the lid temperature set to ≥75°C and run the program: 30 min at 20°C, 30 min at 65°C, hold at 4°C.

3.5.2 Ligation of Adapters

On ice, add the following components directly to the 60 μL reaction mixture from Step 3.5.1 (*see* **Note 9**):

1. Add 2.5 μL of NEBNext EM-seq Adapter.
2. Add 1 μL of NEBNext Ligation Enhancer.
3. Add 30 μL of NEBNext Ultra II Ligation Master Mix.
4. Using a pipettor, mix the entire volume up and down at least ten times. Perform a quick spin to collect the liquid from the walls of the tubes (*see* **Note 10**).
5. Incubate at 20°C for 15 min in a thermal cycler with the heated lid off.

SAFE STOP POINT (samples can be kept overnight in the freezer at −20°C).

3.5.3 Clean-Up of Adapter Ligated DNA

1. Vortex the NEBNext Sample Purification Beads well to resuspend.
2. Add 110 μL of resuspended NEBNext Sample Purification Beads to each sample (*see* **Note 11**). Pipette up and down at least ten times to mix well, and be careful to expel all of the liquid out of the tip during the final mix.
3. Incubate samples for at least 5 min at room temperature.
4. Place the tubes against a magnetic stand to separate the beads from the supernatant (*see* **Note 12**).
5. After 5 min, if the supernatant is clear, carefully remove the supernatant without disturbing the bead pellet and discard it (*see* **Note 13**).
6. Add 200 μL of 80% freshly prepared ethanol to the tubes while on the magnetic stand. Incubate at room temperature for 30 s, then carefully remove and discard the supernatant. Be careful not to disturb the beads that contain DNA targets.
7. Repeat the 80% ethanol wash again for a total of two washes. Be sure to remove all visible liquid after the second wash.
8. Air dry the beads for up to 2 min while the tubes are on the magnetic stand with the lid open (*see* **Note 14**).
9. Remove the tubes from the magnetic stand. Elute the DNA target from the beads by adding 29 μL of Elution Buffer.
10. Mix well by pipetting up and down ten times. Incubate for at least 1 min at room temperature. If necessary, quickly spin the sample to collect the liquid from the sides of the tube.
11. Place the tube on the magnetic stand. After 3 min, or whenever the solution is clear, transfer 28 μL of the supernatant to a new PCR tube.

SAFE STOP POINT (samples can be kept overnight in the freezer at −20°C).

3.5.4 Oxidation of 5-Methylcytosines and 5-Hydroxymethylcytosines

1. Add 100 μL of TET2 Reaction Buffer to one tube of TET2 Reaction Buffer Supplement and mix well. Write the date on the tube (*see* **Note 15**).
2. On ice, add the following components directly to the 28 μL of each adapter ligated sample (*see* **Note 16**).

   a. Add 10 μL of TET2 Reaction Buffer + Supplement mix.
   b. Add 1 μL of Oxidation Supplement.
   c. Add 1 μL of DTT.
   d. Add 1 μL of Oxidation Enhancer.
   e. Add 4 μL of TET2.

3. Mix well by vortexing and then centrifuge briefly to collect the content of the walls of the tube before placing samples on ice.
4. Dilute the 500 mM Fe(II) Solution by adding 1–1249 μL of water (*see* **Note 17**).
5. Combine 5 μL of the diluted Fe(II) solution with 45 μL of the sample.
6. Mix by vortexing, or by pipetting up and down at least ten times, then centrifuge briefly to collect the contents.
7. Incubate at 37°C for 1 h in a thermocycler with the lid temperature set to ≥45°C.
8. Transfer the samples to ice and add 1 μL of Stop Reagent to each sample.
9. Mix well by vortexing, or by pipetting up and down at least ten times, then centrifuge briefly to collect the contents of the walls.
10. Incubate at 37°C for 30 min in a thermocycler with the lid temperature set to ≥45°C.

SAFE STOP POINT (samples can be kept in the thermocycler at 4°C overnight or stored in the freezer at −20°C).

3.5.5 Clean-Up of TET2-Converted DNA

1. Vortex NEBNext Sample Purification Beads to resuspend.
2. Add 90 μL of resuspended NEBNext Sample Purification Beads to each sample. Mix well by pipetting up and down at least ten times (*see* **Note 11**). Be careful to expel all of the liquid out of the tip during the last mix.
3. Incubate samples on the bench top for at least 5 min at room temperature.
4. Place the tubes against a magnetic stand to separate the beads from the supernatant (*see* **Note 12**).
5. After 5 min, or when the supernatant is clear, being careful not to disturb the beads, remove and discard the supernatant (*see* **Note 13**).

6. Add 200 μL of 80% freshly prepared ethanol to the tubes while on the magnetic stand. Incubate at room temperature for 30 s, and then carefully remove and discard the supernatant. Be careful not to disturb the beads that contain DNA targets.
7. Repeat the wash again for a total of two washes. Be sure to remove all visible liquid after the second wash.
8. Air dry the beads for up to 2 min while the tubes are on the magnetic stand with the lid open (*see* **Note 14**).
9. Remove the tubes from the magnetic stand. Elute the DNA from the beads by adding 17 μL of Elution Buffer.
10. Mix well by pipetting up and down ten times. Incubate for at least 1 min at room temperature. If necessary, quickly spin the sample to collect the liquid from the sides of the tube.
11. Place the tube on the magnetic stand for 3 min. Once the supernatant is clear, transfer 16 μL of the supernatant to a new PCR tube.

SAFE STOP POINT (samples can be kept overnight in the freezer at −20°C).

3.5.6 Denaturation of DNA

DNA should be denatured using 0.1 N Sodium Hydroxide.

1. Prepare freshly diluted 0.1 N NaOH.
2. Pre-heat thermal cycler to 50°C with the heated lid set to ≥105°C.
3. Add 4 μL 0.1 N NaOH to the 16 μL of oxidized DNA (from Step 3.5.5).
4. Vortex to mix or by pipetting up and down at least ten times, centrifuge briefly.
5. Incubate at 50°C for 10 min in the pre-heated thermal cycler.
6. Immediately place on ice and allow the sample to fully cool (~2 min) before proceeding.

3.5.7 Deamination of Cytosines with APOBEC (Apolipoprotein B mRNA Editing Enzyme, Catalytic Polypeptide)

1. While samples are on ice, add the following components to the 20 μL of denatured DNA (for multiple reactions, a master mix of the components can be prepared before adding to the denatured DNA).

a. Add 68 μL of nuclease-free water.

b. Add 10 μL of APOBEC Reaction Buffer.

c. Add 1 μL of BSA.

d. Add 1 μL of APOBEC.

2. Mix well by vortexing, or by pipetting up and down at least ten times, then centrifuge briefly to collect the contents.
3. Incubate at 37°C for 3 h and then hold at 4°C in the thermocycler with the lid temperature set to ≥45°C.

3.5.8 Clean-Up of Deaminated DNA (see Note 18)

1. Vortex Sample Purification Beads to resuspend.
2. Add 100 µL of resuspended NEBNext Sample Purification Beads to each sample (*see* **Note 11**). Mix well by pipetting up and down at least ten times. Be careful to expel all of the liquid out of the tip during the last mix.
3. Incubate samples on the benchtop for at least 5 min at room temperature.
4. Place the tubes against a magnetic stand to separate the beads from the supernatant (*see* **Note 12**).
5. After 5 min, once the supernatant is clear, carefully remove and discard the supernatant (*see* **Note 13**).
6. Add 200 µL of 80% freshly prepared ethanol to the tubes while on the magnetic stand. Incubate at room temperature for 30 s, and then carefully remove and discard the supernatant. Be careful not to disturb the beads.
7. Repeat the wash again for a total of two washes. Be sure to remove all visible liquid after the second wash using a 10 µL pipette tip.
8. Air dry the beads for up to 90 s while the tubes are on the magnetic stand with the lid open.
9. Remove the tubes from the magnetic stand. Elute the DNA target from the beads by adding 21 µL of Elution Buffer.
10. Mix well by pipetting up and down ten times. Incubate for at least 1 min at room temperature. If necessary, quickly spin the sample to collect the liquid from the sides of the tube.
11. Place the tube on the magnetic stand. After 3 min, once the supernatant is clear, transfer 20 µL of the supernatant to a new PCR tube.

SAFE STOP POINT (samples can be kept overnight in the freezer at –20°C).

3.5.9 PCR Amplification

On ice, add the following components to each 20 µL sample of deaminated DNA (*see* **Note 19**).

1. Add 5 µL of one EM-Seq Index primer (included with kit) to each sample.
2. Add 25 µL of NEBNext Q5U Master Mix.
3. Mix well by vortexing, or by pipetting up and down at least ten times, then centrifuge briefly to collect the contents of the walls.
4. Place the tube in a thermocycler and perform PCR amplification using the following cycling conditions (*see* **Note 20**): initial denaturation at 98°C for 30 s (one cycle), Denaturation at 98°C for 10 s, Annealing at 62°C for 30 s, and Extension at

65°C for 60 s (12 cycles), Final Extension at 65°C for 5 min (one cycle), hold at 4°C.

SAFE STOP POINT (samples can be kept in the thermocycler at 4°C overnight or stored in the freezer at −20°C).

3.5.10 Clean-Up of Amplified Libraries

1. Vortex Sample Purification Beads to resuspend.
2. Add 45 μL of resuspended NEBNext Sample Purification Beads to each sample (*see* **Note 11**). Mix well by pipetting up and down at least ten times. Be careful to expel all of the liquid out of the tip during the last mix.
3. Incubate samples on the bench top for at least 5 min at room temperature.
4. Place the tubes against a magnetic stand to separate the beads from the supernatant (*see* **Note 12**).
5. After 5 min, or once the supernatant is clear, carefully remove and discard the supernatant (*see* **Note 13**).
6. Add 200 μL of 80% freshly prepared ethanol to the tubes while on the magnetic stand. Incubate at room temperature for 30 s, and then carefully remove and discard the supernatant. Be careful not to disturb the beads that contain DNA targets.
7. Repeat the 80% ethanol wash again for a total of two washes. Be sure to remove all visible liquid after the second wash using a 10 μL pipette tip.
8. Air dry the beads for up to 2 min while the tubes are on the magnetic stand with the lid open (*see* **Note 14**).
9. Remove the tubes from the magnetic stand. Elute the DNA target from the beads by adding 21 μL of Elution Buffer.
10. Mix well by pipetting up and down ten times. Incubate for at least 1 min at room temperature. If necessary, quickly spin the sample to collect the liquid from the sides of the tube.
11. Place the tube on the magnetic stand. After 3 min (or whenever the solution is clear), transfer 20 μL of the supernatant to a new PCR tube.

SAFE STOP POINT (samples can be kept overnight in the freezer at −20°C).

3.6 Library Quantification

1. Use a Bioanalyzer or Tape Station to determine the size distribution and concentration of the libraries. A typical EM-seq library would have a Tape Station trace similar to that shown in Fig. 2.
2. Sequence samples using the preferred Illumina platform with a minimum of pair-ended 100 bp reads for standard-sized libraries. Examples of expected outcomes are summarized in Table 1.

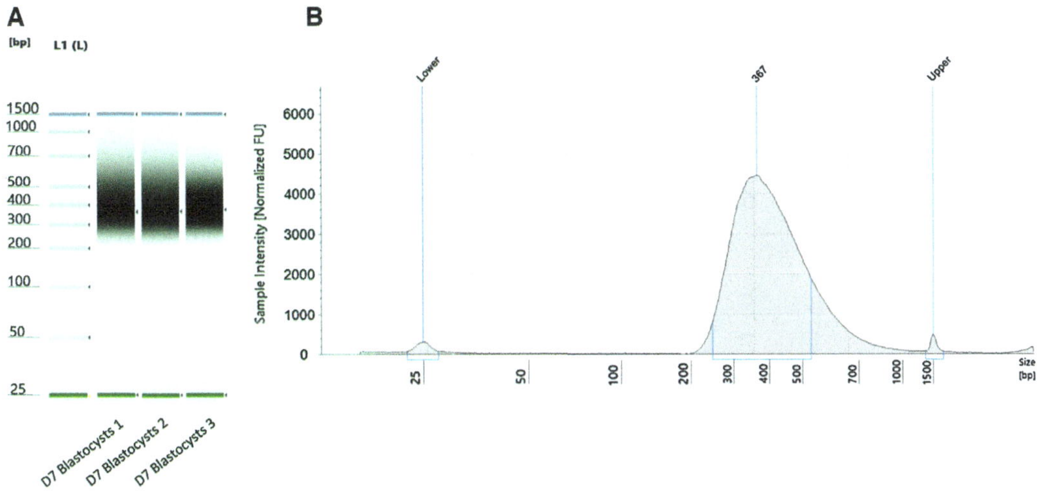

Fig. 2 TapeStation profile of three libraries produced from D7 bovine blastocyst samples. (**A**) TapeStation gel images showing the ladder and three libraries. (**B**) Representative TapeStation electropherogram of "D7 Blastocysts 1" library showing the mean size of the fragments to be 367 bp

Table 1
Example of quality control report of mtDNA sequencing for three libraries produced from D7 bovine blastocysts. The genome of reference was NC_006853.1

Sample name	Startposition	Endposition	Number of reads	Total mtDNA coverage (%)	Mean depth (*X*)
D7 Blastocyst 1	1	16,338	65,012	100	397.419
D7 Blastocyst 2	1	16,338	65,035	100	395.570
D7 Blastocyst 3	1	16,338	64,941	100	399.357

3. Following sequencing, data analysis should be performed according to the goals of the study. For example, sequencing adapters can be removed using Trimmomatic, with read conversion and alignment to the mitochondrial genome performed using Bismark. For identification of differentially methylated cytosines, we have used the MethylKit R package [11].

4 Notes

1. Other DNA extraction kits compatible with mtDNA may be used, but they have not been tested with this protocol.
2. This buffer may be clear or yellow in color.
3. The use of pure PBS is not recommended, as it may cause oocytes or embryos to adhere to the sides of the tubes.

4. This step is not mandatory for proteinase K-treated samples, but it is recommended.
5. If sequencing on NovaSeq, HiSeq or NextSeq: dilute the pUC19 and the unmethylated lambda controls 1:100 using TE buffer.
6. At this time, no other enzymatic methods are currently available. Additionally, kits based on bisulfite conversion are strongly not recommended, as they may lead to bias in methylation detection in mtDNA.
7. NEBNext Ultra II End Prep Reaction Buffer and Enzyme Mix can be pre-mixed ahead of time as a master mix.
8. A small number of bubbles will not interfere with performance.
9. Ligation Enhancer and Ligation Master Mix can be mixed ahead of time and are stable for at least 8 h at 4°C. Do not premix the Ligation Master Mix, Ligation Enhancer and adapter prior to use in the Adapter Ligation Step. Premix adapter and sample, and then add the other ligation reagents.
10. The final mix obtained in Step 3.5.2 is viscous; therefore, it should be mixed carefully. Incomplete mixing of the ligation reaction will result in reduced ligation efficiency. The presence of a small number of bubbles will not interfere with performance.
11. A regular pipettor can be used to dispense the beads into each sample, but a multi-channel pipettor is useful to add/remove the supernatant or ethanol during washes to be faster and more even.
12. If needed, use an elastic band to attach the tube strip to the magnetic stand as firmly as possible.
13. Careful: do not disturb or discard the beads as they contain the DNA targets.
14. Caution: do not over-dry the beads. This may result in lower DNA recovery. Elute the samples when the beads are still dark brown and glossy, but when all visible liquid has evaporated. If the beads turn lighter brown and start to crack, it means they are too dry.
15. The TET2 Reaction Buffer Supplement is a powder. Centrifuge before use to ensure it is at the bottom of the tube. The reconstituted buffer should be stored at −20°C and discarded after 4 months.
16. For multiple reactions, a master mix of the reaction components can be prepared before addition to the sample DNA, since the 5mC/5hmC oxidation is only initiated by the addition of the Fe(II) solution to the reaction.

17. The Fe(II) solution should be used immediately after preparation and discarded after use (do not store it).
18. Caution: The Sample Purification Beads behave differently during the APOBEC clean-up. Do not over-dry the beads.
19. Refer to the manufacturer's manual for barcode pooling guidelines.
20. PCR cycling conditions for libraries prepared from DNA from bovine oocytes or embryos (<10 ng of total DNA). If using higher DNA input, the number of cycles should be adjusted according to the manufacturer's guidelines.

References

1. Vakifahmetoglu-Norberg H, Ouchida AT, Norberg E (2017) The role of mitochondria in metabolism and cell death. Biochem Biophys Res Commun 482(3):426–431. https://doi.org/10.1016/j.bbrc.2016.11.088
2. Fragouli E, Wells D (2015) Mitochondrial DNA assessment to determine oocyte and embryo viability. Semin Reprod Med 33(6):401–409. https://doi.org/10.1055/s-0035-1567821
3. Babayev E, Seli E (2015) Oocyte mitochondrial function and reproduction. Curr Opin Obstet Gynecol 27(3):175–181. https://doi.org/10.1097/GCO.0000000000000164
4. Nishimura Y, Yoshinari T, Naruse K, Yamada T, Sumi K, Mitani H, Higashiyama T, Kuroiwa T (2006) Active digestion of sperm mitochondrial DNA in single living sperm revealed by optical tweezers. Proc Natl Acad Sci U S A 103(5):1382–1387. https://doi.org/10.1073/pnas.0506911103
5. Mercer TR, Neph S, Dinger ME, Crawford J, Smith MA, Shearwood AM, Haugen E, Bracken CP, Rackham O, Stamatoyannopoulos JA, Filipovska A, Mattick JS (2011) The human mitochondrial transcriptome. Cell 146(4):645–658. https://doi.org/10.1016/j.cell.2011.06.051
6. Harvey AJ (2019) Mitochondria in early development: linking the microenvironment, metabolism and the epigenome. Reproduction 157(5):R159–R179. https://doi.org/10.1530/REP-18-0431
7. Lee SR, Han J (2017) Mitochondrial nucleoid: shield and switch of the mitochondrial genome. Oxid Med Cell Longev 2017:8060949. https://doi.org/10.1155/2017/8060949
8. Stoccoro A, Coppede F (2021) Mitochondrial DNA methylation and human diseases. Int J Mol Sci. 22(9). https://doi.org/10.3390/ijms22094594
9. Castegna A, Iacobazzi V, Infantino V (2015) The mitochondrial side of epigenetics. Physiol Genomics 47(8):299–307. https://doi.org/10.1152/physiolgenomics.00096.2014
10. Sirard MA (2019) Distribution and dynamics of mitochondrial DNA methylation in oocytes, embryos and granulosa cells. Sci Rep 9(1):11937. https://doi.org/10.1038/s41598-019-48422-8
11. de Lima CB, Martin H, Pecora Milazzotto M, Sirard MA (2023) Genome-wide methylation profile of mitochondrial DNA across bovine preimplantation development. Epigenetics 18(1):2241010. https://doi.org/10.1080/15592294.2023.2241010
12. Mechta M, Ingerslev LR, Fabre O, Picard M, Barres R (2017) Evidence suggesting absence of mitochondrial DNA methylation. Front Genet 8:166. https://doi.org/10.3389/fgene.2017.00166
13. Dou X, Boyd-Kirkup JD, McDermott J, Zhang X, Li F, Rong B, Zhang R, Miao B, Chen P, Cheng H, Xue J, Bennett D, Wong J, Lan F, Han JJ (2019) The strand-biased mitochondrial DNA methylome and its regulation by DNMT3A. Genome Res 29(10):1622–1634. https://doi.org/10.1101/gr.234021.117
14. Owa C, Poulin M, Yan L, Shioda T (2018) Technical adequacy of bisulfite sequencing and pyrosequencing for detection of mitochondrial DNA methylation: Sources and avoidance of false-positive detection. PLoS One 13(2):e0192722. https://doi.org/10.1371/journal.pone.0192722
15. Hong SR, Shin KJ (2021) Bisulfite-converted DNA quantity evaluation: a multiplex quantitative real-time PCR system for evaluation of bisulfite conversion. Front Genet 12:618955. https://doi.org/10.3389/fgene.2021.618955

Chapter 18

Single-Cell and Single-Embryo RNA Sequencing

Bastien Ducreux, Lucile Ferreux, Julie Firmin, Catherine Patrat and Patricia Fauque

Abstract

The scarcity of available donated human embryos for research and technical difficulties present numerous challenges for human development research. Developments in RNA-seq technologies for small inputs, such as single embryos and single cells, now make human applications possible. This chapter outlines single-embryo RNA-seq, with adaptations noted for single-cell RNA-seq and single embryonic/abembryonic compartment applied to preimplantation embryo analysis, from embryo collection to bioinformatic data processing, and the specificities of applications of these techniques in reproductive biology.

Keywords Single-embryo RNA-seq, Single-cell RNA-seq, Preimplantation embryo, Transcriptome, Trophectoderm, Inner cell mass

1 Introduction

During the first days of development, the human embryo undergoes continuous transcriptomic changes [1]. First, the early phases of embryonic development depend on maternal products stored in the oocyte for appropriate regulation [2]. Shortly after fertilization, the embryo's genome is activated, initiating its own transcriptional program, and embryonic lineage specification is marked by dynamic shifts in gene expression patterns [3–5].

Unraveling the temporal and spatial patterns of gene expression in human embryos is a critical step toward understanding the regulatory changes that govern human embryonic development. Nonetheless, research on human development is hampered by the scarcity of donated human embryos for study, as well as technical challenges. Advances in library preparation techniques have enabled the use of RNA-seq technologies for small inputs, such as single cells. In addition, for the emerging human blastoid and other stem cell-based human embryo models that become an essential asset to

Alexandra J. Harvey and David K. Gardner (eds.), *The Preimplantation Embryo*, Methods in Molecular Biology, vol. 3038,
https://doi.org/10.1007/978-1-0716-5292-3_18,

advance our understanding of human peri-implantation stages, these technological advances are crucial [6].

Over the last decade, collections of single-embryo and single-cell RNA-seq (scRNA-seq) datasets from the human preimplantation embryo have allowed for the creation of an embryonic cell atlas [1, 4, 7–11]. These investigations have revealed genes involved in essential developmental events, including the maternal-zygotic transition and embryonic genome activation, specification of the inner cell mass and trophectoderm, and chromosome X inactivation [1, 2, 5]. It has resulted in significant findings with further therapeutic implications for the etiology of embryo development-related illnesses, as well as for studying the effects of the technologies used in the context of in vitro fertilization.

Indeed, there are various uses and scales at which preimplantation embryos can be studied (Fig. 1). The whole embryo can be used to study the embryonic transcriptome (single-embryo RNA-seq), providing more in-depth sequencing than single-cell approaches. Nonetheless, the single-cell method is ideal for investigating cell heterogeneity and identifying cell type-specific differentiation trajectories during embryonic development. An in-between consists of studying the inner cell mass and the trophectoderm independently by splitting the blastocyst along its abembryonic-embryonic axis (single-TE or single-ICM RNA-seq). This technique optimizes our capacity to record and ascertain the proportionate representation of each cell in the embryo, contrary to the single-cell method. All these techniques address the issue of low input material availability when studying preimplantation embryos.

This chapter describes an adaptation of the original Tang method, used previously by our team [12, 13] (Fig. 2). This method uses an oligo d(T) primer to initiate reverse transcription and subsequent amplification characterizing only the 3′-end of the RNA transcripts, which is generally sufficient for the investigation of cellular

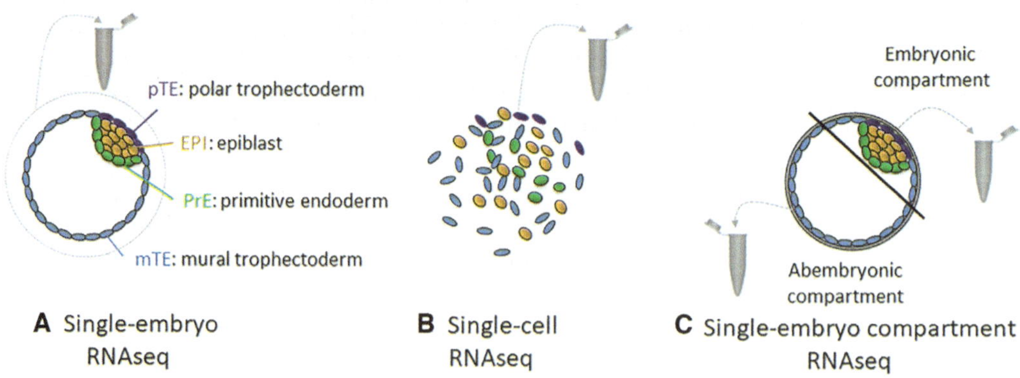

Fig. 1 Embryonic RNA-sequencing: at the single-embryo (**A**), single-cell (**B**), or single-embryo compartment (**C**) scale

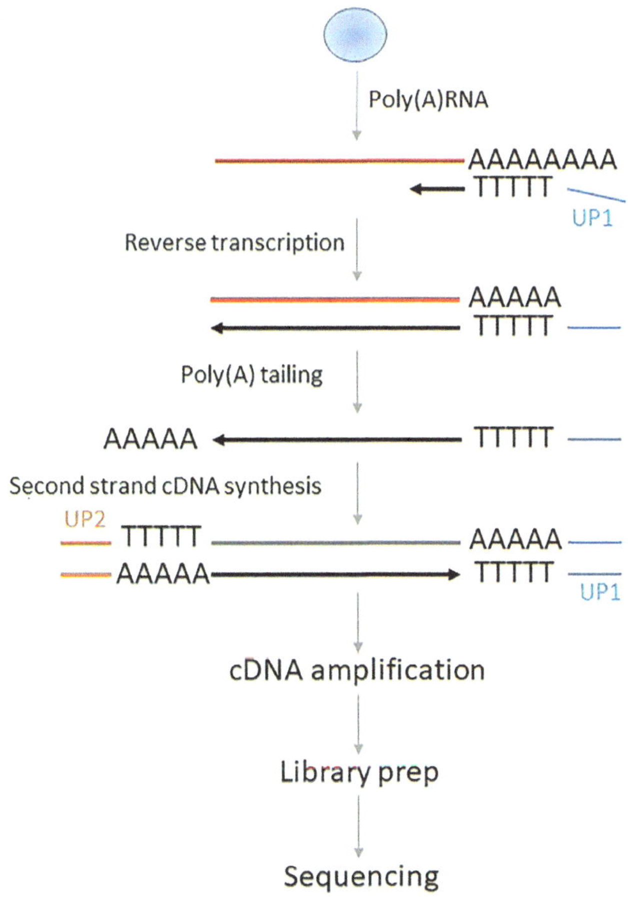

Fig. 2 Schematic comparing Tang's method with that of SMARTseq. mRNAs are reverse-transcribed into cDNAs using a poly(T) primer with anchor sequence (UP1). Poly(A) tails are added to the first-strand cDNAs at the 3′ end, and second-strand cDNAs are synthesized using poly(T) primers with another anchor sequence (UP2). Then cDNAs are evenly amplified by PCR using UP1 and UP2 primers

heterogeneity and the identification of population substructure (*see* **Note 1**). Finally, associated bioinformatic methods that are critical steps in extracting all the information from the isolated RNA of embryonic cells are described.

2 Materials

2.1 Embryo Collection and Dissociation

1. Sterile plastic tissue culture dishes.
2. Stripper® pipettor.
3. Stripper® tips.
4. Commercially available HEPES-buffered culture medium.

5. Hyaluronidase.
6. Acid Tyrode's solution.
7. Digestion medium or Ca^{++}-free medium (e.g., TrypLE™ Express or accutase medium: 0.25% trypsin, 1:1).
8. 0.4% PBS-PVP.
9. 0.2 mL thin-wall PCR tubes, free of RNase, DNase, DNA, and PCR inhibitors, containing freshly prepared lysis buffer.
10. Equipment for isolating anembryonic and embryonic compartments (if desired): an inverted microscope equipped with micromanipulators, a warm plate and a laser.
11. 100 μL pipette tips.
12. Single-channel 100 μL pipettor.

2.2 Cell Lysis

1. GeneAmp® 10× PCR Buffer II and $MgCl_2$.
2. 10% Nonidet P-40 substitute.
3. 0.1 M DTT is included in the SuperScript® III reverse transcriptase kit.
4. dNTP mix (2.5 mM each).
5. SUPERase-In® (20 U/μL).
6. Ambion® RNase Inhibitor (40 U/μL), cloned.
7. Universal primer UP1, 100 μM: 5′-ATATGGATCCGGCGC GCCGTCGACTTTTTTTTTTTTTTTTTTTTTTTT-3′.
8. ERCC spike-in RNAs, 1:100,000 dilution (optional, but recommended).
9. Nuclease-free water.
10. 0.2 mL thin-wall PCR tubes, free of RNase, DNase, DNA, and PCR inhibitors.
11. 10, 200, and 1000 μL pipette tips.
12. Single and multichannel 10, 200, and 1000 μL pipettors.
13. RNase decontamination solution.
14. Mini-spin centrifuge.
15. PCR workstation.

2.3 Reverse Transcription and cDNA Amplification

1. SuperScript® III reverse transcriptase (*see* **Note 2**).
2. Ambion® RNase Inhibitor (40 U/μL), cloned.
3. T4 gene 32 protein (1–10 U/μL).
4. ExoSAP-IT®.
5. GeneAmp® 10× PCR Buffer II and $MgCl_2$.
6. 100 mM dATP.
7. Terminal deoxynucleotidyl transferase, recombinant (15 U/μ L).

8. RNAse-H (2 U/μL).
9. TaKaRa Ex Taq® DNA Polymerase Hot-Start, including 10×ExTaq buffer with $MgCl_2$, Ex Taq® HS and dNTP mix (2.5 mM/each).
10. ino-blocked universal primers, 100 μM: AUP1 (NH2-ATAT GGATCCGGCGCGCCGTCGACTTTTTTTTTTTTTTT TTTTTTTT) and AUP2 (NH2-ATATCTCGAGGGCGCG CCGGATCCTTTTTTTTTTTTTTTTTTTTTTTT).
11. Nuclease-free water.
12. 2, 20, 200, and 1000 μL pipette tips.
13. Single channel 10, 20, 200, and 1000 μL pipettors.
14. 0.2 mL thin-wall PCR tubes, free of RNase, DNase, DNA, and PCR inhibitors.
15. 1.5 mL low retention microcentrifuge tubes, free of RNase, DNase, DNA, and PCR inhibitors.
16. RNAse decontamination solution.
17. Thermocycler.
18. Mini-spin.
19. PCR workstation.
20. 2% agarose gel; dissolve 2.4 g of agarose in 120 mL 1× TBE in a microwave for 1–3 min until it is just bubbling. Add 2.4 μ L SYBR Safe.
21. Gel casting tank and comb.
22. DNA ladder (range 0.5–5 kb).
23. Gel loading dye (e.g., 6× Orange DNA Loading Dye).
24. 1× TAE or 1× TBE running buffer.
25. Electrophoresis running tank and power supply.
26. UV transilluminator.
27. Razor blades.

2.4 Sample Quality Control

1. SYBR-based quantitative PCR kit (e.g., Applied Biosystems SYBR Green Universal Master Mix, consisting of SYBR green dye, dNTPs, Taq polymerase, Passive reference dye, and buffer).
2. Nuclease-free water.
3. Real-time PCR instrument.
4. Housekeeping gene primers (such as *GAPDH*: F:TGCACC ACCAACTGCTTAGC, R:GGCATGGACTGTGGTCATGAG; *TBP*: F: ACCTTACGCTCAGGGCTTGG, R: GCTGTGGGG TCAGTCCAGTG).

5. 96- or 384-well PCR plates and adhesive film (or strip tubes and caps) suitable for use in the real-time PCR instrument to be used.
6. 2, 10, 200, and 1000 μL filter pipette tips.
7. Single channel 2, 10, 200, and 1000 μL pipettors.

2.5 cDNA Purification

1. PCR purification kit (e.g., DNA Clean & Concentrator™, containing DNA binding buffer, DNA elution buffer, Zymo-Spin columns, and collection tubes suitable for 5 μg of sample).
2. Molecular grade ethanol.
3. Agarose gel electrophoresis equipment.
4. Sterile razor blades.
5. Analytical balance.
6. Loading buffer with bromophenol blue dye.
7. 1.5 and 2 mL microcentrifuge tubes, sterile.
8. 20, 200, and 1000 μL pipette tips.
9. Single channel 20, 200, and 1000 μL pipettors.
10. Gel extraction kit (e.g., Zymoclean™ Gel DNA Recovery Kit, containing Agarose Dissolving Buffer, DNA wash buffer, DNA elution buffer, Zymo-Spin columns, and collection tubes).
11. Nuclease-free water.
12. Centrifuge for 1.5–2 mL microcentrifuge tubes.

2.6 Sample Validation and Library Preparation

1. Concentration quantification instruments and buffers (Qubit® or similar technologies), e.g., Qubit dsDNA High Sensitivity Assay containing 1× dsDNA HS working solution and dsDNA HS Standards 1 and 2, or an Agilent Bioanalyzer High Sensitivity DNA kit containing High Sensitivity DNA ladder, DNA Markers, DNA dye concentrate, DNA gel matrix, spin filters, and High Sensitivity microfluidic DNA chips.
2. DNA size distribution quantification instruments and buffers (Tapestation®, Bioanalyzer®, or similar technology), e.g., a High Sensitivity DNA5000 ScreenTape assay (for Tapestation).
3. Nextera® XT DNA Library Prep Kit, containing Amplicon Tagment Mix, Tagment DNA Buffer, Hybridization Buffer, Nextera PCR Master Mix, Resuspension Buffer, Library Normalization Additives 1, Library Normalization Wash 1, Neutralize Tagment Buffer, Library Normalization Beads 1, Library Normalization Storage Buffer 1.

4. Nextera XT Index kit V2, including Index Primers and Replacement caps.
5. Thermocycler.
6. Microplate shaker.
7. Magnetic stand (for 96-well plate).
8. Microplate centrifuge.
9. Vortex.
10. 10, 200, and 1000 μL pipette tips.
11. Single and multichannel 10, 200, and 1000 μL pipettors.
12. 96-well round-bottomed storage plates (0.8 mL).
13. Illumina Purification Beads.
14. Distilled water.
15. Molecular grade Ethanol (200 proof).
16. Low-bind microplates.
17. Microseal film.
18. Molecular grade 1 N NaOH, pH >12.5.
19. Disposable RNase/DNase-free multichannel reagent reservoirs.
20. Ultrapure water.
21. 96-well PCR plates.

2.7 Bioinformatic Analysis

1. A UNIX or UNIX-like terminal, preferably using High-Performance Computing.
2. Open-source bioinformatic tools for data pre-processing:
   FastQC (http://www.bioinformatics.babraham.ac.uk/projects/fastqc/)
   MultiQC (https://seqera.io/multiqc/)
   Trim Galore (https://www.bioinformatics.babraham.ac.uk/projects/trim_galore/)
   STAR (https://github.com/alexdobin/STAR)
   featureCounts (https://subread.sourceforge.net/featureCounts.html).
3. R.
4. Python.

3 Methods

All materials, reagents, and equipment used in this protocol should be used exclusively, where possible, for this technique to minimize contamination. All experiments must be performed in a UV-sterilized hood with laminar flow, and all the surfaces must be free from RNase to prevent degradation of RNA and from DNA to prevent cross-contamination from previous samples (i.e., surfaces and gloves—powder-free—should be cleaned with a RNase decontamination solution until reverse transcription is completed, after which this contamination can be avoided).

3.1 Cell Lysis Buffer Preparation

1. Dilute the UP1 primers to 0.5 mM in cold, nuclease-free water.
2. Thaw all reagents used at 4°C on ice.
3. Prepare lysis buffer master mix based on the number of samples being prepared (see Table 1). The addition of Spike-In RNAs to the cell lysis buffer mix is recommended (*see* **Note 3**).

3.2 Embryo Collection and Dissociation

1. For single embryo analysis, prepare a 60 mm culture dish with 30 μL of pre-warmed (37°C) HEPES-buffered culture medium, 2×30 μL drops of acid Tyrodes and 3×50 μL drops of wash medium (PBS-PVP 0.4%). Collect and transfer embryos (maximum two at a time) through HEPES-buffered medium, then into drops of acidic Tyrode's solution at room temperature (consecutively) to remove the zona pellucida, then transfer to wash medium.
2. If single-cell dissociation is required, prepare a 60 mm culture dish with 2×30 μL drops of pre-warmed (37°C) HEPES-buffered culture medium, 2×30 μL drops of acid Tyrodes, 2×30 μL drops of TrypLE, and 3×50 μL drops of wash medium (PBS-PVP 0.4%). Transfer one embryo to the first drop of pre-warmed HEPES-buffered culture medium, then

Table 1
Cell lysis buffer (1×)

Component	Volume (μL)
2-10× PCR buffer II (without $MgCl_2$)	0.45
3-25 mM $MgCl_2$	0.27
4-10% NP40 (stored at 4°C)	0.225
5-0.1 M DTT	0.225
8-SUPERase-In (20 U/μL)	0.045
9-RNase inhibitor (40 U/μL)	0.045
6-0.5 μM UP1 Primer	0.125
7-dNTP mix (2.5 mM each)	0.09
1-Nuclease-free water	2.575
ERCC Spike-In (1:100,000)	0.1
Total volume	4.15

transfer the embryo through the two consecutive drops of acid Tyrod for zona pellucida removal. Rinse the embryo in the second drop of pre-warmed HEPES-buffered culture medium, then transfer the naked embryo to digestion medium drops (TrypLE solution) at 37°C (using warm plate) for 2–3 min (maximum 5 min) with gentle and careful pipetting to isolate individual cells (soft to be not detrimental) (*see* **Notes 4–6**).

3. Wash the dissociated embryonic cells or single embryos in three consecutive drops of PBS-PVP solution.
4. Transfer individual single-cells (with careful observation during pipetting to ensure that only a single cell is aspirated into the pipette) or single-embryos in a minimum possible amount of PBS-PVP (<1 μL) into PCR tubes containing 4.15 μL of freshly prepared cell lysis buffer. Quickly spin down the tubes and reserve on ice (*see* **Note 7**).
5. Prepare one blank tube by transferring a similar amount (<1 μL) of the last wash drop of PBS to a tube containing the freshly prepared cell lysis buffer (one control tube per run, without a cell/embryo).

3.3 Reverse Transcription and cDNA Amplification

1. Vortex the tubes and maintain on ice.
2. Incubate at 70°C for 90 s in a thermal cycler and hold at 4°C.
3. Prepare reverse transcription mix (Table 2) (*see* **Note 8**).
4. Add 0.45 μL of reverse transcription mix to each tube on ice (*see* **Note 9**).
5. Incubate at 50°C (this temperature is optimal for the SuperScript III, but it should be adapted to the reverse transcriptase kit used, following the chosen manufacturer's recommendations) for 30 min (reverse transcription), 70°C for 15 min (inactivation of the reverse transcriptase), and hold at 4°C. After this step, the first strand cDNAs for all mRNAs are synthesized.
6. Add 1 μL of ExoSAP-IT® to each tube on ice to eliminate free primers not used during reverse transcription, and incubate at

Table 2
Reverse transcription mix (1×)

Component	Volume (μL)
SuperScript III reverse transcriptase (200 U/μL)	0.33
RNase inhibitor (40 U/μL)	0.05
T4 gene 32 protein (1–10 U/μL)	0.07
Total volume	0.45

37°C for 15 min, followed by 80°C for 15 min to inactivate the ExoSAP-IT®, then hold at 4°C. After this step, free UP1 primers are destroyed, leaving 5′ cDNA ends (UP1 sequence) intact.

7. Prepare 3′ poly (A) tailing reaction mix (Table 3).
8. Add 6 μL of 3′ poly (A) tailing reaction mix to each reaction on ice. Incubate at 37°C for 15 min, 70°C for 10 min (inactivation of the TdT enzyme), and hold at 4°C. After this step, the 3′ end of first-stranded cDNAs has a poly (dA) tail.
9. Prepare second-strand cDNA synthesis mix (Table 4).
10. Add 76 μL of second-strand cDNA synthesis mix to each sample on ice. Incubate at 95°C for 3 min, 50°C for 2 min, and 72°C for 10 min, and hold at 4°C. After this step, double-stranded cDNAs are flanked by both UP1 and UP2 sequences.
11. Prepare the PCR preamplification mix (Table 5).

Table 3
3′ poly (A) tailing reaction mix (1×)

Component	Volume (μL)
2-10× PCR buffer II (without $MgCl_2$)	0.6
3-25 mM $MgCl_2$	0.36
4-100 mM dATP (stock solution)	0.06
1-Nuclease-free water	4.58
5-Terminal transferase (15 U/μL)	0.2
6-RNase H (2 U/μL)	0.2
Total volume	6

Table 4
Second-strand cDNA synthesis mix (1×)

Component	Volume (μL)
10× Ex Taq buffer (with $MgCl_2$)	7.6
dNTP mix (2.5 mM each)	7.6
Amine-blockedUP2 (AUP2) (100 μM)	0.76
Nuclease-free water	59.28
TaKaRa Ex Taq HS (5 U/μL)	0.76
Total volume	76

Table 5
PCR preamplification mix (1×)

Component	Volume (μL)
10× Ex Taq buffer (with $MgCl_2$)	7.6
dNTP mix (2.5 mM each)	7.6
Amine-blocked UP1(AUP1) (100 μM)	0.76
Nuclease-free water	59.28
TaKaRa Ex Taq HS (5 U/μL)	0.76
Total volume	76

12. Add 76 μL of the PCR preamplification mix to each tube on ice. As the volume reached is too high for a single tube in the PCR machine (~164 μL), transfer half of each sample (82 μL) into a second PCR tube.
13. Carry out the following PCR preamplification program on both tubes: 95°C for 3 min, then "*N*" cycles of 95°C for 30 s, 67°C for 1 min, and 72°C for 6 min (+12 s each cycle), with a 4°C hold. "*N*" cycles should be tested empirically according to the starting amount of endogenous RNA of the sample. After this step, all cDNAs have been amplified.
14. Quickly spin-down the tubes and pool the two PCR tubes/sample into 1.5 mL low retention microcentrifuge tubes.

3.4 Sample Quality Control

Quality control of the pre-amplified cDNA is performed by qPCR amplification of several housekeeping genes (*see* **Note 10**). This step allows choosing the best samples for further processing, based on the amplification yield.

1. Prepare a 1:20 dilution of pre-amplified cDNA (5 μL of cDNA in 100 μL final volume of nuclease-free water).
2. Prepare a blank sample lacking cDNA alongside embryo samples as a control to detect sample/reagent contamination.
3. Prepare sufficient SYBR green reaction mix (Table 6) to amplify each target in each sample in triplicate.
4. Pipette 1 μL of each diluted sample into triplicate wells of a PCR plate (or individual tubes).
5. Add 9 μL of SYBR green reaction mix to each well.
6. Seal the plate with an optical adhesive cover, then centrifuge briefly to collect the contents at the bottom of each well and eliminate any air bubbles.
7. Run the real-time PCR reactions according to the SYBR green master mix purchased (e.g., enzyme activation at 95°C for

Table 6
SYBR green reaction mix (1×)

Component	Volume (μL)
2× SYBR Green Master Mix	5
Forward and reverse primers	0.5
Nuclease-free water	3.5
Total volume	9

2 min, followed by 40 cycles of denaturation at 95°C for 5 s, followed by combined annealing and extension at 60°C for 30 s, followed by a dissociation curve).

8. Use the melt curve analysis to check for nonspecific amplification.
9. Analyze the results to identify samples with sufficient amplification of the housekeeping genes for further analysis.

3.5 cDNA Purification and Selection

Here, we outline the steps for purifying the PCR amplification product and cDNA of interest outlined using the Zymo™ DNA Clean & Concentrator and Zymoclean™ Gel DNA Recovery Kit. For other kits, purify the remaining undiluted PCR product with a PCR purification kit following the manufacturer's recommendations.

1. If required by the Zymo™ kits purchased, add 24 mL 100% ethanol to the 6 mL of DNA wash buffer (or 96 mL 100% ethanol to the 24 mL DNA wash buffer).
2. Add seven volumes of DNA binding buffer for each volume of DNA sample in a 1.5 mL microcentrifuge tube (e.g., 700 μL buffer to 100 μL sample).
3. Transfer the mixture to a Zymo-Spin Column seated in a collection tube.
4. Centrifuge at 10,000–16,000 × *g* for 30 s then discard the flow-through.
5. Add 200 μL of DNA wash buffer to the column and centrifuge for 30 s. Discard the flow through, then repeat by adding another 200 μL of DNA wash buffer to the column and centrifuge for 30 s.
6. Elute the cDNA in 15–30 μL of nuclease-free water (depending on cell number; *see* **Note 11**). The purified pre-amplified cDNA can be stored at −80°C for 6 months.
7. Prepare a 2% agarose gel and allow it to set in a casting tank with a comb that accommodates the appropriate number of samples to be run.

8. Combine cDNA samples with loading dye (1:6) and load into the set gel using a 10 or 20 μL pipettor and pipette tips. Pipette an appropriate DNA ladder alongside samples.
9. Run the gel at a constant voltage (usually 100–120 V) for ~45 min to 1 h until the dye front reaches the bottom of the gel.
10. Visualize the DNA bands under UV light. Using a fresh razor blade, excise the cDNA band. Transfer the gel slice to a 1.5 mL microcentrifuge tube that has been pre-tared on a balance.
11. Weigh the microcentrifuge with the gel slice in the tube to determine the gel weight.
12. Purify DNA from the agarose gel by adding three volumes of agarose dissolving buffer for each volume of gel slice (e.g., for 100 μL (mg) of agarose gel slice, add 300 μL of agarose dissolving buffer).
13. Incubate at 55°C for a minimum of 10 min. Briefly mix the sample by vortexing or inverting, and ensure the gel slice is completely dissolved.
14. Transfer the dissolved agarose solution to a Zymo-Spin™ Column seated in a collection tube and centrifuge at 10,000–16,000 × *g* for 1 min.
15. Discard the flow-through, then add 200 μL of DNA Wash Buffer to the column. Centrifuge for 30 s, discard the flow-through, then repeat the wash step.
16. Elute the cDNA in 30–50 μL of water (depending on cell number: use 30 μL when there are less than five cells). The purified PCR product can be stored at −80°C for 6 months.

3.6 Sample Validation for Library Preparation

1. Measure the concentration of purified cDNA products using a Qubit® dsDNA kit or equivalent (*see* **Note 12**).
2. Measure the size distribution with a Tapestation®, Bioanalyzer®, or equivalent using a High Sensitivity DNA chip or High Sensitivity DNA5000 ScreenTape assay. A size distribution range of ~500–5000 bp is expected (this will confirm the absence of residual primers which have a size ~100 bp).

3.7 Library Preparation

Library preparation can be performed with any commercial kit. Because the quality of the base (read) is inversely proportional to the length of the DNA strand, the full-length fragments generated after PCR need to be fragmented before being loaded on the flow-cell. For this purpose, several methods can be used, but the one that has gained popularity in the field is the "tagmentation" enzymatic reaction (Tn5 transposase), mainly due to its simplicity and ease of use. Here, we describe the use of the Nextera® XT DNA Library

Preparation Kit method developed by Illumina, in which the sample DNA is simultaneously fragmented and tagged (*see* **Note 13**). The DNA library preparation described here follows the manufacturer's instructions.

1. Prepare samples to a concentration of 1 ng cDNA in 5 μL for the DNA library preparation.
2. Make sure Neutralize Tagment (NT) Buffer is at room temperature and free of precipitate (vortex if necessary).
3. For each library, combine 5 μL of Tagment DNA Buffer (TD) with 2.5 μL of amplified cDNA and 2.5 μL of Amplicon Tagment Mix (ATM) in a low-bind microplate well.
4. Mix gently by pipetting, then briefly centrifuge.
5. Incubate in a pre-heated thermal cycler at 55°C for 5 min.
6. Immediately after incubation, add 2.5 μL of NT Buffer to terminate the tagmentation reaction. Mix samples briefly, then centrifuge and incubate at room temperature (20–25°C) for 5 min to ensure complete neutralization.
7. To each neutralized tagmentation reaction (12.5 μL total volume), add 2.5 μL of i7 index adapter (forward) and 2.5 μL of i5 index adapter (reverse), followed by 7.5 μL of Nextera PCR Master Mix (final volume of 25 μL per library).
8. Mix reactions gently, then centrifuge at 280 × *g* for 1 min at 20°C.
9. Place reactions in a thermal cycler with the following program: 72°C for 3 min, 95°C for 30 s, 12 cycles of: 95°C for 10 s, 55°C for 30 s, 72°C for 30 s, 72°C for 5 min, with a 10°C hold.
10. Maintain amplified libraries on ice or at 4°C until purification.
11. To each 25 μL PCR reaction, add 15 μL of Illumina Purification Beads (IPB; equivalent to a 0.6× bead-to-sample ratio). Mix thoroughly by pipetting, incubate at room temperature for 5 min, then place the plate on a magnetic stand until the supernatant clears (~2 min).
12. Carefully remove the supernatant, then wash the beads twice with 200 μL of freshly prepared 80% ethanol without disturbing the pellet.
13. After the final wash, remove residual ethanol and air-dry the beads on the magnet for 10–15 min, avoiding over-drying.
14. Elute DNA by adding 22.5 μL of Resuspension Buffer (RSB), mix gently, then incubate for 2 min at room temperature before returning to the magnetic stand.
15. Transfer the clarified eluate (~20 μL) containing the purified library to a new low-bind plate well.

16. The resulting libraries can be stored at –20°C until quantification and normalization.
17. Evaluate library quantity and fragment size using a Qubit dsDNA High Sensitivity Assay (for DNA concentration) and an Agilent Bioanalyzer High Sensitivity DNA chip (fragment size profiles). Single-cell libraries typically exhibit a broad size distribution ranging from 300 to 700 bp, consistent with cDNA fragmentation and adapter incorporation.
18. Adjust each library to 4 nM based on molarity (calculated from average fragment size and Qubit quantification).
19. Pool normalized libraries by combining equal volumes (typically 5 μL per library) to generate the multiplexed sequencing pool.
20. Dilute the pooled libraries according to the instrument-specific guidelines (e.g., NovaSeq or Illumina), aiming for a final loading concentration between 1.2 and 1.8 pM, depending on platform and read length.
21. Store final pooled libraries at –20°C and thaw only once before sequencing to minimize freeze–thaw degradation.

3.8 Bioinformatic Analysis

Sequencing results are returned in a .fastq file format. Following a quality control phase, the sequencing reads must be aligned to a reference genome. Gene-by-gene quantification is used to produce an expression matrix, which must be normalized for comparisons between genes and samples/cells for differentially expressed gene identification. Various pipelines can be used; the one presented here, using UNIX and R, serves as an example for single-read sequencing (*see* **Note 14**).

3.8.1 Quality Checks

FastQC is particularly suitable for general use. It presents a color-coded graphical representation of the many measures of interest: Per Base Sequence Quality (Phred score), Per Sequence Quality Scores, Per Base Sequence Content, Sequence Length Distribution, or Presence of adapters. MultiQC is useful to aggregate all quality control files into a single one.

1. Load the FastQC tool into a UNIX or UNIX-like terminal using the following code:

```
mkdir QC
    fastqc *fastq.gz -o QC
    multiqc.
```

3.8.2 Trimming of Adapters and Low-Quality Sequences

It is important to perform adapter cleaning and remove low-quality sequences if QC at the previous step indicated their presence (signified by a warning in the QC report file).

1. Input the following code into your terminal to remove adapters from the.fastq sequencing files:

```
trim_galore *fastq.gz -o /trimmed
```

3.8.3 Alignment

Clean sequences must be mapped to a reference genome using a read aligner tool. In this example, we work with the hg38 human genome version and the STAR aligner tool. This step will generate.bam files, which are a compact and indexable representation of nucleotide sequence alignments.

1. Download the FASTA sequence of the human genome (version hg38; *see* **Note 15**):

```
wget http://hgdownload.soe.ucsc.edu/golden-
Path/hg38/bigZips/hg38.fa.gz
gunzip hg38.fa.gz
```

2. For quantifying the expression of protein-coding genes, you will need a comprehensive annotation file of the genome. You can download the Gencode transcript annotation of the human genome (version hg38) (*see* **Notes 16 and 17**) using the following script:

```
wget https://ftp.ebi.ac.uk/pub/databases/gen-
code/Gencode_human/release_29/gencode.v29.
annotation.gtf.gz
gunzip gencode.v29.annotation.gtf.gz
```

3. Using STAR, generate genome indexes for mapping:

```
STAR --runThreadN 1 --runMode genomeGener-
ate --genomeDir path_to_genome --genomeFasta-
Files path_to_genome/hg38.fa.out --sjdbGTFfile
genome/gencode.v29.annotation.gtf
```

4. Align reads to the reference genome to determine where reads came from (*see* **Note 18**).

```
STAR --readFilesCommand zcat --runThreadN 4 --geno-
meDir path_to_genome --readFilesIn trimmed/*fq.
gz --outSAMtype BAM SortedByCoordinate --run-
Mode alignReads --outFileNamePrefix SAMPLE_
--outFilterMultimapNmax 5000 --outSAMmultNmax 1
--alignEndsType EndToEnd --outFilterMismatchNmax
999 --outFilterMismatchNoverLmax 0.06 --sjdbGTF-
file gencode.v29.annotation.gtf
```

3.8.4 Quantification

featureCounts can be used to determine how many reads correspond to genomic features included in your annotation file (genes, exons, promoters, etc.). Here, we are interested in determining how many reads align with each human gene (*see* **Note 19**).

1. Input the following code to generate a.bam file that can be used to determine the number of reads that mapped to each gene annotated in the.gtf file:

```
featureCounts -g gene_name -M -s 0 -T 8 -C -a
genome/gencode.v29.annotation.gtf -o BAM.count
*.bam
```

2. To create a count matrix suitable for data analysis, format the count file using the following code in your terminal:

```
cat BAM.count | cut -f1,7- | sed 1d > BAM_for-
matted.count
```

3.8.5 Expression Analysis

In order to determine how many reads have mapped to each gene, the aligned and quantified.count files must be converted into a count matrix. There are multiple ways of analyzing embryonic RNA-seq datasets, depending on the study design and the purpose of each research project. Key steps in scRNA-seq analysis generally involve quality control and filtering, normalization, dimensionality reduction, clustering and differential expression.

Here we outline our pipeline for analyzing data from different embryonic stages, identifying differentially expressed genes along embryonic stages, inferring pseudotime for each cell/embryo and visualizing embryonic development using trajectory analysis.

Switching to the R environment, an example dataset of an embryonic dataset obtained by single-cell RNA-seq [7] (available on GEO: GSE36552) that must first be processed as outlined above can be used to practice this pipeline.

1. Start a R session and install the following libraries to be able to run the necessary code lines:

```
BiocManager::install(c("edgeR",      "scater",
"scran", "SingleCellExperiment", "slingshot",
"tradeSeq"))
install.packages(c("ggplot2", "phateR", "readr",
"Seurat","dplyr"))
my_packages <- c("edgeR", "scater", "scran",
"SingleCellExperiment", "slingshot", "tradeSeq",
"ggplot2", "phateR", "readr", "Seurat","dplyr")
lapply(my_packages, library, character.only =
TRUE)
```

2. Import the count matrix and the metadata associated with the samples into your R environment

```
cts <- read_delim("BAM.count", delim = "\t",
escape_double = FALSE, trim_ws = TRUE)
cts <- data.frame(cts)
annotation = cts[,1]
rownames(cts) <- annotation
```

```
cts <- cts[,-1]
Meta_Yan <- read_delim("metadata_Yan.tsv",de-
lim = "\t", escape_double = FALSE, trim_ws =
TRUE)
Meta_Yan <- data.frame(Meta_Yan)
rownames(Meta_Yan) <- Meta_Yan$ID
```

3. Filter the genes to remove lowly expressed genes that can bias the interpretation of results (*see* **Note 20**). We also filter out low-quality cells based on the number of genes expressed (*see* **Note 21**).

```
y <- DGEList(counts=cts, genes=rownames(cts))
isexpr <- rowSums(edgeR::cpm(y)>1) >= 4
cts <- cts[isexpr,]cts <- CreateSeuratOb-
ject(cts, project = "Yan_dataset")
cts <- subset(cts, subset = nFeature_RNA >
5000 & nFeature_RNA < 20000)
```

4. Normalize the read count matrix to make samples comparable.

```
sce <- as.SingleCellExperiment(cts)qclust <-
scran::quickCluster(sce, min.size = 30)
sce <- scran::computeSumFactors(sce, clusters
= qclust)
sce <- scater::logNormCounts(sce)
cts <- as.Seurat(sce, counts = "counts", data
= "logcounts")
cts <- ScaleData(cts, vars.to.regress =
"nCount_RNA", features = rownames(cts))
```

5. At this step, you can apply dimensionality reduction to visualize the similarity of samples in your dataset using a principal component analysis (Fig. 3).

```
cts <- FindVariableFeatures(cts)
```

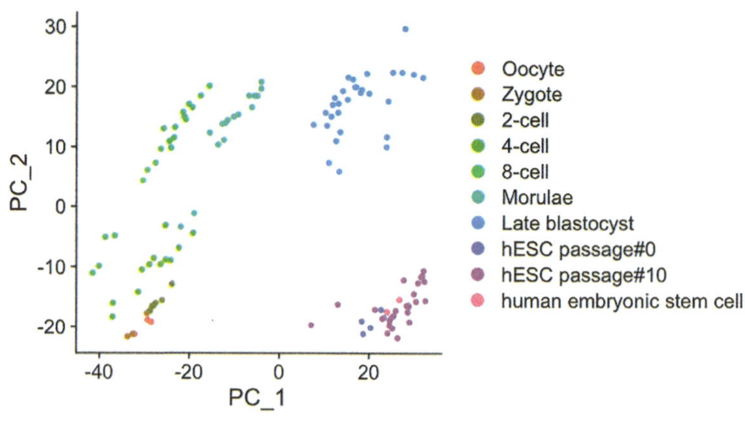

Fig. 3 Principal component analysis of normalized counts

```
cts <- RunPCA(cts)
cts <- AddMetaData(cts, Meta_Yan)
DimPlot(cts, reduction = "pca", group.by=
"Cell")
```

Since each cell stage in this example has been identified by the authors in the metadata file, the gene markers can easily be identified for each stage (Fig. 4). It is possible to distinguish between distinct cell types at each embryonic stage with more comprehensive and sizeable datasets. For instance, at the blastocyst stage, there are three distinct lineages: trophectoderm, primitive endoderm, and epiblast. The user may apply clustering methods to dissect the underlying structure of all cells and group cells with similar transcriptomes.

6. To group cells according to similarities in gene expression, input the following into R:

```
Idents(cts) <- your_clustering
markers <- FindAllMarkers(cts,only.pos=TRUE)
markers %>%
group_by(cluster) %>%
dplyr::filter(avg_log2FC > 1) %>%
slice_head(n = 5) %>%
ungroup() -> top5
DoHeatmap(cts, features = top5$gene)
```

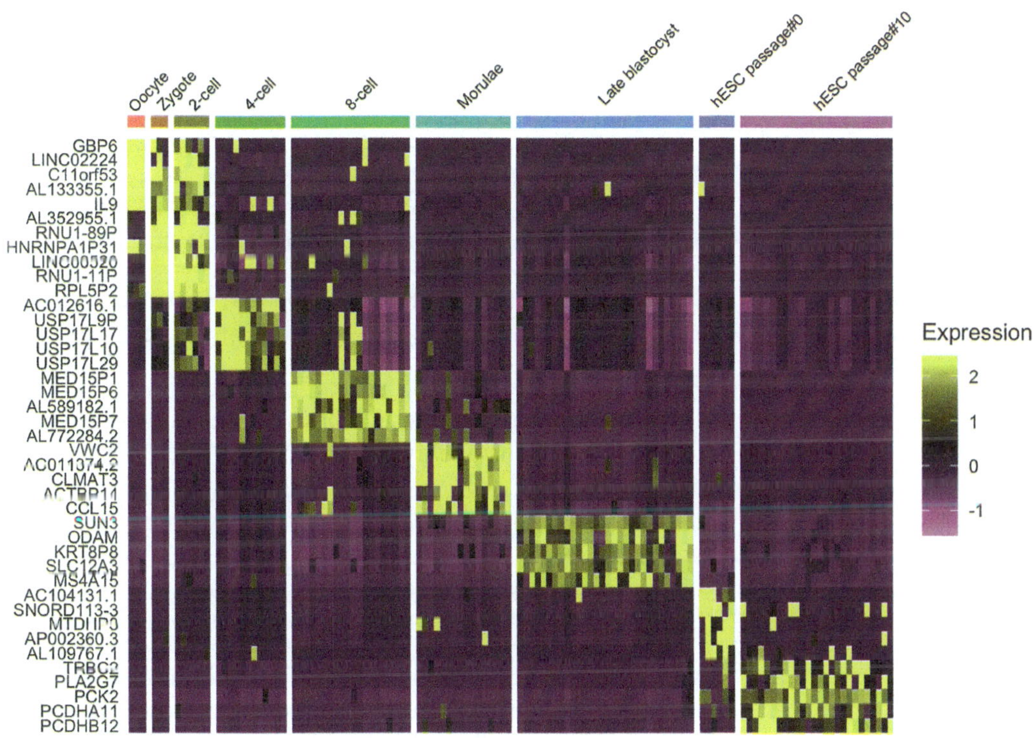

Fig. 4 Heatmap showing the expression of the top five markers for each embryonic stage in the dataset used as an example in this chapter (GSE36552)

7. For comparing the expression of two sets of cells (for example, four-cell vs. two-cell) or embryos (obtained under two or more different settings), specify the desired comparison.

```
markers <- FindMarkers(cts,ident.1=your_group1,
ident.2=your_group2)
```

3.8.6 Trajectory Analysis

To identify genes whose expression changes dynamically and continuously throughout early embryonic development, regardless of embryo stage, PHATE dimensionality reduction, a recently developed method previously applied to human embryonic stem cells [14], can be applied. PHATE is particularly well-suited for early development since its information geometry is based on diffusion dynamics.

1. For this analysis, a Python environment is required, and the path to your Python installation should be provided to be able to run some packages used here.

```
Sys.setenv(RETICULATE_PYTHON =
"your_path_to_python/python")
reticulate::py_discover_config(required_mod-
ule="phate")
```

2. Run the PHATE dimensionality reduction and incorporate the results into the work dataset.

```
data <- cts@assays[["RNA"]]@data
phate_output <- phate(t(data),ndim=3)
clusters_phate <- cluster_phate(phate_output)
phate_output <- as.matrix(phate_output)
colnames(x = phate_output) <- paste0("PHATE_",
1:ncol(x = phate_output))
phate.reduction <- CreateDimReducObject(embed-
dings = phate_output, key = "PHATE_", assay =
"RNA")
cts@reductions$phate <- phate.reduction
sce <- as.SingleCellExperiment(cts, assay =
"RNA")
```

3. Model the developmental trajectories of your samples based on the PHATE dimensionality reduction

```
sce <- slingshot(sce, reducedDim = 'PHATE',
clusterLabels = clusters_phate)
lin <- getLineages(SlingshotDataSet(sce),
clusterLabels = clusters_phate)
crv <- getCurves(lin)
```

4. Extract the pseudotime of each cell. Pseudotime is a metric that could be interpreted as a timing distance between one cell and its precursor cell, and helps identify the ordering of cells along a lineage based on their gene expression profile

```
pseudotime <- slingPseudotime(crv, na = FALSE)
```

```
cellWeights <- slingCurveWeights(crv)
counts <- as.matrix(cts@assays[["RNA"]]@counts)
```

5. To visualize the dynamics of your cells/samples, plot gene expression in the reduced dimensions (Fig. 5).

```
rd <- slingshot::slingReducedDim(crv)
rd <- data.frame(rd)
rd <- merge(rd,Meta_Yan,by.x=0,by.y="ID")

rd$Cell    <-    factor(rd$Cell,level=c("Oo-
cyte","Zygote","2-cell","4-cell","8-cell",
"Morulae","Late blastocyst","hESC passage#0",
"hESC passage#10","human embryonic stem cell"))
ggplot(rd,aes(PHATE_1,PHATE_2))+
geom_point(size=2.5,stroke  =  0,aes(color=
Cell)) +
scale_color_manual(values = c("Oocyte" = "#AE-
CEE3", "Zygote"="#3876B0", "2-cell"="#BAD693",
"4-cell"="#559F51", "8-cell"="#EA9D9D", "Mor-
ulae"="#D23731", "Late blastocyst"="#FF922D",
"hESC passage#0"="#DF70F8", "hESC passage#10"=
"#06A4FF",  "human  embryonic  stem  cell"=
"gray"))+
theme_bw()+
theme(axis.title.x=element_text(size=15))+
theme(axis.title.y=element_text(size=15,vjust=
0))+theme(axis.text.x=element_text(size=10,mar-
gin=margin(3,0,10,0)))+ theme(axis.text.y=ele-
ment_text(size=10,margin=margin(0,3,0,15)))+
theme(legend.title=element_text(size=15))+
theme(legend.text=element_text(size=12))
```

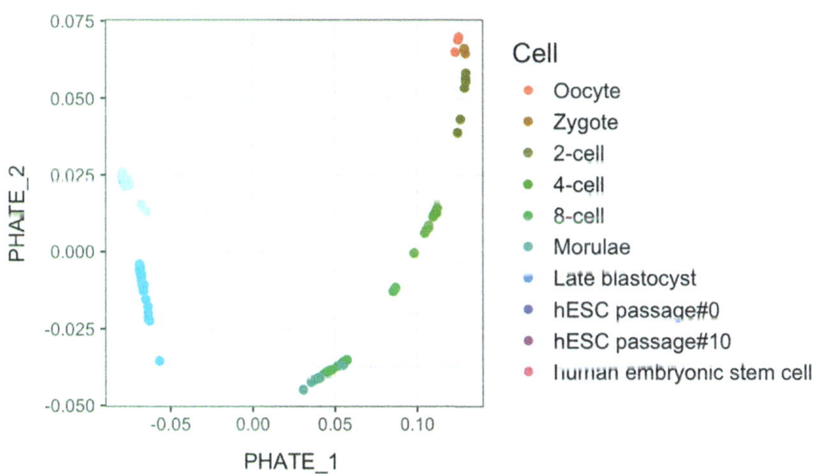

Fig. 5 PHATE dimensionality reduction of normalized counts with samples colored by cell embryonic stage

6. Perform differential analysis of genes along pseudotime after fitting a negative binomial generalized additive model to relate gene expression to pseudotime for each lineage.

```
sce_GAM_all_genes <- fitGAM(counts = counts,
sds = crv, nknots = 6, verbose = FALSE)
assoRes <- associationTest(sce_GAM_all_genes)
```

7. Visualize the gene expression of the identified differentially expressed genes of interest. Here, we outline the method for *IFITM1*, highly expressed in human embryonic stem cells, and

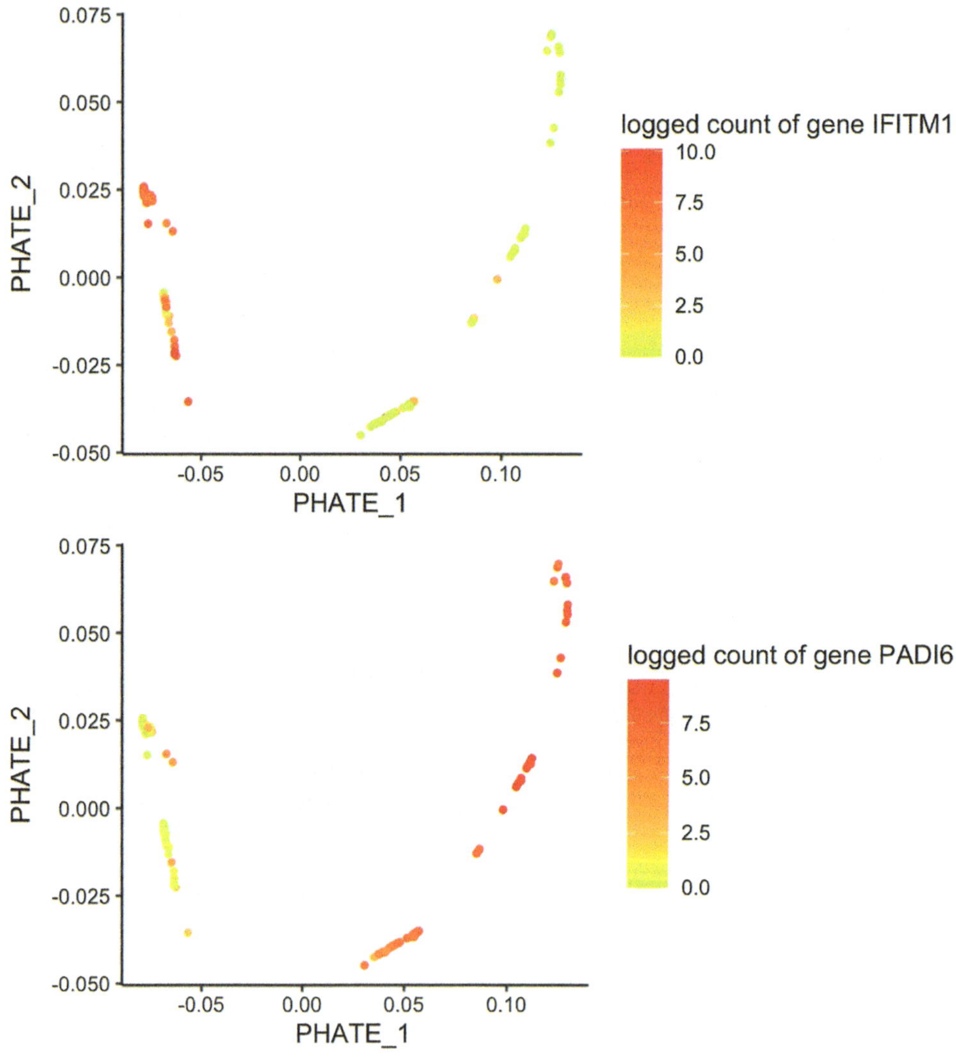

Fig. 6 PHATE dimensionality reduction of normalized counts with samples colored by gene expression of *IFITM1* and *PADI6*

PADI6, a maternal-effect gene (gradually repressed from the oocyte) (Fig. 6, *see* **Notes 22 and 23**).

```
plotGeneCount(crv, counts, gene= "IFITM1")
plotGeneCount(crv, counts, gene= "PADI6")
```

4 Notes

1. With Tang's protocol, the transcriptomes are generated using an oligo d(T) primer to initiate reverse transcription and subsequent amplification; it thus keeps the 3′ skewing. Smart-Seq uses Moloney murine leukemia virus reverse transcriptase to incorporate an oligonucleotide having a PCR-primer sequence at the 5′ end of transcripts. Only these cDNAs are amplified and sequenced, facilitating the analysis of full-length transcripts. A second method exists, the SMART-seq method (Switching Mechanism at the 5′-end of the RNA Transcript; Smart-seq2 protocol), which incorporates an oligonucleotide having a PCR-primer sequence at the 5′ end of transcripts, and only these cDNAs are amplified and sequenced [15].
2. The reverse transcriptase SuperScript® III reverse transcriptase can be replaced by the Promega GoScript™ reverse transcriptase, which has similar performance in terms of sensitivity of transcript detection and transcript length.
3. The use of ERCC Spike-In RNA molecules is optional in this technique. ERCC Spike-In could be used to address technical variation, perform further normalization and compare endogenous RNA quantity between experimental samples. Per reaction, 0.1 μL of a (1:100,000) dilution of ERCC Spike-In can be added in experiments analyzing single-blastomeres.
4. The dissociation can also be obtained by transferring the embryo in drops of pre-warmed Ca^{++}-free media covered by mineral oil set on a 37°C warm plate. The continuous pipetting of the embryo up and down using a pipette mechanically helps the cell dissociation.
5. If the anembryonic and embryonic compartments are to be separately analyzed, separation can be performed by gently pulling the embryonic pole into a holding pipette and then using the laser across the base of the holding pipette to separate the embryonic pole from the anembryonic pole, which remains outside the pipette.
6. When considering cellular identity at the blastocyst stage, the TE differentiates into two compartments: (1) polar TE (pTE), overlying the Inner cell Mass (ICM) and more specifically the

epiblast (EPI), which is primarily responsible for mediating attachment and implantation in the endometrium, and (2) mural TE (mTE), opposing the ICM. As such, in preparation for RNA sequencing, the blastocyst can be separated into two major compartments: one corresponds to the mTE, referred to as the abembryonic compartment, and the other to the ICM/pTE comprising cells of the epiblast (EPI), primitive endoderm-(PrE), and mostly the pTE, collectively referred to as the embryonic compartment. The latter. therefore. corresponds to a mix of cells. Importantly, when the two compartments are analyzed separately, only embryos in which the ICM can be clearly identified under the microscope can be dissected and used for analysis.

7. It is important to point out that the single-cell RNA sequencing of preimplantation embryos is prone to technical issues, and separating technical variation from biological variation is challenging. A single-cell RNA-seq technique measuring a single-cell transcriptome repeatedly, called the MR-seq method, could further contribute to this problem. Splitting the lysate of a single cell into two aliquots prior to performing single-cell RNA-seq analysis could allow for assessing the technical variation by measuring each single cell twice and identifying the differentially expressed genes between these samples [16].

8. The Supersript, RNase inhibitor, and T4 shouldn't be mixed by vortex but only tapped and centrifuged briefly for 30 s at 4°C in a mini-spin.

9. It is recommended to use a PCR cooling block that will be maintained at 4°C during the protocol by setting it on ice. Otherwise, sample tubes should be placed directly on ice.

10. Quality control of the preamplified cDNA is performed by qPCR amplification of housekeeping genes such as *GAPDH* and *TBP*. It is highly recommended to use several housekeeping genes that are expressed at slightly different ranges. Housekeeping genes should be chosen for their quantity and stability of expression across developmental stages. An aliquot of a 1:20 dilution of the preamplified cDNA (5 μL of the PCR product in 100 μL final volume of nuclease-free water) is used for SYBR-based quantitative PCR. The blank sample will be included as a control (e.g., detection of sample/reagents contamination). The qPCR program and reaction mix should be adjusted according to the real-time PCR instrument and kit used.

11. The cDNA elution at the purification step is 15 μL from one cell RNA-seq and up to 30 μL from one embryo.

12. An average concentration range of ~5–10 ng/μL cDNA is typically obtained for cDNA preamplification products of a single blastomere at blastocyst stage and up to 25 ng/μL cDNA at cleavage stage.
13. To avoid cross-contamination when adding or transferring samples or reagent master mixes, change tips between each sample. Open only one index adapter tube at a time to prevent misplacing caps. Remove unused index adapter tubes from the working area; Likewise, for IPB 100 mL bottle resuspension, manually mix the bead reagent container by manual inversion. Inversion mixing constitutes turning a container upside down (rotating bottle 90 degrees every 30 s, and continually inverting for a total of 2 min) and then returning it to its upright position. After this time, visually inspect the inside of the container for any solid material still adhering to the walls. If the container's inner walls are free of any residual material, the bead reagent can be considered fully resuspended. If the reagent is still found on the inner walls, repeat the mixing step for an additional 1 min. Repeat visual inspection of the container walls to ensure complete mixing. If necessary, repeat the mixing step
14. All code lines included in the protocol outlined here require a lot of memory and cannot be run on a local machine. Ideally, access a high-performance computing cluster. Otherwise, free alternatives such as Galaxy exist that do not require computational knowledge and offer free memory for data analysis.
15. For the mouse genome, you can find genome versions at https://hgdownload.soe.ucsc.edu/downloads.html#mouse.
16. While using an annotation is optional, STAR creators highly recommend using annotations whenever they are available.
17. For the determination of repetitive elements, which play a critical role in gene expression regulation, download the FASTA sequence of the human genome (version hg38) with repeat identification from RepeatMasker:

```
wget https://www.repeatmasker.org/genomes/hg38/
RepeatMasker-rm405-db20140131/hg38.fa.out.gz
gunzip hg38.fa.out.gz
```

You can add the repetitive element annotation to the annotation file so that, along with protein-coding genes, you can count the number of reads per repetitive element family.

```
awk '{OFS="\t";if(NR>3){if($9 == "+"){print
$5,"hg38_rmsk","exon",$6,$7,"0","+",".","gene_
name \"TE_"$10"\";",$11}else{print $5,"hg38_
rmsk","exon",$6,$7,"0","-",".","gene_name
\"TE_"$10"\";",$11}}}'
```

```
hg38.fa.out | grep -P "LINE|LTR|SINE" | cut
-f1-9 > hg38.rmsk.gtf
cat hg38.rmsk.gtf gencode.v29.annotation.gtf >
gencode.v29.annotation.rmsk.gtf
```

18. To run STAR for paired-end sequencing, the argument –readFilesIn should provide the two.fastq files at the same time for each sample: e.g., --readFilesIn SAMPLE1_R1.fq.gz SAMPLE1_R2.fq.gz

19. Add –p and --countReadPairs options for paired-end sequencing.

```
featureCounts -g gene_name -M -p --countRead-
Pairs -s 0 -T 8 -C -a genome/gencode.v29.anno-
tation.gtf -o BAM.count *.bam
```

20. The removal of lowly expressed genes should be adapted depending on the number of cells/samples you have. In this dataset, we considered that each gene should have a cpm>1 in at least four samples, and users may modify these values in the code line. Otherwise, edgeR provides the filterByExpr() function, which determines which genes have sufficiently large counts to be retained.

21. There is no universal threshold for filtering cells. Usually, the number of unique genes detected in each cell is informative about the cell quality (notably empty samples). Users may also apply custom thresholds based on mitochondrial RNA proportion or identify outliers using dimensionality reduction or correlation analysis.
You can run the following lines to select only cells with less than 20% of mitochondrial RNA:

```
cts[["percent.mt"]]<-PercentageFeatureSet(cts,
pattern = "^MT-")
cts <- subset(cts, subset = nFeature_RNA >
5000 & nFeature_RNA < 20000 & percent.mt<20)
```

22. Any other gene of interest can be visualized by inputting the relevant gene name.

23. There are other avenues of analysis from scRNA-seq data, including single-nucleotide variant analyses, for example, using toolkits like GATK.

References

1. Petropoulos S, Edsgard D, Reinius B, Deng Q, Panula SP, Codeluppi S, Plaza Reyes A, Linnarsson S, Sandberg R, Lanner F (2016) Single-cell RNA-seq reveals lineage and X chromosome dynamics in human preimplantation embryos. Cell 165(4):1012–1026. https://doi.org/10.1016/j.cell.2016.03.023
2. Sha QQ, Zheng W, Wu YW, Li S, Guo L, Zhang S, Lin G, Ou XH, Fan HY (2020) Dynamics and clinical relevance of maternal mRNA clearance

during the oocyte-to-embryo transition in humans. Nat Commun 11(1):4917. https://doi.org/10.1038/s41467-020-18680-6
3. Vastenhouw NL, Cao WX, Lipshitz HD (2019) The maternal-to-zygotic transition revisited. Development 146(11):dev161471. https://doi.org/10.1242/dev.161471
4. Meistermann D, Bruneau A, Loubersac S, Reignier A, Firmin J, Francois-Campion V, Kilens S, Lelievre Y, Lammers J, Feyeux M, Hulin P, Nedellec S, Bretin B, Castel G, Allegre N, Covin S, Bihouee A, Soumillon M, Mikkelsen T, Barriere P, Chazaud C, Chappell J, Pasque V, Bourdon J, Freour T, David L (2021) Integrated pseudotime analysis of human pre-implantation embryo single-cell transcriptomes reveals the dynamics of lineage specification. Cell Stem Cell 28(9):1625–1640 e1626. https://doi.org/10.1016/j.stem.2021.04.027
5. Asami M, Lam BYH, Ma MK, Rainbow K, Braun S, VerMilyea MD, Yeo GSH, Perry ACF (2022) Human embryonic genome activation initiates at the one-cell stage. Cell Stem Cell 29(2):209–216 e204. https://doi.org/10.1016/j.stem.2021.11.012
6. Kagawa H, Javali A, Khoei HH, Sommer TM, Sestini G, Novatchkova M, Scholte Op Reimer Y, Castel G, Bruneau A, Maenhoudt N, Lammers J, Loubersac S, Freour T, Vankelecom H, David L, Rivron N (2022) Human blastoids model blastocyst development and implantation. Nature 601(7894):600–605. https://doi.org/10.1038/s41586-021-04267-8
7. Yan L, Yang M, Guo H, Yang L, Wu J, Li R, Liu P, Lian Y, Zheng X, Yan J, Huang J, Li M, Wu X, Wen L, Lao K, Li R, Qiao J, Tang F (2013) Single-cell RNA-Seq profiling of human preimplantation embryos and embryonic stem cells. Nat Struct Mol Biol 20(9):1131–1139. https://doi.org/10.1038/nsmb.2660
8. Blakeley P, Fogarty NM, Del Valle I, Wamaitha SE, Hu TX, Elder K, Snell P, Christie L, Robson P, Niakan KK (2015) Defining the three cell lineages of the human blastocyst by single-cell RNA-seq. Development 142(20):3613. https://doi.org/10.1242/dev.131235
9. Xiang L, Yin Y, Zheng Y, Ma Y, Li Y, Zhao Z, Guo J, Ai Z, Niu Y, Duan K, He J, Ren S, Wu D, Bai Y, Shang Z, Dai X, Ji W, Li T (2020) A developmental landscape of 3D-cultured human pre-gastrulation embryos. Nature 577(7791):537–542. https://doi.org/10.1038/s41586-019-1875-y
10. Zhao C, Plaza Reyes A, Schell JP, Weltner J, Ortega NM, Zheng Y, Bjorklund AK, Baque-Vidal L, Sokka J, Trokovic R, Cox B, Rossant J, Fu J, Petropoulos S, Lanner F (2025) A comprehensive human embryo reference tool using single-cell RNA-sequencing data. Nat Methods 22(1):193–206. https://doi.org/10.1038/s41592-024-02493-2
11. Proks M, Salehin N, Brickman JM (2025) Deep learning-based models for preimplantation mouse and human embryos based on single-cell RNA sequencing. Nat Methods 22(1):207–216. https://doi.org/10.1038/s41592-024-02511-3
12. Tang F, Barbacioru C, Wang Y, Nordman E, Lee C, Xu N, Wang X, Bodeau J, Tuch BB, Siddiqui A, Lao K, Surani MA (2009) mRNA-Seq whole-transcriptome analysis of a single cell. Nat Methods 6(5):377–382. https://doi.org/10.1038/nmeth.1315
13. Perez-Palacios R, Fauque P, Teissandier A, Bourc'his D (2021) Deciphering the early mouse embryo transcriptome by low-input RNA-seq. Methods Mol Biol 2214:189–205. https://doi.org/10.1007/978-1-0716-0958-3_13
14. Moon KR, van Dijk D, Wang Z, Gigante S, Burkhardt DB, Chen WS, Yim K, Elzen AVD, Hirn MJ, Coifman RR, Ivanova NB, Wolf G, Krishnaswamy S (2019) Visualizing structure and transitions in high-dimensional biological data. Nat Biotechnol 37(12):1482–1492. https://doi.org/10.1038/s41587-019-0336-3
15. Picelli S, Bjorklund AK, Faridani OR, Sagasser S, Winberg G, Sandberg R (2013) Smart-seq2 for sensitive full-length transcriptome profiling in single cells. Nat Methods 10(11):1096–1098. https://doi.org/10.1038/nmeth.2639
16. Yang L, Ma Z, Cao C, Zhang Y, Wu X, Lee R, Hu B, Wen L, Ge H, Huang Y, Lao K, Tang F (2017) MR-seq: measuring a single cell's transcriptome repeatedly by RNA-seq. Sci Bull (Beijing) 62(6):391–398. https://doi.org/10.1016/j.scib.2017.01.029

Chapter 19

Immunofluorescence Analysis of Epigenetic Markers in Oocytes and Preimplantation Embryos

Marcella Pecora Milazzotto and Jessica Ispada

Abstract

Immunofluorescence (IF) is an essential technique for studying epigenetic modifications in embryos. It offers high spatial resolution for precise visualization of chromatin-associated proteins and modifications involved in gene regulation and epigenetic reprogramming. IF allows multiplexing with multiple fluorophores, enabling the study of protein interactions and co-localization. Widely used in gametes and embryos across species, IF aids in analyzing epigenetic-related protein expression and localization during various stages of development.

Keywords Gamete, Embryo, Epigenetic, Immunofluorescence

1 Introduction

Immunofluorescence (IF), a variation of immunocytochemistry (ICC), relies on the detection of antigens using fluorophore-conjugated antibodies in cells and tissues. This technique was first established by Albert Coons and colleagues in 1941, who pioneered the use of fluorescent antibodies to detect antigens in tissues during their studies on rheumatic fever [1]. When comparing IF with its predecessor, ICC, several advantages of IF become apparent. Firstly, IF provides higher spatial resolution, allowing for more precise identification of subcellular structures and compartments. Secondly, IF enables the simultaneous use of multiple fluorophores to label different antigens. This multiplexing capability makes it easier to study protein interactions and co-localization of proteins and organelles in a more practical and precise manner.

In the context of gametes and embryos, immunofluorescence is widely employed across various species to study protein expression and localization during development, investigate cell fate decisions, and examine tissue organization and morphogenesis at different stages of embryonic development [2–5]. Immunofluorescence is also

Alexandra J. Harvey and David K. Gardner (eds.), *The Preimplantation Embryo*, Methods in Molecular Biology, vol. 3038, https://doi.org/10.1007/978-1-0716-5292-3_19,

a crucial technique for studying epigenetic modifications in embryos, as it allows the visualization of specific proteins or modifications associated with chromatin structure, gene regulation, and epigenetic reprogramming. Epigenetics in embryonic development refers to heritable changes in gene expression that do not involve alterations to the underlying DNA sequence [6]. These modifications include histone modifications, DNA methylation, chromatin remodeling, and noncoding RNA interactions, all of which play vital roles in regulating cellular differentiation and development (reviewed by Milazzotto 2020 and 2023 [7, 8])

Given that the preimplantation embryo undergoes intense epigenetic reprogramming to erase the inherited marks from gametes and initiates a new cycle of reestablishing these marks, immunofluorescence (IF) serves as a practical and accessible tool for identifying these dynamic remodeling profiles.

2 Materials

Prepare all solutions using purified or distilled water and analytical grade reagents. Prepare and store all reagents at room temperature (unless otherwise indicated).

2.1 Equipment

1. Balance.
2. Spatulas and weigh boats.
3. Glass beakers.
4. Graduated cylinders.
5. Magnetic stirrer with hot plate.
6. Magnetic stirrer bars.
7. Filtration system (e.g., syringe and cellulose syringe filter).
8. pH meter.
9. Fume hood.
10. Mechanical pipettors.
11. Glass bottle.
12. Safety equipment (gloves, lab coat, and goggles).
13. Microcentrifuge.
14. Fluorescence microscope (e.g., epifluorescence, confocal, or spinning disk) equipped with the appropriate filters for the excitation and emission wavelengths of the fluorophores used.
15. Image analysis software, such as ImageJ or FIJI.

2.2 Solutions

2.2.1 Sample Preparation Solutions for Immunostaining

1. 1× Phosphate buffered saline (PBS): Weigh 8.4 g of NaCl using a weigh boat and transfer to a glass beaker. Weigh 1.424 g of $Na_2HPO_4*7H_2O$ using a weigh boat and transfer to the glass beaker. Weigh 0.276 g of NaH_2PO_4 using a weigh boat and transfer to the glass beaker. Add 800 mL of Distilled Water to the beaker and mix well using the stirrer and magnetic stirrer bar. Check the pH of the solution. It should be around pH 7.4. If necessary, adjust the pH using small amounts of NaOH or HCl. Transfer the solution to a graduated cylinder and adjust the volume to 1 L. Filter the solution using a filtration system and store the PBS 1× solution at room temperature. Label the container with the date of preparation.
2. 1 N sodium hydroxide (NaOH): Add 800 mL of Distilled Water to an empty beaker (*see* **Note 1**). Weigh 40 g of NaOH pellets using a weighboat. Transfer the NaOH pellets to the beaker containing water and mix well using the stirrer and magnetic stirrer bar. Transfer the solution to a graduated cylinder and slowly adjust the volume to 1 L. Store the solution in an appropriately labeled container in a corrosive storage area.
3. PFA stock solution (8%): Weigh 8 g of paraformaldehyde powder into a weigh boat and transfer it to the glass beaker (*see* **Note 2**). Add 100 mL of distilled water. Place the hot plate in a fume hood turned on (indispensable due to the release of formaldehyde vapors) and heat the mixture to 55–60°C. Stir continuously using the magnetic stirrer bar to aid dissolution. Slowly add drops of 1 N Sodium Hydroxide (NaOH) to the heated mixture until the solution becomes clear (*see* **Note 3**). Filter the clear solution using a filtration system to remove any undissolved particles. Allow the solution to cool to room temperature. Store the 8% paraformaldehyde solution at 2–8°C. Label the container with the date of preparation and concentration.
4. Fixation solution (pH 7.4): Mix equal volumes of the 8% paraformaldehyde solution and 1× PBS. For example, thoroughly mix 25 mL of the 8% paraformaldehyde solution with 25 mL of 1× PBS. Check the pH of the solution. It should be around pH 7.4. If necessary, adjust the pH using small amounts of NaOH or HCl. Store the 4% paraformaldehyde fixation solution at 2–8°C in a properly labeled and dated container.
5. 1% PBS/PVP: Weigh 0.1 g of PVP using a weigh boat and transfer it to a glass beaker. Add 80 mL of 1× PBS to the

beaker and mix well using the stirrer and magnetic stirrer bar. Transfer the solution to a graduated cylinder and adjust the volume to 100 mL. Label the solution and store it at room temperature.

6. Oocyte permeabilization solution (0.5% Triton X-100): Add 99 mL of 1× PBS to a glass bottle. Measure 500 μL of Triton X-100 using a pipettor and add to the glass bottle containing the 1× PBS (*see* **Note 4**). Mix well by inverting the glass bottle up and down, and wait a few minutes until the Triton X-100 is fully dissolved. Label the permeabilization solution and store at 2–8°C.
7. Embryo permeabilization solution (0.2% Triton X-100): Add 99.8 mL of 1× PBS in a glass bottle. Measure 200 μL of Triton X-100 using a pipettor and add to the 1× PBS in the glass bottle. Mix well by inverting the glass bottle up and down, and wait a few minutes until the Triton X-100 is fully dissolved. Label the permeabilization solution and store at 2–8°C.
8. Blocking solution (2% BSA with Triton-X 100): Weigh 2 g of BSA and transfer to a glass bottle. Add 98 mL of 1% PBS to the glass bottle. Measure 50 μL of Triton X-100 using a pipettor and add to the 1× PBS:BSA mix in the glass bottle. Mix well by gently inverting the glass bottle several times and wait a few minutes until the Triton X-100 and BSA are fully dissolved (*see* **Note 5**). Label the blocking solution and store at 2–8°C.
9. Denaturation solution (3 M HCl in 1% PBS/PVP): Add 60 mL of distilled water to a glass bottle. Slowly, add 40 mL of 12 M HCl to the distilled water. Mix well by inverting the glass bottle up and down. Label the denaturation solution and store at 2–8°C.
10. Neutralization solution (100 mM Tris, pH 8.5): Weigh 1.211 g of Tris base. Transfer the weighted Tris base into a glass beaker. Add approximately 90 mL of distilled water to the beaker, then stir the solution using a magnetic stirrer until the Tris base is completely dissolved. Measure the pH of the solution using a pH meter. Gradually add HCl solution dropwise to the Tris solution while continuously stirring until the pH reaches 8.5 (*see* **Note 3**). Once the pH has been adjusted to 8.5, transfer the solution to a graduated cylinder. Add distilled water to the solution until the total volume reaches 100 mL. Transfer the neutralization solution to a properly labeled container and store it at room temperature.

2.2.2 Counterstaining Solutions

1. Hoechst stock solution (5 mg/mL): Using a weighing scale, weigh 5 mg of Hoechst 33342 powder inside a 1.5 mL

microtube. Add 1 mL of 1× PBS to the microtube containing the Hoechst powder. Vortex the solution thoroughly until the Hoechst is fully dissolved. Aliquot and store frozen at −20°C.

2. Nuclear counterstain working solution (5 ng/mL Hoechst): Add 1 μL of Hoechst stock solution to 999 μL of 1% PBS/PVP.
3. Propidium iodide (PI) stock (2.5 mg/mL): Using a weighing scale, weigh 2.5 mg of propidium iodide powder inside a 1.5 mL microtube. Add 1 mL of 1× PBS to the microtube containing the propidium iodide powder. Vortex the solution thoroughly until the propidium iodide powder is fully dissolved. The propidium iodide stock solution can be aliquoted and frozen at −20°C.
4. DNA modification nuclear counterstain working solution (50 μ g/mL PI): Add 10 μL of the propidium iodide stock solution to 490 μL of 1% PBS/PVP.

2.2.3 Mitochondria Colocalization

1. MitoSpy™ Red CMXRos stock solution (*see* **Note 6**): Spin down the vial of lyophilized reagent in a microcentrifuge to ensure the reagent is at the bottom of the vial. Reconstitute MitoSpy™ Red CMXRos by adding 94 μl DMSO (1 mM) to the vial. Aliquot to avoid freeze–thawing cycles and stored at −20°C protected from light (*see* **Note 7**).
2. MitoSpy working solution (50 nM): Dilute MitoSpy™ Red CMXRos stock solution using relevant serum-free culture media for oocytes or embryos (*see* **Note 8**).
3. Primary antibodies specific to the targets of interest (*see* **Notes 9 and 10**).
4. Fluorescently conjugated secondary antibodies (*see* **Note 11**).

2.3 Consumables

1. Clear, round-bottomed 96-well plates.
2. Microcentrifuge tubes.
3. Serological pipettes.
4. Pipette tips.
5. Petri dish (the size will depend on the number of samples).
6. A micropipette that can be used to move oocytes/embryos in 96-well plates.
7. Humidified dark box: a metal or plastic box containing a wet paper napkin or gauze in the bottom that blocks the light and can comfortably fit the 96-well plate.
8. Stereo microscope with a magnification range of 10–20×.
9. Glass slides.
10. Glass coverslips.
11. Antifade mounting medium or glycerol solutions (*see* **Note 12**).

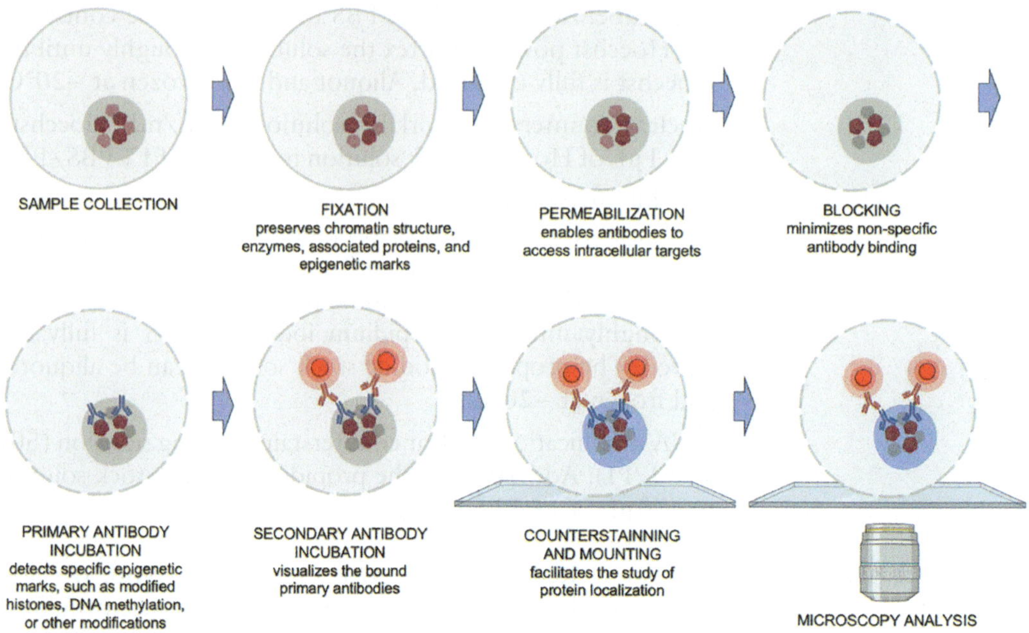

Fig. 1 General workflow of the IF technique in gametes and embryos

3 Methods

The method outlined here is suitable for staining and analyzing DNA methylation and hydroxymethylation (referred to as DNA modifications), posttranslational modifications of histones (referred to as histone PTMs), and associated proteins (as epigenetic-related enzymes) in oocytes and preimplantation embryos. This protocol has been optimized for bovine oocytes and embryos generated in vitro, as well as those obtained in vivo; however, it can be used for a range of species (*see* **Note 13**). Table 1 provides a list of antibodies already used for bovine oocytes and embryos, but the same protocol can also be applied to other antibodies of interest; each primary antibody will require optimization (*see* **Note 13**). Fig. 1 outlines the key steps for performing immunofluorescence in embryos, offering a detailed guide to optimize results while minimizing potential challenges.

3.1 Immunofluorescence of Histone PTMs or Epigenetic-Associated Proteins in Oocytes and Embryos

Handle embryos using a micropipettor to minimize solution carryover. All incubations should be performed in a humidified dark box.

1. Prepare a petri dish for embryo fixation by pipetting a row of 3×70 μL drops of 1% PBS/PVP, then place a 50 μL drop of fixation solution in the center. Complete the dish by pipetting a second row of 70 μL drops of 1% PBS/PVP under the

Table 1
List of antibodies previously tested in bovine oocytes and embryos

Target	1st antibody source	1st antibody cat number	Concentration	Host	2nd antibody	2nd antibody source	2nd antibody cat number	Concentration
TET1	GeneTex, USA	GTX629974	1:50	Mouse	Goat anti-mouse IgG H&L Alexa Fluor® 488	Invitrogen, USA	A-11029	1:250
TET2	Abcam, UK	ab94580	1:100	Rabbit	Donkey anti-rabbit IgG H&L Alexa Fluor® 488	Invitrogen, USA	A-21206	1:350
TET3	GeneTex, USA	GTX121453	1:100	Rabbit	Donkey anti-rabbit IgG H&L Alexa Fluor® 488	Invitrogen, USA	A-21206	1:350
5mC	Cell Signaling, USA	28692S	1:1000	Rabbit	Donkey anti-rabbit IgG H&L Alexa Fluor® 488	Invitrogen, USA	A-21206	1:400

(continued)

Table 1 (continued)

Target	1st antibody source	1st antibody cat number	Concentration	Host	2nd antibody	2nd antibody source	2nd antibody cat number	Concentration
5hmC	Abcam, UK	ab214728	1:200	Rabbit	Donkey anti-rabbit IgG H&L Alexa Fluor® 488	Invitrogen, USA	A-21206	1:400
Anti-histone H3 (acetyl K9)	Abcam, UK	ab12179	1:300	Mouse	Goat anti-mpuse IgG H&L Alexa Fluor® 568	Invitrogen, USA	A-11031	1:300
Anti-trimethylation histone H3 (Lys 27)	Abcam, UK	ab6002	1:300	Mouse	Donkey anti-rabbit IgG H&L Alexa Fluor® 488	Invitrogen, USA	A-21206	1:300

fixation drop (*see* **Note 14**). One petri dish is enough for up to 20 oocytes/embryos.

2. Collect oocyte and embryo stages at relevant times of development following in vitro culture, or in vivo development, according to standard procedures (*see* **Notes 15 and 16**).
3. Wash up to 20 oocytes/embryos through the first 3 drops of 1% PBS/PVP (*see* **Note 17**).
4. Transfer the oocytes/embryos into the central drop of fixation solution and fix for 30 min at room temperature, place the dish in a humidified box to avoid drop dehydration (*see* **Note 18**).
5. Wash the oocytes/embryos through the drops of 1% PBS/PVP. When staining for histone PTMs or other associated proteins, samples can be stored at 2–8°C in PBS/PVP for up to 4 weeks.
6. Prepare rows of a 96-well plate with a 100 μL well of permeabilization solution and a 70–100 μL well of blocking solution. Each row of the 96-well plate is sufficient for processing up to 20 oocytes/embryos.
7. Add oocytes/embryos to each well. Permeabilize oocytes for 1 h and embryos for 30 min at room temperature. Place the plate in the humidified box at room temperature.
8. Transfer the oocytes/embryos with as little media as possible to the well containing the blocking solution. Incubation in blocking solution can be done either at 37°C for 3 h or overnight (maximum of 16 h) at 2–8°C in the humidified box (*see* **Note 19**).
9. Dilute each primary antibody (*see* Table 1) in blocking base and pipette 50–100 μL into wells adjacent to the blocking solution in the 96-well plate. Incubate oocytes/embryos with the prepared primary antibody for 2 h at room temperature in a dark, humidified box (*see* **Notes 20 and 21**, if multiplexing).
10. Pipette 100 μL of blocking solution into two wells adjacent to the primary antibody well. Wash the oocytes/embryos by transferring them to each blocking solution well for 3 min per well
11. Prepare the relevant secondary antibody dilution(s) (*see* Table 1) and pipette 50–100 μL into wells adjacent to the blocking base washes. Incubate samples with secondary antibodies for 1 h at room temperature in the dark (*see* **Notes 20 and 21**, if multiplexing).
12. Pipette 100 μL of blocking solution into two wells adjacent to the secondary antibody and wash the oocytes/embryos through both wells for 3 min per well.

13. Pipette 100 µL of 1% PBS/PVP into two adjacent wells and wash the oocytes/embryos by transferring them twice after the secondary antibody incubation is over. If multiplexing antibodies raised in the same species, repeat Steps 9–13 using the primary and secondary antibodies selected for the second target of interest (*see* **Note 21**).
14. Pipette 50–70 µL of Hoechst working solution into an adjacent well and incubate oocytes/embryos for 20–30 min at room temperature in the dark.
15. Wash the oocytes/embryos two times in wells containing 100 µL of 1% PBS/PVP.
16. Transfer the embryos with the smallest possible volume of medium to 10–20 µL drop of glycerol solution or commercially available mounting medium on a slide. Wait for it to dry a little for 30 s before covering it with a coverslip (*see* **Note 22**). The edges of the coverslip can be sealed with nail polish.
17. Visualize staining on the fluorescence of a confocal microscope (Fig. 2, *see* **Note 23**). Store samples at 4°C protected from light if not visualizing immediately (*see* **Note 24**).

3.2 Immunofluorescence of DNA Modifications for Embryos and Oocytes

Immunofluorescence protocols for DNA modifications, although similar to histone modifications, include a denaturation step to allow the antibody binding to DNA (*see* **Note 25**).

1. Prepare a petri dish for embryo fixation by pipetting a row of 3×70 µL drops of 1% PBS/PVP, then place a 50 µL drop of fixation solution in the center. Complete the dish by pipetting a second row of 70 µL drops of 1% PBS/PVP under the fixation drop. Keep the petri dish at room temperature. Each petri dish is sufficient for one round of up to 20 oocytes/embryos.
2. Collect oocyte and embryo stages at relevant times of development following in vitro culture, or in vivo development, according to standard procedures (*see* **Note 16**).
3. Wash up to 20 oocytes/embryos oocytes/embryos through the 3 drops of 1% PBS/PVP for 3 min per well.
4. Fix oocytes/embryos in fixation solution for 15 min at room temperature (*see* **Note 20**).
5. Wash the oocytes/embryos through the second row of three drops of 1% PBS/PVP for 3 min per well.
6. In a 96-well plate, add 100 µL permeabilization solution and add oocytes/embryos to these wells. Permeabilize for 15 min.
7. Wash the oocytes/embryos through three wells of 70 µL 1% PBS/PVP pipetted into adjacent wells for 3 min per well.

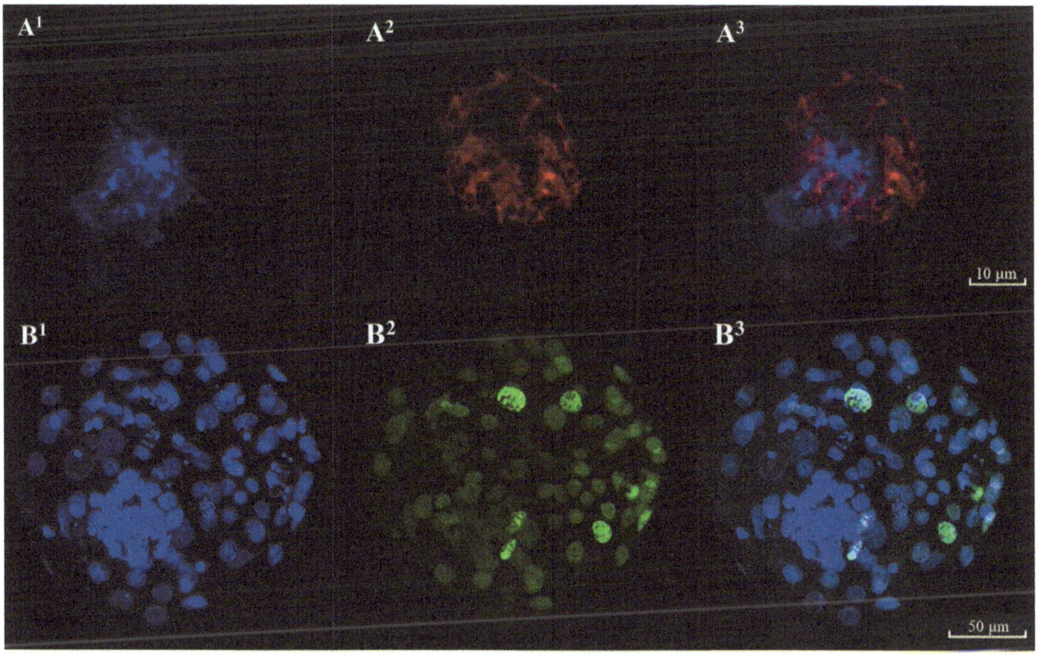

Fig. 2 (**A**) Bovine oocyte collected at the germinal vesicle (GV) stage and (**B**) bovine blastocyst processed using the protocol for histone PTMs or epigenetic-associated proteins in oocytes and embryos. Hoechst 33342 staining of the DNA in the GV (A^1) or blastomeres (B^1), in blue. IF of H3K9ac in the GV (A^2), in red, or blastomeres (B^2), in green. The merged images showing the DNA (blue) and H3K9ac for GV (red—A^3) or blastomeres (green—B^3) were produced using the merge function of ImageJ 1.52q software (U.S. National Institutes of Health, Bethesda, MD, USA). The IgG primary monoclonal antibody (ab12179 Abcam, Cambridge, UK) used against H3K9ac is produced in mice and used at a dilution of 1:300. The secondary antibody is produced in a donkey targeting mouse IgG (ab150109 Abcam, Cambridge, UK), labeled with Alexa Fluor® 488 and used at a dilution of 1:300. Images were obtained with a fluorescence microscope (Leica Microsystems, Wetzlar, HE, Germany, DM16000 B) coupled with fluorescence filters I3 (excitation 450–490 nm/emission 515–550 nm) and L5 (excitation 340–380 nm/emission 435–485 nm). Images were captured by a camera coupled to the microscope (Leica DFC365 FX—Leica Microsystems, Wetzlar, HE, Germany) and processed with LAS X Life Science Software (Leica Microsystems, Wetzlar, HE, Germany). Note that, although the I3 filter corresponds to green fluorescence emission, for the purposes of visual distinction only, in the VG oocytes, the color red was used for visual representation. This change was made solely in the software for imaging purposes, while the actual excitation/emission and fluorescence intensity characteristics remained unaltered. Image credits: Joao Victor Alcantara da Silva and Jessica Ispada

8. Transfer the oocytes/embryos to wells containing 50 μL of denaturation solution and incubate at room temperature for 30 min.
9. Add 100 μL of neutralization solution to the well containing the denaturation solution with the oocytes/embryos inside.
10. Pipette 50 μL of neutralization solution into an adjacent well. Transfer the oocytes/embryos to the well containing neutralization solution and incubate for 10 min at room temperature.

11. Add 100 μL of blocking solution to the well containing the neutralization solution with the oocytes/embryos inside it (*see* **Note 26**).
12. Pipette 100 μL of blocking solution into four adjacent wells. Wash the oocytes/embryos through three wells of blocking solution for 3 min per well.
13. Incubate in a fourth well of blocking solution overnight at 2–8°C (*see* **Notes 19 and 27**).
14. Prepare primary antibody working stock using blocking solution and incubate for 1.5–2 h (depending on the antibody used) at room temperature in a humidified box (*see* **Note 28**).
15. Wash the oocytes/embryos through three wells of 100 μL of blocking solution for 3 min per well.
16. Prepare secondary antibody working dilution(s) with blocking solution and incubate oocytes/embryos for 1 h at room temperature in a humidified dark box. Ideally, all the following steps and incubations should be performed in the dark.
17. Wash the oocytes/embryos through three wells of 100 μL of blocking solution for 3 min per well. Then two times in wells containing 100 μL of 1% PBS/PVP for 3 min per well.
18. Prepare PI solution and incubate oocytes/embryos for 30 min in 50 μL drops at room temperature in the dark.
19. Wash the oocytes/embryos two times in wells containing 100 μL of 1% PBS/PVP for 3 min per well.
20. Transfer the embryos with the smallest possible volume of medium to 10–20 μL drop of glycerol solution or commercially available mounting medium on a slide. Wait for it to dry a little for 30 s before covering it with a coverslip. Seal the edges of the coverslip with nail polish if desired.
21. Visualize staining on the fluorescence of a confocal microscope (Fig. 3, *see* **Note 26**). Store samples at 4°C protected from light if not visualizing immediately.

3.3 Mitochondrial Colocalization or Simultaneous Measurement of Mitochondrial Activity

In addition to visualizing nuclei, the use of dyes targeting specific organelles can be applied to analyze the subcellular localization of epigenetic marks and their associated enzymes.

In the context of epigenetics, mitochondrial counterstains are particularly valuable. They can allow for the precise localization of proteins either inside or outside the mitochondria that potentially regulate the levels of metabolites involved in epigenetic marks modulation [9]. MitoSpy™ Red CMXRos is a cell-permeant fluorescent probe for labeling mitochondria within live cells, which can then

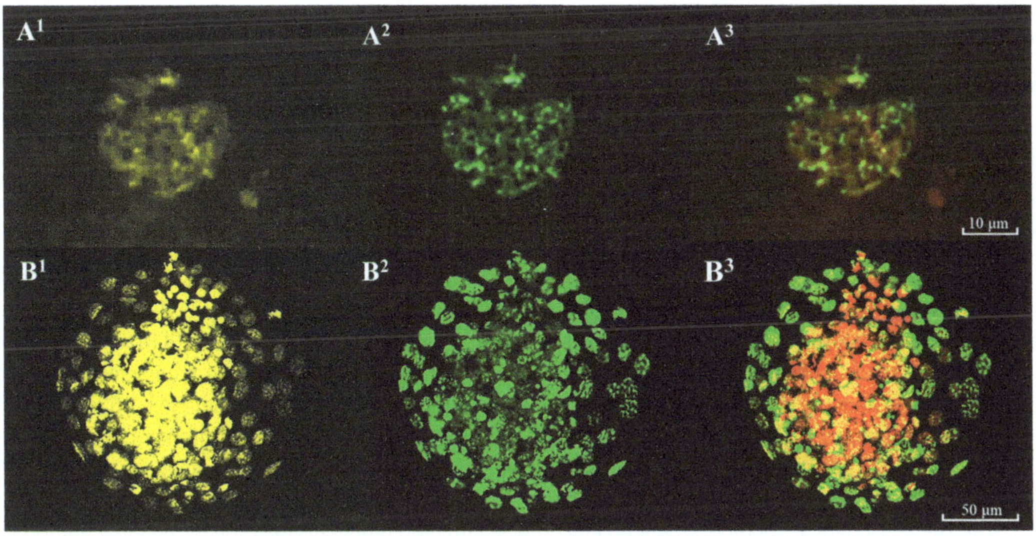

Fig. 3 (**A**) Bovine oocyte collected at the germinal vesicle (GV) stage and (**B**) bovine blastocyst processed using the protocol for DNA modifications for oocytes and embryos. Propidium Iodide staining of the DNA in the GV (A^1) or blastomeres (B^1), in red. Immunofluorescence (IF) of 5mC in the GV (A^2) or blastomeres (B^2), in green. The merged images showing the DNA (red) and 5mC (green) for GV (A^3) or blastomeres (B^3) were produced using the merge function of ImageJ 1.52q software (U.S. National Institutes of Health, Bethesda, MD, USA). The IgG primary monoclonal antibody (39649 Active Motif, Carlsbad, CA, USA) used against the 5mC is produced in mice and was used at a dilution of 1:1000. The secondary antibody is produced in a donkey targeting mouse IgG (ab150109 Abcam, Cambridge, UK), labeled with Alexa Fluor® 488 and used at a dilution of 1:300. Images were obtained with a fluorescence microscope (Leica Microsystems, Wetzlar, HE, Germany, DM16000 B) coupled with fluorescence filters Y3 (excitation 530–560 nm/emission 573–650 nm) and I3 (excitation 450–490 nm/emission 515–550 nm). Images were captured by a camera coupled to the microscope (Leica DFC365 FX—Leica Microsystems, Wetzlar, HE, Germany) and processed with LAS X Life Science Software (Leica Microsystems, Wetzlar, HE, Germany). Image credits: Aldcejam Martins da Fonseca Junior and Jessica Ispada

be fixed and permeabilized for downstream immunofluorescence (Fig. 4). If you wish to colocalize your protein of interest with mitochondria, MitoSpy staining must be performed *prior to* fixation (*see* **Note 6**).

1. The dish with drops needs to be prepared and pre-equilibrated in the incubator at 38.5°C for at least 1 h before use, ideally not more than 2 h. Transfer up to 20 oocytes/embryos to 50 µL drops of 50 nM MitoSpy working solution covered in mineral oil. Incubate for 30 min in the incubator at 38.5°C (or the relevant temperature for your species of interest; *see* **Note 29**).
2. Make enough 70 µL drops of 1% PBS/PVP to wash your samples three times each in a new petri dish; there is no need to cover them with mineral oil. Wash the oocytes/embryos by moving them through the drops.

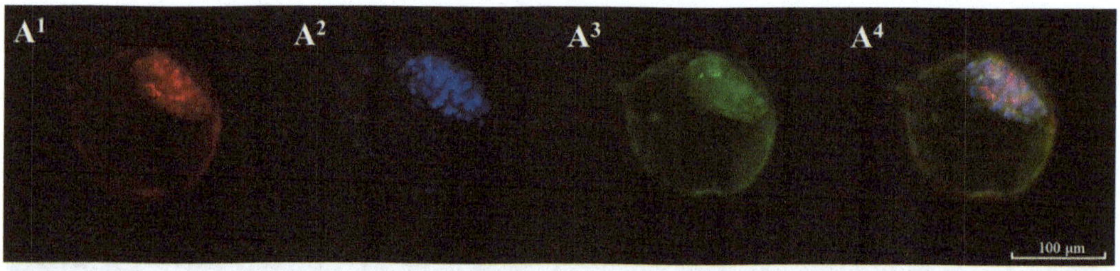

Fig. 4 Bovine blastocyst processed using the protocol for histones PTMs or epigenetic-associated proteins in oocytes and embryos, with the incorporation of the mitochondrial staining with MitoSpy™ Red CMXRos (A^1 in red). The DNA was labeled with Hoechst 33342 (A^2 in blue) and glutamate dehydrogenase (GLUD—A^3 in green) by IF. GLUD is a mitochondrial matrix enzyme that converts glutamate to α-ketoglutarate, a cofactor for histone and DNA demethylations. Merged images (A^4) produced using the merge function of ImageJ 1.52q software (U.S. National Institutes of Health, Bethesda, MD, USA) to visualize the localization of markers of interest (in green) in relation to mitochondria (in red) and DNA (in blue). The IgG primary polyclonal antibody (GTX105765 GeneTex, Irvine, CA, USA) used against the GLUD1 and GLUD2 is produced in rabbit and was used at a dilution of 1:200. The secondary antibody is produced in a donkey targeting rabbit IgG (ab150073 Abcam, Cambridge, UK), labeled with Alexa Fluor® 488 and used at a dilution of 1:350. Images were obtained with a motorized inverted confocal microscope, Leica TCS SP8 AOBS Tandem Scanner Leica (Leica Microsystems, Germany). The Leica software LAS X was used for image acquisition. Image credits: Jessica Ispada

3. Fix up to 20 oocytes/embryos in a 50 μL drop of fixation solution for 30 min.
4. Wash the oocytes/embryos through three drops of ~70 μL 1% PBS/PVP for 3 min per well. Proceed immediately with immunofluorescent staining as outlined above, or store at 2–8°C in 1% PBS/PVP for up to 2 weeks (*see* **Note 30**).

3.4 Image Acquisition and Analysis

Image acquisition for immunofluorescence analysis using fluorescence microscopy or confocal microscopy involves adjusting several key parameters to optimize image quality and ensure repeatability (*see* **Notes 31 and 32**). It is crucial to use consistent image acquisition parameters for all samples and replicates, including the same objective lens for all imaging, to ensure that any detected differences are not artifacts introduced during image acquisition. Optimize parameters with a test sample before starting the actual imaging.

For the analysis of oocyte and embryo images, various methods can be employed depending on the specific goals (*see* **Note 33**).

3.4.1 Analysis of Nuclear-Localized Markers

For a nuclear localized signal, thresholding or manual selection can be used. Thresholding involves setting a pixel intensity cutoff to distinguish the nucleus from the background by adjusting the threshold settings in ImageJ until the desired nuclei are highlighted. Manual selection allows for precise delineation of the nucleus using tools such as the freehand selection tool in ImageJ. Both methods provide effective ways to identify and work with nuclei in oocytes and embryos, and the choice depends on the specific requirements of your analysis and the quality of your images. When delineating the nucleus, it is recommended to use the image containing the stained DNA to define the area that will make the ROI, then use the ROI values to analyze the markers of interest.

1. Open your image in ImageJ.
2. For thresholding, navigate to the menu and select **Image > Adjust > Threshold**. A thresholding window will appear with sliders. Adjust these sliders until the nuclei in your image are highlighted in red.
3. Once satisfied with the selection, click **Apply** to set the threshold. If two or more nuclei are touching or overlapping, use **Process > Binary > Watershed** to separate them into distinct regions.
4. After ensuring that each nucleus is clearly defined, proceed to **Analyze > Analyze Particles**. In the Analyze Particles dialog box, specify the size and circularity parameters to filter for nuclei appropriately.
5. Check the **Display Results** and **Add to Manager** options in this window. This will add each identified nucleus to the ROI Manager as an individual ROI.
6. Alternatively, for manual selection, select the **Oval** or **Freehand Selection Tool** from the toolbar. Then, carefully draw around the nucleus you want to select.
7. Once you have outlined the desired area, go to **Edit > Selection > Add to Manager** (or simply press 't') to save your selection to the ROI manager.

3.4.2 Analysis of Markers Localized to the Cytoplasm

For cytoplasmic evaluations, entire cells, such as oocytes and clearly identifiable blastomeres, can be individually selected using the oval selection tool. Alternatively, randomly defined areas within the cytoplasm can be selected using rectangular or circular selection tools.

For analysis of cytoplasmic or mitochondrial targets, entire cells or regions of cells can be selected:

1. To select entire cells, use the **Oval** or **Freehand Selection Tool** from the toolbar. Draw around the entire cell, such as an oocyte or blastomere. Then, go to **Edit > Selection > Add to Manager** (or press 't') to save the selection. This is especially recommended if the images that will be used for colocalization analysis.
2. For selecting random areas within the cytoplasm, use the **Rectangular** or **Oval Selection Tool** from the toolbar. Draw a rectangle or circle in the cytoplasm area you want to analyze. It is recommended to select at least three different areas per cytoplasm. Go to **Edit > Selection > Add to Manager** (or press

't') to save the selection. Select the same number of regions for all cells/structures analyzed.

3.4.3 Region of Interest (ROI) Analysis

Once the areas for analysis are defined and saved in the regions of interest (ROI) manager, they can be quantified by measuring parameters such as fluorescence intensity. Average intensity can be used to compare samples and determine differences in fluorescence intensity. This can be done using the "Measure" function in ImageJ, which provides a detailed output of selected regions.

1. To quantify fluorescence intensity, start by selecting the parameters you want to measure. Common parameters include average intensity, integrated density, and area.
2. In the ROI Manager, click on **More** and then **Measure** to analyze the selected ROIs.

3.4.4 Colocalization

For colocalization studies, the Coloc2 plugin in FIJI can be used by specifying the channels to be analyzed and interpreting the resulting colocalization metrics, such as Pearson's correlation coefficient and Manders' overlap coefficient, to evaluate the spatial relationship between signals, indicating potential interaction or colocalization of molecules/organelles.

1. To run the Coloc2 plugin, ensure that your images are converted to a binary format by selecting **Image > Type > 8-bit and then Process > Binary > Make Binary.**
2. Next, go to **Plugins > Colocalization > Coloc2.** Select the channels you want to analyze and click **OK** to run the analysis.
3. The output will include metrics such as Pearson's correlation coefficient and Manders' overlap coefficient, which help identify colocalization between stains.

4 Notes

1. The dissolution of NaOH in water is an exothermic reaction (it releases heat), so always add the NaOH to the water and never the other way around.
2. Paraformaldehyde is extremely toxic and potentially carcinogenic, so it must be handled with extreme care. Handle paraformaldehyde powder in a fume hood and wear appropriate personal protective equipment, including gloves, a lab coat, and safety goggles. This includes housing the balance inside the fume hood to prevent inhalation. If you do not have an appropriate setting to make your own fixation solution,

commercially available options can be purchased and diluted with distilled water.

3. When adjusting pH, slowly and carefully add the pH adjuster to avoid overshooting the pH.
4. Triton-X 100 is a viscous liquid. To ensure the correct volume is dispensed, allow sufficient time for the aspirate to accumulate in the tip. Similarly, when expelling, the viscosity can delay the evacuation of Triton-X from the pipette tip, so wait sufficient time to allow this to occur. Rinse the pipette tip by repeated pipetting to ensure transfer of all the Triton-X.
5. Avoid shaking solutions containing BSA, as this breaks up the proteins.
6. MitoSpy can be used to colocalize mitochondria. Alternative fluorescent probes for colocalizing mitochondria, including TMRM and MitoTracker, are commercially available and can be substituted. However, not all mitochondrial probes are suitable for downstream cell fixation. Prepare stock solutions according to the manufacturer's instructions. Working stock concentrations will require optimization.
7. Protect aliquots of MitoSpy stock by wrapping microcentrifuge tubes with foil.
8. Preferably, use the same base media (without serum) in which oocytes/embryos were incubated before staining.
9. There are numerous epigenetic marks that can be quantified and localized within cells using immunofluorescence, including chromatin remodeling proteins, histone modifications, and DNA methylation, as well as proteins that interact with noncoding RNAs and enzymes related to epigenetic modifications. Chromatin remodeling complexes are essential for regulating DNA accessibility and play a critical role in controlling gene expression during development. Proteins associated with these complexes, such as Polycomb group proteins and ATP-dependent chromatin remodelers, can be visualized through immunofluorescence to examine their spatial and temporal distribution in embryos. Histones, the proteins around which DNA is tightly wrapped, undergo various post-translational modifications that significantly impact chromatin structure and gene expression. Common histone modifications that can be analyzed using immunofluorescence include histone H3 acetylation (e.g., H3K27ac and H3K9ac) and histone H3 methylation (e.g., H3K4me3 and H3K27me3). DNA methylation marks, generally involved in the repression of gene expression, imprinting, and X-inactivation, are intensively reprogrammed during preimplantation development. Specific antibodies against 5-methylcytosine (5-mC) and the

intermediate mark 5-hydroxymethylcytosine (5-hmC) can be used to study global DNA methylation reprogramming and the dynamic changes in DNA methylation reprogramming during development.

10. Different types of antibodies are available for detecting proteins of interest. Primary antibodies are categorized into three types: polyclonal, monoclonal, and recombinant. Each type has its advantages and disadvantages (*see* Table 2). Regardless of the target protein, prefer primary antibodies that indicate they have been tested for immunofluorescent application, where available. Likewise, an important consideration when selecting a primary antibody is its host species, as this determines the choice of secondary antibody. This is particularly important when detecting multiple proteins simultaneously, as it helps to avoid cross-reactivity.

11. Secondary antibodies conjugated to fluorophores (e.g., Alexa Fluor, FITC, or Rhodamine), which emit light when excited at specific wavelengths, are used to visualize the bound primary antibodies. Multiple secondary antibodies can be

Table 2
Different types of antibody production

	Polyclonal antibodies	Monoclonal antibodies	Recombinant antibodies
Production	Multiple B cell clones	Single B-cell clone	Recombinant DNA
Epitopes	Multiple from the same antigen	Single B-cell clone	Single
Advantages	(i) Binding to multiple epitopes may enhance the signal	(i) High specificity, avoiding nonspecific binding	(i) High reproducibility
	(ii) Less affected by changes in antigen conformation due to issues in sample preparation	(ii) Less variability in production	(ii) Can be designed for less common species or for searching specific protein variants
	(iii) More likely to be effective across different species		(iii) Production less time-consuming
Disadvantages	(i) Higher background and variability	(i) More affected by changes in antigen conformation due to issues in sample preparation	(i) More expensive
		(ii) Less likely to be effective across different species	

employed simultaneously to detect different epigenetic marks in a multicolor immunofluorescence setup.

When selecting a secondary antibody, it is important to consider not only the (sample) target species, but also the primary antibody host species and isotype. If multiplexing, select different fluorophores to ensure excitation and emission spectra do not overlap. If fluorescent spectra overlap, this will cause bleed-through into other wavelengths.

12. You can opt to use commercial mounting media such as Anti-Fade Fluorescence Mounting Medium—Aqueous, Fluoroshield, or Mounting Medium With DAPI—Aqueous, Fluoroshield, which already contains the DNA stainer DAPI, removing the necessity to use Hoechst 33342 incubation. If using Glycerol solutions, a good concentration is 50% glycerol in 1× PBS.

13. Before analyzing experimental samples, test a range of concentrations of the primary antibody to optimize the staining while minimizing the potential for the secondary to bind nonspecifically, creating background fluorescence. The primary antibody concentration is usually around 1:50 to 1:1000 of antibody:blocking solution. Modifications to antibody concentrations and/or exposure times may also be required when using the same antibody for other species. If nonspecific staining remains a problem, reduce the concentration of the secondary antibody, reduce the time of fixation, or increase the number of washes after antibody incubation, or use normal serum from the same species as the secondary antibody instead of BSA when preparing the blocking solution. Alternatively, if signal intensity is low/absent, increase antibody concentration, reduce light exposure while staining, or analyze samples immediately following preparation. Equally, if using antibodies that have not been tested in your species of interest, check the homology between the antibody's target sequence and that of your species of interest. To avoid repetitive cycles of freezing–thawing and reduce the chances of contamination, it is good practice to dilute the antibody using 1× PBS and aliquot this into microcentrifuge tubes at a volume close to what will be needed for a single experiment.

14. Throughout the experiment, it is crucial to ensure that the cells always remain hydrated. This is especially important when handling samples in microdrops, as they may dry out when not covered with oil.

15. Before initiating immunostaining in gametes and embryos, it is essential to review the literature to confirm the expression levels and intracellular localization of the protein of interest in

these specific cell types. Some proteins may exhibit very low expression levels, making them challenging to detect.

16. Both a positive and a negative control should be included in each staining replicate where possible. Negative controls can be generated by incubating samples with an isotype control in place of the primary antibody or without the secondary antibody (substitute blocking base) at the relevant step.
17. Before fixation, oocytes and embryos may require removal of the zona pellucida (ZP) to achieve optimal immunofluorescence results. This is particularly true if the signal is present at a higher intensity in the ZP and not inside the oocyte/embryo, or is masking the internal signal. In such cases, it is recommended to incubate the samples in a pronase solution before fixation. The pronase concentration recommended for embryos and oocytes is between 0.5 and 1 mg/mL in PBS and the removal can be done in 50 μL drops at 37°C and monitored under a stereo microscope to observe the removal of the ZP, but not the degradation of the oocyte/embryo. Usually, it will take from 0.5 to 5 min to remove the ZP, and once the ZP seems loose in one structure (oocyte or embryo), move the structure up and down a capillary to completely remove it. Quickly and thoroughly wash them in a solution containing 10% FCS at least three times to stop the digestion. Proceed with fixation.
18. Certain epigenetic marks, particularly histone modifications, are sensitive to crosslinking. Therefore, optimizing fixation time and conditions is crucial to maintain the integrity of these modifications. Gametes and embryos are typically fixed using paraformaldehyde or formaldehyde (4% in PBS) to preserve chromatin structure, enzymes, associated proteins, and epigenetic marks. However, some antibodies targeting cytoplasmic proteins may not perform well with formalin-fixed samples; in such cases, methanol fixation may be a suitable alternative. If methanol is used in place of paraformaldehyde, the permeabilization step is unnecessary, as the membranes are already permeabilized by the alcohol.
19. Blocking is performed to minimize nonspecific antibody binding, usually by using serum (preferably from the host species in which the secondary antibody was produced) or bovine serum albumin (BSA). It is important to avoid excessive blocking, as this can obscure the signal of the target protein.
20. The primary antibody incubation step typically requires at least an hour at room temperature. We have found that incubation for 2 h is required to ensure adequate binding.

Alternatively, overnight incubation at 4°C can be performed, for a maximum of 16 h incubation.

21. There are two methods for performing multiplexing, depending on the type of primary antibodies used. If both primary antibodies are raised in the same species (e.g., both are produced in rabbit), you should incubate the samples with one primary antibody at a time. The steps would be: (1) Incubate with the first primary antibody. (2) Wash and incubate with the corresponding secondary antibody. (3) Wash and incubate with the other primary antibody. (4) Wash and incubate with the corresponding secondary antibody. If the primary antibodies are raised in different species (e.g., one in rabbit and the other in rat), you can incubate the samples with both primary antibodies simultaneously. The steps would be: (1) Incubate with both primary antibodies together. (2) Wash and incubate with both corresponding secondary antibodies together.
22. If using commercially available mounting media, there is no need to add spacers, vaseline, or paraffin to the slide to avoid squishing the samples. If using glycerol, one of these options might be required.
23. Fluorescence microscopy, typically confocal microscopy, is utilized to obtain high-resolution images of epigenetic marks in gametes and developing embryos. For early-stage embryos, imaging across various developmental stages is often critical for understanding the dynamic changes in chromatin marks over time. Quantitative analyses, such as measuring fluorescence intensity or evaluating colocalization, provide valuable insights into how epigenetic modifications evolve during gamete maturation and embryonic development.
24. Once samples are processed and slides made, store them at 4°C if not analyzing in the same day. Ideally, samples should not be stored for more than 15 days. If one replicate is stored for a few days, repeat the same storage time for other replicates to reduce variation.
25. The 5-methylcytosine and 5-hydroxymeethylcytosine epitopes in DNA must be exposed for antibody binding, which requires DNA denaturation. In oocytes and embryos, this is typically achieved using HCl. The conditions for HCl-mediated denaturation—such as concentration, incubation temperature, and duration—are critical. Insufficient denaturation can lead to weak and uneven nuclear staining, while excessive denaturation may damage the DNA, producing similar results.
26. Gradual dilution of these solutions avoids abrupt changes in pH and helps reduce oocytes/embryos sticking to the capillary in later steps. If sample loss occurs after neutralization,

increase the amount of blocking solution added to the neutralization solution and aspirate the blocking solution into the pipette to reduce oocyte/embryo binding.

Coat the capillary/pipette in blocking solution to reduce oocyte/embryo binding.

27. Do not exceed 16 h and try to keep the same timing for all replicates.

28. During our tests, we found that the best concentrations were those described in Table 1 and all primary antibodies were incubated for 2 h at room temperature. In general, when using a new antibody, it is important to test it thoroughly under different dilutions and conditions. The recommendations outlined here are suggestions based on our findings.

29. If the intensity of mitochondrial staining is too high, reduce the working solution concentration or incubation time.

30. Do not store blastocysts for later analysis; they should be processed on the same day. Earlier-stage embryos can be stored at 2–8°C in 1% PBS/PVP for up to 2 weeks.

31. The parameters that can be adjusted include excitation intensity, which determines the amount of light used to excite the fluorophores and affects both the brightness and photobleaching of the sample. Exposure time controls how long the detector collects light, influencing the signal-to-noise ratio. Gain adjusts the sensitivity of the detector, impacting the overall brightness and contrast of the image. Offset adjusts the baseline signal level to correct for background fluorescence [10]. When optimizing exposure time, ensure there is sufficient signal without causing photobleaching. For gain and offset, balance signal amplification with background correction.

32. When using a confocal microscope, additional parameters can be adjusted to optimize imaging. The pinhole size controls the thickness of the optical section by adjusting the aperture size. Laser power determines the excitation power of the laser. Scan speed affects the rate at which the laser scans the sample, influencing both image acquisition time and potential photobleaching. Averaging combines multiple scans to reduce noise and enhance image quality. Line/frame accumulation increases the signal by accumulating multiple lines or frames during scanning. Finally, Z-stacking captures optical sections at varying focal planes, enabling 3D sample reconstruction [11, 12]. When adjusting laser power, minimize photodamage while maintaining adequate excitation. Use Z-stacking

whenever possible, especially for thick samples like oocytes and embryos.

33. Once images have been acquired, ImageJ and FIJI are widely used for image analysis in biological research. ImageJ, developed by the National Institutes of Health (NIH), is an open-source image processing program that supports various plugins for different types of analysis [13]. FIJI (Fiji Is Just ImageJ) is a distribution of ImageJ that includes many preinstalled plugins, such as Coloc2, which simplifies colocalization analysis [14]. These tools are popular due to their flexibility, extensive documentation, and active user community, which continuously contributes to their development and improvement. Comprehensive video tutorials are available for common analyses using these software tools.

References

1. Coons AH, Creech HJ, Jones RN (1941) Immunological properties of an antibody containing a fluorescent group. Proc Soc Exp Biol Med 47(2):200–202
2. Jankovics F, Henn L, Bujna Á, Vilmos P, Spirohn K, Boutros M, Erdélyi M (2014) Functional analysis of the Drosophila embryonic germ cell transcriptome by RNA interference. PLoS ONE 9(6):e98579
3. Mihajlović AI, Thamodaran V, Bruce AW (2015) The first two cell-fate decisions of preimplantation mouse embryo development are not functionally independent. Sci Rep 5(1):15034
4. Saiz N, Mora-Bitria L, Rahman S, George H, Herder JP, Garcia-Ojalvo J, Hadjantonakis AK (2020) Growth factor mediated coupling between lineage size and cell fate choice underlies robustness of mammalian development. Elife 9: e56079
5. Skory RM, Moverley AA, Ardestani G, Alvarez Y, Domingo-Muelas A, Pomp O, Hernandez B, Tetlak P, Bissiere S, Stern CD, Sakkas D, Plachta N (2023) The nuclear lamina couples mechanical forces to cell fate in the preimplantation embryo via actin organization. Nat Commun 14(1):3101
6. Gopinathan G, Diekwisch TGH (2022) Epigenetics and early development. J Dev Biol 10(2):26
7. Milazzotto MP, de Lima CB, da Fonseca AM, Dos Santos EC, Ispada J (2020) Erasing gametes to write blastocysts: metabolism as the new player in epigenetic reprogramming. Anim Reprod 17(3):e20200015
8. Milazzotto MP, Ispada J, de Lima CB (2023) Metabolism-epigenetic interactions on in vitro produced embryos. Reprod Fertil Dev 35(2):84–97
9. Gao Z, Li Y, Wang F, Huang T, Fan K, Zhang Y, Zhong J, Cao Q, Chao T, Jia J, Yang S, Zhang L, Xiao Y, Zhou J-Y, Feng X-H, Jin J (2017) Mitochondrial dynamics controls anti-tumour innate immunity by regulating CHIP-IRF1 axis stability. Nat Commun 8(1):1805
10. Dobrucki JW, Kubitscheck U (2017) Fluorescence microscopy. In: Fluorescence microscopy: from principles to biological applications. John Wiley & Sons, Hoboken, Nova Jersey, USA, pp 85–132
11. Jonkman J, Brown CM, Wright GD, Anderson KI, North AJ (2020) Tutorial: guidance for quantitative confocal microscopy. Nat Protoc 15(5):1585–1611
12. Furia L, Pelicci P, Faretta M (2014) Confocal microscopy for high-resolution and high-content analysis of the cell cycle. Curr Protoc Cytom 70(1):7.42.41–47.42.14
13. Schneider CA, Rasband WS, Eliceiri KW (2012) NIH Image to ImageJ: 25 years of image analysis. Nat Methods 9(7):671–675
14. Schindelin J, Arganda-Carreras I, Frise E, Kaynig V, Longair M, Pietzsch T, Preibisch S, Rueden C, Saalfeld S, Schmid B (2012) Fiji: an open source platform for biological-image analysis. Nat Methods 9(7):676–682

Check for updates

Chapter 20

Purification and Characterization of Extracellular Vesicles from Embryo-Conditioned Medium

Krishna Chaitanya Pavani, Ann Van Soom and Katrien Smits

Abstract

Extracellular vesicles (EVs) serve as key communication signals among embryos. They are secreted into the embryo culture medium and hold potential as biomarkers for assessing embryo quality. However, the limited volume of embryo-conditioned medium presents challenges for EV enrichment. This chapter details an effective method for purifying and characterizing EVs from in vitro embryo cultures by using the OptiPrep™ density gradient ultracentrifugation (ODG) method. Additionally, we also provide methods for preparing EV-free embryo culture media.

Keywords Extracellular vesicles, Embryo culture media, EV-free media, EV purification, EV characterization, OptiPrep™ Density gradient

1 Introduction

Extracellular vesicles (EVs) have drawn significant interest in embryonic development research for their potential involvement in facilitating intercellular communication and improving embryo survival, especially in the context of Assisted Reproductive Technologies (ARTs). Earlier studies in embryology primarily focused on soluble autocrine factors such as cytokines, growth factors, and antiapoptotic proteins, which contribute to embryo self-regulation and cell-to-cell communication [1, 2]. However, the discovery of EVs as carriers of various bioactive molecules like mRNAs, microRNAs (miRNAs), and proteins has broadened our understanding of how embryos interact with their microenvironment [3]. These nanoscale vesicles can transport crucial molecular cargo that influences developmental outcomes and may serve as noninvasive biomarkers for embryo quality [4, 5], offering promising applications for ART optimization. Indeed, they have been shown to facilitate embryo–maternal communication by modulating endometrial receptivity, gene expression, and immune responses, thereby enhancing implantation potential and supporting

Alexandra J. Harvey and David K. Gardner (eds.), *The Preimplantation Embryo*, Methods in Molecular Biology, vol. 3038, https://doi.org/10.1007/978-1-0716-5292-3_20,

early pregnancy [3, 6]. Saadeldin et al. demonstrated that EVs from porcine parthenogenetic embryos could enhance blastomere count and blastocyst formation when taken up by cloned embryos [7]. Similarly, studies in bovine embryos have shown that embryo-derived EVs can positively influence the inner cell mass (ICM), a key determinant of viability [8, 9]. Moreover, our team has shown that using serum-free, EV-depleted, or BSA-supplemented media can minimize contamination from other vesicles, enabling more accurate characterization of embryo-specific EVs [9, 10].

Isolated EVs can be utilized for various applications, including high-throughput sequencing of miRNA and mRNA to investigate transcriptomic profiles, mass spectrometry to identify protein cargo, lipidomic analyses, and biomarker discovery for embryo viability and quality assessment [11]. These approaches provide comprehensive insights into the molecular composition of EVs, enabling the exploration of their functional significance and potential applications in embryo culture optimization and assisted reproductive technologies (ART). Further, EVs hold potential as tools for studying paracrine and autocrine signaling mechanisms in embryo–maternal communication, which are critical for implantation and early embryonic development.

Key methods for EV isolation from embryo culture media include differential ultracentrifugation (DU), single-step size exclusion chromatography (SEC), and OptiPrep™ density gradient (ODG) ultracentrifugation. EV isolation techniques vary significantly in terms of cost, equipment requirements, processing time, input material type and volume, and the purity of the isolated EVs (Table 1). While this chapter emphasizes EV extraction from embryo-conditioned media, the described protocol can also be adapted for various sample types, including serum and follicular fluid [9, 10, 12]. Among the three techniques noted above, ODG stands out as an efficient method for isolating EVs from embryo-conditioned media, being easy to perform and offering a balance of purity and recovery (Fig. 1).

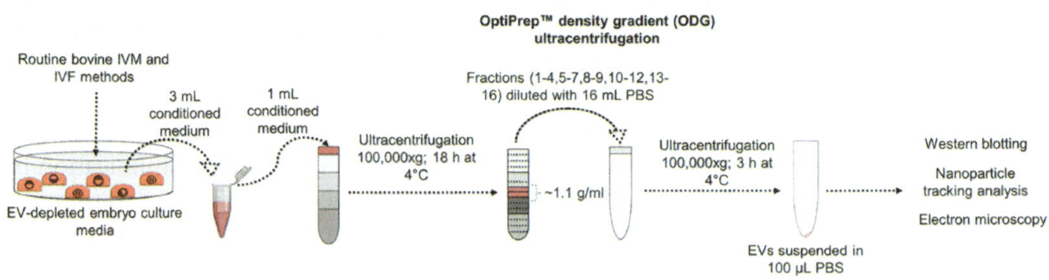

Fig. 1 Schematic representation of extracellular vesicles isolation from bovine embryo conditioned media with three different EV isolation procedures: Optiprep™ density gradient ultracentrifugation (ODG UC). (Adapted from [10])

Table 1

Overview of EV purification techniques, including the mechanism of separation of vesicles from other sample components, type and volume of input, approximate processing time, required equipment, sample purity, and other considerations. (Adapted from [10])

Method	Mechanism of separation	Total input	Processing time	Required equipment	Purity of isolates	Remarks	Approximate price/sample (€)
Differential ultracentrifugation	Sedimentation rate	Raw sample, any volume (>1 mL)	3–6 h	Ultracentrifuge and compatible tubes	Low	May induce vesicle disruption [10, 26–28]	10
Size-exclusion chromatograph (qEV iszon columns)y	Size	>150 µl biofluids and cell culture media	1 h	–	High	May separate vesicles from high-density lipoprotein (HDL) particles [27]	20
OptiPrep ™ Density Gradient Separation	Density	Concentrated volume (>1 mL)	21–22 h	Ultracentrifuge and compatible tubes	High	Co-isolates vesicles and HDL particles [28]	20

To ensure the specific isolation of EVs, multiple characterization techniques, including Western blotting, nanoparticle tracking analysis (NTA), and transmission electron microscopy (TEM), are employed to evaluate EV size, morphology, and specific markers like CD9, CD63, and TSG101 [4, 9, 10]. This chapter, therefore, also includes methods for the characterization of EVs following the requirements of the minimal information for studies of extracellular vesicles (MISEV guidelines [13]).

Finally, biological experiments to assess the function of specific EVs and/or their cargo molecules require the inclusion of EV-depleted media. This chapter also outlines the protocol used for EV depletion of embryo culture media, either containing serum or serum-free, which will deepen our understanding of EV-mediated signaling, paving the way for improved embryo selection methods and better outcomes in assisted reproductive technologies (ART).

2 Materials

Ensure all solutions are made with ultrapure water (e.g., MilliQ water produced by purifying deionized water to reach 18 MΩ cm resistance at 25°C) and use analytical-grade reagents. Store all reagents at room temperature unless otherwise noted. Follow proper waste disposal guidelines for all materials. All media should be filtered before use.

2.1 Preparation of EV-Depleted Culture Media

1. In-house prepared (e.g., SOF [14]) or ready-to-use commercial embryo culture media (e.g., Vitrolife, CooperSurgical, or Stroebech IVF Media).
2. Ultracentrifuge with swing bucket rotor.
3. Polyallomer Centrifuge Tubes: 16.8 mL open-top tubes capable of withstanding forces ≥100,000 × *g*.
4. Pipette controller.
5. Serological pipettes (10 mL).
6. Sterile pipettes and pipette filter tips (10, 100, and 1000 μL).
7. 0.22 μm sterile filters.

2.2 Density Gradient Centrifugation

1. OptiPrep™ Density Gradient Medium (60% w/v Iodixanol).
2. Homogenization medium (HM): 10 mM Tris, 1 mM EDTA, and 0.25 M sucrose. In 500 mL of distilled water, dissolve 42.787 g sucrose, 0.186 g EDTA, and 0.605 g Tris-HCl. Adjust to pH 7.4 with HCl. Store at 4°C for up to 3 weeks.
3. Working buffer (WB): 60 mM Tris-HCL, 6 mM EDTA, 0.25 M Sucrose. In 500 mL of distilled water, dissolve 42.787 g sucrose, 1.116 g EDTA, and 3.633 g Tris-HCl. Adjust to pH 7.4 with HCl. Store at 4 °C for up to 3 weeks.

4. Density gradient working solutions (WS): Mixing OptiPrep™ medium with working buffer (WB) in a 5:1 (OptiPrep™: WB) ratio as follows: WS1: 7 mL OptiPrep™ and 1.4 mL WB; WS2: 14 mL OptiPrep™ and 2.8 mL WB; WS3: 21 mL OptiPrep™ and 4.2 mL WB; WS4: 28 mL OptiPrep™ and 5.6 mL WB; WS6: 42 mL OptiPrep™ and 8.4 mL WB; and WS8: 56 mL OptiPrep™ and 11.2 mL WB.
5. Gradient solutions: For a 5% gradient, mix 0.5, 1, or 1.5 mL WS with 4.5, 9, or 13.5 mL HM, respectively. For a 10% gradient, mix 1, 2, or 3 mL WS with 4, 8, or 12 mL HM. For a 20% gradient, mix 2, 4, or 6 mL WS with 3, 6, or 9 mL HM. Finally, for a 40% gradient, mix 4, 8, or 12 mL WS with 1, 2, or 3 mL HM. These mixtures yield three alternative gradient preparations (Gradient 1, Gradient 2, and Gradient 3, respectively), which can be selected according to experimental requirements (*see* **Note 1**). Store at 4 °C for up to 2 weeks under light protection.
6. Ultracentrifuge with swing bucket rotor.
7. Polyallomer Centrifuge Tubes: 16.8 mL open-top tubes capable of withstanding forces ≥100,000 × *g*.
8. DPBS (Dulbecco's phosphate-buffered saline, no calcium, no magnesium), pH 7.4. Filter sterilize using a 0.22 μm filter.
9. 17 mL sterile polypropylene ultracentrifuge tubes (Open-Top Thinwall Ultra-Clear Tubes).
10. Glassware and plasticware for preparing solutions, mixing, and transferring samples.
11. Pipette controller.
12. Serological pipettes (ranging from 1 to 10 mL) for gradient preparation and fraction handling
13. Sterile pipettes and pipette filter tips (10, 100, and 1000 μL).
14. Low-retention pipette tips (10, 100, and 1000 μL).
15. Cryovials for long-term EV storage at −80°C.

2.3 EV Characterization

2.3.1 Western Blotting

1. 1.5× Laemmli solution: Prepare by combining 3.75 mL of 1 M Tris HCl (pH 6.8), 7.5 mL glycerol (7.5% v/v), 3.0 g sodium dodecyl sulfate (SDS; 3% w/v), and 0.003 g bromo phenol blue (0.003% w/v) (*see* **Note 2**). Add distilled water to a final volume of 100 mL. Stir until fully dissolved and store at 4 °C for up to 3 months.
2. 0.25% bromophenol blue solution: Dissolve 0.25 g of bromophenol blue powder in 100 mL of distilled water. Stir

until the dye is completely dissolved, filter through a 0.22 μ m membrane, and store at room temperature protected from light.

3. *N,N,N′,N′*-tetramethylethylenediamine (TEMED). Store at 4°C (*see* **Note 2**).
4. SDS (0.4% w/v) (*see* **Note 2**).
5. Reducing (denaturation) sample buffer: For each milliliter of buffer required, combine 1 mL 1.5× Laemmli buffer (to impart a negative charge to proteins and glycerol to increase sample density), 50 μL β-mercaptoethanol (to reduce disulfide bonds; *see* **Note 2**), and 50 μL of 0.25% bromophenol blue solution and mix thoroughly. Prepare before use or store in aliquots at −20°C.
6. Non-reducing (non-denaturation) sample buffer: For each milliliter of buffer required, combine 1 mL 1.5× Laemmli buffer and 50 μL of 0.25% bromophenol blue solution and mix until uniform. Store at −20°C in aliquots until use.
7. Thermomixer or heating block (95°C).
8. 30% Acrylamide/bis solution (*see* **Note 2**).
9. Separating Gel buffer (0.4% w/v SDS, 1.5 M Tris, pH 8.8): Dissolve 4 g of SDS (*see* **Note 2**) and 181.7 g of Tris base in approximately 800–900 mL of distilled water. Mix thoroughly to ensure complete dissolution of the reagents before adjusting the pH to 8.8 with concentrated hydrochloric acid (6 N, approximately 25 mL). Add distilled water for a final volume of 1 L. Store in a sealed container at room temperature or at 4°C, depending on the duration and stability required (*see* **Note 3**).
10. Stacking gel buffer (0.4% w/v SDS, 0.5 M Tris, pH 6.8): Dissolve 4 g of SDS and 60.56 g of Tris base in approximately 800–900 mL of distilled water. Adjust the pH to 6.8 with concentrated hydrochloric acid (6 N, approximately 40 mL), then adjust the final volume of the buffer to 1 L with distilled water. Store in a well-sealed container to avoid contamination or pH drift (*see* **Note 3**).
11. 10% APS solution: Dissolve 0.1 g ammonium persulfate (APS) in 1 mL distilled water to obtain a 10% (w/v) solution. Prepare fresh each time before use, as APS rapidly loses activity in solution (*see* **Note 2**).
12. Running buffer (25 mM Tris, 192 mM Glycine, 0.1% SDS): Dissolve 3.0 g Tris base, 14.4 g glycine, and 1 g SDS in distilled water. Store at room temperature.
13. Blotting buffer (25 mM Tris, 192 mM Glycine, 20% Methanol): Dissolve 3.0 g Tris and 14.4 g glycine in

approximately 800 mL distilled water. To this solution, add 200 mL methanol (20%; *see* **Note 2**). Adjust to a final volume of 1 L with distilled water. Mix thoroughly and store at 4°C. Ensure this buffer stays cool before use to prevent overheating during the transfer process.

14. 10% Resolving gel solution; combine 1.70 mL of 30% acrylamide/bis-acrylamide (29:1), 1.30 mL of 1.5 M Tris-HCl buffer pH 8.8 containing 0.4% SDS, 2.10 mL of deionized water, 50 μL freshly prepared 10% ammonium persulfate (APS), and 5 μL TEMED. Mix gently by inversion.
15. 5% Stacking gel solution; combine 30% acrylamide/bis-acrylamide solution (to yield 5% final concentration), 1.0 M Tris-HCl buffer pH 6.8 containing 0.4% SDS, and distilled water to the desired total volume. Immediately prior to pouring, add 50 μL freshly prepared 10% APS and 5 μL TEMED to initiate polymerization.
16. Pipette tips for gel loading.
17. Transfer apparatus (wet or semi-dry blotting system).
18. Nitrocellulose or PVDF membrane.
19. Blot sponges (two per blot).
20. Filter papers (six per blot).
21. SDS-PAGE gel apparatus and power supply.
22. Thermomixer or heating block (95°C).
23. Gel casting cassettes measuring 7.25 cm × 10 cm.
24. 1.5 mm gel casting spacers.
25. Gel comb.
26. Blocking solution: 5% skim milk in TBST or 5% BSA in TBST, as required for specific antibodies (*see* **Note 4**).
27. Ponceau S solution (0.5% w/v): Dissolve 0.5 g Ponceau S powder in 100 mL of 1% acetic acid. Stir until completely dissolved, then filter the solution to remove any undissolved particles. Store at room temperature.
28. Microcentrifuge tubes (1.5, 2, and 5 mL).
29. Pipettes and pipette filter tips (10, 100, and 1000 μL).
30. Falcon™ centrifuge tubes (15 and 50 mL) for sample storage, primary and secondary antibody solution preparation, and fraction handling.
31. Prestained molecular weight standard.
32. Primary antibodies for EV detection: CD63, ApoA-I, and Ago2.
33. Secondary antibodies raised against the primary antibody host (e.g., anti-mouse IgG and anti-rabbit IgG).

34. Humidified chamber for antibody incubation at 4 °C.
35. 1× PBS: Dissolve 8 g NaCl (137 mM), 0.2 g KCl (2.7 mM), 1.44 g Na_2HPO_4 (10 mM), and 0.24 g KH_2PO_4 (1.8 mM) in approximately 800 mL distilled water. Adjust the pH to 7.4 with HCl, then bring the final volume to 1 L with distilled water.
36. PBST (PBS + 0.05% v/v Tween-20): To 1 L of 1× PBS, add 0.5 mL Tween-20 and mix thoroughly. Store at room temperature or at 4°C for long-term use.
37. Chemiluminescence substrate: e.g., WesternBright Sirius.
38. Chemiluminescent imaging system, e.g., Proxima 2850 Imager.
39. Roller.
40. Parafilm.
41. Rocker or platform shaker.

2.3.2 Nanoparticle Tracking Analysis (NTA)

1. NanoSight LM10 microscope (Malvern Instruments Ltd) equipped with a 455 nm laser.
2. NTA Software version 3.2 for video analysis.
3. DPBS (Dulbecco's phosphate-buffered saline, no calcium, no magnesium).

2.3.3 Transmission Electron Microscopy (TEM)

1. Formvar/carbon support film, copper mesh, and electron microscopy grids.
2. Staining solution: 1% uranyl acetate in double-distilled water.
3. UltraPure water.
4. Electron microscope (e.g., JEM 1400 plus and Zeiss EM 109) equipped with an imaging camera.
5. Sterile pipettes and pipette tips (10, 100, and 1000 μL).
6. Parafilm.
7. Fine point tweezers.
8. Filter paper.

3 Methods

The method outlined here has been optimized for the preparation and collection of culture media conditioned by bovine embryos, but it can also be adapted for other species (porcine, human, and murine) with appropriate modifications (*see* **Note 5**).

3.1 Ultracentrifugation of Culture Media for EV Depletion

1. For in-house prepared culture medium, mix components, then filter through a 0.22 μm filter into a new sterile tube, discarding the first 10 drops.
2. To minimize EV contamination, subject in-house or commercial media to ultracentrifugation at 100,000×*g* for 18 h at 4 °C (*see* **Note 6**, Fig. 2).
3. Carefully remove the supernatant, avoiding disturbance of the pellet, and aliquot into 2 mL microcentrifuge tubes, and store

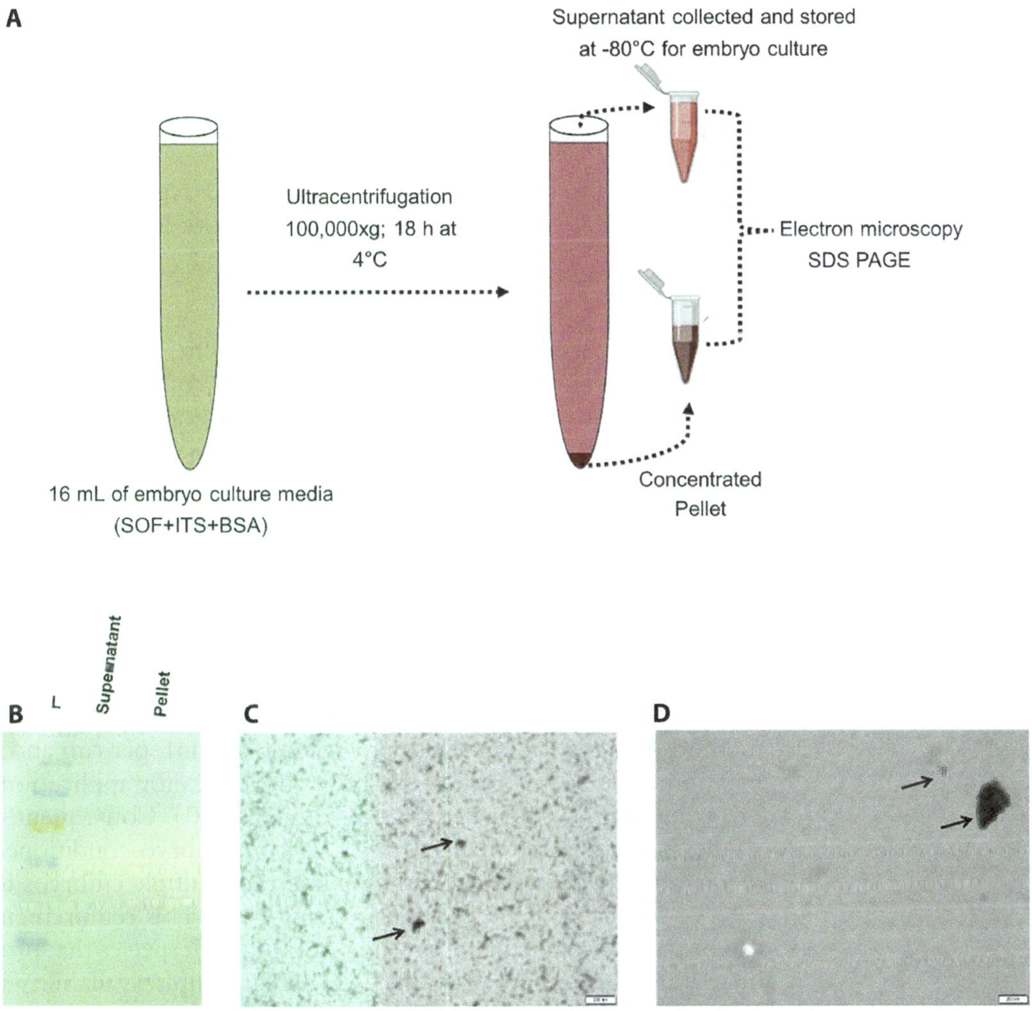

Fig. 2 EV-depleted embryo culture media preparation. (**A**) Ultracentrifugation of embryo culture media (SOF + ITS + BSA) at 100,000 × *g* for 18 h at 4°C. The supernatant is carefully removed to avoid disturbing the pellet B. (**B**) SDS-PAGE gel electrophoresis showing a large smear of protein for the pellet and no protein bands in the supernatant. Negative staining transmission electron microscopy analysis of (**C**) the pellet and (**D**) supernatant of SOF media subjected to ultracentrifugation (scale 200 nm). Amorphous material is indicated in black arrows, but no EVs can be found

at −80°C until further use (*see* **Note 6**). After thawing and warming, the collected supernatant culture medium is ready to use for embryo culture.

3.2 In Vitro Embryo Production and Collection of Conditioned Medium

Culture embryos according to routine in vitro embryo culture methods. The collection method outlined below has been optimized for bovine in vitro embryo production, where embryos are cultured individually in 20 μL droplets or groups of 25 zygotes in 50 μL droplets of ultracentrifuged or EV-depleted embryo culture media. Collection volumes can be adapted for other species (*see* **Note 3**).

1. For each replicate, culture 100 presumptive zygotes in EV-depleted embryo culture medium, placing 25 zygotes per 50 μL droplet (*see* **Note 7**).
2. At 8 dpi, collect 200 μL of embryo-conditioned medium and store it at −80°C (*see* **Note 8**).
3. Perform 15 replicates with a total of 1500 zygotes to obtain 3 mL of embryo conditioned media (*see* **Note 9**).
4. Use 1 mL of pooled conditioned medium for ODG (Fig. 1).

3.3 Density Gradient Centrifugation

Density gradient centrifugation separates vesicles from contaminants based on buoyant density, typically 1.13–1.19 g/mL for EVs [15], allowing effective separation of EVs from contaminants such as protein aggregates and other non-vesicular particles. In this method, solutions of varying densities are layered to create a discontinuous gradient, with the sample layered on top. High-speed ultracentrifugation (100,000 × *g*) is then applied for 16–18 h to separate components by density. This method provides the gold standard for specific isolation of EVs according to the MISEV guidelines.

However, applying density gradient centrifugation to embryo culture media presents specific challenges. A significant limitation is the minimum input volume required for effective separation. The method requires a minimum input volume of 1 mL per run and a maximum of 1.5 mL, which can pose challenges for applications involving limited embryo-conditioned media [10]. Consequently, studies have indicated that isolating EVs from embryo-conditioned media often necessitates pooling media from multiple embryos to achieve sufficient volumes for processing [10]. This requirement can be particularly demanding when working with species that produce limited numbers of embryos or when sample availability is constrained.

1. In a 16.8 mL open-top polyallomer centrifuge tube, layer 4 mL of 40% iodixanol solution at the bottom, followed by 4 mL of 20%, 4 mL of 10%, and finally 3.5 mL of 5% on top.

2. Overlay 1 mL pooled embryo conditioned medium on top of the density gradient.
3. Centrifuge the gradient at 4°C for 18 h at 100,000 × *g* using a swing bucket rotor.
4. After completing centrifugation, carefully divide the gradient into 16 equal fractions of approximately 1 mL each, based on volume rather than visual inspection, to avoid inconsistency. Collect fractions sequentially from the top of the tube using a pipette with a low-retention tip, ensuring that each fraction is removed slowly and precisely to minimize cross-contamination between adjacent layers. Transfer fractions 16.8 mL ultracentrifuge-rated tubes, pooling them as follows: fractions 1–4, 5–7, 8–9, 10–12, and 13–16. Each pooled group represents specific density ranges and should be handled independently to maintain the integrity of the gradient separation.
5. To prepare fractions for ultracentrifugation, adjust each pooled fraction to a final volume of 16 mL by adding the appropriate volume of sterile DPBS. After pooling, the approximate volumes of the fractions are: fractions 1–4 (~4 mL), fractions 5–7 (~3 mL), fractions 8–9 (~1 mL), fractions 10–12 (~2 mL), and fractions 13–16 (~3 mL).
6. Centrifuge at 4°C for 3 h at 100,000 × *g*. Carefully remove the supernatant from each fraction using a pipette, ensuring not to disturb the pellet at the bottom of the tube. Use a low-retention pipette tip to minimize sample loss and contamination. Once the supernatant is completely removed, resuspend the pellets from each fraction in 50 μL of freshly filtered PBS. Gently pipette up and down to ensure complete resuspension of the pellets while avoiding the formation of bubbles, which can affect sample integrity.
7. Store resuspended EVs at −80°C until further characterization or downstream use. Samples may be stored at −80°C for up to 1 year.

3.4 Characterization of EVs According to the MISEV Guidelines

According to the latest MISEV guidelines [13], EVs should be comprehensively characterized by: (1) confirming the presence of EV-specific markers, such as CD63, CD9, or TSG101, and demonstrating the absence of non-EV contaminants like lipoproteins or ribonucleoproteins, using techniques like Western blotting, flow cytometry or enzyme-linked immunosorbent protein assays to ensure sample purity; (2) determining the size distribution and concentration of EVs through nanoparticle tracking analysis (NTA) or comparable methods (e.g., dynamic light scattering), providing quantitative insights into EV populations; and (3) visualizing EV morphology and structure via transmission electron microscopy (TEM) or atomic force microscopy to confirm the characteristic vesicular appearance and integrity of isolated EVs.

Here, we provide the methodology for Western blotting, NTA, and TEM for EV characterization in accordance with the MISEV guidelines. We encourage researchers to submit their EV characterization data to EV-TRACK (https://evtrack.org/; *see* **Note 10**).

3.4.1 Western Blotting

The presence of EVs, as well as sample purity and protein concentration, can be evaluated using western blotting to detect specific vesicle markers. A representative western blot following ODG applied to 1 mL of embryo-conditioned media is shown in Fig. 3A.

1. Begin by cleaning both glass plates thoroughly with distilled water and ethanol, ensuring they are completely free of dust, grease, or chips.
2. Assemble the cassette on a level gel casting stand, placing the short plate in front of the spacer plate to form a sealed sandwich, and secure the assembly with clamps to prevent leakage.
3. Combine all components for a 10% resolving gel in a clean 15 mL tube in the given order, adding APS and TEMED last to initiate polymerization.
4. Using a pipette, carefully pour the resolving gel solution between the glass plates until it reaches approximately two-thirds of the plate height, leaving sufficient space for the stacking gel.
5. Immediately overlay the gel surface with a thin layer of isopropanol or water-saturated n-butanol (about 200–300 μL) to create a smooth interface and prevent oxygen inhibition of polymerization.
6. Leave the gel undisturbed at room temperature for 20–30 min until the gel becomes firm and transparent.
7. Once the resolving gel has polymerized, decant the overlay solvent, rinse the gel surface gently with a small amount of distilled water, and blot dry the top edge with lint-free tissue without disturbing the gel.
8. Next, prepare the 5% stacking gel solution to form the sample loading wells. Pour the stacking gel solution gently on top of the polymerized resolving gel, immediately insert a clean 10-well comb, and check that no air bubbles form beneath the comb teeth. Allow the stacking gel to polymerize completely at room temperature (about 30 min or until firm).
9. After polymerization, remove the comb carefully by pulling it straight upward to form the sample wells. Rinse the wells thoroughly with running buffer to remove any unpolymerized acrylamide. The gel assembly is now ready for sample loading and electrophoresis.

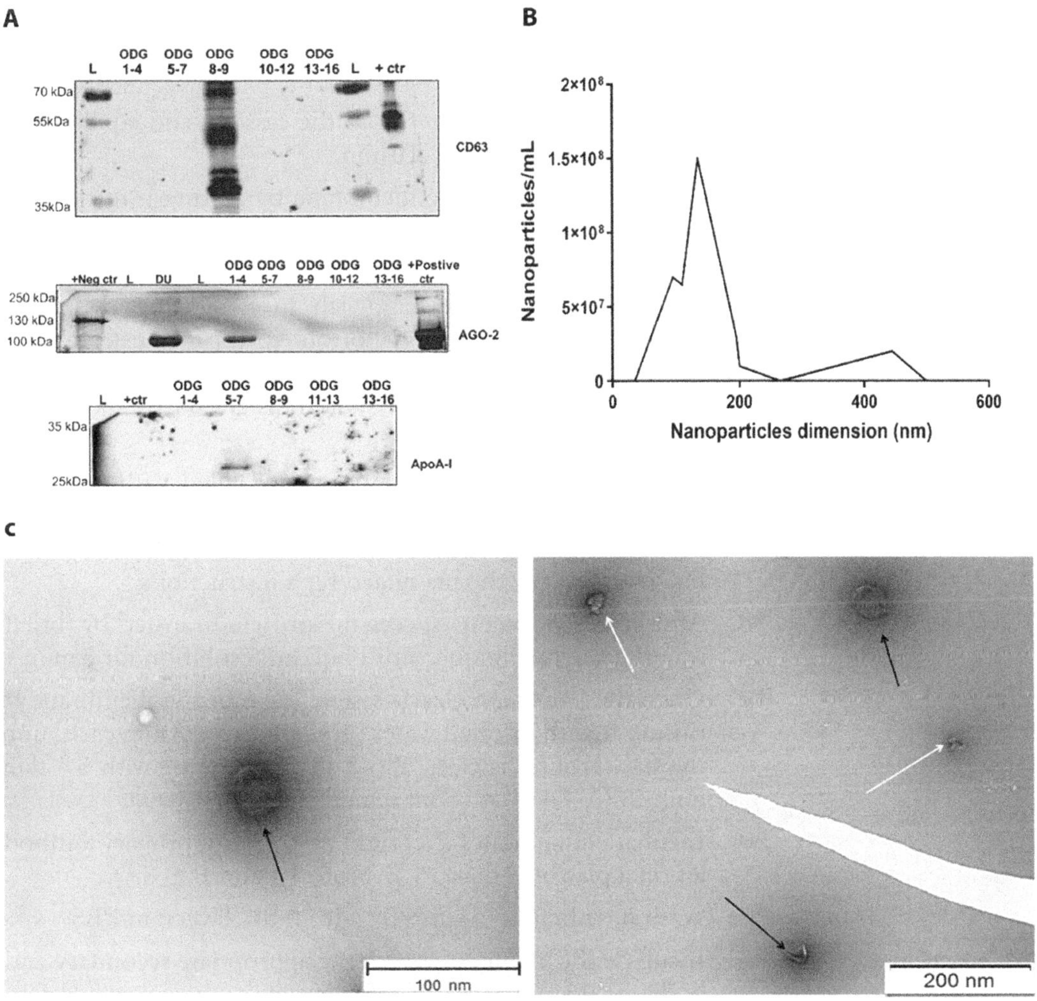

Fig. 3 Characterization of ODG EVs isolated from 1 mL of bovine embryo-conditioned medium. (**A**) Western blot analysis shows the presence of the EV-associated protein CD63 (42 kDa). The non-EV protein ApoA-I (28 kDa) is observed in fractions 5–7, while the non-EV marker Ago-2 (97 kDa) appears in fractions 1–4. (**B**) Nanoparticle tracking analysis (NTA) of ODG-EVs displays a particle size distribution ranging from 40 to 200 nm. (**C**) Transmission electron microscopy images of ODG-EVs, with white arrows indicating protein aggregates and black arrows highlighting EVs. Scale bars represent 100 and 200 nm. Abbreviations: DU differential ultracentrifugation, ODG 1–4 OptiPrep™ density gradient fractions 1 to 4 (similar format for 5–7, 8–9, 10–12, 13–16), +ctr follicular fluid-derived EVs isolated by ODG, L protein ladder, +Neg ctr EVs derived from HEK 293 T lysate, +Positive ctr pure plasma. (Adapted from [10])

10. Suspend 25 μL of EV sample in 5 μL of reducing buffer or nonreducing buffer (*see* **Note 11**).
11. Boil samples at 95°C for 5 min in a thermomixer or heated dry tube rack.
12. Load each sample into individual wells of the gel along with a protein molecular weight marker or ladder in a separate well.

13. Run the gel at a constant voltage (e.g., 120–150 V) in SDS-PAGE running buffer until the dye front reaches the bottom of the gel.
14. Carefully remove the gel from the cassette and equilibrate it in blotting buffer for 5–10 min.
15. Prepare a nitrocellulose membrane by cutting it to the same size as the gel (*see* **Note 12**) and soaking it in transfer buffer for 5–10 min to ensure proper wetting.
16. Assemble the transfer sandwich in a blotting cassette in the following order (from bottom to top): transfer cassette sponge (1×), filter paper (3×), SDS-PAGE gel, nitrocellulose membrane, filter paper (3×), and transfer cassette sponge (1×) (*see* **Note 13**).
17. Place the cassette into a transfer tank filled with prechilled transfer buffer and perform protein transfer at a constant current (e.g., 250 mA) for 1–2 h at 4°C or using a semidry transfer system as per the manufacturer's instructions.
18. After transfer, verify successful protein transfer by briefly staining the membrane with Ponceau S solution for 5 min.
19. Once the bands are clearly visible, destain the membrane by rinsing it with distilled water three times for 3 min each, until the background is clear. Block the membrane with 5% skim milk/BSA TBST at room temperature for 45 min.
20. Incubate membranes overnight at 4°C with primary antibodies on a platform shaker (*see* **Note 14** and Fig. 3A).
21. Wash membranes extensively with 0.5% Tween in PBS.
22. Incubate the membranes with the appropriate secondary antibody for 45 min at room temperature on a rocker/platform shaker to ensure even binding.
23. Prepare the working detection reagent by mixing equal volumes of the two chemiluminescent substrate components (1:1 ratio) immediately before use, following the manufacturer's instructions. Use sufficient volume to fully cover the membrane surface—typically 0.1 mL of mixed substrate per cm^2 of membrane. Gently pipette or pour the mixed substrate evenly over the membrane placed protein-side up on a clean, flat surface (*see* **Note 15**).
24. Incubate for 1–2 min at room temperature to allow the enzymatic reaction between horseradish peroxidase (HRP) and the substrate to initiate light emission.
25. Carefully remove any excess substrate and immediately place the membrane in the chemiluminescence imaging system for signal detection (*see* **Note 16**).

3.4.2 EV Particle Concentration Determination by Nanoparticle Tracking Analysis (NTA)

Physical characterization of EVs can be performed using NTA. A representative NTA of isolated EVs from 1 mL of embryo-conditioned media prepared by ODG is shown in Fig. 3B.

1. Dilute 10 μL of EV suspension with sterile DPBS to achieve a 1:50 or 1:100 dilution, depending on the sample concentration. The optimal dilution should ensure that particle concentration falls within the linear range of the NTA instrument (typically 10^6–10^9 particles/mL).
2. Analyze samples with a NanoSight LM10 microscope equipped with a 455 nm laser.
3. Record three 60-s videos per sample at detection threshold 3 and camera level 13.
4. Use NTA Software version 3.2 for video analysis. First, open the NTA v3.2 software and load the recorded video files ("captures") for your sample.
5. Set up measurement parameters: enter the sample dilution factor and, if applicable, the viscosity of the medium (*see* **Note 17**).
6. Adjust the focus and camera level so that particles are clearly visible, limiting saturation (e.g., no more than ~20% of pixels saturated), ensuring uniform distribution across the field.
7. Set a detection threshold: this determines how bright a particle must be to be detected (*see* **Note 18**).
8. Select the number of video captures (frames/video) to analyze and execute particle analysis (*see* **Note 19**).
9. Once processing is complete, export the data (size distribution, concentration, etc.) as graphs (PDF format) and summary and raw data (CSV format) for further analysis.

3.4.3 Transmission Electron Microscopy (TEM)

Physical EV characterization can be achieved using TEM. Figure 3C illustrates the size of isolated EVs from 1 mL of embryo-conditioned media following ODG.

1. Place a 15 μL drop of the non-diluted EV sample onto a clean piece of Parafilm.
2. Gently place the carbon-coated side of a formvar/carbon-coated copper grid onto the drop, ensuring the grid is fully immersed in the sample. Allow the sample to adsorb onto the grid surface for 10 min.
3. Using fine-point tweezers, carefully lift the grid from the sample drop and gently blot the edge of the grid with filter paper to remove excess liquid.
4. Transfer the grid to a 15 μL drop of ultrapure water placed on Parafilm for a brief rinse (10–15 s) to remove excess buffer or

contaminants. Repeat this rinse step once more with a fresh drop of ultrapure water.

5. Immediately transfer the grid to a 15 μL drop of 1% uranyl acetate staining solution for 1 min, ensuring the entire grid surface is covered by the stain.
6. Carefully lift the grid from the staining drop and blot the edge with filter paper to remove excess stain, leaving a thin, even film.
7. Allow the grids to air dry completely in a dust-free environment before imaging.
8. Analyze the prepared grids using a transmission electron microscope at an accelerating voltage of 80 kV.
9. Capture multiple fields of view at varying magnifications using a digital camera system, such as a Quemasa camera, to obtain high-resolution images to visualize the structural integrity and morphology of the EVs.
10. Analyze the TEM images to confirm the presence and morphology of extracellular vesicles. Parameters typically assessed include (i) size range, measured across multiple vesicles to ensure consistency with expected EV dimensions (typically 30–200 nm for exosomes and up to 1000 nm for microvesicles), (ii) shape and structural integrity, with EVs generally appearing as round or cup-shaped vesicles, and (iii) membrane integrity and contrast following uranyl acetate staining. Quantitative size measurements can be performed using imaging software to generate histograms of particle size distribution, while qualitative assessment ensures that the vesicles observed are distinct from potential contaminants such as protein aggregates or debris.

4 Notes

1. In general, gradients with higher OptiPrep™ (iodixanol concentration (20%–40%)) yield cleaner but lower-recovery fractions, whereas shallower gradients (5%–10%) maximize EV yield but may include minor protein contamination. The choice of gradient depends on the input volume, expected vesicle density, and purity requirements of the sample. For small input volumes (<1 mL) or low EV concentrations, Gradient 1 (lower density range) provides better recovery of lighter, embryo-derived EVs with minimal loss. Gradient 2 is suited for standard preparations where both exosome- and microvesicle-sized particles are anticipated. Gradient 3 (higher density range) is recommended when higher purity

is required, for example, to separate EVs from protein aggregates, lipoproteins, or other dense particles.

2. When handling chemicals such as β-mercaptoethanol, acrylamide, TEMED, APS, methanol, and SDS, always wear appropriate PPE, including a lab coat, safety goggles, and nitrile gloves. Perform all procedures involving volatile or toxic reagents (e.g., β-mercaptoethanol, acrylamide, TEMED, and methanol) inside a fume hood to avoid inhalation exposure. Avoid direct skin or eye contact and dispose of chemical waste according to institutional safety guidelines.
3. For preparation of both the separating and stacking gel buffers, the use of high-purity reagents and freshly prepared distilled or deionized water is essential to ensure reproducibility. It is recommended to monitor pH carefully with a calibrated pH meter at room temperature, as temperature fluctuations can influence the readings. Slow addition of hydrochloric acid is advised to prevent overshooting the target pH. For long-term storage, filtration of the buffers can be considered to prevent microbial growth and maintain buffer clarity.
4. The choice of blocking reagent depends on the antibody type and antigen sensitivity. In general, 5% skim milk in TBST is suitable for most total protein antibodies and non-phosphorylated targets, while 5% BSA in TBST is preferred for antibodies that recognize conformational or posttranslationally modified epitopes, as milk contains casein, a phosphoprotein that can interfere with such detection. For EV-specific antibodies, we recommend using 5% BSA for rabbit-derived antibodies against CD9 and CD63 to minimize nonspecific binding and preserve epitope integrity, whereas 5% skim milk is appropriate for mouse-derived antibodies such as TSG101, which recognize linear epitopes and are less sensitive to milk components. The Jackson ImmunoResearch Western Blotting Guide (2024) https://www.jacksonimmuno.com/secondary-antibody-resource/immuno-techniques/western-blotting-guide/ serves as a useful guide for those new to working with antibodies.
5. While this protocol was optimized for bovine embryos, it can be adapted for other mammalian species, including porcine, ovine, caprine, equine, and even human, for the isolation and characterization of extracellular vesicles (EVs) from embryo culture media (reviewed by [11]). The primary adjustments involve modifying the embryo-to-medium ratio, droplet volume, and pooling strategies based on the typical embryo yield and EV concentration in each species. For human embryo culture systems, several studies have demonstrated that EVs can be recovered from spent human embryo culture media,

offering potential biomarkers for embryo quality and implantation success [16, 17]. Whether working with livestock or human embryos, pilot experiments are strongly recommended to optimize culture conditions (e.g., pH, osmolarity, and EV-depletion methods) to ensure robust embryo development and sufficient EV recovery.

6. Fetal bovine serum (FBS) is known to contain extracellular vesicles (EVs), which can act as contaminants [18, 19]. To eliminate these contaminants, the media should be ultracentrifuged to remove EVs. The ultracentrifuged embryo culture media (EV-depleted media) can be aliquoted and stored at −80°C for up to 3–4 weeks. Before use, thaw the media and allow it to equilibrate to ensure optimal conditions for embryo culture.
7. Pooling of embryo culture media can vary based on experimental design. Working with individual or single-droplet embryo culture has the advantage of pooling the media based on embryo quality. For EVs isolation from individual embryo culture, pooling multiple samples is required to achieve a final volume of >1 mL, as the concentration of EVs in medium conditioned by individual embryos is very low [4, 20]. Working with group-cultured embryos requires relatively less pooling of conditioned culture media (>500 μL) [21]. Initially, we started off with a volume of 500 μL of conditioned medium, by pooling the media from 250 embryos cultured in groups of 25 in 50 μL droplets. Preliminary data, however, identified a low EV particle concentration ($2.9 \pm 5.8 \times 10^7$/mL) which was not in the optimal range (3×10^8–1×10^9/mL) as required by the NTA Software. Ideally, the minimum concentration should be 3–10×10^8 particles/mL, which is within the linear range [22]. Above this concentration, the screen becomes crowded, and it may be difficult to discern small EVs in a polydispersed sample. Analyzing concentrations of EVs below 2×10^8 particles/mL results in reduced precision [23] and also difficulty in detecting EVs in electron microscopy, as well as EV-specific marker protein in western blotting. Hence, in this book chapter, we prefer pooling of group culture media to a final volume of 1 mL for isolation by ODG UC.
8. When collecting embryo-conditioned media from culture droplets, there is a risk of inadvertently drawing in oil from the droplet's surrounding layer. To minimize or prevent oil contamination, first remove (aspirate) the oil from around the droplet. Remove the oil prior to pooling of the conditioned media. This procedure helps ensure that any carryover of oil into the collected media is significantly reduced or eliminated.

9. Embryo culture duration and media handling significantly impact EV yield and quality. Studies have shown that EVs secreted during early developmental stages, such as cleavage or morula, differ in concentration and molecular cargo compared to the blastocyst stage [9, 20]. Shorter culture durations may limit EV collection, while prolonged culture allows for developmental stage-specific EV analysis. For example, EV concentrations are higher in media conditioned by embryos with impaired developmental potential, such as those degenerating post-morula stage [21]. Changing media mid-culture versus consistent media use also affects EV secretion patterns. A uniform culture medium may better reflect continuous EV secretion dynamics, whereas media replacement could lead to artificial EV spikes or dilution [4, 19]. In human and murine embryo cultures, medium adjustments are often tailored to specific developmental stages, which might also affect EV dynamics. For accurate EV collection, it is recommended to use EV-depleted media, collect samples prior to medium changes, and adhere to species-specific protocols to standardize EV isolation and characterization [9, 18, 19, 24]. Earlier studies of embryo-derived EVs have demonstrated that using volumes of conditioned medium of 750 μL (derived from a pool of 750 embryos [7]) or 500 μL (derived from 10 embryos of group culture medium [21]) or even 50 μL (derived from six embryos [16]) may be sufficient for EV isolation with ultracentrifugation. These pioneering studies highlighted the presence of embryo-derived EVs, but they do not comply with the current MISEV guidelines, as particle concentration and EV-specific antibody identification were performed on conditioned media prior to ultracentrifugation. Hence, for ODG UC or DU, we recommend using a higher volume of >1 mL of embryo-conditioned media to fulfill all EV characterization requirements recommended by MISEV [13].
10. The EV-TRACK platform (https://evtrack.org/) offers significant advantages for tracking extracellular vesicles (EVs), especially in embryo-derived EV research [25]. It promotes transparent reporting, ensuring that critical experimental parameters are documented and accessible. This transparency facilitates reproducibility and enhances the rigor of EV studies by standardizing data reporting. Researchers benefit from the community-driven knowledge base that aggregates detailed methodology, allowing for improved consistency in EV isolation, characterization, and interpretation. Using EV-TRACK helps researchers avoid common pitfalls, thus providing a reliable foundation for embryo-derived EV

studies and advancing insights into EV-mediated communication in reproductive biology.

11. When preparing samples for Western blotting, it is important to select reducing or nonreducing sample buffers according to the antibody being used. Some antibodies (for example, CD63) only recognize epitopes in the native (non-reduced) form of a protein, whereas others detect epitopes that are exposed only under reducing conditions (for example, CD9 and TSG101) after disulfide bonds are broken. Using the wrong buffer may result in a weak or absent signal, or the detection of different protein isoforms or subunits. Therefore, always consult antibody datasheets or perform pilot blots to determine whether reducing or non-reducing conditions are optimal for each specific target.
12. After running SDS-PAGE, cut a small piece (triangular or square notch) from the upper left-hand corner of the gel. This cut serves as a reference mark to maintain consistent orientation of both the gel and the membrane (e.g., matching molecular weight marker lanes, ensuring correct placement of samples).
13. During transfer assembly, align the upper left corner of the gel with the corresponding corner of the membrane. Some protocols alternatively use the lower right corner of the membrane after transfer for marking; if your lab follows such a convention, clearly document and consistently use that. Ensure there are no air bubbles between the gel and the nitrocellulose membrane, as this can interfere with protein transfer. Remove bubbles by gently pushing them to the edges of the gel using a roller.
14. Data for one EV-specific antibody (CD63) and two non-EV-specific antibodies (ApoA-I and AGO-2 [4, 9, 10]) are presented to illustrate protocol optimization in line with MISEV 2024 guidelines [13]. However, there are other common EV markers, including tetraspanins such as CD9, TSG101, and CD81, and proteins such as HSP70 and flotillin. Selecting appropriate antibodies for detecting extracellular vesicles (EVs) in embryo culture media is challenging due to low sample volumes and species specificity. In our optimization studies, we found that rabbit-derived antibodies exhibit high cross-species reactivity, performing well across various applications. For instance, rabbit antibodies have been shown to recognize antigens in both mouse and human samples effectively. This broad reactivity makes them valuable tools in EV research involving multiple species. The non-EV-specific markers ApoA-I and AGO-2 were specifically chosen due to their high abundance in embryo-conditioned media [9].

15. A plastic tray or plastic cling wrap serves as a suitable surface for incubation. When using cling wrap, ensure there are no folds or bubbles, as this can interfere with introducing artefacts in chemiluminescent binding.
16. Optimize exposure time empirically, beginning with a short exposure (10–30 s for strong signals and sensitive substrates). If the signal appears weak, increase exposure time gradually (e.g., 1 min, 5 min, or longer) until the protein bands are clearly visible but not saturated. Save both low- and high-exposure images for accurate qualitative and quantitative analysis.
17. If the sample contains components such as serum, culture medium with high protein concentration, or other additives that increase solution thickness or resistance to flow, it is considered viscous. Viscosity affects particle movement and thus influences size and concentration calculations. If you are unsure, compare the sample visually or by pipetting against pure water; viscous samples will flow more slowly and appear thicker. When in doubt, consult the instrument manual or manufacturer's viscosity reference chart to input an appropriate value before starting analysis.
18. Select a threshold low enough to include true particles, but high enough to exclude background noise. The same threshold must be used across all videos for the same sample to allow comparability.
19. Multiple short captures are better than one long video to get robust statistics. The software tracks particle motion (Brownian motion) within each video, then computes size distributions, modal/mean particle size, concentration (particles/mL), and optionally scatter (intensity). Ensure consistency between replicate captures by checking for low background noise, minimal drift or vibration, and sufficient particle tracks in each video.

References

1. Paria BC, Dey SK (1990) Preimplantation embryo development in vitro: cooperative interactions among embryos and role of growth factors. Proc Natl Acad Sci U S A 87(12):4756–4760. https://doi.org/10.1073/pnas.87.12.4756
2. Watson AJ, Hogan A, Hahnel A, Schultz GA, Wiemer KE (1992) Expression of growth factor ligand and receptor genes in the preimplantation bovine embryo. Mol Reprod Dev 31(2):87–95. https://doi.org/10.1002/mrd.1080310202
3. Wydooghe E, Vandaele L, Heras S, De Sutter P, Deforce D, Peelman L, De Schauwer C, Van Soom A (2017) Autocrine embryotropins revisited: how do embryos communicate with each other in vitro when cultured in groups?. Biol Rev 92(1):505–520. https://doi.org/10.1111/brv.12241
4. Pavani KC, Meese T, Pascottini OB, Guan X, Lin X, Peelman L, Hamacher J, Van Nieuwerburgh F, Deforce D, Boel A, Heindryckx B, Tilleman K, Van Soom A, Gadella BM, Hendrix A, Smits K (2022) Hatching is modulated by

microRNA-378a-3p derived from extracellular vesicles secreted by blastocysts. Proc Natl Acad Sci U S A 119(12):e2122708119. https://doi.org/10.1073/pnas.2122708119
5. Fazeli A, Godakumara K (2024) The evolving roles of extracellular vesicles in embryo-maternal communication. Commun Biol 7(1):754. https://doi.org/10.1038/s42003-024-06442-9
6. Pavani KC, Alminana C, Wydooghe E, Catteeuw M, Ramírez MA, Mermillod P, Rizos D, Van Soom A (2017) Emerging role of extracellular vesicles in communication of preimplantation embryos in vitro. Reprod Fertil Dev 29(1):66–83. https://doi.org/10.1071/RD16318
7. Saadeldin IM, Kim SJ, Choi YB, Lee BC (2014) Improvement of cloned embryos development by co-culturing with parthenotes: a possible role of exosomes/microvesicles for embryos paracrine communication. Cell Reprogram 16(3):223–234. https://doi.org/10.1089/cell.2014.0003
8. Qu P, Qing S, Liu R, Qin H, Wang W, Qiao F, Ge H, Liu J, Zhang Y, Cui W, Wang Y (2017) Effects of embryo-derived exosomes on the development of bovine cloned embryos. PLoS ONE 12(3):e0174535. https://doi.org/10.1371/journal.pone.0174535
9. Pavani KC, Hendrix A, Van Den Broeck W, Couck L, Szymanska K, Lin X, De Koster J, Van Soom A, Leemans B (2018) Isolation and characterization of functionally active extracellular vesicles from culture medium conditioned by bovine embryos in vitro. Int J Mol Sci 20(1):38. https://doi.org/10.3390/ijms20010038
10. Pavani KC, Lin X, Hamacher J, Broeck WVD, Couck L, Peelman L, Hendrix A, Van Soom A (2020) The separation and characterization of extracellular vesicles from medium conditioned by bovine embryos. Int J Mol Sci 21(8):2942. https://doi.org/10.3390/ijms21082942
11. Saadeldin IM, Pavani KC, Gnagnarelli J, Ehab S, Assiri AM, Van Soom A (2025) Unlocking a decade of research on embryo-derived extracellular vesicles: discoveries made and paths ahead. Stem Cell Rev Rep 21(3):698–708. https://doi.org/10.1007/s12015-025-10844-5
12. Asaadi A, Dolatabad NA, Atashi H, Raes A, Van Damme P, Hoelker M, Hendrix A, Pascottini OB, Van Soom A, Kafi M, Pavani KC (2021) Extracellular vesicles from follicular and ampullary fluid isolated by density gradient ultracentrifugation improve bovine embryo development and quality. Int J Mol Sci 22(2):578. https://doi.org/10.3390/ijms22020578
13. Welsh JA, Goberdhan DCI, O'Driscoll L, Buzas EI, Blenkiron C, Bussolati B, Cai H, Di Vizio D, Driedonks TAP, Erdbrügger U, Falcon-Perez JM, Fu Q-L, Hill AF, Lenassi M, Lim SK, Mahoney MG, Mohanty S, Möller A, Nieuwland R, Ochiya T, Sahoo S, Torrecilhas AC, Zheng L, Zijlstra A, Abuelreich S, Bagabas R, Bergese P, Bridges EM, Brucale M, Burger D, Carney RP, Cocucci E, Crescitelli R, Hanser E, Harris AL, Haughey NJ, Hendrix A, Ivanov AR, Jovanovic-Talisman T, Kruh-Garcia NA, Ku'ulei-Lyn Faustino V, Kyburz D, Lässer C, Lennon KM, Lötvall J, Maddox AL, Martens-Uzunova ES, Mizenko RR, Newman LA, Ridolfi A, Rohde E, Rojalin T, Rowland A, Saftics A, Sandau US, Saugstad JA, Shekari F, Swift S, Ter-Ovanesyan D, Tosar JP, Useckaite Z, Valle F, Varga Z, van der Pol E, van Herwijnen MJC, Wauben MHM, Wehman AM, Williams S, Zendrini A, Zimmerman AJ, Consortium M, Théry C, Witwer KW (2024) Minimal information for studies of extracellular vesicles (MISEV2023): from basic to advanced approaches. J Extracell Vesicles 13(2):e12404. https://doi.org/10.1002/jev2.12404
14. Wydooghe E, Vandaele L, Piepers S, Dewulf J, Van den Abbeel E, De Sutter P, Van Soom A (2014) Individual commitment to a group effect: strengths and weaknesses of bovine embryo group culture. Reproduction 148(5):519–529. https://doi.org/10.1530/rep-14-0213
15. van der Pol E, Böing AN, Harrison P, Sturk A, Nieuwland R (2012) Classification, functions, and clinical relevance of extracellular vesicles. Pharmacol Rev 64(3):676. https://doi.org/10.1124/pr.112.005983
16. Giacomini E, Vago R, Sanchez AM, Podini P, Zarovni N, Murdica V, Rizzo R, Bortolotti D, Candiani M, Viganò P (2017) Secretome of in vitro cultured human embryos contains extracellular vesicles that are uptaken by the maternal side. Sci Rep 7(1):5210. https://doi.org/10.1038/s41598-017-05549-w
17. Makieva S, Giacomini E, Scotti GM, Lazarevic D, Pavone V, Ottolina J, Bartiromo L, Schimberni M, Morelli M, Alteri A, Minetto S, Tonon G, Candiani M, Papaleo E, Viganò P (2024) Extracellular vesicles secreted by human aneuploid embryos present a distinct transcriptomic profile and upregulate MUC1 transcription in decidualised endometrial stromal cells. Human Reprod Open 2024(2):hoae014. https://doi.org/10.1093/hropen/hoae014
18. Eitan E, Zhang S, Witwer KW, Mattson MP (2015) Extracellular vesicle–depleted fetal bovine and human sera have reduced capacity to support cell growth. J Extracell Vesicles 4:26373. https://doi.org/10.3402/jev.v4.26373
19. Shelke GV, Lässer C, Gho YS, Lötvall J (2014) Importance of exosome depletion protocols to eliminate functional and RNA-containing extracellular vesicles from fetal bovine serum.

J Extracell Vesicles 3:24783. https://doi.org/10.3402/jev.v3.24783

20. Dissanayake K, Nõmm M, Lättekivi F, Ressaissi Y, Godakumara K, Lavrits A, Midekessa G, Viil J, Bæk R, Jørgensen MM, Bhattacharjee S, Andronowska A, Salumets A, Jaakma Ü, Fazeli A (2020) Individually cultured bovine embryos produce extracellular vesicles that have the potential to be used as non-invasive embryo quality markers. Theriogenology 149:104–116. https://doi.org/10.1016/j.theriogenology.2020.03.008
21. Mellisho EA, Velásquez AE, Nuñez MJ, Cabezas JG, Cueto JA, Fader C, Castro FO, Rodríguez-Álvarez L (2017) Identification and characteristics of extracellular vesicles from bovine blastocysts produced in vitro. PLoS ONE 12(5):e0178306. https://doi.org/10.1371/journal.pone.0178306
22. Dragovic RA, Gardiner C, Brooks AS, Tannetta DS, Ferguson DJP, Hole P, Carr B, Redman CWG, Harris AL, Dobson PJ, Harrison P, Sargent IL (2011) Sizing and phenotyping of cellular vesicles using Nanoparticle Tracking Analysis. Nanomedicine 7(6):780 788. https://doi.org/10.1016/j.nano.2011.04.003
23. Gardiner C, Ferreira YJ, Dragovic RA, Redman CWG, Sargent IL (2013) Extracellular vesicle sizing and enumeration by nanoparticle tracking analysis. J Extracell Vesicles 2(1):19671. https://doi.org/10.3402/jev.v2i0.19671
24. Wydooghe E, Heras S, Dewulf J, Piepers S, Van den Abbeel E, De Sutter P, Vandaele L, Van Soom A (2014) Replacing serum in culture medium with albumin and insulin, transferrin and selenium is the key to successful bovine embryo development in individual culture. Reprod Fertil Dev 26(5):717–724 https://doi.org/10.1071/RD13043
25. Consortium E-T, Van Deun J, Mestdagh P, Agostinis P, Akay Ö, Anand S, Anckaert J, Martinez ZA, Baetens T, Beghein E, Bertier L, Berx G, Boere J, Boukouris S, Bremer M, Buschmann D, Byrd JB, Casert C, Cheng L, Cmoch A, Daveloose D, De Smedt E, Demirsoy S, Depoorter V, Dhondt B, Driedonks TAP, Dudek A, Elsharawy A, Floris I, Foers AD, Gärtner K, Garg AD, Geeurickx E, Gettemans J, Ghazavi F, Giebel B, Kormelink TG, Hancock G, Helsmoortel H, Hill AF, Hyenne V, Kalra H, Kim D, Kowal J, Kraemer S, Leidinger P, Leonelli C, Liang Y, Lippens L, Liu S, Lo Cicero A, Martin S, Mathivanan S, Mathiyalagan P, Matusek T, Milani G, Monguió-Tortajada M, Mus LM, Muth DC, Németh A, Nolte-'t Hoen ENM, O'Driscoll L, Palmulli R, Pfaffl MW, Primdal-Bengtson B, Romano E, Rousseau Q, Sahoo S, Sampaio N, Samuel M, Scicluna B, Soen B, Steels A, Swinnen JV, Takatalo M, Thaminy S, Théry C, Tulkens J, Van Audenhove I, van der Grein S, Van Goethem A, van Herwijnen MJ, Van Niel G, Van Roy N, Van Vliet AR, Vandamme N, Vanhauwaert S, Vergauwen G, Verweij F, Wallaert A, Wauben M, Witwer KW, Zonneveld MI, De Wever O, Vandesompele J, Hendrix A (2017) EV-TRACK: transparent reporting and centralizing knowledge in extracellular vesicle research. Nat Methods 14:228–232. https://doi.org/10.1038/nmeth.4185
26. Nordin JZ, Lee Y, Vader P, Mäger I, Johansson HJ, Heusermann W, Wiklander OPB, Hällbrink M, Seow Y, Bultema JJ, Gilthorpe J, Davies T, Fairchild PJ, Gabrielsson S, Meisner-Kober NC, Lehtiö J, Smith CIE, Wood MJA, Andaloussi SEL (2015) Ultrafiltration with size-exclusion liquid chromatography for high yield isolation of extracellular vesicles preserving intact biophysical and functional properties. Nanomedicine 11(4):879–883. https://doi.org/10.1016/j.nano.2015.01.003
27. Böing AN, van der Pol E, Grootemaat AE, Coumans FAW, Sturk A, Nieuwland R (2014) Single-step isolation of extracellular vesicles by size exclusion chromatography. J Extracell Vesicles 3:23430. https://doi.org/10.3402/jev.v3.23430
28. Yuana Y, Levels J, Grootemaat A, Sturk A, Nieuwland R (2014) Co-isolation of extracellular vesicles and high-density lipoproteins using density gradient ultracentrifugation. J Extracell Vesicles 3:23262. https://doi.org/10.3402/jev.v3.23262

Check for updates

Chapter 21

Modeling Early Human Embryo Implantation Using Human Endometrial Organoids and Embryo Surrogate Co-cultures for microRNA Transfection

Wei Zhou, Michaela Sacco and Evdokia Dimitriadis

Abstract

In vitro models of human implantation offer a powerful approach to studying the molecular and cellular events critical for a successful pregnancy. In this chapter, we describe a method that combines human endometrial epithelial organoids with human embryo surrogate co-cultures to model the early stages of implantation. Endometrial epithelial organoids, derived from primary tissue, are co-cultured with human cytotrophoblast progenitor cells to simulate the uterine microenvironment during implantation. This model enables a focused study of embryo adhesion and key signaling pathways, offering insights into the molecular mechanisms underlying implantation and reproductive disorders such as infertility.

Keywords Human embryo implantation, In vitro model, Endometrial organoids, Embryo adhesion, Infertility, Cellular interactions, Molecular mechanisms

1 Introduction

In vitro human implantation models have made a tremendous impact on the reproductive sciences, providing invaluable insights into the complex processes underlying implantation. Traditionally, these models have relied on a range of cellular systems, including single-cell types, cell lines, and primary human cells. More recently, the development of endometrial organoid cultures has further advanced the study of embryo–endometrial interactions that are critical for successful embryo implantation, offering a more physiologically relevant model for investigating the uterine microenvironment during implantation [1, 2]. Previous studies have used organoid culture systems to study other aspects of the endometrium, such as menstruation and the hormonal regulation of tissue remodeling [3–5]. However, these models typically lack the ability to fully simulate the embryonic component of implantation.

Alexandra J. Harvey and David K. Gardner (eds.), *The Preimplantation Embryo*, Methods in Molecular Biology, vol. 3038,
https://doi.org/10.1007/978-1-0716-5292-3_21,

Building on previous models, the co-culture system detailed in this chapter integrates human-derived endometrial organoids with human embryo surrogates or embryos [6, 7]. This multilayered approach allows for the study of key implantation events, such as embryo attachment, cell signaling, and critical regulators of embryo implantation in real-time. The method described here builds upon previous work with organoid culture systems and embryo culture techniques [8–10].

In this method, we detail a co-culture model using human endometrial organoid-derived epithelial cell monolayers to mimic luminal epithelial cells, along with human cytotrophoblast progenitor cells (HTPCs) as embryo surrogates [11]. Endometrial epithelial organoids are hormonally treated to simulate the receptive mid-secretory phase of the menstrual cycle (*see* **Note 1**) and are subsequently fragmented to generate epithelial monolayers that model the luminal surface of the endometrium. Embryo surrogates are generated using low-attachment U-shape culture wells to form spherical embryo-sized surrogates in parallel, facilitating the spheroid adhesion assay as the endpoint functional analysis.

Using microRNA (miR) miR-23b-3p as an example [7], organoid-derived epithelial monolayers have been shown to recapitulate the in vivo changes of miR-23b-3p expression in the luminal epithelium, transitioning from the non-receptive proliferative phase to the receptive mid-secretory phase following hormone treatment. We are able to overexpress miR-23b-3p in fertile organoid-derived epithelial monolayers, test its function on embryo surrogate adhesion, and identify its gene target to confirm the in vivo relevance.

2 Materials

Prepare all cell culture medium under sterile conditions.

2.1 Culture Reagents

1. Advanced DMEM/F12.
2. N-2 supplement.
3. B-27 supplement minus vitamin A.
4. 50 mg/mL Primocin.
5. 125 mM *N*-Acetyl-L-cysteine.
6. 200 mM L-glutamine.
7. 50 μg/mL recombinant human EGF.
8. 25 μg/mL recombinant human Noggin.
9. 8 μg/mL recombinant human Rspondin-1.
10. 50 μg/mL recombinant human FGF-10.
11. 50 μg/mL recombinant human HGF.
12. 1 mM A83-01.

13. 10 μM Nicotinamide.
14. DMEM/F12 + 1× GlutaMAX-l.
15. Fetal bovine serum (FBS).
16. 10 μg/mL bFGF.
17. 20 mM SB431542.
18. Dulbecco's phosphate-buffered saline (DPBS).
19. Penicillin/streptomycin.
20. DPBS with 1% penicillin–streptomycin; to 99 mL DPBS add 1 mL penicillin/streptomycin.
21. 40 mg/mL Collagenase III.
22. 10 mg/mL DNase I.
23. Calcium chloride.
24. Matrigel (growth factor reduced) basement membrane matrix.
25. Plate coating Matrigel; dilute 1 volume of neat Matrigel with 9 volumes of Advanced DMEM/F12.
26. TrypLE Express.
27. Estradiol (E2) stock: 10 μM prepared in 100% ethanol.
28. Medroxyprogesterone acetate (MPA) stock: 1 mM prepared in 100% ethanol.
29. Cyclic adenosine monophosphate (cAMP) stock 1 mM prepared in sterile water.
30. 0.4% trypan blue stain.
31. Sterile nuclease-free water.

2.2 Culture Media

1. Endometrial dissociation buffer: DPBS containing 0.73 mg/mL collagenase III, 25 μg/mL DNase I, and 0.36 mM calcium chloride.
2. Organoid expansion medium: Advanced DMEM/F12, 1× N-2 supplement, 1× B-27 supplement minus vitamin A, 100 μg/mL Primocin, 1.25 mM *N*-Acetyl-L-cysteine, 2 mM L-glutamine, 50 ng/mL recombinant human EGF, 100 ng/mL recombinant human Noggin, 80 ng/mL recombinant human Rspondin-1 (*see* **Note 2**), 100 ng/mL recombinant human FGF-10, 50 ng/mL recombinant human HGF, 500 nM A83-01, 10 nM Nicotinamide. Filter through a 0.22 μm filter. Medium can be stored at 4°C for up to 2 weeks.
3. Estrogen supplemented organoid expansion medium; organoid expansion medium supplemented with 10 nM E2.
4. Receptive phase medium: organoid expansion medium supplemented with 10 nM E2, 1 μM MPA, and 1 μM cAMP.

5. Epithelial monolayer culture medium: organoid expansion medium supplemented with 10 nM E2, 1 μM MPA, and 1 μM cAMP.
6. Advanced DMEM/F12 supplemented with 20% FBS.
7. Embryo surrogate culture medium base: Prepare 500 mL stock medium by adding 450 mL DMEM/F12 + GlutaMAX, 50 mL FBS. Filter sterilize the solution using a 0.22 μm filter.
8. Complete embryo surrogate culture medium: to prepare the complete medium, supplement the embryo surrogate culture medium base medium with 10 ng/mL bFGF and 10 μM SB431542. The complete medium can be stored at 4°C and used for up to 7 days.
9. 0.5% gelatin substrate solution: Warm a 100 mL bottle of 2% gelatin in a 37°C water bath for 30 min. Add 300 mL of sterile water, sterile-filter the solution, and store at 4°C until use.

2.3 Transfection Reagents and Media

1. Opti-MEM (Reduce serum medium).
2. Transfection medium base: Primocin-free organoid expansion medium supplemented with 10 nM E2, 1 μM MPA, and 1 μM cAMP.
3. Transfection medium: add 30 μL transfection mix and 120 μL transfection medium base.
4. Lipofectamine™ RNAiMAX Transfection Reagent.
5. 1 μM stock mIR mimic of interest prepared in sterile nuclease-free water (e.g., miR-23b-3p).
6. Scramble control miR.

2.4 Consumables

1. 0.22 μm filters.
2. 40 μm cell strainers.
3. 1 mL pipette tips.
4. 200 μL pipette tips.
5. 10 μL pipette tips.
6. 1.5 mL microcentrifuge tubes.
7. 15 mL falcon tubes.
8. 50 mL falcon tubes.
9. 12-well plates.
10. T25 flasks.
11. 96-well flat-bottom plates.
12. 96-well round bottom, low-attachment plates.
13. Sterile scissors.
14. Countess Cell Counting Chamber Slides.

2.5 Equipment

1. Shaking water bath set to 37°C.
2. Humidified incubator set at 5% CO_2 and 37°C.
3. Centrifuge equipped with 1.5 mL centrifuge tube and 15 mL centrifuge rotors.
4. Countess Automated Cell Counter.
5. Inverted light microscope.
6. 10, 200, and 1000 μL pipettors.

3 Methods

3.1 Establishment of Endometrial Organoids

1. Collect an endometrial biopsy by dilation and curettage (*see* **Note 3**).
2. Wash endometrial tissue two to three times in DPBS with 1% penicillin–streptomycin. Break down tissue via mechanical mincing with scissors and enzymatic digestion with 2 mL of endometrial dissociation buffer in a 37°C shaking water bath for 30–60 min depending on tissue size.
3. Filter digested tissue through a 40 μM cell strainer into a fresh 50 mL Falcon tube to separate epithelial cells and undigested fragments from digested stromal and immune cell solutions.
4. Reverse wash epithelial components off the filter and transfer to a 1.5 mL microcentrifuge tube. Centrifuge epithelial components at 600 × *g* for 6 min at room temperature.
5. Aspirate the supernatant, gently flick microcentrifuge tube to loosen the cell pellet, then transfer to ice for 2–3 min. Resuspend epithelial components in 15× volume:volume of ice-cold Matrigel (*see* **Note 4**) using a 1 mL pipette tip with approximately 1 mm cut off the end using sterile scissors.
6. Plate Matrigel containing epithelial cells at three domes per well in a 12-well plate at a ratio of 20–25 μL of Matrigel + cell suspension per dome. Incubate organoids at 37°C and 5% CO_2 for 15 min to allow the domes to set.
7. Overlay domes with 750 μL of organoid expansion medium and return to an incubator at 37°C and 5% CO_2. Replace medium every 2–3 days (organoids should form after 48 h of culturing). Passage every 7–10 days as appropriate (*see* **Note 5**).

3.2 Passaging of Endometrial Organoids

1. Without removing the culture medium, gently detach the Matrigel domes into the medium by scraping them back and forth using a 1 mL pipette tip. Collect and transfer the contents of each well into a 1.5 mL microcentrifuge tube, centrifuge at 600 × *g* for 6 min at room temperature.

2. Carefully remove the supernatant (*see* **Note 6**), add 150 μL precool (4°C) Advanced DMEM/F12 and mechanically dissociate organoids by gently pipetting 150 times using a 200 μL pipette. Add 850 μL precooled (4°C) Advanced DMEM/F12 to the tube and centrifuge at 600 × *g* for 6 min at room temperature.
3. Carefully remove the supernatant (*see* **Note 6**), add 150 μL precool (4°C) Advanced DMEM/F12 and mechanically dissociate organoids by gently pipetting 80 times using a 200 μL pipette. Add 850 μL precooled (4°C) advanced DMEM/F12 to the tube and centrifuge at 600 × *g* for 6 min at room temperature.
4. Carefully remove the supernatant. The pellet at this stage should contain very little Matrigel. If Matrigel is still present, estimate the volume.
5. Gently flick microcentrifuge tube to loosen the cell pellet, then transfer to ice for 2–3 min. Resuspended epithelial components in 15× volume:volume of ice-cold Matrigel (*see* **Note 4**), minus the estimated volume of any residual Matrigel which remained in the pellet, with a 1 mL pipette tip with approximately 1 mm cut off the end using sterile scissors.
6. Plate Matrigel containing epithelial cells at three domes per well in a 12-well plate at a ratio of 20–25 μL of Matrigel per dome. Incubate organoids at 37°C and 5% CO_2 for 15 min to allow the domes to set.
7. Overlay domes with 750 μL of organoid expansion medium and return to an incubator at 37°C and 5% CO_2. Replace medium every 2–3 days (organoids should form after 48 h of culturing). Passage every 7–10 days as appropriate (*see* **Note 5**).

3.3 Generation of Epithelial Cell Monolayers from Endometrial Organoids

1. Passage organoids into a 12-well plate, plating three Matrigel domes per well and overlaying with 750 μL organoid expansion medium. Culture for 4 days at 37°C and 5% CO_2 to allow organoid formation; no medium change is required.
2. Pretreatment with E2: Remove the culture medium from each well, then add 750 μL of fresh organoid expansion medium containing 10 nM E2 and incubate for 2 days.
3. Induce the receptive phase by replacing 750 μL of fresh receptive phase medium. Culture for 4 days with a medium change on day two (*see* **Note 7**, Fig. 1). When changing the medium (750 μL), carefully avoid disturbing the Matrigel domes in each well by slowly pipetting the medium against the side of the well.
4. Following hormone treatment, without removing the culture medium, gently detach the Matrigel domes by scraping

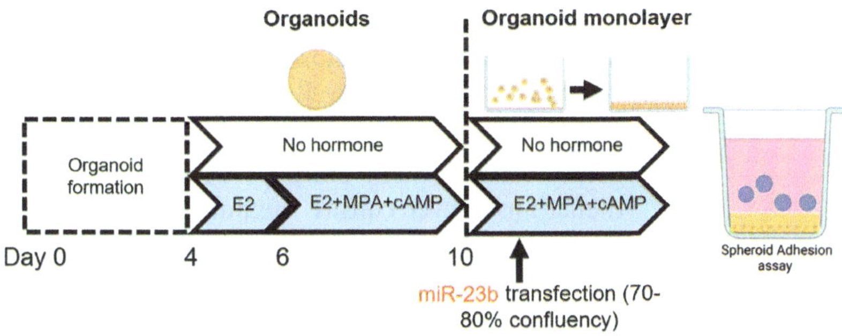

Fig. 1 Establishment of human endometrial organoid epithelial monolayers. Hormone-treated organoids are used to generate epithelial monolayers for spheroid adhesion assays

them back and forth using a 1 mL pipette tip. Collect and transfer the contents of each well (3 domes) into a 1.5 mL microcentrifuge tube, centrifuge at 600 × *g* for 6 min at room temperature.

5. Carefully remove the supernatant (*see* **Note 6**), add 150 μL precooled (4°C) Advanced DMEM/F12, and mechanically dissociate organoids by gently pipetting 150 times using a 200 μL pipette. Add 850 μL precooled (4°C) advanced DMEM/F12 to the tube and centrifuge at 600 × *g* for 6 min at room temperature.
6. Carefully remove the supernatant as much as possible, then digest organoid fragments in each microcentrifuge tube using 300 μL prewarmed TrypLE Express for 20 min at 37°C (pipette up and down to resuspend at 10 min intervals). Stop digestion by adding 700 μL Advanced DMEM/F12 supplemented with 20% FBS, then centrifuge at 600 × *g* for 6 min. Wash the cell pellet once with 1 mL precooled Advanced DMEM/F12, then centrifuge at 600 × *g* for 6 min at room temperature.
7. Meanwhile, use plate coating Matrigel (*see* **Note 4**) to coat culture wells (96-well flat-bottom plate) for epithelial monolayers by diluting Matrigel in precooled Advanced DMEM/F12. Use reverse pipetting with 50 μL to evenly coat the well bottom, remove excess Matrigel, and incubate at 37°C for 15 min.
8. Resuspend organoid epithelial cells in 300 μL epithelial monolayer culture medium.
9. Combine 10 μL cell suspension with 10 μL 0.4% trypan blue stain in a microcentrifuge tube and mix by pipetting up and down a few times. Gently pipette 20 μL of solution into the half-moon-shaped loading area of a Countess cell counting chamber slide. Transfer the chamber slide to the Countess to determine cell viability and concentration.

10. Seed 1×10^5 cells per well in a 96-well plate precoated with Matrigel. Add epithelial monolayer culture medium to a final volume of 150 µL per well (*see* **Note 8**).
11. The following day, remove non-adherent and dead cells by gently washing with 100 µL DPBS. Replace with fresh epithelial monolayer culture medium and continue culturing at 37°C and 5% CO_2 until cells reach 70%–80% confluency, typically achieved within 2 days post-seeding, before proceeding with downstream applications.

3.4 miR Transfection in Organoid Epithelial Cell Monolayers

Here we use miR-23b-3p as an example for miR transfection [7]. The required concentration of the miR mimic (synthetic miR) may vary depending on the specific case (*see* **Note 9**, Fig. 2).

1. Once the epithelial cell monolayer reaches 70%–80% confluency, prepare the miR-23b-3p transfection mix. For a 100 µL reaction, label two nuclease-free, sterile 1.5 mL microcentrifuge tubes as "Lipofectamine" and "miR-23b-3p". Add 50 µL Opti-MEM to each tube. To the Lipofectamine tube, add 3 µL Lipofectamine™ RNAiMAX reagent and mix gently. To the miR-23b-3p tube, add 1 µL miR-23b-3p stock and mix gently. Transfer the miR-23b-3p solution to the Lipofectamine tube, mix gently by pipetting, then incubate at room temperature for 5 min.

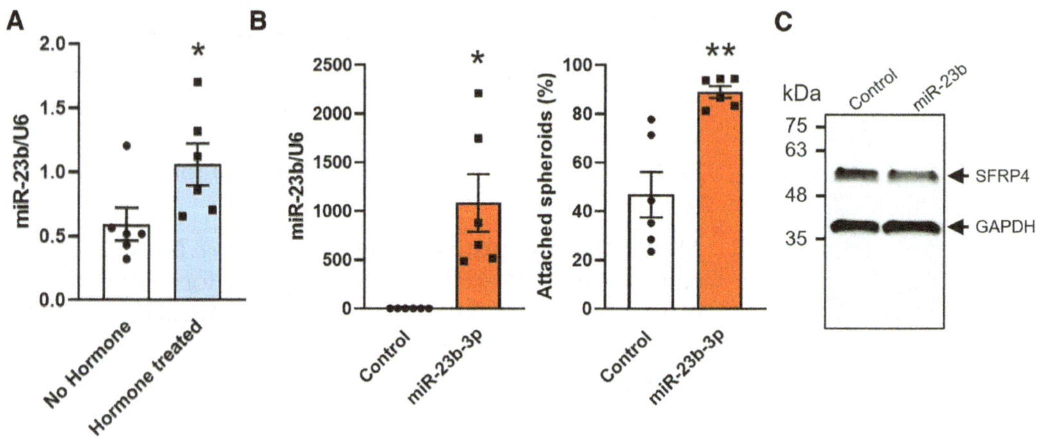

Fig. 2 The effects of miR-23b-3p overexpression on spheroid adhesion. (**A**) miR-23b-3p expression is significantly increased in fertile organoid-derived epithelial monolayers following E2 + MPA + cAMP treatment compared to controls. (**B**) miR-23b-3p overexpression is confirmed by qPCR ($n = 6$). Its overexpression in fertile organoid-derived epithelial monolayers significantly increases HTPC spheroid adhesion. (**C**) Immunoblotting reveals that overexpression of miR-23b-3p in fertile organoid-derived epithelial monolayers reduces its target SFRP4 expression compared to control. GAPDH is used as a loading control ($n = 4$). Data presented as mean ± SEM. *$P < 0.05$, **$P < 0.01$. E2 estrogen, MPA medroxyprogesterone acetate, cAMP cyclic adenosine monophosphate. (The figure has been modified from Barton et al. [7])

2. Prepare another set of tubes for the scramble control, replacing the miR mimic with a non-targeting, randomized nucleotide sequence at the same final concentration (*see* **Note 10**).
3. During the 5-min incubation, gently wash the epithelial monolayer twice with 150 μL DPBS, ensuring that the attached cells are not disturbed. To achieve this, position the pipette tip against the side of the culture well and slowly release the DPBS.
4. Remove DPBS and add transfection medium. For each well of a 96-well plate, add 150 μL of transfection medium. Incubate the cells for 24 h at 37°C and 5% CO_2, then replace with fresh epithelial monolayer medium. Incubate for a further 48 h at 37°C and 5% CO_2 before performing the spheroid adhesion assay. At this stage, organoid monolayers should reach 100% confluence.

3.5 Generation of Embryo Surrogates

HTPCs, derived from blastomeres of cleavage-stage human embryos [11], are used to generate embryo surrogates (*see* **Note 11**). An initial optimization study has determined that 2500 cells form a compact spheroid approximately the same size as human blastocysts [12].

1. HTPCs are maintained in embryo surrogate medium in a T25 flask pre-coated with 0.5% gelatin substrate solution. To coat, 2 mL of gelatin solution is added into a T25 flask, incubated at 37°C for 1 h, aspirated, and then HTPCs are plated at 1×10^6 cells in 5 mL medium per flask.
2. Once a confluency of 80%–100% is achieved (typically within 1 week) remove complete embryo surrogate culture medium and rinse the T25 flask with 1 mL of DPBS to eliminate residual FBS. Incubate HTPCs in 1 mL prewarmed TrypLE Express at room temperature for less than 3 min, monitoring under a microscope for cell detachment. To neutralize the reaction, add 3 mL embryo surrogate culture medium. Transfer cell suspension to a 15 mL falcon tube and centrifuge at 500 × *g* for 5 min at room temperature. Resuspend the cell pellet in 4 mL of medium and determine cell viability and concentration using the Countess. In our experiments, cell viability consistently exceeds 90%.
3. Transfer a total of 2500 HTPCs in 150 μL of complete embryo surrogate culture medium into each well of a 96-well round-bottom, ultra-low attachment plate to form one spheroid. To calculate the required number of spheroids, prepare 18–22 spheroids for each well of epithelial monolayer available (*see* **Note 12**).
4. Culture HTPCs at 37°C and 5% CO_2 for 48 h to form spheroids. Check the morphology of each spheroid and exclude any

misformed spheroids (*see* **Note 13**) (generally <5%) from the spheroid adhesion experiments (*see* **Note 14**).

3.6 Spheroid Adhesion Assay to Model Embryo Adhesion

1. Using a 1 mL pipette tip with approximately 1 mm cut off the end with sterile scissors to allow spheroids to pass through without damage, transfer 18–22 spheroids per 15 mL Falcon tube and centrifuge at 200 × *g* for 8 min at room temperature. Prepare as many 15 mL Falcon tubes as needed to match the number of wells of the epithelial cell monolayer.
2. Carefully remove as much supernatant as possible and gently resuspend spheroids in 150 μL epithelial monolayer culture medium using a 200 μL pipette tip with approximately 1 mm cut off the end. Gently mix to separate the individual spheroids, which should be visible without magnification.
3. Aspirate the epithelial monolayer culture medium and transfer spheroids from each tube into a corresponding well of the 96-well epithelial cell monolayer culture plate (18–22 spheroids per well). Repeat for all wells and allow spheroids to settle for 5–10 min (Fig. 3).
4. Using an inverted light microscope, count and record the number of spheroids per well as the time 0 spheroid count. Incubate the co-cultures at 37°C for 4 h. After incubation, gently wash the co-cultures once with DPBS by pipetting slowly along the side of the well to remove non-adherent spheroids (*see* **Note 15**). To prevent cell drying, perform the washing step one well at a time. Count the number of attached spheroids and calculate the percentage of adhesion relative to the time 0 spheroid count (*see* **Note 16**).
5. Following the adhesion assay, monolayers can be collected for RNA and protein analysis (*see* **Note 17**). To prepare cells for collection, wash each well twice with 150 μL DPBS, applying increased force to remove any remaining spheroids while ensuring the monolayer remains undisturbed. If some spheroids remain attached, they can be removed by using a 10 μL pipette tip to flick them off the epithelial cell monolayer.

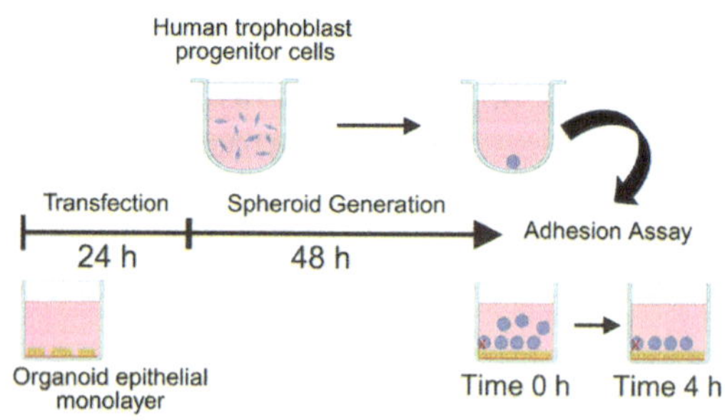

Fig. 3 Schematic representation of the spheroid adhesion assay

4 Notes

1. Endometrial organoids may be used to model other phases of the menstrual cycle using differential hormone treatment protocols. Treatment with E2 for 8 days models the proliferative phase.
2. The human recombinant Rspondin-1 concentration required for organoid expansion media is specific to BioTechne. Growth may not be as robust using other sources, requiring optimization. Likewise, the concentration of Collagenase III required for tissue digestion is specific to Worthington. Endometrial tissue digestion may be insufficient to effectively separate epithelial fragments from stromal cells; using a different supplier or a different concentration may be required.
3. The establishment of organoids requires fresh tissue collected from a clinic, which must be processed immediately. A minimum tissue size equivalent to that of a pea is required to ensure the successful isolation of sufficient endometrial epithelial fragments.
4. Matrigel should be thawed at 4°C until a liquid-like viscosity is achieved; the time to thaw is dependent on the aliquot volume. During Matrigel use, aliquots should be kept on ice whenever not being handled to prevent unwanted warming and associated setting.
5. Organoid size defines the passaging interval. Generally, endometrial organoids should be passaged every 7–10 days and allowed to reestablish within fresh Matrigel droplets, even if organoids are not clearly visible within the Matrigel drop. The nutrients in Matrigel support organoids' growth for a maximum of 14 days; however, overgrowth may compromise cell viability and quality. When overgrown (>500 μm), organoids could fold in on themselves and appear "dark" such that viability is then compromised.
6. Organoids are cultured in Matrigel, and after centrifugation, the Matrigel forms a tail above the pellet, which makes it challenging to remove supernatant close to the cell pellet. Change the angle of the pipette to aspirate the supernatant from a distance, away from the Matrigel tail, and perform this step slowly to help remove the supernatant.
7. It is recommended to include a non-receptive control. This can be achieved by preparing additional organoids and treating them with E2 alone or no hormone treatment, following the same protocol in "Generation of epithelial cell monolayers from endometrial organoids." When investigating specific targets, it is important to determine whether their expression

differences between non-receptive and receptive organoid monolayers reflect the changes observed in the luminal epithelium between the proliferative (non-receptive) and mid-secretory (receptive) phases of human endometrium, as we demonstrated with miR-23b-3p (Fig. 2A) [7].

8. Each well of a 12-well plate (containing 3 domes of organoids) yields enough cells for approximately 6–10 wells of a 96-well plate, depending on organoid density. Ensure even plating for reproducible monolayer formation. It can be challenging to achieve even monolayers in a 96-well plate. The cross-shaking technique does not work well for small wells. Instead, it is recommended to mix the cell suspension within each well and allow it to settle for ~10 min before transferring the plate to the incubator for culture. Avoid sudden movements, as they can cause cells to cluster in the center. It is also recommended to generate extra wells as a backup.
9. For miR transfection, it is essential to confirm its overexpression by qPCR and, ideally, the downregulation of known miR targets by qPCR/immunoblotting [7]. The optimal concentration of each miR for transfection may vary.
10. Lipofectamine™ RNAiMAX Transfection Reagent is used as it is specifically designed for in vitro delivery of microRNA and siRNA into cells. If alternative transfection reagents are used, the protocol should be adjusted accordingly based on the manufacturer's specifications.
11. HTPCs were provided by Professor Susan Fisher and Dr. Olga Genbacev (University of California, San Francisco) and are therefore not readily available. As an alternative, HTR-8/SVneo trophoblast cells (available via ATCC) can be used to generate spheroids following the same workflow as for HTPCs. For each spheroid, 2000 cells are seeded per well in standard HTR-8/SVneo growth medium (RPMI 1640 supplemented with 10% FBS); all remaining procedures are identical to those described for HTPCs.
12. Prepare more spheroids than required (at least 5% extra), as some may be lost due to malformation and during transfer. A highly confluent T25 flask contains enough cells to produce between 500 and 2000 spheroids depending on cell viability.
13. A properly formed spheroid should be round in shape, approximately the size of a human blastocyst (0.1–0.2 mm).
14. The spheroids and transfected epithelial monolayers must be prepared simultaneously for the spheroid adhesion assay. Begin Step 2 of "Generation of embryo surrogates" on the same day as the transfection medium change for epithelial monolayers.

15. Do not count spheroids attached to the edges of the well, as their firm attachment may be caused by physiological contact with the wall, rather than the adhesive capacity of the epithelial cells. When counting attached spheroids at 4 h, it is important to distinguish floating spheroids, which can be achieved by gently shaking the well and observing their movement relative to the culture surface.
16. Following potential spheroid loss during transfer and exclusion of spheroids attached to the edges of a well, a minimum of 10 spheroids per well is required to initiate the adhesion assay, as lower spheroid numbers increase variability in adhesion percentage. In controls, adhesion typically reaches ~50%.
17. Organoid epithelial cells from a single well of a 96-well plate are sufficient to isolate over 300 ng RNA for qPCR. To extract protein for quantification and immunoblotting, at least two wells of cells are required.

Acknowledgments

This work was supported by a National Health and Medical Research Council of Australia Ideas Grant (APP2038919) to E.D. and an Australian Research Council Discovery Project to E.D. and W.Z. (DP250103322).

References

1. Turco MY, Gardner L, Hughes J, Cindrova-Davies T, Gomez MJ, Farrell L, Hollinshead M, Marsh SG, Brosens JJ, Critchley HO (2017) Long-term, hormone-responsive organoid cultures of human endometrium in a chemically defined medium. Nat Cell Biol 19(5):568–577
2. Boretto M, Cox B, Noben M, Hendriks N, Fassbender A, Roose H, Amant F, Timmerman D, Tomassetti C, Vanhie A (2017) Development of organoids from mouse and human endometrium showing endometrial epithelium physiology and long-term expandability. Development 144(10):1775–1786
3. Hewitt SC, Wu S-p, Wang T, Ray M, Brolinson M, Young SL, Spencer TE, DeCherney A, DeMayo FJ (2022) The estrogen receptor α cistrome in human endometrium and epithelial organoids. Endocrinology 163(9):bqac116
4. Hewitt SC, Wu S-P, Wang T, Young SL, Spencer TE, DeMayo FJ (2022) Progesterone signaling in endometrial epithelial organoids. Cells 11(11):1760
5. Garcia-Alonso L, Handfield L-F, Roberts K, Nikolakopoulou K, Fernando RC, Gardner L, Woodhams B, Arutyunyan A, Polanski K, Hoo R (2021) Mapping the temporal and spatial dynamics of the human endometrium in vivo and in vitro. Nat Genet 53(12):1698–1711
6. Zhou W, Van Sinderen M, Rainczuk K, Menkhorst E, Sorby K, Osianlis T, Pangestu M, Santos L, Rombauts L, Rosello-Diez A, Dimitriadis E (2024) Dysregulated miR-124-3p in endometrial epithelial cells reduces endometrial receptivity by altering polarity and adhesion. Proc Natl Acad Sci U S A 121(41):e2401071121. https://doi.org/10.1073/pnas.2401071121
7. Barton S, Zhou W, Santos LL, Menkhorst E, Yang G, Teh WT, Ang C, Lucky T, Dimitriadis E (2023) miR-23b-3p regulates human endometrial epithelial cell adhesion implying a role in implantation. Reproduction 165(4):407–416
8. Mardon, H, Grewal, S, Mills, K, 2007. Experimental models for investigating implantation of the human embryo, In: Semin Reprod Med. © Thieme Medical Publishers, Stuttgart, Germany, vol 06pp. 410–417

9. Carver J, Martin K, Spyropoulou I, Barlow D, Sargent I, Mardon H (2003) An in-vitro model for stromal invasion during implantation of the human blastocyst. Hum Reprod 18(2):283–290
10. Dimova T, Alexandrova M, Vangelov I, You Y, Mor G (2004) The modeling of human implantation and early placentation: achievements and perspectives. Hum Reprod Update 31(2):133–163
11. Zdravkovic T, Nazor KL, Larocque N, Gormley M, Donne M, Giritharan G, Bernstein HS, Wei G, Hebrok M, Zeng X (2015) Human stem cells from single blastomeres reveal pathways of embryonic or trophoblast fate specification. Development 142(23): 4010–4025
12. Evans J, Walker KJ, Bilandzic M, Kinnear S, Salamonsen LA (2020) A novel "embryo-endometrial" adhesion model can potentially predict "receptive" or "non-receptive" endometrium. J Assist Reprod Genet 37(1):5–16

Chapter 22

Targeted Degradation of Mouse Embryo Proteins Using Trim-Away

Thomas Nolte, Steffen Israel and Michele Boiani

Abstract

Trim-Away is an antibody-based method for the degradation of endogenous cellular proteins without prior genetic manipulation. From the effect that protein degradation has on the phenotype, one can infer the function of the protein-coding gene. Among the strengths of Trim-Away, the authors highlight the possibility to target embryonic proteins that have not yet been produced or are not yet functionally required at the time of antibody delivery. Conversely, a limitation of Trim-Away is the possibility that while the target protein is being degraded, de novo translation may replenish it, whereby the two processes offset each other, and the phenotype is inconspicuous. Appropriate controls, as described, are therefore essential for the correct interpretation of Trim-Away results. With these considerations in mind, the authors provide guidance on how to set up a Trim-Away experiment in fertilized mouse oocytes, using microinjection as the delivery method and the epithelial CADHERIN protein as an example.

Keywords Antibody, E-CADHERIN, Mouse, Oocyte, Protein of interest (POI), *Trim21* mRNA, TRIM 21 protein

1 Introduction

Traditionally, nucleic acid targeting via gene locus knockout and RNA knockdown has been the method of choice to inactivate gene expression and thereby understand the function of those genes through the phenotypic consequences of inactivation. However, the peculiarity of oocytes and embryos makes these methods of gene targeting difficult. Oocytes mature inside the ovary for weeks or months, depending on the species, and store proteins, some of which are long-lived and are required during embryogenesis. Put simply, mature oocytes are not in a steady state of gene expression, whereby an intervention on the DNA or mRNA would immediately have an impact on the protein; such effects may take time. This delay creates two problems: (1) In general, the time between gene locus knockout and phenotypic assessment allows for compensatory mechanisms and secondary phenotypes to ensue [1], which

Alexandra J. Harvey and David K. Gardner (eds.), *The Preimplantation Embryo*, Methods in Molecular Biology, vol. 3038, https://doi.org/10.1007/978-1-0716-5292-3_22,

can confound the interpretation of a protein's function [2]; (2) in particular cases, extremely long-lived proteins persist for a long time after the knockout [3]. Moreover, genes can be functionally required at multiple stages in development, and therefore, the knockout may only reveal the first stage at which the gene function is required, while the later functions may go unnoticed. Compensatory mechanisms and secondary phenotypes are less of an issue in RNAi-mediated and morpholino-mediated knockdowns, because these methods operate faster than DNA knockout. However, RNA-based knockdowns are almost powerless toward the protein that has already been translated, except in cases where the protein has a short half-life. As a result, this protein can compensate for and mask the phenotype that the RNA knockdown may cause. These considerations converge on the need to target the gene product directly at the end of the gene expression cascade, that is, the protein.

Immunoglobulin proteins can be exploited to recognize and bind other proteins, such that inactivation or degradation of the bound protein can be achieved. While there are various classes of immunoglobulins, here we mainly consider those of class G (IgG), commonly known as antibodies. In nature, antibodies have existed for as long as there has been multicellularity. In medicine, antibodies have been exploited since the late 1800s in the context of vaccination [4]. In laboratory science, antibodies have been produced industrially thanks to hybridoma technology since 1975 [5]. In embryology, the possibility of taking advantage of antibodies was recognized four decades ago, when antibodies were either microinjected [6–8] or co-incubated with embryos [9–12] to inactivate the function of proteins of interest (POI). However, it was not until 2017 that this basic approach evolved significantly [13], making it possible to not only bind and mask a POI, but also to trigger its intracellular degradation. To this end, scientists hijacked the natural defense system that is built into cells. This system hinges on the RING E3 ligase and cytosolic antibody receptor known as 'Tripartite motif containing 21', TRIM21, which binds with high affinity to the Fc (Fragment, crystallizable) domain of antibodies [14–17].

Normally, TRIM21 becomes engaged when pathogens, e.g., viruses, bypass the frontline defense of the immune system and manage to enter cells as immunoglobulin-coated pathogens. These are recognized and bound by TRIM21 in the cytosol and directed to the proteasome for destruction [15, 18]. It was reasoned that in place of the pathogen, there could be an intracellular protein, so that the system could be "hijacked" to degrade a particular protein of interest (POI), provided the levels of TRIM21 and antibody inside the cell were high enough. This methodology was first reported by Clift and colleagues in 2017 in mouse oocytes, and as a result of the intervention, which was to wipe away the protein–antibody–TRIM21 complex via the proteasome, the method was nicknamed

"Trim-Away" [13] (*see* **Note 1**). In mechanistic terms, it is not yet clear how exactly TRIM21 works, with some studies proposing that the TRIM21 protein itself is ubiquitinated on lysines [15] and other studies proposing that the POI, but not TRIM21, is ubiquitinated on lysines [19] (*see* **Note 2**). Uncertainties about the mechanism aside, Trim-Away has nonetheless been demonstrated to be effective in several cell types in addition to oocytes, including HEK293T cells [20], N2A cells [21], and HeLa cells [22].

Protein degradation with TRIM21 and the specific antibody occurs within minutes after application [13]. Trim-Away is therefore an attractive method for studying maternal proteins and early developmental processes, without the necessity of prior elimination through knockout models, such as CRISPR–Cas, or downregulation through RNA interference. However, oocytes and embryos do not express *Trim21* and *IgG* genes at significant levels. Therefore, both TRIM21 and the antibody of choice must be delivered experimentally. Two established methods for delivering the antibody and TRIM21 into oocytes or embryos are microinjection [13, 23] and electroporation [24, 25]. Recent endeavors have been aimed at creating strains of mice that express a *Trim21* transgene, thereby producing significant levels of protein within oocytes [26]. Although the antibody remains to be introduced in this instance, the availability of *Trim21*-expressing oocytes is advantageous since the *Trim21* mRNA does not have to be synthesized in the laboratory and microinjected. At present, when it comes to oocytes and early blastomeres, the method of choice is microinjection, and TRIM21 is typically provided not as a protein but as precursor mRNA (including a sequence that encodes a fluorescent tag such as Cherry). *Trim21* mRNA can either be delivered separately prior to the antibody or together with the antibody (chemical cocktail). Since TRIM21 and the antibody can be stable in the cytoplasm for some time (TRIM21 has a half-life of 75–263 h; IgG light and heavy chains have half-lives of 56 and 26–251 h, respectively; [27]), this stability can be exploited to target POIs that appear later in preimplantation development, as shown in the case of embryonic POIs that appear at the morula–blastocyst stage, such as TEAD4 [28].

In this chapter, the authors describe methods to generate *Trim21* mRNA, and its co-microinjection with an antibody of choice (Fig. 1) to inactivate E-CADHERIN [29] (*see* **Note 3**), the product of gene *Cdh1* in mice, which produces an easily recognizable phenotype (*see* **Note 4**), noted by some as a maternal-effect gene [30]. Here, the authors show that E-CADHERIN is amenable to inactivation using Trim-Away, with the expected cellular phenotype. However, caution must be exercised. Although TRIM21 has a higher binding affinity for the antibody, it can nonetheless bind other molecules as well (Fig. 2), and this should be kept in mind when designing experimental controls and interpreting results. With reference to this and other cautionary aspects, the authors describe the strengths and limitations

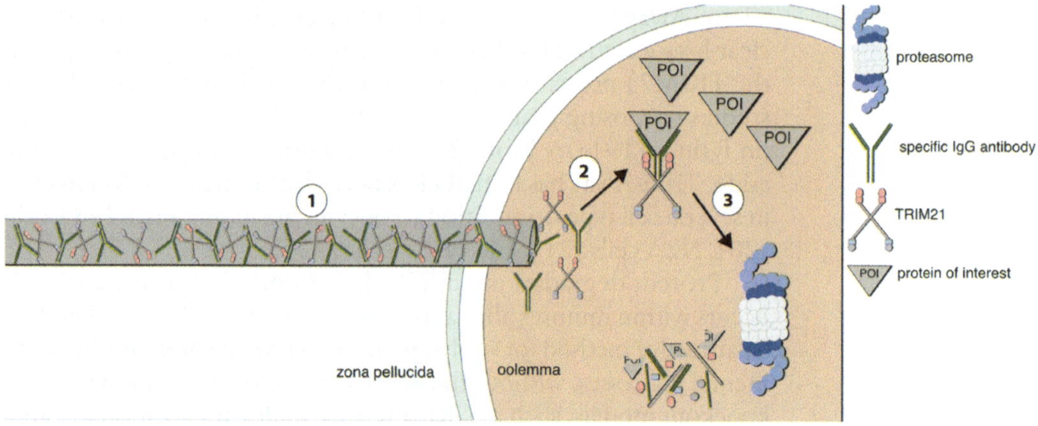

Fig. 1 Schematic principle of Trim-Away. TRIM21 protein or its coding mRNA is microinjected (1) into the freshly fertilized oocyte, or zygote, along with a specific IgG antibody directed against a POI. After microinjection, the antibody, TRIM21, and the POI form a complex (2), which is then ubiquitinated and destroyed in the proteasome (3). Abbreviations: IgG immunoglobulin G, POI protein of interest

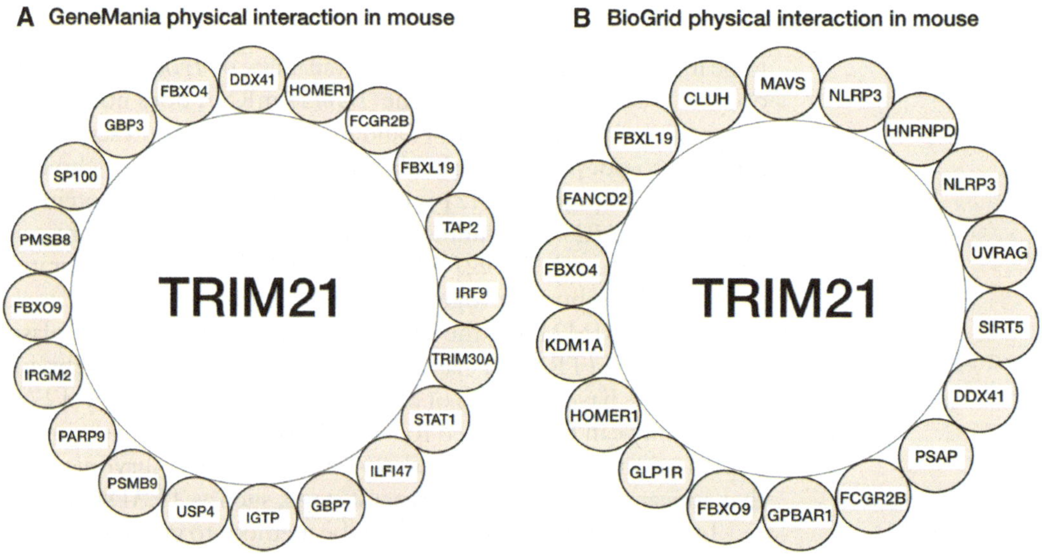

Fig. 2 Molecular interactions of TRIM21 with other proteins. TRIM21 not only binds to the Fc region of the antibody, but also interacts with a variety of cytosolic proteins, as revealed by queries (November 2024) to GeneMania (**A**) and BioGRID (**B**) for both physical and high-throughput interactions. This network is important to keep in mind when examining the effects of Trim-Away, as the effect observed may be due to the interactions of TRIM21 with other proteins rather than with the antibody

of Trim-Away based on their own experience in studying embryonic proteins in mice, using off-the-shelf antibodies delivered into zygotes by microinjection [23, 28, 31]. Apart from these cautionary notes, use of Trim-Away is limited only by the availability of antibodies to the POI and by the reader's imagination.

2 Materials

For a successful Trim-Away experiment conducted in mouse embryos, TRIM21 protein or its mRNA must be delivered into oocytes along with a specific antibody that targets an intracellular POI. Chemical materials and reagents (e.g., mRNA and antibody) must be prepared in advance and aliquoted ready for use. By way of example, the authors outline the application of Trim-Away to inactivate E-CADHERIN [29] in mouse zygotes (Fig. 3).

2.1 Chemical Materials and Instruments for Off-the-Shelf Antibody Purification

1. Based on the POI's amount (Fig. 4), structure and subcellular location (*see* **Note 5**), and accessibility (*see* **Note 6**), identify a suitable antibody (*see* **Note 7**), e.g., anti-E-CADHERIN (BD Biosciences).
2. Denatured antibody for the POI to serve as a negative control (*see* **Note 8**).
3. Amicon Ultra-0.5 150 kDa centrifugal filter devices.
4. MilliQ water.
5. NanoDrop One/One^{C} microvolume spectrophotometer.
6. Low-protein-binding 2 mL microcentrifuge tubes.
7. 1000 μL pipette tips and single channel pipettor.
8. Centrifuge capable of accommodating 2 mL tubes and a max speed of 21,000 × *g*.

2.2 Chemical and Microbiological Materials for Trim21 mRNA Synthesis (see Note 9)

1. pGEMHE-*mCherry-Trim21* plasmid (Addgene Plasmid #105522).
2. Bacteria (*E. coli* Top 10 strain).
3. Water bath.
4. LB (Luria–Bertani) medium.
5. Ampicillin antibiotic.
6. LB+ Ampicillin (100 μg/mL) agar plates.
7. Spreading bar.
8. Bunsen burner.
9. Dry oven set to 37°C.
10. 1.5 mL microcentrifuge tubes.
11. Sterile 12 mL tube for bacterial growth.
12. 200 μL RNase/DNase-free PCR tubes.
13. Plasmid Miniprep Kit (e.g., Zymo Research containing P1 buffer, P2 buffer, P3 buffer, Endo-Wash buffer, Plasmid-Wash buffer, DNA elution buffer, Zymo-Spin™ IIN collection tubes, and Zymo-Spin™ IIN columns).
14. 95% ethanol.

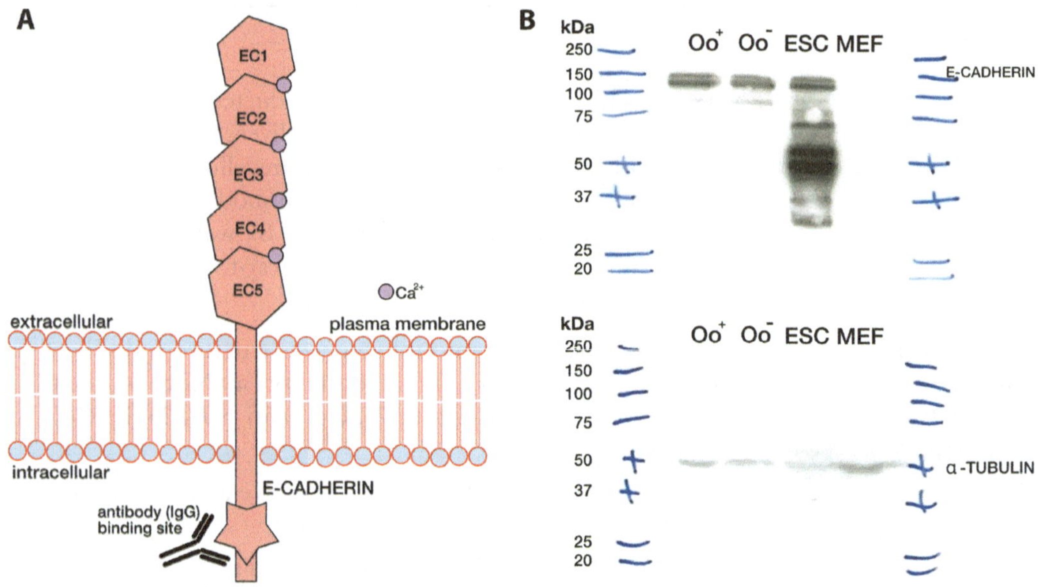

Fig. 3 Schematic structure of E-CADHERIN protein and test of antibody specificity. The antibody used in this Trim-Away example binds the intracellular domain of E-CADHERIN. (**A**) The structure of mature mouse E-CADHERIN consists of 5 CADHERIN repeats (EC1–EC5) in the extracellular domain, one transmembrane domain, and an intracellular domain, for a total of 728 amino acids (additional 23 and 133 amino acids form the signal peptide and the propeptide, respectively, which are cleaved to generate the mature protein). The EC repeats bind Calcium ion and account for its cell adhesion properties. (**B**) The antibody (BD Biosciences cat no. 610182) was checked for specificity by Western blot using lysates of cells where E-CADHERIN is expected to be present (OO+, OO−, ESCs) and lysates of cells where E-CADHERIN is not expected to be present (MEFs). The anti-E-CADHERIN antibody produces a 120 kDa band in oocytes, flanked by a slightly larger (130 kDa) band that most likely is the precursor of E-CADHERIN (including the signal peptide and propeptide). In ESCs, there are multiple bands in addition to the 120–130 kDa doublet, whose nature is unclear. These additional bands are not due to the secondary antibody, which is the same for both anti-E-CADHERIN and anti-α-TUBULIN (loading control). In MEFs, there is no E-CADHERIN band because the protein is known not to be expressed. The loading control produces a band at 50 kDa in all samples, as expected from the use of α-TUBULIN antibody (cat. no. T6199, Sigma-Aldrich Chemie GmbH). Abbreviations: *OO+*, oocytes with intact zona pellucida; *OO−*, oocytes with zona pellucida removed; *ESCs* mouse embryonic stem cells, *MEFs* mouse embryonic fibroblasts

15. Single-channel pipettors and disposable pipette tips.
16. Restriction enzymes (SwaI).
17. MilliQ water.
18. PCR CleanUp Kit (e.g., Takara Bio NuceloSpin Gel and PCR CleanUp kit containing Binding buffer NTI, Wash buffer NT3, Elution buffer NE, NucleoSpin® Gel and PCR clean-up columns, and 2 mL collection tubes).
19. mMessage mMachine T7 Transcription kit, containing Nuclease-free water, 2× NTP/CAP, 10× reaction buffer, RNA polymerase enzyme mix, and Turbo DNAse.
20. 200 μL RNase/DNase-free PCR tube.

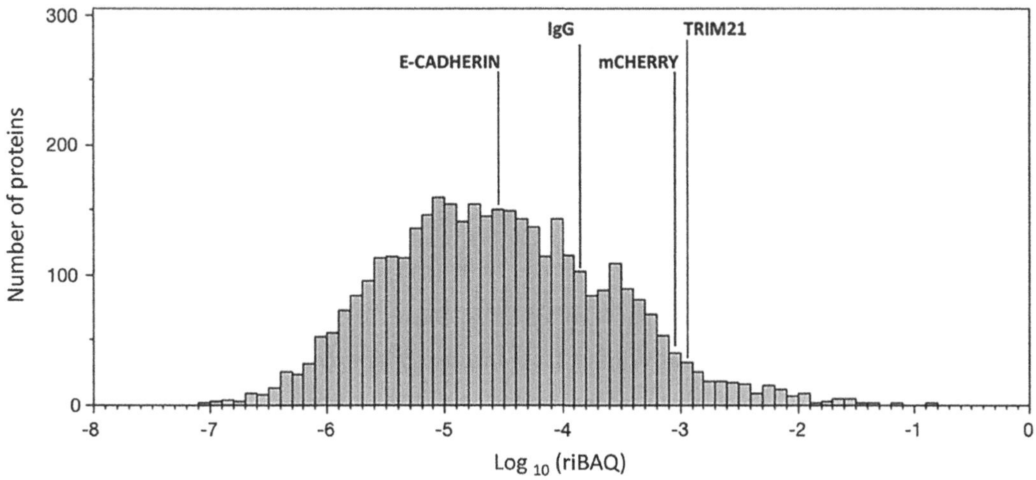

Fig. 4 Relative abundance of POI and Trim-Away cocktail reagents. Zygotes were microinjected with the Trim-Away cocktail, or with IgG antibody alone (1 mg/mL) or *mCherry-Trim21* mRNA alone (0.15 mg/mL) and cultured as described. Twenty-four hours later, embryos were collected at the 2-cell stage (approximately 200 embryos per sample) and processed for mass spectrometric analysis, as per our previously published pipeline [28]. Protein abundances were calculated using the iBAQ algorithm (for more information, see [28]). iBAQ values were normalized to the sum of the iBAQ values, thereby generating relative iBAQ (riBAQ) values, which are roughly comparable to molarities. Shown is the genome-wide distribution of the riBAQ values on a logarithmic scale, and the number of proteins with those riBAQ values. The full set of data plotted is available at https://doi.org/10.6084/m9.figshare.28788236.v1. The amount of TRIM21 and IgG antibody is in excess of the amount of E-CADHERIN, which suggests that it is possible to degrade all POI, in the case of E-CADHERIN. This, however, may or may not be the case for other POIs, depending on their abundance and on how much reagent is microinjected. Abbreviations: *iBAQ* intensity-based absolute quantification, riBAQ relative iBAQ, *IgG* immunoglobulin G, *POI* protein of interest

21. Quick-RNA Microprep Kit (e.g., Zymo Research, containing RNA Lysis buffer, RNA Prep buffer, RNA Wash buffer, DNase/RNase-free water, DNase (lyophilized), DNA Digestion buffer, Zymo-Spin® IC columns, and collection tubes).
22. 100% ethanol.
23. NanoDrop spectrophotometer.
24. PCR thermal cycler.
25. Rotating/shaker incubator set to 37°C.

2.3 Instruments for Microinjection of Mouse Zygotes and Embryos

The authors have been using the following setup for many years.

1. Inverted microscope fitted with 4× and 40× DIC (Nomarski) ELWD objectives and suitable light sources for bright field and fluorescence.
2. Narishige MMO-4 Micromanipulator.
3. Piezo drill (e.g., PMM-150, Prime Tech Ltd).

4. Gilmont GS-1200 2 mL micrometer syringe for hydraulic control of the holding and injection micropipettes.
5. Borosilicate glass capillaries: outer diameter, 1 mm; inner diameter, 0.78 mm.
6. Elemental mercury (Hg) and a microliter syringe (Hamilton 701 N) for loading the microinjection capillary with Hg (*see* **Note 10**).
7. Microforge (e.g., F-1000 de Fonbrune).
8. Capillary puller (e.g., Sutter P-97).
9. Humidified CO_2 incubator set to 37°C and 6% CO_2.
10. Fluorescent tracer (OGDB, Oregon Green™ 488 Dextran beads; 70 kDa MW).
11. Glass coverslips (at least 24 × 50 mm).

2.4 Homemade Media and Plasticware for Zygote Collection/Microinjection and Embryo Culture

1. Glass pipette for isolation and manipulation of mouse oocytes and preimplantation embryos (*see* **Note 11**).
2. Zygote collection and microinjection medium (Hepes-buffered Chatot, Ziomek, Bavister (H-CZB) medium) supplemented with 0.2% polyvinylpyrrolidone (PVP) 40 kDa: KH_2PO_4, 161 mg/L; $CaCl_2$ $2H_2O$, 251 mg/L; $MgSO_4$, 142 mg/L; NaCl, 4760 mg/L; Hepes (free acid), 4760 mg/L; KCl, 360 mg/L; EDTA (di-Sodium salt), 40 mg/L; Glutamine, 146 mg/L; Sodium lactate, 3509 mg/L; Sodium pyruvate, 30 mg/L; Glucose, 1000 mg/L; $NaHCO_3$, 420 mg/L; Phenol red, 5 mg/L; PVP 40 kDa, 2000 mg/L (*see* **Note 12**).
3. Hyaluronidase dissolved in H-CZB medium at 50 International Units (I.U.)/mL.
4. Embryo culture medium (potassium (K)-supplemented simplex optimized medium containing amino acids (KSOMaa): $KH_2PO_4$48 mg/L; $CaCl_2$ $2H_2O$, 251 mg/L; $MgSO_4$ $7H_2O$, 49 mg/L; NaCl, 5552 mg/L; KCl, 186 mg/L; EDTA (di-Sodium salt), 4 mg/L; Glutamine, 146 mg/L; Sodium lactate, 1121 mg/L; Sodium pyruvate, 22 mg/L; Glucose, 36 mg/L; $NaHCO_3$, 2100 mg/L; Phenol red, 5 mg/L; bovine serum albumin, 2000 mg/L; nonessential free amino acids, 0.5×; essential free amino acids, 0.5×; Gentamicin sulfate, 50 mg/L (*see* **Note 12**).
5. OGDB (*see* **Note 13**) dissolved in MilliQ water at 5 mg/mL (stock solution).
6. Micromanipulation chamber, made from a 60 mm plastic dish with a 20 mm hole drilled in the middle and a round coverslip glued on top (using Amazing Goop Marine Goop Glue).
7. Four-well culture plates.
8. 35 mm plastic culture dishes.

9. Surgical scissors.
10. Surgical tweezers.
11. Warming plate.
12. Zygotes (fertilized oocytes) collected from B6C3F1 female mice 8–10 weeks old mated to CD1 stud males >12 weeks old.

3 Methods

Unless otherwise specified, all methods are carried out at room temperature (RT). In the authors' laboratory, the micromanipulation room is permanently heated to 27°C (81 °F) via central air conditioning.

3.1 Antibody Purification

Since commercial antibodies may contain preservatives such as sodium azide, it is essential to clean the antibody before use in live cells. Antibodies can be purified via Amicon (150 kDa) or protein G columns. The authors prefer to use Amicon as outlined below.

1. Transfer ~0.1 mg antibody (*see* **Note 14**) and up to 500 μL MilliQ water into the column and centrifuge at 4°C for 15 min at 14,000 × *g*.
2. Discard the flow through. Place the column into a fresh collection tube and add 400 μL MilliQ water to the column and gently mix.
3. Centrifuge for 15 min at 4°C and 14,000 × *g*.
4. Discard flow through. Repeat the centrifugation again for 15 min at 4°C and 14,000 × *g*.
5. After centrifugation, place the column upside down in a new collection tube and centrifuge for 2 min at 4°C at 1000 × *g* to retrieve the purified antibody.
6. Measure the purified antibody concentration using a NanoDrop spectrophotometer (wavelength 280 nm).
7. Make aliquots (4 μL in low protein binding tubes) for storage at 4°C until use (*see* **Note 15**).
8. The same cleaned antibody directed against the POI can also be used as a negative control, provided it is denatured. Given the denaturation, the antibody should not produce any cellular effect in association with TRIM21 (*see* **Note 8**). Using heat as a denaturing agent, the temperature should be above 71°C [32]. For Trim Away, the authors denature the antibody by heating at 94°C for 30 min directly before use, using a PCR thermal cycler.

3.2 *Trim21* mRNA Synthesis

Trim21 mRNA is synthesized in vitro from a bacterial plasmid. For this, the authors use a pGEMHE-*mCherry-Trim21* plasmid, which is transformed and amplified using *E. coli*. Here we describe an optimized protocol where preculture preparation uses the Zymo Research Plasmid Miniprep-Classic and Takara PCR CleanUp kits, with mRNA synthesis carried out using an Invitrogen mMessage mMachine™ T7 Transcription Kit and resultant *mCherry-Trim21* mRNA isolated using a Zymo Quick-RNA Microprep Kit.

1. For transformation, thaw 1 μL plasmid (pGEMHE-*mCherry-Trim21*) and competent *E. coli* Top10 cells on ice. Transfer 1 μL plasmid to the competent *E. coli* Top10 cells and incubate for 30 min on ice.
2. After incubation, heat-shock the cells for 90 s at 42°C and then place them directly back on ice. Add 1 mL cold (4°C) LB medium to the cells and incubate for 60 min at 37°C.
3. After incubation, transfer 50 μL cells to LB+Amp agar plates. Sterilize a spreader bar using ethanol, then briefly pass it through the flame of a Bunsen burner. Allow the bar to cool before spreading the cells across the agar plates. Incubate the plates overnight in a dry oven at 37°C.
4. Prepare precultures of *E. coli* Top 10 pGEMHE-mCherry-Trim21 by placing 5 mL LB+ Ampicillin (100 μg/mL) into a sterile 12 mL tube. Select a colony of *E. coli* Top10 pGEMHE-mCherry-Trim21 using a sterile and disposable tip and add it to the tube. Incubate the preculture overnight at 37°C, on a shaker at 120 revolutions per minute.
5. For Plasmid wash buffer preparation as part of the Plasmid Miniprep-Classic kit, add 96 mL 95% ethanol to the concentrated Plasmid wash buffer.
6. Using a fresh 1.5 mL tube, centrifuge the precultures for 30 s at max speed and discard the supernatant.
7. Then add 200 μL P1 buffer and resuspend the pellet by pipetting or using a vortex.
8. For cell lysis, add 200 μL P2 buffer, invert 2–4 times, and proceed to the next step within 1–2 min.
9. Neutralize the reaction by adding 400 μL P3 buffer, mix thoroughly but do not vortex, and incubate it for 2 min at RT and 11,000–16,000 × *g* (Author's recommendation: 13,000 × *g*).
10. Centrifuge the mixture at RT for 2 min. Meanwhile, transfer the Zymo-Spin™ IIN column into the Zymo-Spin™ IIN collection tube.
11. Without disturbing the pellet, transfer the supernatant into the column and centrifuge for 30 s at 11,000–16,000 × *g*, then discard the flow through.

12. Add 200 μL Endo wash buffer, centrifuge for 30 s at 11,000–16,000 × *g*, discard the flow through.
13. Add 400 μL Plasmid wash buffer, centrifuge for 1 min at 11,000–16,000 × *g* and place the column into a fresh 1.5 mL tube.
14. Add 20 μL DNA elution buffer or MilliQ water directly onto the membrane and centrifuge for 30 s at 11,000–16,000 × *g*.
15. Apply the flow through to the column and centrifuge again.
16. Measure the concentration of plasmid using a NanoDrop spectrophotometer. Sufficient plasmid DNA is essential for the subsequent linearization process. A concentration of at least 250 ng/μL is recommended.
17. Linearize the plasmid by restriction enzyme digest (RE-digest) by combining 1500 ng plasmid with 1 μL RE-buffer, 2 μL RE (SwaI; fast digest), with the reaction made up to 10 μL with MilliQ water. Let the reaction take place on a PCR thermal cycler for 50 min at 37°C (*see* **Note 16**).
18. Purify linearized plasmids using a PCR CleanUp Kit. First, add NTI buffer to the sample in a ratio of 2:1 and mix by pipetting.
19. Transfer the sample onto the NucleoSpin® Gel and PCR clean-up column and centrifuge at 11,000 × *g* for 30 s. Discard the flow through.
20. Wash the membrane by adding 700 μL NT3 Wash buffer and centrifuge again at 11,000 × *g* for 30 s. Discard the flow through (*see* **Note 17**).
21. To dry the membrane, centrifuge at 11,000 × *g* for 1 min. Add 15 μL NE or MilliQ water to the membrane. Incubate for 1 min at RT and centrifuge at 11,000 × *g* for 1 min.
22. Apply the flow through to the column and centrifuge again.
23. Measure the concentration using a NanoDrop spectrophotometer. In order to obtain sufficient plasmid DNA for the subsequent mRNA synthesis, it is essential to achieve a concentration of at least 50 ng/μL.
24. First, thaw frozen mMessage mMachine T7 Transcription Kit reagents, placing the enzyme mix on ice.
25. Assemble a 20 μL reaction as follows (pipetting order to be followed strictly): Pipette template DNA (~800 ng) into a fresh 200 μL RNase/DNase-free PCR tube. Then add 10 μL 2× NTP/CAP, 2 μL 10× reaction buffer, and finally RNA polymerase enzyme mix (*see* **Note 18**).
26. After mixing the tube gently, centrifuge briefly to collect the reaction mixture at the bottom of the tube.

27. Incubate the reaction mixture for 2 h at 37°C in a PCR thermal cycler.
28. After incubation, add 1 μL Turbo DNase for each 20 μL reaction mix and incubate for 15 min at 37°C in a PCR thermal cycler.
29. Purify *mCherry-Trim21* mRNA following manufacturers' kit instructions (Quick-RNA Microprep Kit) by first adding 96 mL 100% ethanol to the concentrated RNA wash buffer.
30. Add 100% ethanol in a ratio of 1:1 to the RNA synthesis reaction mix and mix by pipetting.
31. Place a Zymo-Spin® IC column into a collection tube and transfer the sample onto the membrane of the column, then centrifuge at 10,000–16,000 × *g* (author's recommendation is 13,400 × *g*) for 30 . Discard the flow through.
32. Optionally, wash the column with 400 μL RNA wash buffer, centrifuge, and discard the flow-through.
33. Mix 5 μL DNase I and 35 μL DNA digestion buffer in a fresh nuclease-free tube and transfer the mix onto the membrane.
34. Incubate for 15 min at RT.
35. Add 400 μL RNA wash buffer, centrifuge for 30 s at 10,000–16,000 × *g* and discard the flow through.
36. Then add 700 μL RNA wash buffer, centrifuge for 30 s at 10,000–16,000 × *g* and discard the flow through.
37. Add another 400 μL RNA wash buffer and centrifuge for 2 min at 10,000–16,000 × *g*.
38. Transfer the column into a fresh nuclease-free 1.5 mL tube and add 15 μL Dnase/RNase-free water. Centrifuge for 30 s at 10,000–16,000 × *g*.
39. Measure eluted RNA concentration using a NanoDrop spectrophotometer (wavelength 280 nm). RNA can be used directly or stored at −80°C. A minimum concentration of 1 mg/mL is required in order to be able to apply a sufficiently high dilution factor when preparing the Trim-Away cocktail for microinjection.
40. Test the *mCherry-Trim21* mRNA by microinjecting it alone into zygotes (embryos should fluoresce red after 6 h and should produce 80%–90% blastocysts; losses should only be due to the microinjection *per se*, which is invasive, not the mRNA) (*see* **Note 19**).

3.3 Mouse Zygote Collection

1. Inject 10 I.U. pregnant mare serum gonadotropin per mouse at 4 pm, followed 48 h later by injection of 10 IU human chorionic gonadotropin per mouse. If there are unforeseen events and the experiment needs to be postponed, it is also possible

to increase the interval from 48 to 72h [33]. Mate females to stud males right after the human chorionic gonadotropin injection (*see* **Note 20**). The authors typically inject 10 females and obtain 7–8 copulatory plugs on average.

2. At least 6h in advance, pre-equilibrate 10mL of KSOMaa medium in a humidified CO_2 incubator.
3. Sacrifice mice by cervical dislocation (*see* **Note 20**).
4. Cut the dorsal skin and peritoneum with scissors and expose the ovaries using tweezers.
5. Dissect oviducts and place them in 1mL of H-CZB medium in a 35mm petri dish.
6. Tear oviducts to release cumulus-enclosed zygotes.
7. Gently transfer the cumuli to a 100μL drop of hyaluronidase solution using a Gilson pipette.
8. Let the hyaluronidase digest the cumulus cells for 10min at 30°C on a warming plate.
9. Place cumulus-free zygotes in 500μL KSOMaa in 4-well plates without oil overlay. Absence of an oil overlay serves the purpose of letting ammonium, which is embryo-toxic [34], more easily escape as gas in the air.

3.4 Mixing the Trim-Away Cocktail for Microinjection

Typically, the authors inject the reagents altogether in the form of a cocktail (Table 1). However, on first use, it is advisable to test the *mCherry-Trim21* mRNA on its own, especially when a new batch of in vitro transcribed mRNA has been made, to ensure that the embryos fluoresce red and that they do not present any obvious abnormality (e.g., low blastocyst rate) (*see* **Note 19**).

1. Combine the mRNA, purified antibody (or denatured antibody; *see* **Note 8**), and fluorescent tracer (Table 1).

3.5 Microinjection of Antibody and Trim21 mRNA Cocktail in Mouse Zygotes (Trim-Away)

1. Ensure the micromanipulator is set up and ready to operate (*see* **Note 21**).
2. Deposit a 4–5μL drop of Trim-Away cocktail on a rectangular coverslip on the stage of the micromanipulator. The coverslip should be long enough (i.e., 24×50 or 24×60mm) to rest on the edges of the hole in the round plate that lies on top of the objectives' revolver.
3. Prepare the micromanipulation chamber with an elongated drop (300–400μL) of H-CZB medium.
4. Dip the tip of the injection capillary in the microdrop of Trim-Away cocktail, and aspirate by applying negative pressure with the Gilmont GS-1200 syringe.
5. Once the capillary has been loaded, lift the injection pipette, remove the coverslip, replace it with the micromanipulation

Table 1
Composition of the Trim-Away Cocktail

Component	Final concentration (mg/mL)
Purified *mCherry-Trim21* mRNA (*see* **Note 19**)	0.15–0.20
Purified antibody (*see* **Note 7**)	≥1
OGDB	0.01–0.02
milliQ water	As needed to obtain the above concentrations in a final volume of 5 μL

chamber, and lower the injection and holding pipettes into the chamber containing the elongated drop of H-CZB medium (*see* **Note 22**).

6. Using a pipette, transfer 10–30 zygotes into the H-CZB medium in the south sector of the micromanipulation chamber (*see* **Note 23**).
7. Pick up each single zygote with the holding capillary and roll it using the injection capillary until the 2nd polar body is positioned at 6 or 12 o'clock, i.e., out of the injector pipette's trajectory.
8. Position the tip of the injection capillary against the zona pellucida. Apply two to four piezo pulses to pierce through the zona pellucida (*see* **Notes 24 and 25**).
9. Insert the capillary through the hole made by the piezo and cross the zygote diameter without applying any pulse. Once the opposite side of the zygote is reached, apply one single piezo pulse to puncture the oolemma at the capillary's tip. Successful penetration of the oolemma is indicated by a rapid relaxation of the oolemma itself (*see* **Note 24**).
10. Release a small volume of Trim-Away cocktail inside the egg (*see* **Note 26**), until the perivitelline space has almost disappeared.
11. Withdraw the capillary quickly but gently.
12. Move injected zygotes one by one (after each injection) to the north sector of the micromanipulation chamber.
13. Allow zygotes to recover for 5–10 min before removing them from the micromanipulation chamber.

3.6 Culture of Trim-Away Zygotes

KSOMaa medium and culture dish should be prepared in advance, and culture dishes must be placed in a humidified CO_2 incubator for equilibration for at least 1 h prior to use.

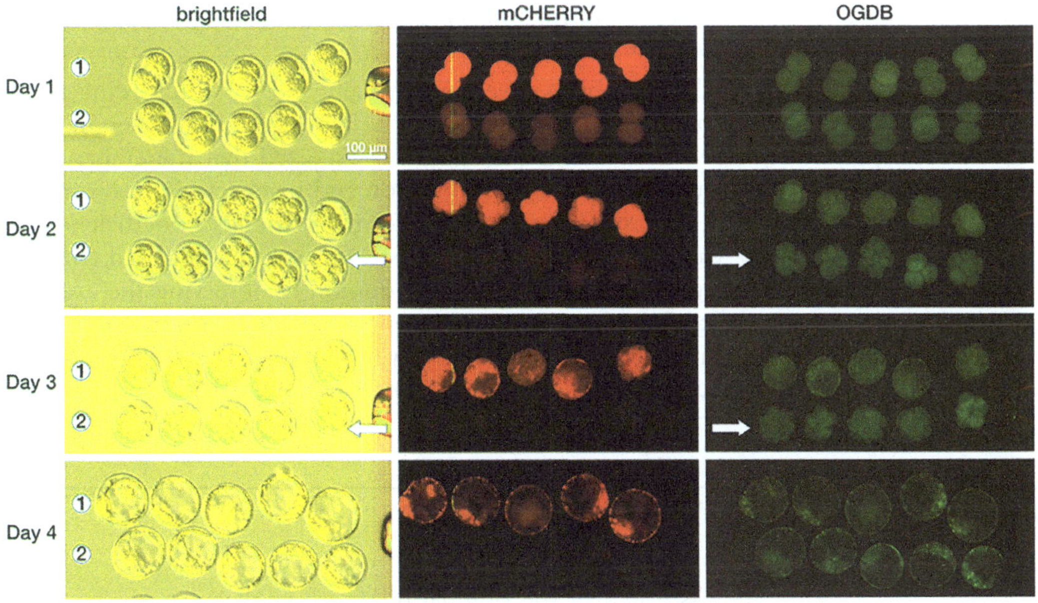

Fig. 5 Developmental phenotype (embryonic days 1–4) of mouse zygotes subjected to Trim-Away of E-CADHERIN. Zygotes were microinjected with the Trim-Away cocktail on day 0 (not shown) and imaged on consecutive days on the stage of the micromanipulator fitted with an excitation source for fluorescence (568 nm for mCHERRY; 488 nm for OGDB; same exposure time for all pictures). In each insert, the upper row (1) shows zygotes that were injected with the Trim-Away cocktail containing heat-inactivated (denatured) anti-E-CADHERIN antibody, the bottom row (2) shows zygotes that were injected with the Trim-Away cocktail containing anti-E-CADHERIN antibody in its native form. The red channel confirms that mCHERRY-TRIM21 was produced but not degraded (upper row, (1)) or produced and degraded (bottom row, (2)) in the cytoplasm, depending on the state of the antibody. The green channel of OGDB serves the purpose of showing that the volumes injected were similar across embryos and treatments. The bright field channel shows that on day 2 the embryos in which the Trim-Away reaction did not take place were compacted at the same time while the "trimmed" embryos were uncompacted (white arrow); on day 3 the embryos in which the Trim-Away reaction did not take place had begun to cavitate at the same time when the "trimmed" embryos were still morulae (white arrow); the "trimmed" embryos demonstrate catch up on day 4 forming blastocysts, thereby confirming the absence of irreversible toxicity of the procedure. Abbreviations: OGDB Oregon Green Dextran beads

1. Prepare the four-well plate with 500 µL of KSOMaa per well in each well.
2. Transfer the microinjected zygotes from the micromanipulation chamber to the first well of the four-well plate using a glass pipette.
3. After 30 min, remove zygotes that have lysed (*see* **Note 27**). Move surviving zygotes to the second well to be washed and counted, then they are transferred to the third well for culture to the blastocyst stage (if this can be reached, depending on the phenotype elicited by the Trim-Away of the POI;

blastocyst formation is expected in the case of E-CADHERIN; *see* **Note 28**).

4. Check embryo development daily and score according to the appropriate grading system (*see* **Note 29**). In the case of Trim-Away of E-CADHERIN, it is expected that the embryos will have difficulties at compaction and trophectoderm formation, but most of them should form blastocysts nonetheless—just with a delay (same blastocyst rate as unmanipulated controls, e.g., 80%–90%; Fig. 5). In addition to observing the expected cellular phenotype, it is appropriate to confirm—at least at the beginning of a Trim-Away series—that the POI has been degraded, by immunoblotting (*see* **Note 30**).

4 Notes

1. As noted in the original publications that introduced Trim-Away to the field of oocyte and early embryo biology [13, 35], the major advantages of this technique over other protein degradation methods are that it can be applied to any endogenous protein without prior modification; that it uses off-the-shelf antibodies that are commercially available; and that it can reduce the lag time between gene locus knockout and phenotypic effect (the longer the time, the higher the possibility of adaptive or secondary effects [1]). These advantages help the field advance, and one certainly does better with Trim-Away than without it. However, there are also a number of limitations of which the field is not yet fully aware, and which the authors realized soon after entering this field as novices at the time—it was 2019 [28].
2. It is unclear whether TRIM21 and the antibody are really degraded in the proteasome only after they have bound the POI, or also independently. The authors have observed, thanks to TRIM21 tagged with mCHERRY fluorescent peptide, that the fluorescence declined—albeit slowly—after microinjection of Trim-Away reagents even if there was no target protein, such as in the case of wild type mouse zygotes microinjected with monoclonal anti-GFP antibody or isotype IgG (Boiani, unpublished). There might be a possibility that the dimer TRIM21-antibody is degraded in the proteasome irrespective of the POI, which means that Trim-Away could induce a phenotype (nonspecific) even if the POI was not targeted.
3. E-CADHERIN is a transmembrane protein that operates as a Calcium-dependent cell adhesion molecule (hence the name

Cadherin, from "Calcium adhering"). It is expressed in various cell types, including early embryos of mice and other mammalian species, mouse embryonic stem cells (ESCs), but not mouse embryonic fibroblasts (MEFs) [36, 37] (Fig. 3). E-CADHERIN is expressed throughout mouse preimplantation development, localizing at cell–cell contact sites from the 8-cell stage onward, and is crucial for compaction at the 8-cell stage and for trophectoderm epithelium formation and blastocyst cavity development [38]. Embryos lacking E-CADHERIN fail to maintain compaction and exhibit severe developmental abnormalities prior to implantation [39–42], which lend themselves to being easily visualized under a stereomicroscope without the need for molecular assays. The authors note that cell-adhesion molecules were inactivated already in the 1980s by simply co-incubating the embryos with the antibodies [9–12], since one part of the molecule is exposed outside the cell. To the author's knowledge, degradation of cell adhesion molecules from inside using Trim-Away has not been reported to date.

4. It is recommended that when performing the Trim-Away method for the first time in mammalian embryos (e.g., mouse), a protein is tested that has a known phenotype, and this phenotype is easy to visualize without needing molecular assays . Therefore, a Trim-Away pilot experiment should be conducted on a POI that is known a priori to have an effect when removed, and this effect can be seen under a stereomicroscope. If the results of this pilot are as expected, then one can move on to the study of other POIs with more confidence.

5. For Trim-Away to work successfully after microinjection, part of the protein must be exposed to the cytosol and accessible from the cytosolic side. Indeed, this is the case of E-CADHERIN (Fig. 3), and commercial antibodies are available that recognize the cytosolic part of the molecule. Moreover, it needs to be considered whether the POI may undergo de novo synthesis during embryogenesis. If this is the case, then the amount degraded by Trim-Away may be regenerated, reducing the effectiveness of the Trim-Away method.

6. In vitro, the application of Trim-Away is hampered by several factors, the most severe being the large size of antibodies (150 kDa) that can hinder tissue penetration and accessibility to certain epitopes in conjunction with the impermeability of the cell membrane [28, 36]. In somatic cells, scientists are developing nanobody fragments of smaller size (15 kDa) that can be delivered by lentiviral transfection [37], but lentiviral transfection is not ideal for treating oocytes or blastomeres, which, moreover, are surrounded by the zona pellucida. In order to overcome the zona and the oolemma, the Trim-Away

reagents are typically delivered by microinjection, except in a few cases where membrane permeabilization was used [24]. However, if the POI is localized in the cell nucleus, this poses an additional barrier. It is not trivial for antibodies to reach inside the nucleus, except during mitosis, when the nuclear membrane transiently disappears. An Fc-nanobody strategy was developed to facilitate targeting of nuclear proteins in interphase [13]. If different POIs are localized in the same subcellular compartment, and the antibodies are similar to each other in the rates at which they bind and dissociate from their targets, then POIs present in small amounts are easier to remove than POIs present in larger amounts. Therefore, knowing the amount of POI is useful to calibrate the amount of antibody and *Trim21* mRNA in the Trim-Away reaction (Fig. 4).

7. The authors recommend rabbit antibodies, if available, because the affinity of TRIM21 is higher for rabbit IgG than for IgG of most other species [38]. If available for the POI, monoclonal rabbit antibodies (RabMAb®) are probably the best. A large proportion of antibodies from commercial vendors and in academic laboratories are not sufficiently characterized in terms of quantitative binding properties against specific and nonspecific targets. There are studies revealing that, on average, 50% of commercial antibodies do not produce expected binding results as advertised, and the success rate varies from 0% to 100% for different vendors [39]. Some commercial antibodies for antigens expressed in oocytes and embryos have been tested by immunoblot on brain lysates. Since Trim-Away critically depends on the antibody and its quality, it is essential to check it thoroughly before embarking on a Trim-Away experiment. If the Western blot shows multiple bands, then it is recommended to look for another antibody, since the use of such an antibody would mean that the Trim-Away is not specific, i.e., it degrades something other than the POI. For these reasons, the authors recommend checking the antibodies before use, to make sure these recognize a single band of the expected molecular weight in Western blot analysis performed with lysate of a tissue where the target protein is known to be expressed. This recommendation has been put into practice by the authors (Fig. 3). The anti-E-CADHERIN antibody used in this chapter recognizes only one band in oocytes and no band in mouse fibroblasts, just as it should be, whereas in embryonic stem cells, there are multiple bands that are puzzling. As further control, pull-down experiments with the antibody are also valuable.
8. To properly interpret the results of a Trim-Away experiment, suitable controls are essential. Resist the temptation to stop

the tuning of your Trim-Away experiment as soon as you see a phenotype after the microinjection of the Trim-Away cocktail (antibody directed against POI plus TRIM21 protein or *Trim21* mRNA). It might be a nonspecific phenotype – a false positive. Once inside the cell, the antibody is faced with the possibility of cross-reactions with other proteins, which can lead to spurious phenotypes (false positives). Therefore, adequate experimental controls for antibody specificity are necessary. Of note, the injection of the antibody alone (without *Trim21* mRNA or TRIM21 protein) can cause a phenotype already *per se* in the absence of POI degradation. In their study [23] addressing the intracellular presence and role of ZP3 in preimplantation mouse embryos, for example, the authors showed that already the ZP3 antibody alone led to a phenotype after it was microinjected in zygotes—without *Trim21* mRNA. The authors recommend performing Trim-Away with the denatured antibody, or with an antibody directed against an irrelevant protein (such as "green fluorescent protein," GFP, which is not present in wild-type oocytes and embryos). As well as the risk of false positives, there is also the risk of false negatives. The reason for this is not just the trivial case of an antibody not being good at binding the POI. There is also a possibility that protein neo-synthesis can partly replenish or even restore the amount of POI that has been degraded by Trim-Away, thereby masking the phenotype. Therefore, depending on the POI, it is advisable to assess potential neo-synthesis of the POI by embryos with suitable methods. It is likely that this means doing mass spectrometry analysis of embryos that have incorporated isotopic (but not radioactive) amino acids, as reported by the authors [23].

9. Clearly, having a plasmid, one can produce mRNA at will in bacteria, while producing protein is more challenging (due to posttranslational modifications, which require eukaryotic cells). TRIM21 can be applied directly as a protein, which can be purchased from a number of vendors. However, commercial TRIM21 proteins are typically available without a fluorescent tag, and injecting proteins with the piezo drill can be problematic, as proteins can form an amalgam with Hg.

10. CAUTION: Hg is toxic. It can be replaced with Fluorinert™ (e.g., FC-77), but the energy transfer of Fluorinert is not as high as that of Hg, which makes it more difficult to use the piezo system.

11. Mouth pipetting is not recommended for occupational health and safety reasons. However, it is still widely used in non-clinical laboratory settings, including the authors' laboratory, because it gives excellent control over pressure/movement

inside the pipette, and because it is ergonomic and comfortable over long time periods. Alternatively, manual glass pipettes with an appropriate diameter can be used.

12. The authors prepare concentrated stocks of the culture media, dissolving the chemicals (all except albumin and gentamicin in the case of KSOMaa) in one quarter of the final volume (i.e., the amounts indicated for 1 L are dissolved in 0.25 L). The stocks are sterile-filtered, aliquoted, and stored at 4°C for up to 6 months. Albumin and gentamicin are added after the dilution, before equilibration of KSOMaa for final use.

13. The green-fluorescent dextran beads are suited for the combined use with *mCherry-Trim21* mRNA. These beads can be replaced with beads of another color, as long as their color is distinguishable from that of mCHERRY.

14. Too much antibody clogs the column and reduces the flow through. Approximately 35 μL eluate is obtained with an input of 50 μg antibody. If the antibody is too concentrated, then more eluate will come out at the end because the column is clogged. Although this eluate may still be used, the best thing to do is to repeat the procedure, loading the column with a lower amount of antibody.

15. In the authors' experience, these storage conditions are safe for about 3 months. Freezing may allow for longer storage, but there is a risk that the antibody is damaged and precipitates.

16. After linearization, a test gel is recommended to check if the plasmid was cut or not. The authors recommend a 1% agarose gel. During mRNA synthesis, it is important to strictly follow the outlined order for pipetting. Also, a DNase digestion step is recommended. After mRNA synthesis, add 1 μL DNAse to the sample and incubate for 15 min at 37°C.

17. A second wash step is recommended.

18. To prevent coprecipitation of the 10× reaction buffer and template DNA, assemble the reaction at RT.

19. Not only can the antibody alone cause a phenotype, but TRIM21 or its mRNA can also have side- or even toxic effects in mouse embryos [28] if injected in too high a concentration, and this can lead to spurious phenotypes (false positives). The authors recommend not exceeding a *Trim21* mRNA concentration of 0.2 mg/mL. This concentration, when translated in the cytoplasm, produces a high amount of TRIM 21 protein (Fig. 4). It is often overlooked that TRIM21, in addition to being an antibody receptor, is also part of the NF-kB, AP-1, and IRF signaling pathway [18] and has various interaction partners as revealed by database searches in GeneMania and

BioGrid [40, 41] (Fig. 2). Toxicity and unspecific phenotypes can also result from overloading of the proteasome by the ternary complex antibody-POI-TRIM21 [42]. There is also the opposite risk, namely, that the concentrations of antibody and TRIM21 in the ooplasm are below the concentration of POI, leading to incomplete depletion of the POI, in which case a wrong conclusion can be drawn that the POI has no effect.

20. Adhere to the local regulations concerning the use of experimental animals for scientific purposes.

21. In the authors' laboratory, micromanipulation is performed using a manual micromanipulator with Nomarski optics. The microinjection and the holding capillaries are mounted on the left and right sides, respectively, of the microscope arms. Pressure in the microinjection circuit is operated with water. Pressure in the holding circuit is operated with mineral oil. The injection capillary is essentially an ICSI needle, with an inner diameter (ID) of approximately 6–8 μm at the tip, which is cut bluntly using a microforge. The capillary is bent with an angle of approximately 20° at a distance of 8–10 mm from the tip. The microinjection needle is back-loaded with a 2–3 μL volume of Hg using a Hamilton micrometer syringe and pushed to the tip using water pressure. Accurate setting of the injection capillary is very important for the success of the Trim-Away microinjection. Good working condition and effectiveness of the microinjection pipette are crucial. During injections, the pipette can become sticky or even clog, due to the fact that the antibody protein and Hg form an amalgam. This event will affect not only zygote survival but also embryo development after microinjection. It is recommended to test the efficacy of the piezo system at drilling the zona pellucida (*see* **Notes 24 and 25**) before front-loading the microinjection capillary with the Trim-Away cocktail and starting to inject the zygotes.

22. It needs to be ensured that there is a minimal but constant flow of cocktail coming out of the tip, to prevent aspiration and thereby the cocktail inside the capillary from being diluted with H-CZB medium.

23. The appropriate number of zygotes depends on the operator's skills. Each group should be processed within 15 min.

24. Using PMM-150 as a piezo drill, the authors recommend intensity 10, speed 3 for piercing the zona; intensity 4, speed 3 for penetrating the oolemma. Depending on the amount of Hg and how much Trim-Away cocktail volume has been front-loaded into the pipette, the energy transfer efficiency may vary, and the piezo intensity may need to be adjusted.

25. When you are not experienced with the use of piezo, it should be tested thoroughly beforehand to ensure that the energy transfer from Hg to the tip of the microinjection needle is sufficient. The needle must be able to pierce through the zona already before front-loading with the Trim-Away cocktail. After loading, the energy transfer will be weaker, due to the fluid occupying the space between the front of the Hg and the tip of the pipette. Therefore, ensure that sufficient energy transfer is available before loading the needle, and that sufficient energy transfer is still available after front-loading the needle. This is tested by trying on discarded oocytes: the microinjection needle should be able to penetrate the zona with 2, 3, or a maximum of 4 piezo pulses. If penetration requires more pulses, the first thing to try is to increase the intensity of the piezo setting. If this is already maxed out, one can try disassembling and reassembling the pipette in the piezo drill shaft.
26. Most investigators keep the microinjected volume to a minimum, typically in the vicinity of 10 pL per mouse oocyte. The authors inject substantially larger volumes, up to 100 pL, based on the preliminary studies and calibrations described in [28]. This larger volume allows for maximizing the amounts of Trim-Away reagents (*Trim21* mRNA and antibody), which is particularly important when aiming to target abundant proteins.
27. Under optimal settings, the immediate survival rate of the zygotes right after microinjection, as indicated by membrane resealing and lack of lysis, is >90%. Decline of performance after many injections is normal, and survival rates of 70%–80% are still acceptable. If the immediate survival rate of the zygotes falls below 70%, then it means that the pipette is no longer suited for its task and should be replaced without hesitation.
28. Protein synthesis of TRIM21 within the zygote requires a certain amount of time (approximately 6 h), and the amount of translated product places a limit on the amount of POI that can be removed via Trim-Away. For this reason, it is recommended to know in advance in what quantity the POI is present, what half-life it has, and whether it is synthesized de novo, before embarking on a Trim-Away experiment. To this end, there are published resources in which the proteome of mouse oocytes and embryos has been quantified [28, 33]. In addition, in the present work, the authors also quantified the antibody (IgG) injected (Fig. 4).
29. On day 1 after the Trim-Away cocktail injection, the embryos should be 2-cell, and in the late evening (8 pm), they should begin to divide to 4-cells. On day 2 after injection, the

embryos should be 8-cell, non-compacted in the morning, compacted in the early afternoon. On day 3 after injection, the embryos should be 16-cell morulae, which begin to cavitate in the afternoon. On day 4 after injection, the embryos should be blastocysts.

30. Confirmation of protein depletion after Trim-Away can be achieved by immunofluorescence (ensuring that the primary antibody used for detection is different from the antibody used for Trim-Away) or by immunoblotting. If there is no option but to use the same antibody that was used for Trim-Away, then primary antibodies can be directly labeled with fluorophores prior to use, thereby avoiding the problem that the secondary antibody would not distinguish between the two identical primary antibodies. If the primary antibody was raised in a mouse, then this is not an ideal situation for confirming protein depletion in mouse embryos, as the secondary anti-mouse antibody could recognize other epitopes in addition to the primary antibody.

Acknowledgments

The authors thank all present and former members of the Boiani team. Hannes Drexler is gratefully acknowledged for the mass spectrometry data presented in Fig. 4. This work was supported by the Deutsche Forschungsgemeinschaft (DFG grant BO-2540/8-1 to M.B.).

References

1. El-Brolosy MA, Stainier DYR (2017) Genetic compensation: a phenomenon in search of mechanisms. PLoS Genet 13(7):e1006780. https://doi.org/10.1371/journal.pgen.1006780
2. Camlin NJ (2025) Protein-targeting reverse genetic approaches: the future of oocyte and preimplantation embryo research. Mol Hum Reprod 31(2):gaaf008. https://doi.org/10.1093/molehr/gaaf008
3. Smoak EM, Stein P, Schultz RM, Lampson MA, Black BE (2016) Long-term retention of CENP-A nucleosomes in mammalian oocytes underpins transgenerational inheritance of centromere identity. Curr Biol 26(8):1110–1116. https://doi.org/10.1016/j.cub.2016.02.061
4. Keller MA, Stiehm ER (2000) Passive immunity in prevention and treatment of infectious diseases. Clin Microbiol Rev 13(4):602–614. https://doi.org/10.1128/CMR.13.4.602
5. Kohler G, Milstein C (1975) Continuous cultures of fused cells secreting antibody of predefined specificity. Nature 256(5517):495–497. https://doi.org/10.1038/256495a0
6. Buehr M, Lee S, McLaren A, Warner A (1987) Reduced gap junctional communication is associated with the lethal condition characteristic of DDK mouse eggs fertilized by foreign sperm. Development 101(3):449–459. https://doi.org/10.1242/dev.101.3.449
7. Emerson JA (1988) Disruption of the cytokeratin filament network in the preimplantation mouse embryo. Development 104(2):219–234. https://doi.org/10.1242/dev.104.2.219
8. Lee S, Gilula NB, Warner AE (1987) Gap junctional communication and compaction during preimplantation stages of mouse development. Cell 51(5):851–860. https://doi.org/10.1016/0092-8674(87)90108-5

9. Johnson MH, Maro B, Takeichi M (1986) The role of cell adhesion in the synchronization and orientation of polarization in 8-cell mouse blastomeres. J Embryol Exp Morphol 93:239–255
10. Shirayoshi Y, Okada TS, Takeichi M (1983) The calcium-dependent cell-cell adhesion system regulates inner cell mass formation and cell surface polarization in early mouse development. Cell 35(3 Pt 2):631–638. https://doi.org/10.1016/0092-8674(83)90095-8
11. Yoshida-Noro C, Suzuki N, Takeichi M (1984) Molecular nature of the calcium-dependent cell-cell adhesion system in mouse teratocarcinoma and embryonic cells studied with a monoclonal antibody. Dev Biol 101(1):19–27. https://doi.org/10.1016/0012-1606(84)90112-x
12. Fleming TP, McConnell J, Johnson MH, Stevenson BR (1989) Development of tight junctions de novo in the mouse early embryo: control of assembly of the tight junction-specific protein, ZO-1. J Cell Biol 108(4):1407–1418. https://doi.org/10.1083/jcb.108.4.1407
13. Clift D, McEwan WA, Labzin LI, Konieczny V, Mogessie B, James LC, Schuh M (2017) A method for the acute and rapid degradation of endogenous proteins. Cell 171(7):1692–1706 e1618. https://doi.org/10.1016/j.cell.2017.10.033
14. Li X, Yang L, Chen S, Zheng J, Zhang H, Ren L (2023) Multiple roles of TRIM21 in virus infection. Int J Mol Sci 24 (2):1683. https://doi.org/10.3390/ijms24021683
15. Mallery DL, McEwan WA, Bidgood SR, Towers GJ, Johnson CM, James LC (2010) Antibodies mediate intracellular immunity through tripartite motif-containing 21 (TRIM21. Proc Natl Acad Sci U S A 107(46):19985–19990. https://doi.org/10.1073/pnas.1014074107
16. Yoshimi R, Chang TH, Wang H, Atsumi T, Morse 3rdHC, Ozato K (2009) Gene disruption study reveals a nonredundant role for TRIM21/Ro52 in NF-kappaB-dependent cytokine expression in fibroblasts. J Immunol 182(12):7527–7538. https://doi.org/10.4049/jimmunol.0804121
17. Zeng J, Santos AF, Mukadam AS, Osswald M, Jacques DA, Dickson CF, McLaughlin SH, Johnson CM, Kiss L, Luptak J, Renner N, Vaysburd M, McEwan WA, Morais-de-Sa E, Clift D, James LC (2021) Target-induced clustering activates Trim-Away of pathogens and proteins. Nat Struct Mol Biol 28(3):278–289. https://doi.org/10.1038/s41594-021-00560-2
18. McEwan WA, Tam JC, Watkinson RE, Bidgood SR, Mallery DL, James LC (2013) Intracellular antibody-bound pathogens stimulate immune signaling via the Fc receptor TRIM21. Nat Immunol 14(4):327–336. https://doi.org/10.1038/ni.2548
19. Mevissen TET, Prasad AV, Walter JC (2023) TRIM21-dependent target protein ubiquitination mediates cell-free Trim-Away. Cell Rep 42(2):112125. https://doi.org/10.1016/j.celrep.2023.112125
20. Chatterjee P, Ponnapati M, Kramme C, Plesa AM, Church GM, Jacobson JM (2020) Targeted intracellular degradation of SARS-CoV-2 via computationally optimized peptide fusions. Commun Biol 3(1):715. https://doi.org/10.1038/s42003-020-01470-7
21. Lin CY, Chen SH, Tsai CL, Tang YH, Wu KY, Chao A (2021) Intracellular targeting of STIP1 inhibits human cancer cell line growth. Transl Cancer Res 10(3):1313–1323. https://doi.org/10.21037/tcr-20-3333
22. Pozzi S, Codron P, Soucy G, Renaud L, Cordeau PJ, Dutta K, Bareil C, Julien JP (2020) Monoclonal full-length antibody against TAR DNA binding protein 43 reduces related proteinopathy in neurons. JCI Insight 5 (21):e14042. https://doi.org/10.1172/jci.insight.140420
23. Israel S, Seyfarth J, Nolte T, Drexler HCA, Fuellen G, Boiani M (2023) Intracellular fraction of zona pellucida protein 3 is required for the oocyte-to-embryo transition in mice. Mol Hum Reprod 29 (11):gaad038. https://doi.org/10.1093/molehr/gaad038
24. Gerri C, McCarthy A, Alanis-Lobato G, Demtschenko A, Bruneau A, Loubersac S, Fogarty NME, Hampshire D, Elder K, Snell P, Christie L, David L, Van de Velde H, Fouladi-Nashta AA, Niakan KK (2020) Initiation of a conserved trophectoderm program in human, cow and mouse embryos. Nature 587(7834):443–447. https://doi.org/10.1038/s41586-020-2759-x
25. Huang J, Ru G, Sun J, Sun L, Li Z (2022) Elevated RIF1 participates in the epigenetic abnormalities of zygotes by regulating histone modifications on MuERV-L in obese mice. Mol Med 28(1):17. https://doi.org/10.1186/s10020-022-00446-z
26. Mehlmann LM, Uliasz TF, Yee SP, Kaback D, Lowther KM (2024) Generation and Characterization of a TRIM21 overexpressing mouse line. Genesis 62(5):e23616. https://doi.org/10.1002/dvg.23616
27. Mathieson T, Franken H, Kosinski J, Kurzawa N, Zinn N, Sweetman G, Poeckel D, Ratnu VS, Schramm M, Becher I, Steidel M, Noh KM, Bergamini G, Beck M, Bantscheff M, Savitski MM (2018) Systematic analysis of protein turnover in primary cells. Nat Commun 9(1):689. https://doi.org/10.1038/s41467-018-03106-1
28. Israel S, Casser E, Drexler HCA, Fuellen G, Boiani M (2019) A framework for TRIM21-mediated protein depletion in early mouse

embryos: recapitulation of Tead4 null phenotype over three days. BMC Genomics 20(1):755. https://doi.org/10.1186/s12864-019-6106-2
29. Vestweber D, Gossler A, Boller K, Kemler R (1987) Expression and distribution of cell adhesion molecule uvomorulin in mouse preimplantation embryos. Dev Biol 124(2):451–456. https://doi.org/10.1016/0012-1606(87)90498-2
30. Kim KH, Lee KA (2014) Maternal effect genes: findings and effects on mouse embryo development. Clin Exp Reprod Med 41(2):47–61. https://doi.org/10.5653/cerm.2014.41.2.47
31. Israel S, Drexler HCA, Fuellen G, Boiani M (2021) The COP9 signalosome subunit 3 is necessary for early embryo survival by way of a stable protein deposit in mouse oocytes. Mol Hum Reprod 27 (8):gaab048. https://doi.org/10.1093/molehr/gaab048
32. Vermeer AW, Norde W (2000) The thermal stability of immunoglobulin: unfolding and aggregation of a multi-domain protein. Biophys J 78(1):394–404. https://doi.org/10.1016/S0006-3495(00)76602-1
33. Taher L, Israel S, Drexler HCA, Makalowski W, Suzuki Y, Fuellen G, Boiani M (2021) The proteome, not the transcriptome, predicts that oocyte superovulation affects embryonic phenotypes in mice. Sci Rep 11(1):23731. https://doi.org/10.1038/s41598-021-03054-9
34. Lane M, Gardner DK (2003) Ammonium induces aberrant blastocyst differentiation, metabolism, pH regulation, gene expression and subsequently alters fetal development in the mouse. Biol Reprod 69(4):1109–1117. https://doi.org/10.1095/biolreprod.103.018093
35. Clift D, So C, McEwan WA, James LC, Schuh M (2018) Acute and rapid degradation of endogenous proteins by Trim-Away. Nat Protoc 13(10):2149–2175. https://doi.org/10.1038/s41596-018-0028-3
36. Sui B, Wang M, Cheng C, Zhang Q, Zhang J, Fan D, Xu P (2021) Nanogel-facilitated protein intracellular specific degradation through Trim-Away. Adv Funct Mater 31 (30):2010556. https://doi.org/10.1002/adfm.202010556
37. Chen G, Kong Y, Li Y, Huang A, Wang C, Zhou S, Yang Z, Wu Y, Ren J, Ying T (2021) A promising intracellular protein-degradation strategy: TRIMbody-Away technique based on nanobody fragment. Biomolecules 11 (10):1512. https://doi.org/10.3390/biom11101512
38. Ramos AF, Fernandes LA, Batista F, de Souza Vieira B, Thompson M, Mattos JJ, Marques MRF, de Lourdes Borba Magalhaes M, da Silva GF (2022) TRIM21 chimeric protein as a new molecular tool for multispecies IgG detection. J Genet Eng Biotechnol 20(1):111. https://doi.org/10.1186/s43141-022-00396-3
39. Perkel JM (2014) The antibody challenge. Biotechniques 56(3):111–114. https://doi.org/10.2144/000114143
40. Warde-Farley D, Donaldson SL, Comes O, Zuberi K, Badrawi R, Chao P, Franz M, Grouios C, Kazi F, Lopes CT, Maitland A, Mostafavi S, Montojo J, Shao Q, Wright G, Bader GD, Morris Q (2010) The GeneMANIA prediction server: biological network integration for gene prioritization and predicting gene function. Nucleic Acids Res 38(Web Server issue):W214–W220. https://doi.org/10.1093/nar/gkq537
41. Oughtred R, Rust J, Chang C, Breitkreutz BJ, Stark C, Willems A, Boucher L, Leung G, Kolas N, Zhang F, Dolma S, Coulombe-Huntington J, Chatr-Aryamontri A, Dolinski K, Tyers M (2021) The BioGRID database: a comprehensive biomedical resource of curated protein, genetic, and chemical interactions. Protein Sci 30(1):187–200. https://doi.org/10.1002/pro.3978
42. Namba S, Moriya H (2024) Toxicity of the model protein 3xGFP arises from degradation overload, not from aggregate formation. J Cell Sci 137 (11):jcs261977. https://doi.org/10.1242/jcs.261977

Check for updates

Chapter 23

Somatic Cell Nuclear Transfer in Sheep and Approaches to Improve Efficiency

Renata Blocher, Ying Liu, Tayler Patrick and Irina A. Polejaeva

Abstract

Sheep somatic cell nuclear transfer (SCNT) is widely used in both agricultural and biomedical research, especially for generating animal models of human disease. Here, we present the most accessible method of sheep oocyte collection for use in subsequent SCNT, using commercial in vitro maturation and culture media to increase SCNT efficiency. We describe a detailed procedure of sheep SCNT and SCNT animal production, as well as outline two methods to determine embryo quality: in vitro culture until blastocyst formation and blastocyst inner cell mass and trophectoderm cell staining.

Keywords Somatic cell nuclear transfer, Sheep, Ovum pick up

1 Introduction

The history of somatic cell nuclear transfer (SCNT) goes back to 1996, when Dolly the sheep was the first mammal cloned from an adult somatic cell [1]. Since Dolly, many animal species have been cloned, including pigs [2], cows [3], horses [4], goats [5], and others. SCNT in cows and horses has mainly been used to preserve superior genetics, including high-value beef and dairy cattle [6], bucking bulls, and performance horses (cutting, jumping, dressage, etc.). SCNT in domestic species like pigs, goats, and sheep allows the use of animals for pharmaceutical therapeutics, animal models of human diseases, and xenotransplantation [7–11]. Additional applications for SCNT include reviving endangered or extinct species through interspecies cloning [12, 13] or reproducing a deceased pet, a service offered by companies such as ViaGen Pets.

SCNT involves taking a mature oocyte in the metaphase stage of meiosis II, removing the first polar body and the metaphase plate, injecting a somatic cell under the oocyte's zona pellucida and fusing the two, then activating the subsequent zygote to start dividing. The oocyte provides the necessary cytoplasmic components, such

Alexandra J. Harvey and David K. Gardner (eds.), *The Preimplantation Embryo*, Methods in Molecular Biology, vol. 3038, https://doi.org/10.1007/978-1-0716-5292-3_23,

as maternal RNAs, proteins, mitochondria, and other organelles, to support embryonic development. The somatic cell provides the DNA for the resultant embryo [14]. Therefore, all resulting offspring will be genetically identical to the somatic cells used during SCNT. To this day, SCNT remains highly inefficient at 0.1%–16.0%, requiring large numbers of oocytes to create live offspring (reviewed in refs. [15, 16]). SCNT inefficiency stems largely from the inability to reset the somatic cell epigenome to embryonic status, which is necessary for proper embryonic development to the blastocyst stage, and in part from in vitro oocyte maturation, meaning not all oocytes that reach metaphase of meiosis II undergo the cytoplasmic maturation needed to support subsequent embryonic development. These problems contribute to changes in gene expression throughout embryonic development, which is further impacted by the increased environmental stress associated with in vitro procedures [17, 18].

Good-quality oocytes are essential for successful SCNT. The age of oocyte donors is important in determining oocyte developmental potential. After in vitro fertilization (IVF), zygotes from prepubertal sheep have a significantly decreased in vitro blastocyst rate compared to those from adult sheep, but perform similarly in terms of pregnancy rate and number of lambs born following embryo transfer [19]. Oocyte quality and in vitro developmental potential can be improved by using mature animals for ovariectomy, although this approach increases animal care costs [20]. Another way researchers aim to increase SCNT efficiency is by modifying the in vitro oocyte maturation and embryo culture media. Serum-containing media, which are routinely used for in vitro procedures, have been associated with increased offspring health complications, such as large offspring syndrome (LOS) [21, 22]. Use of a commercially available serum-free medium increased the 90-day pregnancy rate and full-term rate, almost doubling our SCNT efficiency [23]. The largest effects of media were observed in oocytes derived from prepubertal sheep, but SCNT efficiency was increased when using the commercial media with oocytes from both prepubertal and adult sheep [23]. These results indicate that in vitro maturation and culture media have a substantial impact on the in vivo development of SCNT embryos, more so than oocyte donor age.

Here, we describe detailed procedures of sheep SCNT, including using commercial in vitro maturation and culture media instead of an “in-house” prepared media to increase SCNT efficiency. Animal production, including embryo transfers, pregnancy monitoring, lamb delivery, and postnatal care, is also described in this chapter. Finally, we outline two methods used to assess SCNT embryo quality: blastocyst rate and blastocyst cell staining.

2 Materials

All materials were readily available from major scientific supply companies, such as Thermo Fisher Scientific. This method has been optimized using IVF Bioscience oocyte maturation and embryo culture media. Other commercially available, serum-free maturation and culture media need to be tested before use with this SCNT method.

2.1 Abattoir Ovary Collection

1. Warm Saline Solution: 0.9% NaCl in Milli-Q Ultrapure Water.
2. Insulated Cooler.
3. Colander.
4. 100 mL Beaker.
5. Sharp Dissecting Scissors.
6. Scalpel Handle and Blades (#15).
7. Tweezers.
8. 50 mL Conical Centrifuge Tube.

2.2 Oocyte Collection

1. 50 mg/mL Heparin Sodium Stock.
2. BSA Fraction V, Fatty Acid Free.
3. 1% Penicillin/Streptomycin Solution.
4. Fetal Bovine Serum (FBS).
5. Medium 199 (M199) with Earle's Balanced Salts, L-Glutamine, and HEPES.
6. Oocyte Collection Medium: M199 supplemented with 1% Penicillin/Streptomycin Solution, 0.3% BSA Fraction V, 1% FBS, and 0.06% Heparin Sodium Stock.
7. 100 mm Petri Dish.

2.3 In Vitro Maturation (IVM)

1. Commercially available IVM Medium (e.g., BO-IVM™).
2. Four-well culture dishes.
3. 50 mL Conical Centrifuge Tube.
4. 100 mm Square Petri Dish with Grid.
5. 35 mm Round Petri Dish.
6. Sterile Transfer Pipettes.
7. CO_2 Incubator.

2.4 Cell Culture

1. Cryopreserved Cells.
2. 15 mL Conical Centrifuge Tube.
3. Dulbecco's Modified Eagle's Medium (DMEM)/High Glucose.
4. Cell Starving Medium: DMEM supplemented with 0.5% FBSand 1% Penicillin/Streptomycin Solution.

5. Dulbecco's Phosphate-Buffered Saline (DPBS) without Calcium and Magnesium.
6. TrypLE™ Express.
7. Cell Culture Medium: DMEM supplemented with 15% FBS and 1% Penicillin/Streptomycin Solution.
8. Four-well culture plates.
9. Trigas Incubator.
10. Water bath.
11. HEPES-buffered SOF (HSOF; Table 1).

2.5 SCNT

1. Light Mineral Oil (e.g., Fujifilm).
2. 0.6 mg/mL Hyaluronidase Stock.
3. Commercially available In Vitro embryo Culture (IVC) Medium (e.g., BO-IVC™).
4. Injection Pipette Washing Medium: HSOF supplemented with 10% polyvinylpyrrolidone (PVP) w/v.
5. Cytochalasin B (CB) Medium: HSOF supplemented with 10 µg/mL CB.
6. Enucleation Medium: HSOF supplemented with 10 µg/mL CB and 45 mM sucrose.

Table 1
Components found in 200 mL of HEPES-buffered synthetic oviductal fluid (HSOF)

Component	Cat #	Concentration (final)
Cell culture water	SH30529	Q.S. to 200 mL
NaCl	S5886	99.7 mM
KCl	P5405	8.03 mM
KH_2PO_4	P5655	1.19 mM
$NaHCO_3$	S5761	5 mM
$CaCl_2 \bullet 2H_2O$	C7902	1.71 mM
$MgCl_2 \bullet 6H_2O$	M2393	0.49 mM
$CH_3COCOONa$	P4562	0.4 mM
Glutamax	35050-061	0.20 mM
Pen/strep	15070-063	1%
Na lactate acid	L7900	0.16%
$C_6H_{12}O_6$	G6152	1.499 mM
$C_8H_{18}N_2O_4S$ (Hepes)	H6147	20 mM
FBS	SH30396	2%

7. Inverted Microscope equipped with a Micromanipulator.
8. BTX Electro Cell Manipulator (ECM) with connector cables: set to 2.0 kV/cm, 40 μs, two pulses.
9. BTX 0.5 mm Microslide Chamber.
10. Milli-Q Ultrapure Water.
11. 20 μm Outer diameter (OD) Injection Pipettes (made in-house; *see* **Note 1**).
12. 100 μm OD Holding Pipettes (made in-house; *see* **Note 1**).
13. Trigas Incubator.
14. Timer.
15. Pipettors.
16. Pipette Tips.
17. Sheep Fusion Medium (Table 2).
18. 6-dimethylamino purine (6-DMAP).
19. Cycloheximide (CHX).
20. Activation Medium: IVC medium supplemented with 2 mM 6-DMAP and 10 μg/mL CHX.
21. Ionomycin Medium: HSOF supplemented with 5 μM ionomycin.
22. 35 mm Culture Dishes.
23. Micropipette.

2.6 In Vitro Culture (IVC)

1. 35 mm Culture Dishes.
2. Light Mineral Oil (e.g., Fujifilm).
3. IVC Medium.
4. Trigas Incubator.
5. Pipettors.
6. Pipette Tips.

Table 2
Components found in 200 mL of Sheep Fusion Media

Component	Cat #	Concentration (final)
Embryo grade water	W1503	Q.S. to 200 mL
D-sorbitol	194742	309.3 mM
BSA	A8806	0.10%
$Ca(CH_3CO_2)_2 \cdot H_2O$	402850	0.1 mM
$(CH_3COO)_2Mg \cdot 4H_2O$	M5661	0.5 mM
$C_8H_{18}N_2O_4S$ (Hepes)	H6147	0.5 mM

7. Micropipette.

2.7 Recipient Synchronization and Embryo Transfers

1. Lubricant.
2. Vaginal Sponges containing 40 mg Flurogestone Acetate (e.g., SYNCRITE).
3. Luteolytic Prostaglandin Solution (e.g., Estrumate®).
4. 2 mL Syringe with an 18 G Needle.
5. Surgery Table.
6. 22 Blade on a #4 Scalpel Handle.
7. Atraumatic Forceps (e.g., DeBakey or Iris).
8. Metezenbaum and Mayo Scissors.
9. Backhaus Towel and Allis Tissue Clamps.
10. Rat Tooth Forceps.
11. Needle Drivers.
12. Hemostats.
13. Surgical Drape.
14. Anti-psychotic Drug, such as Acepromazine (0.05 mg/kg).
15. Xylazine Sedative Solution, such as AnaSed LA (0.03 mg/kg).
16. Flunixin Analgesic, such as Prevail (2.2 mg/kg).
17. Ampicillin For Injectable Suspension, such as Polyflex (10 mg/kg).
18. Ketamine, e.g., Zetamine (3 mg/kg).
19. Cuffed Endotracheal Tube with Bite Block.
20. Isoflurane Anesthesia Machine with Fluriso.
21. Heartbeat Monitor.
22. 70% Isopropyl Alcohol.
23. Surgical Scrub Solution: 4% Chlorhexidine.
24. 4×4 Gauze Pads and Stainless Steel Bowl with Saline.
25. 10% Povidone-iodine (e.g., Betadine). Embryo Transfer Medium.
26. Round-bottom Embryo Glass Dish.
27. 14 cm Open-Ended Tomcat Catheter with Adapter (e.g., Covidien) and 2 mL Syringe.
28. Stereomicroscope.
29. Sterile Saline.
30. CO_2 Incubator.
31. Portable Incubator, if required for transportation.

32. Thermal Beads are prewarmed to 38°C, if required for transportation.
33. Electric Wool Shears and Shaver with a 40 Blade.
34. #1 Polydioxanone Veterinary Sutures on a Reverse Cutting Needle.
35. Sterile 35 W Wound Staples.
36. Aluminum Bandage Spray (e.g., AluShield).
37. 1.0 mL Cryovials.
38. Pipettors.
39. Pipette Tips.
40. Four-well Dishes.
41. Gonadotropin Releasing Hormone (e.g., Fertagyl).

2.8 Pregnancy Checks and Lambing

1. Transabdominal Ultrasound Machine.
2. Lubricant.
3. Dexamethasone.

2.9 Blastocyst Cell Staining

1. Phosphate-Buffered Saline with Calcium and Magnesium (PBS+).
2. PB1+: PBS+ supplemented with 0.035 mg/mL Sodium Pyruvate, 1 mg/mL Glucose, 1% Penicillin/Streptomycin Solution, and 3 mg/mL BSA Fraction V.
3. Pronase (3.3 mg/mL).
4. Fix Washing Medium: DPBS supplemented with 0.25% BSA and 0.01% polyvinyl alcohol.
5. 4% Paraformaldehyde.
6. Permeabilization Medium: DPBS supplemented with 1% Triton X-100.
7. Washing Medium: DPBS supplemented with 0.05% Tween-20.
8. Blocking Medium: DPBS supplemented with 2% BSA.
9. Primary Antibody: Mouse anti-CDX-2 (BioGenex Cat#: MU392A-UC).
10. Primary Antibody Diluent: HK941.
11. Secondary Antibody: Goat anti-mouse Alexa Fluor 488.
12. DNA Stain: 20 μg/mL Hoechst 33342.
13. Mounting medium (e.g., Permafluor Mountant).
14. Glass Slides.
15. 18 mm² Glass Coverslips.
16. 96-well Plate with Lid.

17. Pipettors.
18. Pipette Tips.
19. 0.5 mL Microcentrifuge Tubes.
20. Aluminum Foil.
21. 100 mm Dish.

3 Methods

3.1 Abbatoir Ovary Collection

This protocol outlines the use of oocytes obtained from an abattoir, as this is the most readily available option for the majority of laboratories. However, oocytes can also be obtained from superstimulated animals through ovariectomy after culling or laparoscopic ovum pickup (*see* **Note 2**).

1. Prepare oocyte collection medium and store it at 4°C for no more than 2 weeks. On the day of oocyte acquisition, warm the oocyte collection medium to 38°C thoroughly before use.
2. Collect sheep ovaries from a local abattoir and transport them in 0.9% saline warmed to 30°C to a laboratory within 4 h of collection (*see* **Note 3**).
3. Rinse ovaries in the colander with 0.9% saline. Trim excess tissue and fat from the ovaries with scissors, and place them into a beaker with 40 mL of warmed oocyte collection medium.

3.2 Oocyte Collection

1. Place approximately 40 mL of oocyte collection medium in a 100 mm petri dish.
2. Hold each ovary in the petri dish with the tweezers. Release cumulus-oocyte complexes (COCs) from ovaries by slicing the entire surface with a scalpel blade.
3. Pour the medium containing COCs into a 50 mL tube and allow the sediment to settle for 5 min.
4. Place 20 and 3 mL of pre-warmed oocyte collection medium into a square 100 mm search plate and a 35 mm round petri dish, respectively (*see* **Note 4**).
5. Place the settled pellet from the 50 mL tube into the 100 mm search plate using a transfer pipette. Search for COCs under a stereomicroscope and place them into the round 35 mm collecting dish at room temperature (*see* **Note 5**).
6. Wash all COCs four times in oocyte collection medium in groups of 40 oocytes.

3.3 Oocyte In Vitro Maturation

1. Place 500 μL of IVM medium in each well of a four-well dish and allow at least 2 h of equilibration at 38.5°C and 5% CO_2 in air.
2. Wash all COCs through 1–2 wells of equilibrated IVM medium.
3. Place up to 40 oocytes per well of equilibrated 500 μL IVM medium.
4. Incubate at 38.5°C and 5% CO_2 in air for 21–22 h (Fig. 1).

3.4 Donor Cell Preparation

Thaw donor cells 3–5 days before nuclear transfer (*see* **Note 6**).

1. Quickly thaw one vial of cells stored in liquid nitrogen in a water bath at 37°C.
2. Transfer cell suspension and an equal volume of fresh cell culture medium to a 15 mL tube and centrifuge at 406 × *g* for 5 min to form a cell pellet.
3. Remove the supernatant and resuspend the pellet in 500 μL of cell culture medium.
4. Seed cells at various concentrations in a four-well dish in a final volume of 500 μL/well (*see* **Note 7**).
5. Remove and replace culture medium every 48 h.
6. Replace culture medium with cell starvation medium 24 h before SCNT.
7. Immediately before SCNT cell injection, rinse the cells with DPBS(−) then incubate them with 250 μL TrypLE™ Express for 5 min at 38.5°C and 5% CO_2 until the cells detach from the dish, rechecking at additional 1-min intervals until detachment is observed (*see* **Note 8**).
8. Add 500 μL of cell culture medium to inactivate the TrypLE, and transfer the cell suspension into a 15 mL tube.

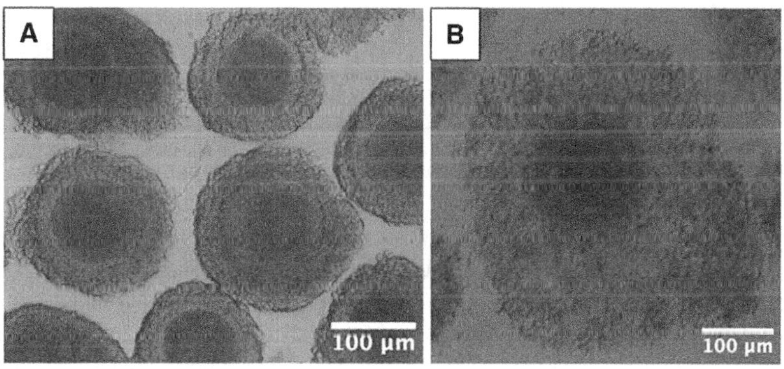

Fig. 1 Representative images (taken at 80x zoom) of (**A**) COCs before IVM and (**B**) a COC after IVM

9. Centrifuge at 406 × *g* for 5 min to form a cell pellet.
10. Remove the supernatant and gently resuspend the pellet in 60–80 μL HSOF (*see* **Note 9**).

3.5 SCNT

A flow diagram of the whole SCNT procedure, including oocyte in vitro maturation, donor cell preparation, embryo in vitro culture, and embryo transfer, can be found in Fig. 2. A more precise timeline of each step can be found in Fig. 3.

1. Prepare the following for each SCNT group in a flow hood, all under 2–3 mL of light mineral oil in a 35 mm culture dish:
   a. 1× post-fusion dish: seven drops of 40 μL IVC medium.
   b. 1× activation dish: seven drops of 40 μL activation medium.
   c. 1× CB dish: seven drops of 40 μL CB medium.

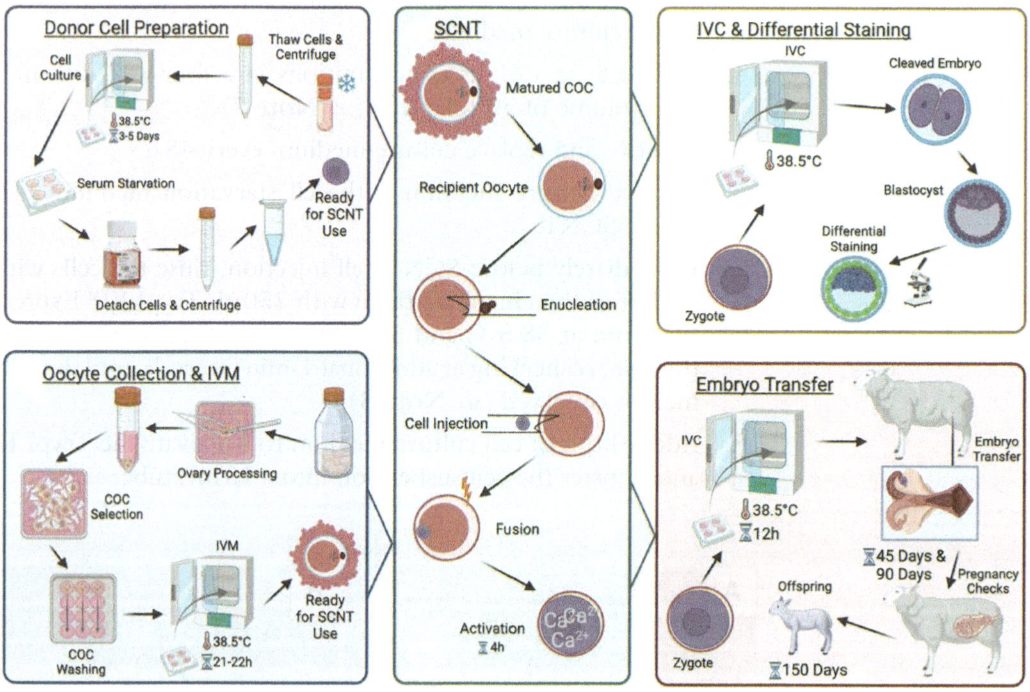

Fig. 2 Flow diagram of the sheep somatic cell nuclear transfer procedure. (**A**) Donor cell preparation, including cell culture, serum starvation for 24 h, and cell detachment for SCNT use. (**B**) Oocyte collection and IVM, including ovary processing, COC selection and washing, and IVM. (**C**) SCNT process, including enucleation, cell injection, fusion, and activation. (**D**) IVC and differential staining, including IVC of embryos, with cleavage rate checked on day 2 and blastocyst rate checked on day 7. Differential staining can be performed several months after blastocyst fixation. (**E**) Embryo transfer, including IVC, embryo transfer into estrus synchronized recipient ewes, pregnancy checks at days 45 and 90, and lambing approximately 150 days after embryo transfer. *SCNT* somatic cell nuclear transfer, *IVM* in vitro maturation, *COC* cumulus-oocyte complex, *IVC* in vitro culture. (Figure created in BioRender)

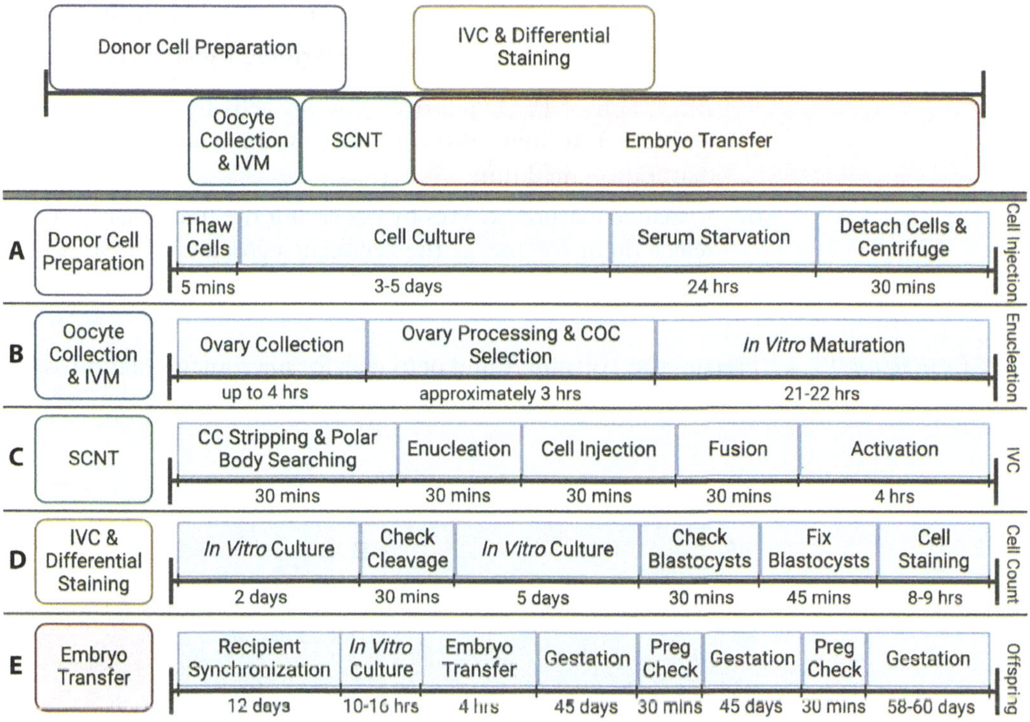

Fig. 3 A precise timeline of each step of the sheep somatic cell nuclear transfer procedure. (**A**) Donor cell preparation timeline, including cell culture, serum starvation, and cell preparation on the day of SCNT. (**B**) Oocyte collection and IVM timeline, including ovary processing and COC selection, as well as in vitro maturation. (**C**) SCNT timeline, including CC stripping and polar body searching, enucleation, cell injection, fusion, and activation. (**D**) IVC and differential staining timeline, including in vitro culture with a cleavage check on day 2 and a blastocyst check on day 7, fixing of blastocysts with paraformaldehyde, and differential cell staining, which can be performed several months after blastocyst fixing if storing at 4°C. (**E**) Embryo transfer timeline, including recipient synchronization, in vitro culture, embryo transfer, gestation, and pregnancy checks. *SCNT* somatic cell nuclear transfer, *IVM* in vitro maturation, *COC* cumulus-oocyte complex, *CC* cumulus cells, *IVC* in vitro culture. (Figure created in BioRender)

   d. 1× culture dish: seven drops of 40 μL IVC medium.

2. Prepare the following four-well dishes for post-fusion and post-activation washing, each well containing 400 μL of IVC medium under 400 μL of light mineral oil:
   a. 1× post-fusion wash dish (one well used for each group)
   b. 1× post-activation wash dish (one well used for each group).
3. Place all dishes into an incubator set to 38.5°C with 5% O_2, 90% N_2, and 5% CO_2. Allow them to equilibrate for approx. 2 h (*see* **Note 10**).
4. Prepare and pre-warm the following to 37°C in a water bath:

a. 15 mL Sheep Fusion medium in a conical centrifuge tube.

b. 25 mL HSOF in a conical centrifuge tube.

5. After 21 h of IVM, remove cumulus cells by gentle pipetting with a 1:1 volume of 0.6 mg/mL hyaluronidase stock:oocyte maturation medium.

6. Visualize mature oocytes by observing the first polar body and select them for use as the recipient cytoplasts for SCNT (*see* **Note 11** and Fig. 4).

3.5.1 Enucleation and Cell Injection

Prepare the 100 mm round petri dish for micromanipulation as outlined in Fig. 5. Wash the injection pipette in the injection washing

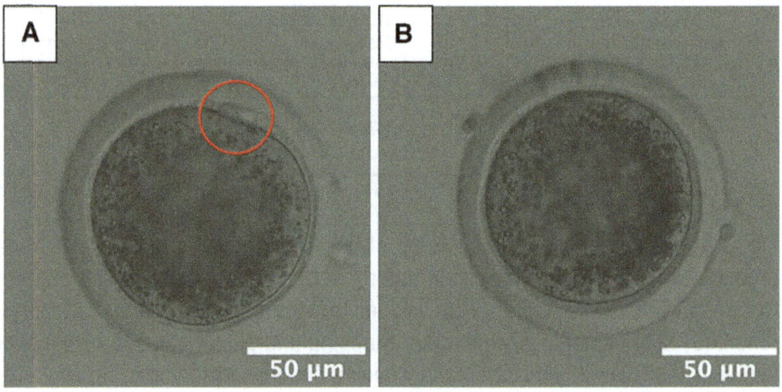

Fig. 4 Representative images (taken at 160x zoom) of (**A**) a mature (with a polar body circled) oocyte and (**B**) an immature oocyte

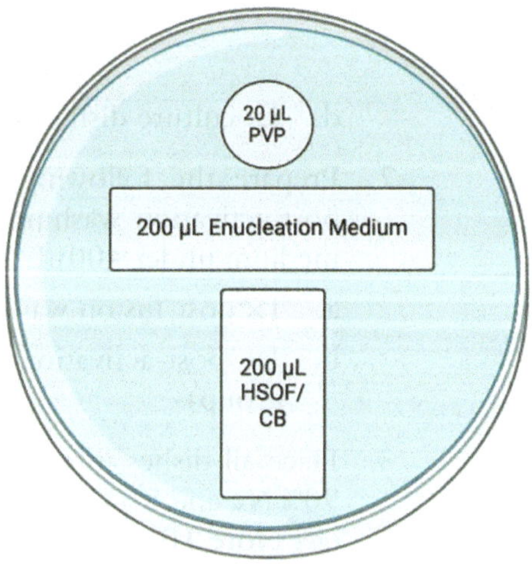

Fig. 5 Micromanipulation dish layout. (Figure created in BioRender)

medium containing 10% PVP prior to starting by aspirating the liquid approximately 7–8 times and ejecting.

1. Place oocytes (up to as many as may be processed in a 30-min time period) into the enucleation medium drop using a micropipette.
2. Immobilize an oocyte with a holding pipette (*see* **Note 12**).
3. Use the injection pipette to pierce the zona pellucida and remove the first polar body along with 10%–20% of the cytoplasm. Repeat Steps 3–4 for each oocyte (*see* **Note 13**).
4. Return enucleated oocytes to IVM medium and incubate at 38.5°C and 5% CO_2 for 30 min.
5. Gently pipette the donor cell suspension to mix well and place 5 μL of the cell suspension into the vertical HSOF/CB drop. Move enucleated oocytes to the same drop (*see* **Note 13**).
6. Pick up donor cells with the injection pipette and insert one cell into the perivitelline space of each enucleated oocyte while it is immobilized by the holding pipette (*see* **Note 14**). Once all oocytes are injected, place them into HSOF dish 1 (described below; *see* **Note 13**).

3.5.2 Fusion

1. Rinse the BTX microslide chamber in milli-Q ultrapure water and allow it to dry completely for approximately 15 min.
2. Connect the cables from the BTX ECM to the wires of the chamber and place the chamber onto the microscope stage (*see* **Note 15**).
3. Prepare two HSOF wash dishes (HSOF 1 and HSOF 2) and one fusion wash dish (Fusion) by placing 3 mL HSOF or 3 mL sheep fusion medium, respectively, in three 35 mm culture dishes. These are kept at room temperature (*see* **Note 4**).
4. Rinse the BTX microslide chamber with 600 μL of sheep fusion medium.
5. Replace with 600 μL of fresh sheep fusion medium.
6. Move up to 20 injected oocytes at a time from HSOF 1 into the Fusion dish (*see* **Note 16**). Allow the oocytes to sink to the bottom of the dish for approximately 10 s.
7. Place four to six injected oocytes at a time (from the Fusion dish) between the electrodes of the fusion chamber with the injected cell facing either the 12 or 6 o'clock position. Apply two DC electric fusion pulses of 2.0 kV/cm for 40 μs.
8. Remove fused oocyte-cell couplets and place them into HSOF 2. Repeat Step 7 until all injected oocytes in the Fusion dish have had an electric current applied.
9. Repeat Steps 5–8 until all injected oocytes have undergone the fusion process (*see* **Note 17**).

10. Wash all couplets from HSOF 2 in the equilibrated post-fusion wash well, then transfer into the equilibrated post-fusion dish, washing through all drops and incubating in the final drop at 38.5°C with 5% O_2, 90% N_2, and 5% CO_2. Evaluate fusion after 30 min by visualizing any unfused couplets, which will have an apparent somatic cell remaining between the zona pellucida and the oocyte membrane (Fig. 6), and removing them.

3.5.3 Activation

1. Move all fused embryos to the equilibrated CB dish and incubate at 38.5°C with 5% O_2, 90% N_2, and 5% CO_2 for 45 min.
2. Set a timer to 5 min. Wash embryos through two smaller drops of ionomycin medium (approx. 250 μL) and incubate in a third larger drop (approx. 500 μL) until the timer chimes.
3. Wash embryos through four 1 mL HSOF drops and one of the equilibrated post-activation wash wells.
4. Remove the equilibrated activation dish from the incubator and wash embryos through all seven drops, leaving them in the final drop at 38.5°C with 5% O_2, 90% N_2, and 5% CO_2 for 4 h.

3.6 In Vitro Culture (IVC)

1. After 4 h of activation, move and wash all surviving embryos through all drops of the IVC dish and culture at 38.5°C with 5% O_2, 90% N_2, and 5% CO_2 for 7 days, until blastocyst formation (*see* **Note 18**).
2. Replace 20 μL of BO-IVC™ medium every 2 days during in vitro culture with fresh, pre-equilibrated medium (minimum equilibration time of 2 h at 38.5°C with 5% O_2, 90% N_2, and 5% CO_2).
3. For SCNT embryo transfer/animal production, culture embryos for approximately 10–16 h. For in vitro embryo

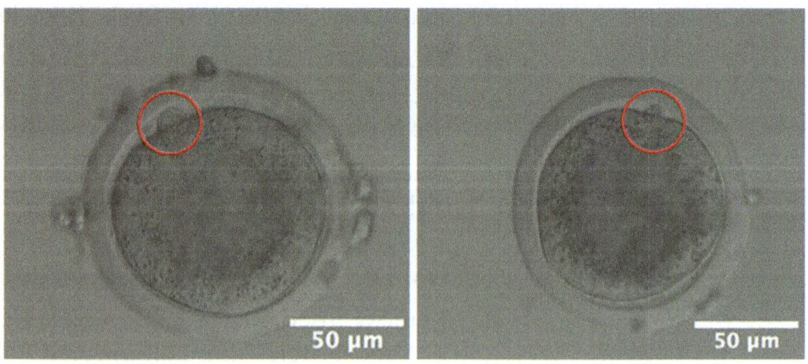

Fig. 6 Representative images (taken at 160x zoom) of unfused couplets, which have a clearly identifiable somatic cell (circled) between the zona pellucida and the oocyte membrane

assessment, culture embryos for 7 days and check the cleavage rate and the blastocyst rate on day 2 and day 7, respectively (*see* **Note 19**).

3.7 Recipient Synchronization and Embryo Transfers

1. Using proper disinfection and lubrication, insert one vaginal sponge into each recipient ewe's vagina and leave it in for 10 days (*see* **Note 20**).
2. Inject 500 µg of cloprostenol, a synthetic prostaglandin F2α (e.g., Estrumate) intramuscularly at the time of vaginal sponge removal (*see* **Note 21**).
3. Load SCNT embryos for transport to the surgical facility location by completely filling a 1 mL cryovial with equilibrated IVC medium, gently pipetting all embryos into the cryovial, and securing the cap. Place vial(s) into a portable incubator set to 38°C.
4. On site, remove cryovial(s) from the portable incubator and pipette all media and embryos into a four-well dish. Place the dish into a CO_2 incubator.
5. Prior to embryo transfer surgery, administer an anti-psychotic drug, a xylazine sedative solution, a flunixin analgesic, and ampicillin (e.g., Acepromazine, AnaSed LA, Prevail, and Polyflex) to each animal. Wait approximately 15 min until the ewe is calm enough to handle, then administer Ketamine (e.g., Zetamine) to complete pre-surgery anesthesia.
6. Insert the endotracheal tube with a bite block. Shave the wool off the stomach. Place the ewe onto the surgery table, secure the limbs, and apply disinfectant to the ewe's stomach, scrubbing by alternating the surgical scrub solution (4% Chlorhexidine) and 70% isopropyl alcohol for at least three rounds. Spray the ewe's stomach with povidone-iodine (10% betadine). Hook up the Isofluorane anesthesia machine to the endotracheal tube and the heartbeat monitor on the ear or tongue of the ewe.
7. The embryo transfer surgery follows all sterile procedures, using the backhaus towel clamps to secure the drape to the ewe's body and the allis tissue clamps to hook the drape over and around the rear hoofs to prevent contamination.
8. Make a small cut in the drape with mayo scissors to produce a surgical field. Make an incision 10–15 cm cranial to the udder and find the uterus, cutting tissue with Metzenbaum scissors if necessary.
9. Move SCNT embryos into a round-bottom dish with 2 mL of embryo transfer medium (*see* **Note 22**).
10. Remove the uterus from the body cavity, making sure not to dry it out by covering it with gauze pads soaked in sterile

saline, and measure the dominant follicle size on the ovaries (*see* **Note 23**).

11. Draw up all embryos with embryo transfer media into an open-end catheter and deposit them into the oviduct closest to the ovary with the dominant follicle. Check that all embryos went into the oviduct by flushing the catheter with DPBS in a round-bottom glass embryo dish under a stereomicroscope.
12. Place the uterus back into the body cavity and pour saline into the body cavity to prevent future uterine adhesions. Suture the linea alba and the subcutaneous tissue using needle drivers and rat tooth forceps, and staple the skin together.
13. Apply disinfectant and an aluminum bandage to the wound, take out the endotracheal tube, and allow the ewe to wake up from anesthesia in a padded room.

3.8 Pregnancy Checks and Lambing

1. Check pregnancy status by visualizing the uterus at day 45 and day 90 using transabdominal ultrasonography.
2. Recipients are allowed to lamb naturally (*see* **Note 24**).

3.9 Blastocyst Cell Staining

1. Visualizing on a stereomicroscope in a 100 mm dish, wash day 7 blastocysts through a 300 μL drop of PB1+, then incubate in a 30 μL drop of pronase for 2–5 min to remove the zona pellucida (ZP). Next, wash through three 300 μL PB1+ drops until all of the ZP is gone.
2. In a four-well dish, wash ZP-free blastocysts in 500 μL of fix washing medium, incubate in 500 μL of 4% PFA for 20 min, and wash twice, for approximately 30 s each time, in 500 μL of fix washing medium. Store fixed embryos in a microcentrifuge tube containing 200 μL of fix washing medium at 4°C until used for staining (*see* **Note 25**).
3. In a 96-well plate, permeabilize fixed blastocysts in 100 μL permeabilization medium for 1 h (*see* **Note 26**), followed by washing once in 100 μL of washing medium for 5 min.
4. Block non-specific binding by incubating blastocysts in 100 μ L of blocking medium for 1 h.
5. Transfer blastocysts to 100 μL of the primary antibody (diluted 1:100 with its diluent) for 1 h then wash in 100 μL washing medium three times for 15 min each.
6. Dilute secondary antibody 1:100 in blocking solution and transfer blastocysts to 100 μL for 1 h. Cover with foil and keep in the dark for the remaining steps, including this one.
7. Wash stained blastocysts in 100 μL washing medium three times for 15 min each.

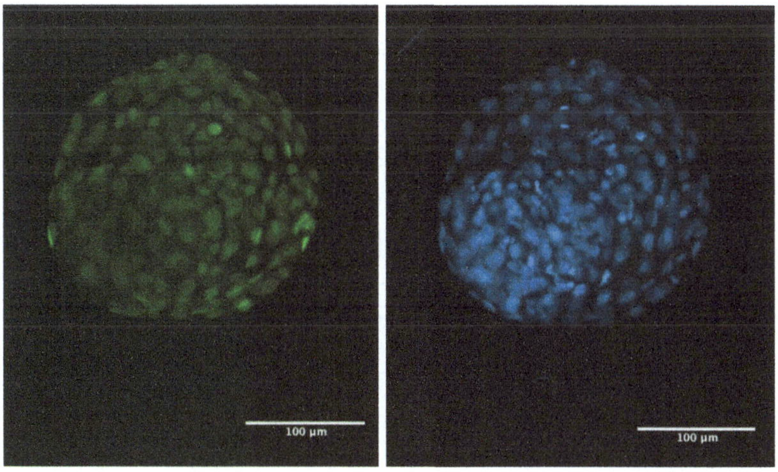

Fig. 7 Representative images of trophectoderm cells (Green) and all cells (Blue) in day 7 sheep blastocysts. Green staining is achieved using a mouse anti-CDX2 primary antibody and an AF488 secondary antibody. Blue stain is Hoechst 33342

8. Stain all cellular DNA in 100 µL Hoechst 33342 diluted in washing medium to achieve 20 µg/mL for 10 min.
9. Pipette each blastocyst into 5 µL of mounting medium placed on slides under a cover glass.
10. Take images with the DAPI (blue filter) and AF488 (green filter) settings on a fluorescent microscope (*see* **Note 27**; for representative images, see Fig. 7).

4 Notes

1. Micropipettes for injection and holding are available for purchase through commercial suppliers (e.g., Origio or TPC). In-house prepared injection pipettes are pulled with a Sutter Instrument micropipette puller, beveled at a 45° angle on a Narishige microgrinder and subsequently shaped with a Narishige micro-forge. Holding pipettes excludes a bevel.
2. Ovarian superstimulation can be achieved over the course of 7 days through the use of a CIDR and FSH injections. Ovariectomy after ovarian superstimulation in sheep allows for the harvesting of oocytes from adult animals, while laparoscopic ovum pick up negates the need for animal culling. Please contact the authors if you are interested in either of these oocyte collection methods.

3. All ovary collections for cloning were performed during the sheep breeding season, which is started by shorter day length (i.e., October through January in the Northern hemisphere).
4. All embryo and oocyte handling is performed in warm rooms kept at a steady 28°C.
5. For abattoir-derived COCs, select oocytes with three or more layers of compact cumulus cells and homogenous cytoplasm.
6. Cells used for SCNT procedures in our laboratory are sheep fetal fibroblasts or skin fibroblasts from young and adult animals. Cells may be acquired from culturing a small tissue biopsy from a living donor.
7. We usually seed our cells at 40,000–60,000 cells/mL and use cell culture medium to dilute accordingly.
8. Harsher enzymes are not generally necessary, as they could cause additional trauma to the cells. However, 1× trypsin EDTA may also be used. Gentle agitation of the dish aids in speeding up cell detachment.
9. The amount of HSOF used for resuspension will be adjusted based on the number of cells present. Use less HSOF if there are fewer cells or more to dilute higher-density cell suspensions. The desired density, once resuspended, is around 28,000 cells/mL. This adds 150–200 cells with every 5 μL, allowing the selection of the best cells during cell injection. Once resuspended in HSOF, cells should be used for injection immediately or kept at 4°C for up to 2 h.
10. Avoid splashing oil on the lids/rims of the dishes as this will seal them and prevent gas exchange.
11. Oocyte maturation is assessed by the presence of a first polar body under a stereo-microscope (metaphase of meiosis II). Oocyte selection for SCNT includes consistent cytoplasm and good morphology to increase the developmental potential of subsequent embryos.
12. The first polar body should be at the 4–5 o'clock position when the oocyte is immobilized.
13. It is important to handle only the number of oocytes that may be processed within a 30-min time period for all steps of SCNT, including fusion. For an experienced embryologist, this may be 50 oocytes, whereas for a novice, it may only be ten.
14. Select round, healthy cells for injection to increase the chances of fusion and embryo development.
15. To secure the chamber to the stage, tape may be used on the cables.

16. The number of embryos moved from HSOF 1 to the Fusion dish will depend on the embryologist's fusion skill, ideally taking 5–10 min per group. If experienced, fusion of 20 embryos may take 5 min, whereas for a novice it may take 15 min. Therefore, a novice may want to only move five embryos at a time to avoid prolonged exposure time to sheep fusion medium.
17. Replace the sheep fusion medium in the chamber every 15–20 min or between groups. HSOF 1, Fusion dish, and HSOF 2 may be reused until all injected oocytes have gone through the fusion process, if the fusion step is completed within 2 h.
18. Surviving embryos may be identified by visualizing an intact and solid plasma membrane. Lysed embryos will appear "grainy" or "sandy" and should be removed (Fig. 8).
19. Cleavage rate = the number of cleaved embryos/total number of embryos. Blastocyst rate = the number of blastocysts/the number of cleaved embryos. Blastocyst rate can be used as one assessment for oocyte quality because only the embryos that come from high-quality oocytes develop to this stage.
20. Recipient ewes are usually 2–5 years old and have lambed before. This ensures they are mature and can carry offspring to term.
21. Estrus should occur 36–48 h after sponge removal. Ovulation should occur 12–24 h after estrus.
22. For SCNT embryos derived from abattoir oocytes, we transfer 13–18 embryos into each recipient ewe.
23. The uterus should be kept covered by saline-soaked gauze to prevent drying out. The ideal ovarian status for embryo transfer in sheep is a 6–12 mm follicle or the presence of a corpus hemorrhagicum, which indicates recent ovulation. Recipients that exhibit a 5 mm dominant follicle usually receive 100 μg of

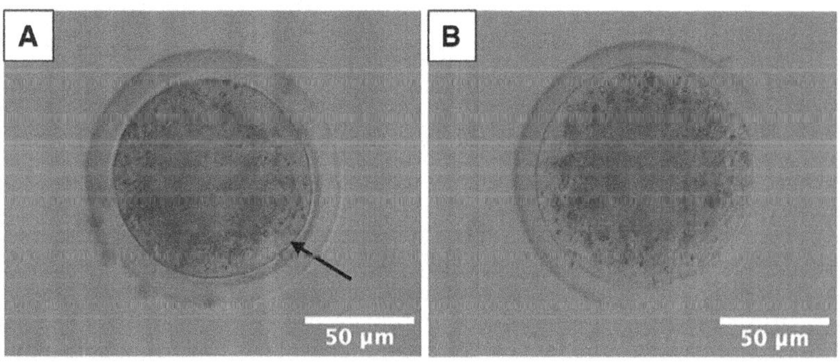

Fig. 8 Representative images (taken at 160x zoom) of (**A**) a surviving embryo and (**B**) a lysed embryo. The arrow is pointing to the oolemma, which cannot be seen in the lysed embryo

gonadotropin-releasing hormone intramuscularly at the time of embryo transfer.

24. If recipients have not lambed naturally by gestation day 148, induce with 15 mg of dexamethasone intramuscularly 24 h before cesarean section surgery. For postnatal care of SCNT animals, refer to the 5th Edition IETS Manual: Chapter 11A [24].
25. Fixed embryos can be stored in solution at 4°C for several months to 1 year prior to staining.
26. All steps from permeabilization to DNA staining should be performed in a 96-well plate on a rocker at room temperature. For staining in a 96-well plate, a p10 pipettor set to 4 μL may be utilized to move embryos. From the secondary antibody onward, all steps should be performed in the dark to prevent photobleaching.
27. Hoechst 33342 (DAPI) stains all cells, and CDX2 (AF488) will selectively stain trophectoderm cells. Inner cell mass numbers can be calculated by subtracting the number of trophectoderm cells from the total number of cells.

References

1. Wilmut I, Schnieke AE, McWhir J, Kind AJ, Campbell KHS (1997) Viable offspring derived from fetal and adult mammalian cells. Nature 385:810–813
2. Polejaeva IA, Chen SH, Vaught TD, Page RL, Mullins J, Ball S et al. (2000) Cloned pigs produced by nuclear transfer from adult somatic cells. Nature 407(6800):86–90
3. Cibelli JB, Stice SL, Golueke PJ, Kane JJ, Jerry J, Blackwell C et al. (1998) Cloned transgenic calves produced from nonquiescent fetal fibroblasts. Science 280(5367):1256–1258
4. Cortez JV, Hardwicke K, Cuervo-Arango J, Grupen CG (2023) Cloning horses by somatic cell nuclear transfer: Effects of oocyte source on development to foaling. Therio 203:99–108
5. Baguisi A, Behboodi E, Melican DT, Pollock JS, Destrempes MM, Cammuso C et al. (1999) Production of goats by somatic cell nuclear transfer. Nat Biotech 17:456–461
6. Polejaeva IA, Broek DM, Walker SC, Zhou W, Walton M, Benninghoff AD, Faber DC (2013) Longitudinal study of reproductive performance of female cattle produced by somatic cell nuclear transfer. PLoS One 8(12):e84283
7. Polejaeva I, Rutigliano H, Wells K (2016) Livestock in biomedical research: history, current status and future prospective. Reprod Fert Dev 28(2):112–124
8. Polejaeva IA (2021) 25th Anniversary of cloning by somatic cell nuclear transfer: generation of genetically engineered livstock using somatic cell nuclear transfer. Reprod Suppl 162(1):F11–F22
9. Phelpfs CJ, Koike C, Vaught TD, Boone J, Wells KD, Chen SH et al. (2003) Production of alpha 1,3-galactosyltransferase-deficient pigs. Science 299(5605):411–414
10. Wu H, Fan Z, Brandsrud M, Meng Q, Bobbitt M, Regouski M et al. (2019) Generation of H7N9-specific human polyclonal antibodies from a transchromosomic goat (caprine) system. Sci Rep 9(1):366
11. Fan Z, Perisse IV, Cotton CU, Regouski M, Meng Q, Domb C, et al. (2018) A sheep model of cystic fibrosis generated by CRISPR/Cas9 disruption of the CFTR gene. JCI Insight 3(19):e123529
12. Loi P, Ptak G, Barboni B, Fulka J, Cappai P, Clinton M (2001) Genetic rescue of an endangered mammal by cross-species nuclear transfer using post-mortem somatic cells. Nat Biotech 19(10):962–964
13. Adams L, Liu Y, Polejaeva IA (2024) Current status of interspecies somatic cell nuclear transfer and meta-analysis of the effects of phylogenetic distance on embryonic and fetal development. Mammal Rev 54(4):387–403. https://doi.org/10.1111/mam.12352

14. Evans MJ, Gurer C, Loike JK, Wilmut I, Schnieke AE, Schon EA (1999) Mitochondrial DNA genotypes in nuclear transfer-derived cloned sheep. Nat Genet 23:90–93
15. Long CR, Westhusin ME, Golding MC (2014) Reshaping the transcriptional frontier: epigenetics and somatic cell nuclear transfer. Mol Reprod Dev 81:183–193
16. Zhang X, Gao S, Liu X (2021) Advance in the role of epigenetic reprogramming in somatic cell nuclear transfer-mediated embryonic development. Stem Cells Int 2021:6681337
17. Leibfried-Rutledge ML, Critser ES, Eyestone WH, Northey DL, First NL (1987) Development potential of bovine oocytes matured in vitro or in vivo. Biol Reprod 36:376–383
18. Caetano LC et al. (2023) In vivo and in vitro matured bovine oocytes present a distinct pattern of single-cell gene expression. Zygote 31:31–43
19. O'Brien JK, Catt SL, Ireland KA, Maxwell WMC, Evans G (1997) In vitro and in vivo developmental capacity of oocytes from prepubertal and adult sheep. Therio 47(7):1433–1443
20. Ferrer-Roda M, Izquierdo D, Gil A, Oliveira MEF, Paramio M-T (2024) Oocyte competence of prepubertal sheep and goat oocytes: an assessment of large-scale chromatin configuration and epidermal growth factor receptor expression in oocytes and cumulus cells. Int J Mol Sci 25(8):4474
21. Thompson JG, Gardner DK, Pugh PA, Mcmillan WH, Tervit HR (1995) Lamb birth weight is affected by culture system utilized during in vitro pre-elongation development of ovine embryos. Biol Reprod 53(6):1385–1391
22. Lazzari G, Wrenzycki C, Herrmann D, Duchi R, Kruip T, Niemann H et al. (2002) Cellular and molecular deviations in bovine in vitro-produced embryos are related to the large offspring syndrome. Biol Reprod 67(3):767–775
23. Liu Y, Stott R, Regouski M, Fan Z, Viotti Perisse I, Patrick T et al. (2024) A retrospective analysis of sheep generated by somatic cell nuclear transfer. Therio 227:102–111
24. Chavatte-Palmer P, Lee R, Oback B, Fermin LM, Stott R, Sweat A et al (2024) Chapter 11a Health and welfare of cloned ruminants. In: IETS Manual, 5th edn. vol. 1. pp 151–169. https://www.iets.org/Publications/IETS-Manual

Chapter 24

Clustered Regularly Interspaced Palindromic Repeats (CRISPR) Applied to Gametes and Embryos

Munuse C Savash Ishanzadeh and Dagan Wells

Abstract

Preimplantation genetic testing (PGT) facilitates the identification of embryos affected by specific types of genetic abnormalities. However, PGT does not seek to treat the genetic abnormality, rather it is an embryo selection tool, employing a strategy of detection and exclusion. Until recently, the notion that mutations and aneuploidies could be corrected in gametes, or in embryos produced using in vitro fertilization (IVF), seemed improbable. However, rapid progress in the evolution of gene editing technologies may make this a realistic possibility in the near future. Not only would such an approach help to avoid the discard of human embryos, which some find challenging from ethical or religious perspectives, but it would also increase the number of embryos considered suitable for transfer, potentially leading to higher pregnancy rates than achieved in PGT cycles. This chapter considers the use of genome editing applied to human preimplantation embryos, describing a protocol that can be used to inactivate genes for research purposes, and which might, in the future, allow for the correction of pathogenic mutations.

Keywords CRISPR-Cas9, Genome editing, Human embryo, Double strand breaks, Genome sequencing

1 Introduction

Genome editing has been utilized since the 1990s; however, early tools such as zinc finger nucleases (ZFNs) and transcription activator-like effector nucleases (TALENs) were challenging to design and had practical limitations that restricted their use [1, 2]. The development of clustered regularly interspaced palindromic repeats (CRISPR) greatly simplified and enhanced the ability of scientists to interact with the genome. CRISPR is a simple and efficient tool, consisting of two components: a guide RNA (gRNA), which is an RNA sequence complementary to a chosen target DNA sequence; and a CRISPR-associated protein 9 (Cas9) endonuclease, an enzyme which creates a double strand break (DSB) in the DNA by incising both strands precisely three bases away from a protospacer activation motif (PAM) (Fig. 1) [3]. PAM sites are characterized by the

Alexandra J. Harvey and David K. Gardner (eds.), *The Preimplantation Embryo*, Methods in Molecular Biology, vol. 3038, https://doi.org/10.1007/978-1-0716-5292-3_24,

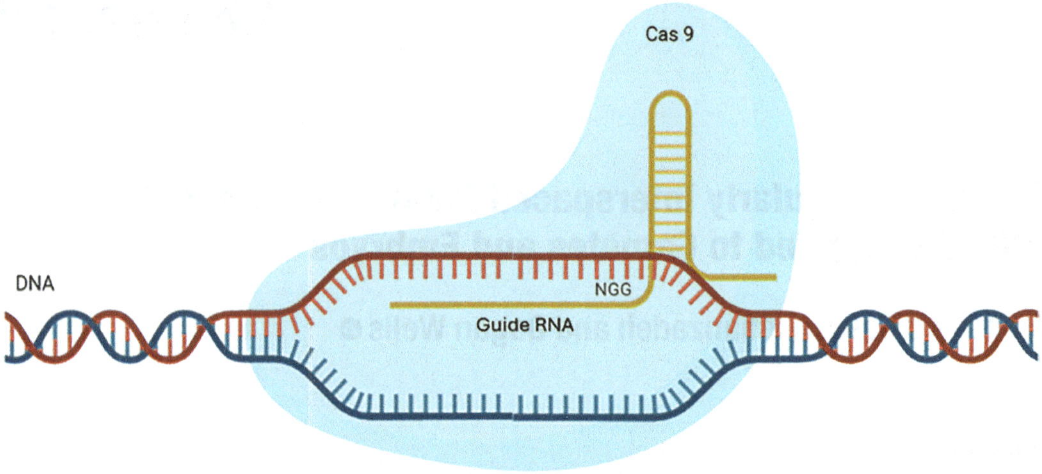

Fig. 1 Illustration of a CRISPR complex performing an edit of a DNA. The Cas9 protein (blue) is guided to the target DNA sequence by the guide RNA (orange). The guide RNA binds to the complementary DNA strand, allowing the Cas9 endonuclease to recognize and cleave the DNA at a specific site in the vicinity of a protospacer adjacent motif (PAM) sequence (NGG) resulting in a double strand break [24]

sequence NGG and need to be located directly downstream of the gRNA target sequence for the Cas9 endonuclease to bind. The gRNA consists of CRISPR RNA (crRNA) and transactivating CRISPR RNA (tracrRNA), the former having a complementary sequence to the target site and the latter helping to facilitate the interaction between crRNA and the Cas9 endonuclease by acting as a scaffold [4]. In the laboratory, the crRNA can be synthesized and its sequence defined so the Cas9 targets the intended genomic locus.

Since CRISPR-Cas9 involves the introduction of a DSB, it is essential that the treated cells have the capacity to repair the induced damage. While DNA repair is crucial in all cells, it is particularly important in preimplantation embryos. Failed or abnormal repair, leading to the presence of mutations or unresolved damage, can impair embryo viability or cause genetic abnormalities. Eukaryotic cells have evolved multiple repair pathways to ensure the maintenance of cellular genetic health and survival [5]. Two of the principal repair pathways used by cells are homology directed repair (HDR) and non-homologous end joining (NHEJ). HDR is a mechanism that involves rebuilding the site of DNA damage using a closely related sequence as a template. In the natural situation, this sequence is located on the homologous chromosome. The intact copy of the affected genomic site, including flanking sequences, is copied across, thus repairing the DSB. This is known as interhomolog homologous recombination (IH-HR). However, in an experimental context, an artificial DNA construct with a sequence that closely matches the area containing the DSB, but with some changes with respect to the native

DNA sequence, can be introduced into cells. The construct may then be used as a template for repair, incorporating the desired alterations of the sequence into the genome during the process of HDR. In the context of therapy, this approach could be used to substitute a mutant sequence (present in the cell) for that of the wild-type (present in the artificial construct), i.e., to "correct" a mutation.

NHEJ does not utilize a template for repair. Instead, the broken ends of DNA are directly joined (ligated) together. This mechanism is highly error-prone and typically introduces insertions or deletions (indels) of a few nucleotides at the site of repair [6]. Clearly, NHEJ is not helpful if the intention is to remove a mutation from the genome. However, this mode of repair is of great value when the aim is to inactivate a gene, since the indels introduced frequently disrupt the reading frame of the targeted gene, leading to loss of function of the protein produced.

In most cell types, checkpoints interrupt the progression of the cell cycle when damage occurs, thus preventing replication of DNA until repair has been completed. Little is known about the activity of DNA repair pathways during preimplantation embryo development; however, it appears that cell cycle checkpoints are unusually relaxed during the first few mitoses following fertilization [7]. Furthermore, the high incidence of segmental aneuploidy observed during early developmental stages (~24.3% of human preimplantation embryos have some affected cells) is suggestive of a deficiency in DSB repair during the first few days of developement [8]. It is, therefore, reasonable to hypothesize that early human embryos may be particularly vulnerable to genotoxic damage, and that DSBs induced by CRISPR-Cas9 might be problematic. It is likely that DNA repair remains relatively inefficient until the activation of the embryonic genome, around the 4–8 cell stage in human embryos. Repair mechanisms and repair capacity need to be carefully considered when planning an experiment using CRISPR-Cas9 or any other genome editing technology that causes breakage of DNA strands, as the specific mode of repair will influence how the DNA sequence is altered at the targeted site, while the efficiency of repair may influence the viability of the treated embryos.

CRISPR-Cas9 gene editing has frequently been applied to research involving mammalian embryos, especially the mouse. However, extrapolation of the findings to human development is not always valid due to differences in timings of key developmental events, and variation in the molecular mechanisms governing important cellular pathways. Indeed, in the first study to deploy CRISPR-Cas9 for the study of human embryos, inactivation of *OCT4* in human and mouse oocytes at the time of fertilization resulted in differing consequences in the two species investigated [9]. While this research succeeded in increasing knowledge of early embryonic development, it also revealed some unintended outcomes of CRISPR-Cas9 editing, including failed or inappropriate repair of DSBs in some cells—a potentially worrisome finding if genome editing at the preimplantation stage were to be considered for clinical use [10]. In further pre-clinical work in human embryos, CRISPR-Cas9 reagents were injected along with a single sperm during intracytoplasmic sperm injection (ICSI), and the researchers involved claimed successful correction of an inherited mutation in more than half of the resulting embryos [11].

Genome editing studies using human embryos have usually involved injection of CRISPR-Cas9 reagents either during ICSI or at the zygote (1-cell) stage [9, 11]. In theory, this approach seems highly desirable. The delivery of reagents can be virtually guaranteed by microinjection, and the edits introduced should be present throughout the embryo, since all cells will be descended from the injected oocyte/zygote. However, there may be difficulties as this early developmental stage appears to be the time when DNA repair processes are at their least efficient. Several studies involving genome editing applied to human preimplantation embryos have reported abnormal DNA repair, complete failure of repair, or have been unable to provide a definitive evaluation of whether the targeted site has been edited due to an inability to distinguish certain repair outcomes from failed/abnormal repair [10–13].

A full discussion of ethical considerations around germline genome editing is beyond the scope of this chapter. However, it must be acknowledged that substantial concerns have been raised concerning the safety and ethics of heritable genome editing. In some quarters, there has been unease over the potential misuse of the technology. There remain many unresolved questions, such as concerns over the potential for genome editing to be used for enhancement rather than cure, costs of the technology, and equity of access, the possibility of potential impacts on the way society views people living with inherited conditions, to name but a few. More detailed discussion of the appropriateness of germline genome editing has been published elsewhere [14–16].

The method described in this chapter represents an optimized protocol that has been used successfully in genome editing experiments involving human embryos. The creation and validation of gRNAs is described, followed by information on the application of CRISPR-Cas9 to cells from embryos, and protocols for evaluating the outcome of genome editing.

2 Materials

2.1 Guide Design and Cas9 Complex

1. Guide RNAs for the gene or target of interest (*see* **Note 1**).
2. Cas9 enzyme.
3. RNP complex comprised of guide RNA and Cas9 enzyme.

2.2 Validation and application

1. Oragene Saliva self-collection/swab kit.
2. Qiamp DNA mini kit, containing Buffer AL, Proteinase K, Buffer AW1, Buffer AW2, Buffer AE.
3. Qubit 1× dsDNA HS assay kit containing 1× dsDNA HS working solution, 1× dsDNA HS Standard 1, and 1× dsdDNA Standard 2.
4. Qubit assay tubes or tube strips.
5. Platinum II Taq Hot-Start kit containing Platinum II PCR Buffer and Platinum II Taq Hot-Start.
6. dNTPs.
7. Nuclease-free water.
8. CutSmart Buffer.
9. UltraPure Agarose powder.
10. 10× TBE.
11. 1× TBE: dilute 10× TBE 1:10 with nuclease-free water.
12. Agarose ultrapure powder
13. GelRed Nucleic Acid Gel Stain.
14. 10× BlueJuice Gel Loading Buffer.
15. 100 bp low-scale DNA ladder.
16. Embryo culture media such as Genea, Cooper, Fuji, or Vitrolife.
17. Quick calf intestinal phosphatase.
18. Primary Template-directed Amplification (PTA) (Bioskryb Genomics),
19. Nanopore ligation sequencing V14-PCR barcoding kit (SQK-LSK114), with the following reagents required from the kit: Ligation adapter (LA), Ligation Buffer (LNB), Long fragment buffer (LNB), Sequencing buffer (SB), Elution buffer (EB), Library beads (LIB), Library solution (LIS), Flow cell flush (FCF) and Flow cell tether (FCT).
20. Ultra II End-prep reaction buffer.
21. Ultra II End-prep enzyme mix.
22. AMPure XP beads.
23. Molecular grade ethanol.
24. 80% ethanol.
25. Nanopore EXP-PBC096 kit containing Barcode adapters 1-96 and corresponding barcodes 1-96
26. Blunt/TA Ligase Master mix, containing T4 DNA ligase in an optimized reaction buffer.
27. NEBNext Quick T4 DNA ligase.

28. TWIST Library Preparation Enzymatic Fragmentation (EF) Kit containing a ligation mastermix, Equinox Library amp mastermix.
29. TWIST Universal Adapter System/TWIST Long Read Library Preparation and Standard Hyb v2 Enrichment containing TWIST universal adapter, TWIST UDI primers, and DNA purification beads.
30. LongAmp® Taq 2× Master Mix is comprised of a concentrated solution with an optimised enzyme.
31. Twist Hybdrisation reagents containing hybridisation mix and hybridisation enhancer.
32. TWIST custom panel.
33. TWIST universal blockers.
34. TWIST wash buffers containing wash buffers 1 and 2.
35. Streptavidin Dynabeads.
36. Acid Tyrode's.
37. Tapestation DNA D1000 screentape.
38. Tapestation DNA D1000 reagents (containing sample buffer and ladder).

2.3 Equipment and Consumables

1. Pipette tips.
2. Manual pipettors.
3. 0.2 mL PCR tubes.
4. 1.5 mL sterile Eppendorf tubes.
5. Heat block suitable for 1.5 mL Eppendorf tubes.
6. Vortex.
7. Primers flanking gRNA.
8. PCR thermocycler.
9. Qubit fluorometer.
10. Tapestation.
11. Gel casting tray and 15-well combs.
12. Gel loading tips.
13. Electrophoresis unit.
14. Nanopore MinION.
15. Nanopore flow cells.
16. Magnetic stand.
17. Centrifuge.
18. Minicentrifuge.
19. Micromanipulator with injection and holding pipettes.

3 Methods

3.1 Guide Design and Cas9 Complexing and Validation

1. Define the gene or target of interest and download the target sequence from NCBI, https://www.ncbi.nlm.nih.gov (*see* **Note 1**).
2. The CRISPR guides are designed using online tools such as Benchling https://www.benchling.com/crispr (*see* **Note 2**). Once registered, the page will contain a sidebar menu, click + > CRISPR > CRISPR guides.
3. Insert the gene name to import the sequence. Once loaded, the type of design can be selected (single or paired guides and guide length; *see* **Note 3**) and the target region highlighted. Potential guides will then be generated and presented for review (*see* **Note 4**).
4. Load the IDT website (https://www.idtdna.com/) and navigate to Products & Services > CRISPR gene editing > Guide RNAs > Cas9 Guide RNAs > Use your own design. The designed guide can be pasted and then purchased (*see* **Note 5**).
5. An RNP complex can be created by combining the gRNA and the enzyme. When using a single-guide RNA (sgRNA) dilute the sgRNA in PCR grade water to the desired concentration (~1–3 μM). Dilute the Cas9 enzyme in Cas9 nuclease reaction buffer to the desired concentration (~1–3 μM) (*see* **Note 6**). Combine the diluted sgRNA and diluted Cas9 at a 1:1 ratio and incubate at room temperature for 10–20 min.

3.1.1 PCR gRNA Validation

Once the RNP complex has been generated, validation should be carried out to ensure the complex will create a DSB at the target location. One method of validation involves amplifying a DNA fragment including the targeted site using a polymerase chain reaction (PCR) and demonstrating that the amplicons can be successfully cleaved.

1. Design oligonucleotide primers flanking the region encompassed by the guide RNA using any primer design software package (e.g., Primer3, https://www.primer3plus.com/) (*see* **Note 7**).
2. To obtain genomic DNA start by collecting a buccal sample using a swab kit.
3. Using a QIAamp DNA Mini kit, combine 400 μL of the buccal sample with 20 μL of protease K solution and 400 μL of Buffer AL in a sterile 1.5 mL Eppendorf tube and vortex.
4. Place the solution in a heat-block set to 56°C for 10 min.
5. Following incubation, add 400 μL of molecular grade ethanol and mix well by vortexing for 30 secs.

6. Transfer 700 μL of the mixture to a QIamp mini spin column already placed in a 2 mL collection tube and centrifuge for 1 min at 8000 rpm.
7. Discard the flow through and add the remaining sample mixture to the column. Centrifuge at 8000 rpm for 1 min.
8. Discard the flow through and transfer the spin column to a clean 2 mL collection tube. Add 500 μL of Buffer AW1 to the spin column and centrifuge at 8000 rpm for 1 min.
9. Discard the flow through and transfer the spin column to a clean 2 mL collection tube. Add 500 μL of Buffer AW2 and centrifuge at 14,000 rpm for 3 min.
10. Discard the flow through, transfer the spin column to a clean 2 mL collection tube, and centrifuge for another minute at 14,000 rpm.
11. Place the spin column in a labelled 1.5 mL Eppendorf tube. Add 100 μL of Buffer AE, incubate for 1 min at room temperature, then centrifuge the spin column at 8000 rpm for 1 min.
12. Quantify extracted DNA using a Qubit fluorimeter using the Qubit 1× dsDNA HS assay kit. Add 10 μL of each standard to a Qubit tube, and 1–20 μL of each sample to a Qubit tube. Genomic DNA samples can be stored at 4°C for later use.
13. Add Qubit™ 1× dsDNA working solution to each tube to a final volume of 200 μL.
14. Vortex vigorously for 3–5 s.
15. Incubate tubes at room temperature for 2 min then read samples on the Qubit fluorometer.
16. To conduct a PCR, prepare a 25 μL DNA amplification reaction in 0.2 mL PCR tubes comprising 6 μL reaction buffer (1×), 0.6 μL of 10 mM dNTP mixture, 0.6 μL of pooled primers (each at 10 mM), 1 unit of Platinum II Taq polymerase, and 10–20 ng/μL of genomic DNA, the final volume adjusted to 25 μL with nuclease-free water (*see* **Note 8**).
17. Cycle reaction through the following series of incubations in a PCR thermocycler: 94°C for 2 min, then 35 cycles of denaturation at 94°C for 15 s, primer annealing at 58°C for 15 s, primer extension at 68°C for 15 s, followed by hold at 4°C (*see* **Note 9**).
18. Determine the concentration of the PCR reaction using a Qubit fluorimeter using the Qubit 1× dsDNA HS assay kit. Add 10 μL of each standard to a Qubit tube, and 1–20 μL of each sample to a Qubit tube. Genomic DNA sSamples can be stored at 4°C for later use.13.Add Qubit™ 1× dsDNA working solution to each tube tofor a final volume of 200 μ L.14.Vortex vigorously for 3–5 s.15.Incubate tubes at room temperature for 2 min then read samples on the Qubit fluorometer.

19. Combine 1 μL of 100 nM of the resulting PCR product with 7 μL of nuclease-free water, 1 μL of 10× NEB CutSmart buffer.
20. Incubate the reaction at 37°C for 60 min in a PCR thermocycler (*see* **Note 10**).
21. Inactivate the enzyme by incubating the sample at 65°C for 10 min.
22. Add 1 μL of RNP complex to the sample and incubate at 80°C for 10 min.
23. Following incubation the sample should be ran on an agarose gel by electrophoresis.
24. Prepare a 1.5% agarose gel by combining 100 ml of 1× TBE and 1.5 g of UltraPure agarose gel and heat in a microwave until the powder is dissolved (ensure to be careful with the molten agarose as it can cause burns). Once slightly cool, add 10 μL of GelRed and swirl the flask gently to mix.
25. Prepare a gel cassette by placing 2 × 15-well combs into a gel casting tray, then pour the warm 1.5% agarose mix into the tray. Allow to solidify (can take up to 20 min). Remove the combs once solidified and place them into an electrophoresis tank filled with 1× TBE up to the maximum mark.
26. Combine 3 μL of CRISPR-Cas9 treated sample with 0.5 μL of DNA gel Loading Dye (6×) and load into wells of a 1.5% agarose gel.
27. A DNA ladder should be run alongside the samples as a control, allowing estimation of the size of the amplified DNA fragments (e.g., 5 μL of 100 bp Low Scale DNA Ladder) (*see* **Note 11**).
28. Run at 120 V for 40 min on a gel electrophoresis system.
29. View the gel under UV light. Successful editing will be confirmed by the presence of two bands on the gel from the sample.

3.1.2 Nanopore Ligation Sequencing V14-PCR Barcoding

A second method for validating the specificity of the guide RNA is to use Nanopore sequencing. Samples are sequenced via Nanopore ligation sequencing V14-PCR barcoding (SQK-LSK114 with EXP PBC096) as detailed below (the original protocol was slightly altered to minimise reagent use). As such, any sites cleaved by CRISPR-Cas9 will be revealed. This can also be a useful way of identifying any off target editing (*see* **Note 12**).

1. Dephosphorylate the 5′ ends of fragments within a genomic DNA sample (extracted buccal swab) by combining 1–10 μg of DNA with 3 μL of 10× NEB CutSmart Buffer in a 0.2 mL tube and make up to a total reaction volume of 27 μL with nuclease-free water.

2. Add 3 μL of Quick calf intestinal phosphatase (5000 units/mL)
3. Incubate the sample at 37°C for 10 min, then 80°C for 2 min (*see* **Note 13**).
4. Add 3 μL of RNP complex and incubate at 80°C for 10 min.
5. Combine 30 μL of amplified product with 4.5 μL of Ultra II End-prep reaction buffer and 1.8 μL of Ultra II End-prep enzyme mix.
6. Mix the reaction well by vortexing, spin, and incubate in a thermocycler at 20°C for 5 min followed by 65°C for 5 min.
7. Purify the reaction using AMPure XP beads. Add 36 μL of AMPure XP beads to the sample and mix by vortexing and spin down for 2 s in a minicentrifuge. Incubate at room temperature for 5 min.
8. Prepare 500 μL of 80% ethanol (per sample). After incubation, spin down the sample and place on a magnetic stand until the supernatant is clear and a bead pellet has formed on the side of the tube. Remove and discard the supernatant while keeping the tube on the magnetic stand.
9. With the tube on the magnetic stand, add 200 μL of 80% ethanol without disturbing the pellet and immediately remove and discard the ethanol. Repeat this step so that the bead pellet has been washed twice.
10. Spin down the tube and place it back on the magnetic stand, and carefully remove any residual ethanol. Remove the tube from the magnetic stand and allow the bead pellet to dry for 1 min then add 12 μL of nuclease-free water, vortex, and centrifuge. Incubate for 2 min. Place the tube back on the magnetic stand and after 1 min, elute the clear supernatant.
11. Transfer 10 μL of the eluted end-prepped DNA to a clean PCR tube.
12. Add 6.6 μL of barcode adapter (provided in a 96-well plate as part of the EXP-PBC096 kit) and 24.9 μL Blunt/TA Ligase Master mix.
13. Mix well by vortex, then centrifuge for 5 s with a minicentrifuge.
14. Incubate the reaction at room temperature for 10 min.
15. Purify the mixture with 13 μL of AMPure XP beads, mix by vortexing, then spin down for 2 s in a minicentrifuge. Incubate at room temperature for 5 min.
16. Place samples on a magnetic stand until a bead pellet has formed. Remove and discard the supernatant, keeping the tube on the magnetic stand.
17. Wash the bead pellet twice by adding 200 μL of 80% ethanol to each sample on the magnetic stand without disturbing the

pellet, followed by immediately removing and discarding the ethanol.

18. Remove the tube from the magnetic stand and allow the bead pellet to dry for 1 min then add 7 μL of nuclease-free water, vortex, and centrifuge. Incubate for 2 min. Place the tube back on the magnetic stand and after 1 min, elute the clear supernatant.
19. In a clean 0.2 mL PCR tube, combine 3 μL of adapter ligated DNA with 1 μL of barcode (provided in a 96-well plate as part of the EXP-PBC096 kit), 1 μL of water, and 5 μL of LongAmp Taq 2× Master Mix.
20. Mix well by vortexing, then centrifuge for 5 s in a minicentrifuge.
21. Incubate the mixture in a thermocycler using the following conditions: 95°C for 3 min, 95°C for 15 s, 62°C for 15 s, followed by 18 cycles of 65°C for 5 min, 65°C for 5 min and hold at 4°C.
22. Following incubation, purify the mixture by adding 4 μL of AMPure XP beads and mixing by vortexing. Spin samples down for 2 s in a minicentrifuge. Incubate at room temperature for 5 min.
23. Place samples on a magnetic stand until a bead pellet has formed. Remove and discard the supernatant, keeping the tube on the magnetic stand.
24. Wash the bead pellet twice by adding 200 μL of 80% ethanol to each sample on the magnetic stand without disturbing the pellet, followed by immediately removing and discarding the ethanol.
25. Remove the tube from the magnetic stand and allow the bead pellet to dry for 1 min then add 8 μL of nuclease-free water, vortex, and centrifuge. Incubate for 2 min. Place the tube back on the magnetic stand and after 1 min, elute the clear supernatant.
26. Transfer 5 μL of the elution to a clean PCR tube and quantify with a Qubit Fluorometer.
27. Pool samples to create a final volume of 50 μL (*see* **Note 14**).
28. Combine the 50 μL sample pool with 7 μL of Ultra II End-prep Reaction buffer and 3 μL of Ultra II End-prep Enzyme mix.
29. Mix well by vortexing, then centrifuge for 5 s with a mini-centrifuge. Incubate at 20°C for 5 min followed by 65°C for 5 min in a PCR thermocycler.
30. Following incubation, purify the sample with 60 μL of AMPure XP beads and mix by vortexing. Spin samples down

for 2 s in a minicentrifuge. Incubate at room temperature for 5 min.

31. Place samples on a magnetic stand until a bead pellet has formed. Remove and discard the supernatant, keeping the tube on the magnetic stand.
32. Wash the bead pellet twice by adding 200 μL of 80% ethanol to each sample on the magnetic stand without disturbing the pellet, followed by immediately removing and discarding the ethanol.
33. Remove the tube from the magnetic stand and allow the bead pellet to dry for 1 min then add 61 μL of nuclease-free water, vortex, and centrifuge. Incubate for 2 min. Place the tube back on the magnetic stand and after 1 min, elute the clear supernatant.
34. Combine 60 μL of the elution with 25 μL of ligation buffer (supplied in Nanopore ligation sequencing kit), 10 μL of NEBNext Quick T4 DNA ligase, and 5 μL of Ligation Adapter (supplied in Nanopore ligation sequencing kit).
35. Thoroughly mix the mixture by vortexing then incubate at room temperature for 10 min.
36. Purify the mixture with 40 μL of AMPure XP beads and incubate at room temperature for 5 min. Place the tube on the magnetic stand until the supernatant is clear. Remove and discard the supernatant without disturbing the pellet.
37. Add 250 μL of long fragment buffer (from the Nanopore ligation sequencing kit) to the tube while keeping it on the magnetic stand. Remove the tube from the stand, vortex and spin for 2 s. Return the tube to the magnetic stand and wait 5 min or until the supernatant is clear. Remove and discard the supernatant. Repeat this step another two times so the bead pellet has been washed three times.
38. Add 15 μL of elution buffer to the bead pellet, mix, centrifuge for 5 s on a minicentrifuge then incubate at room temperature for 10 min.
39. Following incubation, place the tube on the magnetic stand until the supernatant is clear. Transfer 12 μL of supernatant to a clean PCR tube.
40. Prepare flow cell priming mix by adding 30 μL of Flush Tether directly to a tube of Flush Buffer (supplied in the Nanopore ligation kit) and mix thoroughly by vortexing.
41. Open the MinION lid and place the flow cell under the clip. Open the flow cell priming port cover and insert a pipette tip fitted to a P1000 manual pipette set to 200 μL.

42. Turn the pipette wheel until the dial shows 220–230 μL or until a small amount of yellow buffer coming out of the flow cell is visible. Discard the tip.
43. Load 800 μL of priming mix into the priming port, avoiding bubbles, and incubate for 5 min.
44. During incubation, prepare the library by adding 37.5 μL of Sequencing Buffer and 25.5 μL of Nanopore Loading beads (LB) (mixed immediately before) to the 12 μL of eluted sample.
45. Once incubation of the flow cell is complete, open the SpotON sample port cover and add a further 200 μL of the priming mix to the priming port (do <u>not</u> add priming mix to the SpotON sample port).
46. Mix the sample and load the entire volume (75 μL) into the SpotON port in a dropwise manner.
47. Close both ports and replace the MinION Lid.
48. Start the sequencer after inputting the relevant information on the sample. Samples should be sequenced for up to 48 h, until the number of sequence reads originating at the targeted site exceeds 20.
49. Once the run has terminated, the reads are aligned and trimmed manually in Epi2Me Labs software using Minimap2 (https://nanoporetech.com/products/analyse/epi2me). Inspect the aligned reads (in bam files) for edits visually in Integrative Genomic Viewer (IGV; https://igv.org/) (*see* **Note 15**).

3.2 Introducing CRISPR-Cas 9 into Living Cells

Several methods of transfection exist, including chemical transfection, various physical transfection techniques, and the use of viral vectors. Ensuring the most effective delivery method is used given a particular situation is crucial [17] (*see* **Note 16**). When applied to oocytes and preimplantation embryos CRISPR-Cas9 has most often been introduced by microinjection at the time of fertilization (*see* **Note 17**).

1. Using a micromanipulation setup, aspirate RNP complex along with a sperm and inject as per a standard intracytoplasmic sperm injection (ICSI) protocol (*see* **Note 18**). Following fertilization via ICSI, culture the embryo for the appropriate amount of time until it has reached the target developmental stage for analysis (*see* **Note 19**).

3.2.1 Disaggregation of CRISPR-Treated Embryos to Test for Editing in Individual Cells

1. Remove the zona pellucida by briefly incubating the embryo in a 20 μL drop of Acid Tyrode's Solution for 5-30 s (Pronase can also be used).

2. Observe embryos under the microscope and as soon as the zona pellucida can be seen to have dissolved, transfer to a drop of medium (e.g., calcium-magnesium-free buffer G-PGD) for embryo biopsy.
3. Gently pipette the embryo to separate the cells (*see* **Note 20**).
4. In order to remove DNA contaminants that could invalidate downstream testing, it is important that the isolated blastomeres be thoroughly washed through drops of wash buffer. Prepare a 60 mm culture dish with three 20 μL drops of clean embryo culture media.
5. Maintaining sterile technique, wash each cell through the three drops, then place into a 0.2 mL PCR tubes (*see* **Note 21**).
6. Isolated cells can be stored at −80°C until processing.

3.2.2 Whole Genome Amplification

To maximize the information that can be obtained from individual cells, it is desirable to perform whole genome amplification (WGA), increasing the amount of DNA available for analysis.

1. Methods such as multiple displacement amplification (MDA) (e.g., REPLI-g Single cell kit; Qiagen), Primary Template-directed Amplification (PTA) (Bioskryb Genomics), and various others are available for this purpose (*see* **Note 22**). PTA being the most efficient.
2. Following WGA, analyses of the samples can be conducted using one or more of the methods outlined below, revealing the outcome of the CRISPR-Cas9 treatment.

3.2.3 Low Pass Genome Sequencing

A common way to evaluate editing of a cell sample is to use PCR to amplify a DNA fragment that includes the targeted site and then to sequence it via Nanopore sequencing. The detection of an altered DNA sequence is indicative of successful targeting. Repair via NHEJ is revealed by the presence of indels at the targeted site, while HDR is demonstrated by the observation of specific nucleotide changes only found within a supplied template. If any heterozygous mutations or polymorphisms were present at the targeted site or in flanking sequences, loss of heterozygosity (LOH) may be observed (i.e., sites that were heterozygous before editing are subsequently found to be homozygous). This can potentially be explained by repair using IH-HR (*see* **Note 23**). Low-coverage whole genome sequencing can reveal whether failed or abnormal repair has led to chromosome loss or certain large-scale duplications or deletions affecting the chromosome bearing the target site (e.g., segmental aneuploidy) (*see* **Note 24**). Following cell lysis and whole genome amplification, the amplified DNA is then used to prepare libraries for next-generation sequencing (*see* **Note 25**).

1. Design oligonucleotide primers flanking the region encompassed by the edit using any primer design software package (e.g., Primer3, https://www.primer3plus.com/) (*see* **Note 7**).
2. Perform a PCR on the edited sample amplified with PTA with the Q5® High-Fidelity DNA Polymerase (New England Biolabs M0492). Prepare a 50 μL reaction in a 0.2 ml PCR tube containing 1× Q5 reaction buffer, 200 μM dNTPs, 0.5 μ M of forward and reverse primer, 0.5 μL of Q5 High-fidelity DNA polymerase, 2 μL of DNA. Make up the final volume with nuclease-free water. Mix the sample and spin with a bench-top centrifuge for 5 s. Incubate the sample at the following conditions: 94°C for 30 s, 94°C for 30 s, 55–62°C for 45 s, 35 cycles of 72°C for 45 s, 72°C for 45 s and hold at 4°C.
3. Combine 30 μL of amplified product with 4.5 μL of Ultra II End-prep reaction buffer and 1.8 μL of Ultra II End-prep enzyme mix.
4. Mix the reaction well by vortexing, spin, and incubate in a thermocycler at 20°C for 5 min followed by 65°C for 5 min.
5. Purify the reaction using AMPure XP beads. Add 36 μL of AMPure XP beads to the sample and mix by vortexing and spin down for 2 s in a minicentrifuge. Incubate at room temperature for 5 min.
6. Prepare 500 μL of 80% ethanol (per sample). After incubation, spin down the sample and place on a magnetic stand until the supernatant is clear and a bead pellet has formed on the side of the tube. Remove and discard the supernatant while keeping the tube on the magnetic stand.
7. With the tube on the magnetic stand, add 200 μL of 80% ethanol without disturbing the pellet and immediately remove and discard the ethanol. Repeat this step so that the bead pellet has been washed twice.
8. Spin down the tube and place it back on the magnetic stand, and carefully remove any residual ethanol. Remove the tube from the magnetic stand and allow the bead pellet to dry for 1 min then add 12 μL of nuclease-free water, vortex and centrifuge. Incubate for 2 min. Place the tube back on the magnetic stand and after 1 min, elute the clear supernatant.
9. Transfer 10 μL of the eluted end prepped DNA to a clean PCR tube.
10. Add 6.6 μL of barcode adapter (provided in a 96-well plate as part of the EXP-PBC096 kit) and 24.9 μL Blunt/TA Ligase Master mix.
11. Mix well by vortex, then centrifuge for 5 s with a minicentrifuge.

12. Incubate the reaction at room temperature for 10 min.
13. Purify the mixture with 13 μL of AMPure XP beads, mix by vortexing, then spin down for 2 s in a minicentrifuge. Incubate at room temperature for 5 min.
14. Place samples on a magnetic stand until a bead pellet has formed. Remove and discard the supernatant, keeping the tube on the magnetic stand.
15. Wash the bead pellet twice by adding 200 μL of 80% ethanol to each sample on the magnetic stand without disturbing the pellet, followed by immediately removing and discarding the ethanol.
16. Remove the tube from the magnetic stand and allow the bead pellet to dry for 1 min then add 7 μL of nuclease-free water, vortex, and centrifuge. Incubate for 2 min. Place the tube back on the magnetic stand and after 1 min, elute the clear supernatant.
17. In a clean 0.2 mL PCR tube, combine 3 μL of adapter ligated DNA with 1 μL of barcode (provided in a 96-well plate as part of the EXP-PBC096 kit), 1 μL of water, and 5 μL of LongAmp Taq 2× Master Mix.
18. Mix well by vortexing, then centrifuge for 5 ss in a minicentrifuge.
19. Incubate the mixture in a thermocycler using the following conditions: 95°C for 3 min, 95°C for 15 s, 62°C for 15 s, followed by 18 cycles of 65°C for 5 min, 65°C for 5 min and hold at 4°C.
20. Following incubation, purify the mixture by adding 4 μL of AMPure XP beads and mixing by vortexing. Spin samples down for 2 s in a minicentrifuge. Incubate at room temperature for 5 min.
21. Place samples on a magnetic stand until a bead pellet has formed. Remove and discard the supernatant, keeping the tube on the magnetic stand.
22. Wash the bead pellet twice by adding 200 μL of 80% ethanol to each sample on the magnetic stand without disturbing the pellet, followed by immediately removing and discarding the ethanol.
23. Remove the tube from the magnetic stand and allow the bead pellet to dry for 1 min then add 8 μL of nuclease-free water, vortex, and centrifuge. Incubate for 2 min. Place the tube back on the magnetic stand and after 1 min, elute the clear supernatant.
24. Transfer 5 μL of the elution to a clean PCR tube and quantify with a Qubit Fluorometer.

25. Pool samples to create a final volume of 50 μL (*see* **Note 14**).
26. Combine the 50 μL sample pool with 7 μL of Ultra II End-prep Reaction buffer and 3 μL of Ultra II End-prep Enzyme mix.
27. Mix well by vortexing, then centrifuge for 5 s with a mini-centrifuge. Incubate at 20°C for 5 min followed by 65°C for 5 min in a PCR thermocycler.
28. Following incubation, purify the sample with 60 μL of AMPure XP beads and mix by vortexing. Spin samples down for 2 s in a minicentrifuge. Incubate at room temperature for 5 min.
29. Place samples on a magnetic stand until a bead pellet has formed. Remove and discard the supernatant, keeping the tube on the magnetic stand.
30. Wash the bead pellet twice by adding 200 μL of 80% ethanol to each sample on the magnetic stand without disturbing the pellet, followed by immediately removing and discarding the ethanol.
31. Remove the tube from the magnetic stand and allow the bead pellet to dry for 1 min then add 61 μL of nuclease-free water, vortex, and centrifuge. Incubate for 2 min. Place the tube back on the magnetic stand and after 1 min, elute the clear supernatant.
32. Combine 60 μL of the elution with 25 μL of ligation buffer (supplied in Nanopore ligation sequencing kit), 10 μL of NEBNext Quick T4 DNA ligase, and 5 μL of Ligation Adapter (supplied in Nanopore ligation sequencing kit).
33. Thoroughly mix the mixture by vortexing then incubate at room temperature for 10 min.
34. Purify the mixture with 40 μL of AMPure XP beads and incubate at room temperature for 5 min. Place the tube on the magnetic stand until the supernatant is clear. Remove and discard the supernatant without disturbing the pellet.
35. Add 250 μL of long fragment buffer (from the Nanopore ligation sequencing kit) to the tube while keeping it on the magnetic stand. Remove the tube from the stand, vortex and spin for 2 s. Return the tube to the magnetic stand and wait 5 min or until the supernatant is clear. Remove and discard the supernatant. Repeat this step another two times so the bead pellet has been washed three times.
36. Add 15 μL of elution buffer to the bead pellet, mix, centrifuge for 5 s on a minicentrifuge then incubate at room temperature for 10 min.

37. Following incubation, place the tube on the magnetic stand until the supernatant is clear. Transfer 12 µL of supernatant to a clean PCR tube.
38. Prepare flow cell priming mix by adding 30 µL of Flush Tether directly to a tube of Flush Buffer (supplied in the Nanopore ligation kit) and mix thoroughly by vortexing.
39. Open the MinION lid and place the flow cell under the clip. Open the flow cell priming port cover and insert a pipette tip fitted to a P1000 manual pipette set to 200 µL.
40. Turn the pipette wheel until the dial shows 220–230 µL or until a small amount of yellow buffer coming out of the flow cell is visible. Discard the tip.
41. Load 800 µL of priming mix into the priming port, avoiding bubbles, and incubate for 5 min.
42. During incubation, prepare the library by adding 37.5 µL of Sequencing Buffer and 25.5 µL of Nanopore Loading beads (LB) (mixed immediately before) to the 12 µL of eluted sample.
43. Once incubation of the flow cell is complete, open the SpotON sample port cover and add a further 200 µL of the priming mix to the priming port (do not add priming mix to the SpotON sample port).
44. Mix the sample and load the entire volume (75 µL) into the SpotON port in a dropwise manner.
45. Close both ports and replace the MinION Lid.
46. Start the sequencer after inputting the relevant information on the sample. Samples should be sequenced for up to 48 h, until the number of sequence reads originating at the targeted site exceeds 20.
47. Upon the completion of the sequencing run, Fastq files must obtain a quality score of seven or higher (determined as the threshold by the manufacturer)
48. Demultiplex and trim sequences using the Porechop tools (Oxford Nanopore Technologies). Execute Porechop with the command **porechop -i input.fastq.gz -b demultiplexed/--discard_middle --threads 8**, which identifies and removes adapter and barcode sequences from both ends of reads while discarding chimeric reads containing internal adapters.
49. Use **the --native_barcodes** option to separate reads into barcode-specific FASTQ files, with unclassified reads stored separately. The resulting high-quality, adapter-free reads can then be used for downstream analyses.

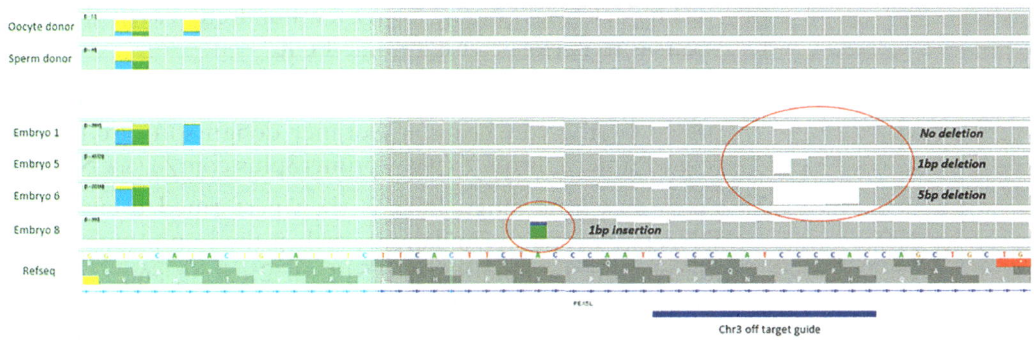

Fig. 2 Representative image showing sequencing data from four embryos that underwent genotyping using nanopore sequencing. Embryo 1 presents an intact sequence compared to the reference sequence, whereas embryos 5 and 6 displayed deletions of 1 bp and 5 bps, respectively. Embryo 8 however showed the occurrence of a single base pair insertion

50. Align sequences to the human genome reference build GRCh38 in Minimap2, using the default parameters.
51. Sort and index the BAM files in Samtools and then split into bins of 5 Mbp across all chromosomes.
52. Count the number of reads falling within each bin in the Genome Analysis Tool Kit (Broad Institute) using the CountReads function. Calculate the proportion of reads attributable to each bin by dividing this number by the total number of mapped reads. The proportions for each 5 Mb region are compared to reference sets created following analysis of several karyotypically normal male and normal female DNA samples (Fig. 2).

3.3 Deeper Analysis of Targeted Sites

3.3.1 PCR Amplification and Gel Electrophoresis

The interpretation of the editing outcomes following CRISPR-Cas9 can be challenging, especially at the level of individual cells. While LOH can be indicative of IH-HR, it can also occur if PCR only succeeded in amplifying one of the two copies of the targeted region, leaving the other copy undetectable. This can occur if one of the two primer annealing sites is separated from the other by a large-scale genomic rearrangement (e.g., an inversion or translocation), if the region is subject to a large deletion eliminating one or both annealing sites, or if there is an unrepaired DSB between the primers. All of these possibilities have been observed at appreciable frequencies after CRISPR-Cas9 treatment of human preimplantation embryos, but most of these undesirable outcomes are invisible to standard PCR-based methods and can only be distinguished from IH-HR if a battery of genetic methods is deployed, including low-pass genome sequencing and sequence capture. The most straightforward means of revealing deletions ranging from a few hundred to a few thousand base-pairs is to perform a long-range PCR.

1. Design oligonucleotide primers flanking the region encompassed by the edit using any primer design software package (e.g., Primer3, https://www.primer3plus.com/) (*see* **Note 7**).
2. Perform a long-range PCR is then performed using special DNA polymerase formulations, such as the LongAmp® Taq 2× Master Mix by combining 7.5 μL of 2× long amp Taq with 0.6 μL of 10 μM of both forward and reverse primers. Make up the reaction to 14.5 μL with nuclease-free water. Add 0.5 μL of DNA, vortex for 5 s and spin on a benchtop centrifuge for 5 s.
3. Indicative reaction conditions are initial denaturation at 94°C for 30 s, then 35 cycles of denaturation at 94°C for 30 s, annealing at 55–62°C for 45 s, and extension at 72°C for 8 min, followed by a final extension at 72°C for 8 min (*see* **Note 26**).
4. DNA amplification is followed by electrophoresis analysis. Prepare 1.5% agarose by combining 100 ml of 1× TBE and 1.5 g of UltraPure agarose gel and heat in a microwave until the powder is dissolved. Once slightly cool, add 10 μL of GelRed and swirl the flask gently to mix.
5. Prepare a gel cassette by placing 2 × 15-well combs into a gel casting tray, then pouring the warm 1.5% agarose mix into the tray. Allow to solidify (can take up to 20 min). Remove the combs once solidified and place them into an electrophoresis tank filled with 1× TBE up to the maximum mark.
6. Combine 3 μL of CRISPR-Cas9 treated sample with 0.5 μL of DNA gel Loading Dye (6×) and load into wells of a 1.5% agarose gel.
7. A DNA ladder should be run alongside the samples as a control, allowing estimation of the size of the amplified DNA fragments (e.g., 5 μL of 100 bp Low Scale DNA Ladder) (*see* **Note 11**).
8. Run at 120 V for 40 min on a gel electrophoresis system.
9. Bands of a lower molecular weight than expected indicating the presence of amplicons containing large deletions.

Alternatively, the PCR amplicons can be subjected to long-read sequencing, revealing the precise breakpoints of any deletions (e.g., the Nanopore ligation sequencing V14-PCR barcoding; SQK-LSK114 with EXP-PBC096).

3.3.2 Sequence Capture

In instances where loss of heterozygosity has been observed, with the polymorphism/mutant alleles from one parent remaining undetected after sequencing of the targeted site, one possible explanation is that repair of the DSB has involved IH-HR. This is a reasonable hypothesis, especially if long-range PCR has failed to detect any deletions, and low-pass sequencing has not revealed any chromosome loss or segmental aneuploidy affecting the targeted region. However, repair using IH-HR, which is a potentially desirable outcome of CRISPR-Cas9 treatment if the aim is to remove a heterozygous mutation, cannot be confirmed without further analyses. For this purpose, *sequence capture* can be performed, a process that concentrates DNA from a defined genomic region prior to sequencing. The method allows interrogation of large areas of interest, extending over thousands of base pairs of the genome, without using PCR. This approach can identify certain aberrant forms of repair that are invisible to the other methods described in this chapter, including those producing large inversions and balanced translocations.

Sequence capture involves the combination of two protocols: TWIST Library Preparation EF2.0 with Enzymatic Fragmentation and TWIST Universal Adapter System/TWIST Long Read Library Preparation and Standard Hyb v2 Enrichment protocols and Nanopore Ligation sequencing gDNA (SQK-LSK114) protocol (*see* **Note 27**).

1. Warm the PTA sample at 37°C in a thermocycler for 5 min or until no precipitate is visible. Transfer 5 μL of warmed MDA embryo sample to a 0.2 ml tube and combine with 35 μL of nuclease-free water. Clean with AMPure XP purification beads at a 1× ratio and mix by vortexing. Spin samples down for 2 s in a minicentrifuge. Incubate at room temperature for 5 min.
2. Place samples on a magnetic stand until a bead pellet has formed. Remove and discard the supernatant, keeping the tube on the magnetic stand.
3. Wash the bead pellet twice by adding 200 μL of 80% ethanol to each sample on the magnetic stand without disturbing the pellet followed by immediately removing and discarding the ethanol.
4. Remove the tube from the magnetic stand and allow the bead pellet to dry for 1 min then add 20 μL of nuclease-free water, vortex, and centrifuge. Incubate for 2 min. Place the tube back on the magnetic stand and after 1 min, elute the clear supernatant.
5. Quantify the elution on a Qubit Fluorometer using a Qubit dsDNA HS Assay.
6. In a separate PCR tube, prepare a dilution of 100 ng in a final volume of 41 μL.
7. Program a thermocycler to the following conditions: hold at 4°C, followed by a 7-min incubation at 30°C, 65°C for 30 min and hold at 4°C. Start the program to pre-chill the thermocycler.

8. Combine 4 μL of Frag/AT buffer and 6 μL of Frag/AT enzyme (from the TWIST Library Preparation EF Kit) with the prepared sample. Vortex the mixture, spin and place the reaction in the pre-chilled thermocycler and continue the program.
9. Run all samples on Tapestation via D1000 high sensitivity (the maximum number that can be run in at once is 16).
10. Add 5 μL of TWIST universal adapters (from the TWIST Universal Adapter System) to each tube containing the dA-tailed fragmented DNA and vortex.
11. Add 20 μL of ligation master mix (from the TWIST Library Preparation EF Kit), then mix, spin, and incubate at 20°C for 15 min in a thermocycler with the heated lid off or set at minimum.
12. Following incubation, purify the mixture by adding 78 μL of AMPure XP beads and mixing by vortexing. Spin samples down for 2 s in a minicentrifuge. Incubate at room temperature for 5 min.
13. Place samples on a magnetic stand until a bead pellet has formed. Remove and discard the supernatant, keeping the tube on the magnetic stand.
14. Wash the bead pellet twice by adding 200 μL of 80% ethanol to each sample on the magnetic stand without disturbing the pellet, followed by immediately removing and discarding the ethanol.
15. Remove the tube from the magnetic stand and allow the bead pellet to dry for 1 min then add 17 μL of nuclease-free water, vortex, and centrifuge. Incubate for 2 min. Place the tube back on the magnetic stand and after 1 min, elute the clear supernatant.
16. Remove 15 μL of the clear supernatant and place it in a clean PCR tube.
17. Add 10 μL of TWIST UDI primer into a 96-well plate (from the TWIST Universal Adapter System) and 25 μL of 2× Equinox Ligation amp mix (from the TWIST Library Preparation EF Kit) with each sample and mix well. Incubate samples with the following programme: 98°C for 45 s, 10 cycles of 98°C for 15 s, 60°C for 30 s, 72°C for 30 s, a final extension of 72°C for 1 min and hold at 4°C.
18. Following incubation, purify the mixture by adding 50 μL of AMPure XP beads and mixing by vortexing. Spin samples down for 2 s in a minicentrifuge. Incubate at room temperature for 5 min.

19. Place samples on a magnetic stand until a bead pellet has formed. Remove and discard the supernatant, keeping the tube on the magnetic stand.
20. Wash the bead pellet twice by adding 200 μL of 80% ethanol to each sample on the magnetic stand without disturbing the pellet, followed by immediately removing and discarding the ethanol.
21. Remove the tube from the magnetic stand and allow the bead pellet to dry for 1 min then add 22 μL of nuclease-free water, vortex, and centrifuge. Incubate for 2 min. Place the tube back on the magnetic stand and after 1 min, elute the clear supernatant.
22. Place 20 μL of the elution in a clean PCR tube. Prepare a 1:10 dilution of the 20 μL elution and analyse using Qubit quantification and Tapestation.
23. Using the quantification values of each sample, calculate how many samples can be pooled together to create a sample concentration of 4000 ng. Referring to our results of around 25 ng/μL for each sample, we generally pool 3/4 samples in a single pool.
24. Heat the hybridisation mix (from the TWIST Hybridisation Reagents) at 65°C for 10 min or until the precipitate dissolves.
25. Prepare a probe solution by combining 20 μL of hybridisation mix, 4 μL of TWIST custom panel (designed and validated by Twist Ltd specifically for your target region) and 4 μL of water at room temperature.
26. Prepare a dried index library by adding 36 μL of AMPure purification beads to a clean 1.5 mL microcentrifuge tube containing the sample pool, mix and incubate at room temperature for 5 min.
27. Place the tube on a magnetic stand then remove and discard the clear supernatant. Wash the bead pellet twice with 200 μL of 80% ethanol.
28. Spin the tube with a bench top centrifuge for 5 s and place it back on the magnetic stand. Remove and discard any residual ethanol.
29. Retain the bead pellet and add 5 μL of blocker solution and 7 μ L of universal blocker (from the TWIST Universal Blockers) to the dried index library pool and mix well to re-suspend the bead pellet.
30. Heat the probe solution at 95°C for 2 min then immediately cool on ice for 5 min.
31. While the probe solution cools on ice, incubate the index library pool at 95°C for 5 min.

32. Equilibrate both solutions at room temperature for 5 min.
33. Vortex the probe solution, spin, and transfer to the index library pool and mix well. Add 30 μL of hybridisation enhancer (from the TWIST Hybridisation Reagents) to the sample and spin to ensure there are no bubbles. Incubate the hybridisation mixture at 70°C for 16 h in a thermocycler with lid temperature at 85°C.
34. For each sample, 30 min prior to the completion of the hybridisation incubation, prepare 800 μL of binding buffer equilibrated to room temperature, 225 μL of wash buffer 1 heated to 68°C, and 700 μL of wash buffer 2 equilibrated to 48°C (all found in the TWIST Wash buffers kit).
35. Place 100 μL of streptavidin beads in a 1.5 mL microcentrifuge tube along with 200 μL of binding buffer and mix well.
36. Place the tube on a magnetic stand for 1 min then remove and discard the clear supernatant. Repeat the wash twice more, for a total of three washes. Add a final 200 μL of binding buffer and mix with the beads. Incubate the mixture at 68°C for at least 10 min.
37. Once hybridisation is complete, while remaining in the thermocycler, transfer the hybridisation mix to the preheated streptavidin beads. Rapid transfer is crucial to avoid a decrease in temperature of the hybridisation mix.
38. Incubate tubes containing the streptavidin beads and hybridisation mix at 68°C for 5 min.
39. Place the tube on a magnetic stand for 1 min then remove and discard the clear supernatant.
40. Remove the tube from the magnetic stand, add 200 μL of 68°C wash buffer 1, mix and incubate at 68°C for 5 min.
41. Transfer the sample to a clean tube and place it on the magnetic stand for 1 min. Remove and discard the clear supernatant.
42. Add 200 μL of wash buffer 2, mix well to resuspend the beads, and incubate at 48°C for 5 min.
43. Place the tube on the magnetic stand, then remove and discard the clear supernatant. Repeat the wash another two times. After the final wash, spin the tube and place it on the magnetic stand to remove any traces of the supernatant.
44. Remove the tube from the magnetic stand, add 22 μL of water, mix well, resuspending the beads, and place on ice. The sample is hereafter referred to as a streptavidin slurry.
45. Combine 3 μL of amplification primers and 25 μL of Equinox library amp 2× mix with the slurry. Mix the sample and incubate at the following thermocycler conditions: 94°C for 2 min, 17 cycles of 98°C for 10 s, 58.8°C for 30 s and 72°C

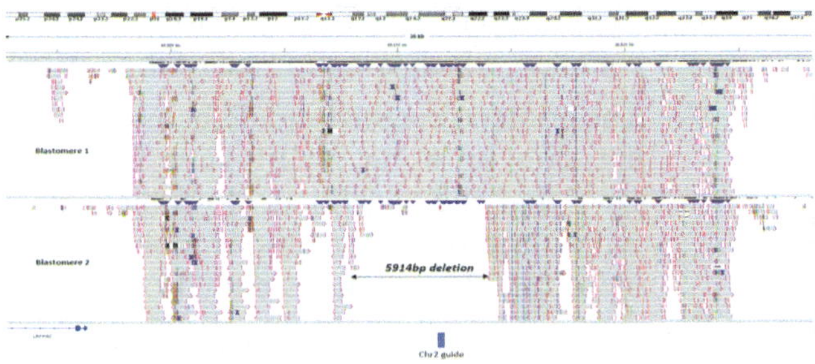

Fig. 3 Representative image showing sequencing data from two blastomeres that underwent sequence capture. Blastomere 1 presents an intact chromosome, whereas blastomere 2 shows a large deletion (a gap in the DNA sequence) of almost 6000 base pairs

for 3 min. Followed by a final extension of 72°C for 3 min and hold at 4°C.

46. Purify the sample with 50 μL of beads and elute in 42 μL of water.
47. Remove 40 μL of the elution and process according to the Nanopore ligation sequencing V14-PCR barcoding protocol as indicated in section 3.2.3. Quantify the remaining elution using a Qubit fluorometer.
48. View the data on IGV to identify the presence and size of any deletions present, as illustrated in Fig. 3.

4 Notes

1. Ensembl gene browser, https://www.ensembl.org/index.html, is an alternative site for downloading DNA sequences.
2. Other online tools are also available, such as CHOPCHOP, CRISPOR, and GuideScan.
3. While the length of a CRISPR guide can vary, 20 bp generally works well.
4. In order to select the most ideal gRNA, it is important to ensure there is a PAM site at the end of the guide identified as a sequence of "NGG" when using Cas9 and a GC content (40%–60%) that matches to your target in the seed region (8–12 bp to the PAM).
5. If guides are ordered through other companies, such as Benchling, they may be inserted within a plasmid and require amplification and purification, as detailed in [9]. RNP complexes do not need to be generated within plasmids.

6. The most efficient way of utilizing CRISPR is with an RNP complex, which is composed of the sgRNA and Cas enzyme.
7. The primers designed should be assessed for off-target annealing sites by using NCBI blast https://blast.ncbi.nlm.nih.gov/ and the likelihood of problematic hairpin formations can be assessed using various tools, such as Oligocal http://biotools.nubic.northwestern.edu/OligoCalc.html
8. Volumes for the reaction buffer and Taq will vary depending on the manufacturer. The primer pool is generated by combining the forward and reverse primers (typically supplied separately at concentrations of 100 μM) to a final concentration of 10 μM for each primer.
9. The optimal annealing temperature is likely to differ depending on the length and sequence of the primers, and the extension temperature may vary depending on the specific Taq formulation used for PCR (refer to manufacturer's recommendations).
10. Temperature for incubation varies depending on the enzyme used, e.g., Cas12 requires incubation at 37–42°C.
11. This PCR-based validation provides a basic assessment of the ability of the guide to target the chosen site. If the amplicon is successfully cleaved, two bands will be present on the gel, their combined size being equivalent to the full-length amplicon. A band on the gel the same size as the PCR amplicon indicates the presence of DNA fragments that were not cleaved by CRISPR-Cas9.
12. Validation combining dephosphorylation and Nanopore sequencing is more complicated than the simple assessment of the ability to cleave a PCR amplicon containing the targeted site but has the added advantage that it not only confirms successful cleavage of the intended target but also reveals sites susceptible to off-target editing. Essentially, it provides a snapshot of CRISPR-Cas9 cleavage events across the entire genome.
13. Cleavage by CRISPR-Cas9 exposes a 5′ phosphate group in the otherwise dephosphorylated DNA, to which sequencing adapters can be ligated.
14. For samples with a concentration of > 10 ng/μL, 2 μL is taken forward to make a single pool; therefore, 25 samples can be pooled together to create a final volume of 50 μL, whereas samples with a concentration of less than 10 ng/μL but greater than 5 ng/μL, 4 μL is required to be added to the pool. Concentrations less than 5 ng/μL are deemed unsuccessful, and the barcoding step should be repeated (Section

3.1.2, Steps 19–26). If the concentration of the sample remains low, the entire process is repeated.

15. In theory, the only sites in the genome to which sequencing adapters can ligate are those cleaved during CRISPR-Cas9 treatment. Although a significant number of reads will be seen scattered across the genome, any sites cleaved by CRISPR-Cas9 will have a large number of sequence reads associated with them, all beginning from the same nucleotide position.
16. The efficiency with which the CRISPR-Cas9 reagents are delivered into cells, otherwise known as transfection efficiency, is of critical importance [17]. Transfection efficiency is dependent on several factors, including cell membrane composition, cell cycle phase, cellular health, and the transfection method used [18]. Primary cells tend to contain membranes that are poorly permeable, which often prevents the CRISPR-Cas9 from entering the nucleus. There are higher rates of transfection efficiency in dividing, healthy cells compared to more mature or stressed cells.
17. If applying genome editing to oocytes or zygotes, microinjection is likely to be the most efficient strategy for introducing the necessary reagents, since successful transfer can be virtually guaranteed. However, microinjection is likely to be impractical at later embryonic stages due to the technical challenge of injecting cells that become progressively smaller in size as preimplantation development proceeds, and the rapidly increasing number of cells that would require injection. Alternative introduction methods are available, such as electroporation, or with the use of liposomes or other nanoparticles.
18. The amount of RNP complex to aspirate varies according to the ICSI protocol of transfer but is typically 4–12 picolitres.
19. It is also essential to leave a minimum of 24 h to allow the CRISPR-Cas9 to interact with and edit the DNA of the cell(s). Embryo culture after transfection/microinjection typically proceeds according to standard protocols. To date, many experiments performed using human embryos have involved disaggregation into single cells after a small number of mitoses (at least 24 h after microinjection/transfection), followed by characterization of the editing in individual cells [9, 10].
20. Disaggregation is useful, allowing assessment of individual cells, thereby facilitating detection of mosaic editing outcomes. With no zona pellucida, and in the absence of calcium and magnesium ions, cell junctions break down, and the constituent cells of the embryo can be separated by gentle pipetting. However, this approach is likely to be impractical after the cleavage stage as the increasing number of cells in

the embryo means the cost associated with analysis becomes prohibitive.

21. It is recommended that the final volume in the tube, including the cell(s), should not exceed 2 μL. Keeping the volume small allows for maximum flexibility for subsequent analyses (avoiding excessive dilution of enzymatic reactions).

22. Amplification of DNA from single cells is important as the amount of genetic material available is extremely low (5–10 pg). Alternative whole genome amplification methods include MDA, MALBAC and SurePlex each having different pros and cons. Until recently, MDA has been the most widely used method and has generally delivered satisfactory results. However, PTA offers certain technical advantages and is now our preferred technology. As kits for these methods evolve rapidly, it is always best to refer to the manufacturer's instructions.

23. While PCR will provide the least amount of information, it can still provide a useful insight into the outcome of editing at many sites.

24. Loss of part of the chromosome, extending from the targeted site to the end of the chromosome, or loss of the entire chromosome, prevents PCR amplification across the targeted site, as one (or both) of the two primer annealing sites is no longer present. If the homologous chromosomal copy is intact, all sequence data will be derived from that chromosome only, and there will appear to be a loss of heterozygosity. Without cytogenetic analysis, this may be wrongly classified as IH-HR.

25. Any NGS platform should be suitable, but one possibility is to perform nanopore sequencing using the Ligation Sequencing kit SQK-LSK-114 kit (Oxford Nanopore Technologies) with sequencing performed using a MinION device (Oxford Nanopore Technologies). The number of mapped reads suitable for downstream analysis being approximately 200,000 per sample.

26. Precise annealing temperatures depend on the design of one's target-specific primers. Also note that the extension temperature may differ if another polymerase is used.

27. The use of TWIST enrichment protocol maximizes the number of sequence reads at site the targeted by CRISPR-Cas9, and for thousands of base-pairs either side, without the use of PCR. This allows detection of large deletions, translocations, and inversions that cannot be detected when PCR is used due to the removal of primer annealing sites or because their location is altered [19–23].

References

1. Mirza Z, Karim S (2019) Advancements in CRISPR/Cas9 technology-focusing on cancer therapeutics and beyond. Semin Cell Dev Biol 96:13–21
2. Joung JK, Sander JD (2013) TALENs: a widely applicable technology for targeted genome editing. Nat Rev Mol Cell Biol 14:49–55
3. Redman M, King A, Watson C, King D (2016) What is CRISPR/Cas9?. Arch Dis Child Educ Pract Ed 101:213–215
4. Karvelis T, Gasiunas G, Miksys A, Barrangou R, Horvath P, Siksnys V (2013) crRNA and tracrRNA guide Cas9-mediated DNA interference in streptococcus thermophilus. RNA Biol 10:841–851
5. Hanahan D, Weinberg RA (2011) Hallmarks of cancer: the next generation. Cell 144:646–674
6. Takata M, Sasaki MS, Sonoda E, Morrison C, Hashimoto M, Utsumi H, Yamaguchi-Iwai Y, Shinohara A, Takeda S (1998) Homologous recombination and non-homologous end-joining pathways of DNA double-strand break repair have overlapping roles in the maintenance of chromosomal integrity in vertebrate cells. Embo J 17:5497–5508
7. Khokhlova EV, Fesenko ZS, Sopova JV, Leonova EI (2021) Features of DNA repair in the early stages of mammalian embryonic development. Genes (Basel) 11:1138
8. Babariya D, Fragouli E, Alfarawati S, Spath K, Wells D (2017) The incidence and origin of segmental aneuploidy in human oocytes and preimplantation embryos. Human Reprod 32:2549–2560
9. Fogarty NME, McCarthy A, Snijders KE, Powell BE, Kubikova N, Blakeley P, Lea R, Elder K, Wamaitha SE, Kim D et al (2017) Genome editing reveals a role for OCT4 in human embryogenesis. Nature 550:67–73
10. Alanis-Lobato G, Zohren J, McCarthy A, Fogarty NME , Kubikova N, Hardman E, Greco M, Wells D, Turner JMA , Niakan KK (2021) Frequent loss of heterozygosity in CRISPR-Cas9-edited early human embryos. Proc Natl Acad Sci U S A 118(22):e2004832117. https://doi.org/10.1073/pnas.2004832117
11. Ma H, Marti Gutierrez N, Park S-W, Wu J, Lee Y, Suzuki K, Koski A, Ji D, Hayama T, Ahmed R et al (2017) Correction of a pathogenic gene mutation in human embryos. Nature 548:413–419
12. Kang X, He W, Huang Y, Yu Q, Chen Y, Gao X, Sun X, Fan Y (2016) Introducing precise genetic modifications into human 3PN embryos by CRISPR/Cas-mediated genome editing. J Assist Reprod Genet 33:581–588
13. Liang P, Xu Y, Zhang X, Ding C, Huang R, Zhang Z, Lv J, Xie X, Chen Y, Li Y et al (2015) CRISPR/Cas9-mediated gene editing in human tripronuclear zygotes. Protein Cell 6:363–372
14. Kubikova N, Keefe DL, Wells D, Oktay KH, Feinberg EC (2023) Should we use CRISPR gene editing in human embryos?. Fertil Steril. 120:737–744
15. Gyngell C, Douglas T, Savulescu J (2017) The ethics of germline gene editing. J Appl Philos 34:498–513
16. Rubeis G, Steger F (2018) Risks and benefits of human germline genome editing: an ethical analysis. Asian Bioeth Rev 10:133–141
17. Liu W, Li L, Jiang J, Wu M, Lin P (2021) Applications and challenges of CRISPR-Cas gene-editing to disease treatment in clinics. Precis Clin Med 4:179–191
18. Chong ZX, Yeap SK, Ho WY (2021) Transfection types, methods and strategies: a technical review. PeerJ 9:e11165
19. Cimadomo D, Rienzi L, Capalbo A, Rubio C, Innocenti F, García-Pascual CM, Ubaldi FM, Handyside A (2020) The dawn of the future: 30 years from the first biopsy of a human embryo. The detailed history of an ongoing revolution. Human Reprod Update 26:453–473
20. Lea RA, Niakan KK (2019) Human germline genome editing. Nat Cell Biol 21:1479–1489
21. Jinek M, Chylinski K, Fonfara I, Hauer M, Doudna JA, Charpentier E (2012) A programmable dual-RNA-guided DNA endonuclease in adaptive bacterial immunity. Science 337:816–821
22. Callaway E (2016) UK scientists gain licence to edit genes in human embryos. Nature 530:1
23. Guo C, Ma X, Gao F, Guo Y (2023) Off-target effects in CRISPR/Cas9 gene editing. Front Bioeng Biotechnol 11:1143157
24. Ran FA, Hsu PD, Wright J, Agarwala V, Scott DA, Zhang F (2013) Genome engineering using the CRISPR-Cas9 system. Nat Protoc 8:2281–2308

Chapter 25

Pronuclear Transfer and Maternal Spindle Transfer in the Mouse

Ross Stocker, Jeffrey R. Mann and John Carroll

Abstract

Maternal Spindle Transfer (MST) and Pronuclear Transfer (PNT) are micromanipulation techniques that allow the transfer of the nuclear material from one oocyte or zygote to the cytoplasm of another. The application of these techniques has provided novel insights into nuclear-cytoplasmic interactions necessary for development, as well as fundamental discoveries such as genomic imprinting. Now, after extensive pre-clinical research, PNT and MST are starting to be used clinically for Mitochondrial Donation (also known as mitochondrial replacement therapy) and for patients with a history of repeated IVF failure. Here, we describe the methods for PNT and MST currently used in the mouse model. The techniques are readily adaptable for application to other mammalian species, including humans.

Keywords Pronuclear transfer, Maternal spindle transfer, Mouse, Zygote, Oocyte, Mitochondrial disease, Sendai virus

1 Introduction

McGrath and Solter were the first to record live births after PNT in mammalian zygotes [1]. PNT has since been used to investigate functional differences between parental genomes leading to the discovery of genomic imprinting [2-5]) nuclear-cytoplasmic interactions [6, 7], embryonic genome activation [8], and the regulation of pre-implantation development [9–11]. The MST technique was introduced after PNT [12] and has been used in mice to investigate the nuclear and cytoplasmic contributions to reduced oocyte quality after transferring immature oocyte nuclei [13] and in maternal aging [14].

In the last decade, PNT has been tested in some clinical IVF settings in efforts to improve in vitro fertilization and embryo development [15, 16]. These studies have been small and exploratory, and

Supplementary Information The online version contains supplementary material available at https://doi.org/10.1007/978-1-0716-5292-3_25

Alexandra J. Harvey and David K. Gardner (eds.), *The Preimplantation Embryo*, Methods in Molecular Biology, vol. 3038, https://doi.org/10.1007/978-1-0716-5292-3_25,

there is insufficient evidence at this stage to suggest broader use. Very recently, MST has been used in a cohort of IVF patients who had undergone repeated failure of fertilization or preimplantation development in multiple IVF attempts [17]. While the numbers remain small, this study reported that the transfer of spindles from patient oocytes to enucleated donor oocytes improved rates of fertilization and preimplantation development and resulted in the birth of six babies. These early but encouraging findings in patients with repeated failures of IVF cycles suggest that MST may be a viable approach to increasing the chance of these patients having a genetically related baby [16]. It should be noted that in regulated jurisdictions such as the UK and Australia, the use of nuclear transfer techniques for infertility treatment is illegal.

In addition to fundamental discovery research and possible future application in the treatment of some forms of infertility, arguably the most exciting impact of MST and PNT is in preventing the transmission of mitochondrial genetic disease [18, 19]. The transfer of nuclear material from an egg or zygote carrying affected mitochondria to an enucleated donor egg has the potential to prevent transmission of mitochondrial genetic disease [20–23]. Both techniques have been used in a number of species, and at the time of going to press only one baby has been born after MST for mitochondrial genetic disease [24]. While preclinical studies indicate much hope for the clinical application of MST and PNT for mitochondrial donation, ongoing research in animal and preclinical models will continue. These studies will focus on improving the success rates of the techniques, minimizing carry over of mitochondria, and understanding if there are any long-term effects of these micromanipulation procedures.

The first key feature of PNT involves preparing the zygotes for manipulation by transiently inhibiting two key elements of the cytoskeleton: actin filaments and microtubules. This enables the pronuclei to be extracted as a karyoplast without egg lysis. In PNT, the karyoplast consists of the pronuclei surrounded by a small amount of cytoplasm and intact plasma membrane. In contrast to PNT, MST only involves inhibition of actin filaments, while the microtubules need to be maintained to support normal spindle structure and chromosomal alignment—the spindle-complex [25]. This inhibition of actin allows for the spindle-complex to be removed from the oocyte as a karyoplast, which in MST is comprised of the spindle surrounded by a small amount of cytoplasm and intact plasma membrane. In both PNT and MST, this karyoplast is then exposed to a fusogen, usually HVJ-E, purified from inactivated Sendai virus, before being transferred into the perivitelline space of the recipient enucleated zygote or oocyte. Fusion of the karyoplast and enucleated cytoplasm normally ensues within 30 min.

While inactivated Sendai virus has emerged as the preferred method of achieving fusion, electrofusion can offer a viable alternative in mouse and human models [12, 13, 26–29]. Preferences of inhibitors vary among the literature with cytochalasin B, cytochalasin

D, and latrunculin A being the most commonly used reagents for actin inhibition, and with nocodazole and colcemid being favored for microtubule inhibition [1, 9, 10, 12, 13, 17, 20, 29–33].

Visualisation of the spindle complex is essential for performing MST. However, in murine and human oocytes, the metaphase-II spindle is not readily visible under brightfield, phase contrast, or differential interference contrast microscopy [34]. Advances in using polarized light to interact with the birefringent properties of the metaphase-II spindle have enabled the visualization of the spindle-complex without needing invasive techniques such as immunostaining or exogenous dyes [35]. Software associated with spindle-viewing systems can also provide increased contrast in real-time between the spindle-complex and cytoplasm. The adoption of micro-biopsy lasers and piezo-actuated drilling has allowed easier penetration through the zona pellucida, replacing methods such as the application of acid Tyrode's to thin the zona, or the use of spiked enucleation pipettes for manual penetration [36, 37].

The pronuclear transfer and maternal spindle transfer methods, present in scientific research for over 40 and 30 years, respectively, continue to develop and mature, especially as their usage in clinics becomes more prevalent. Here we share our most up-to-date methods.

2 Materials

2.1 Animals

1. 4–6 week old female mice.
2. Stud male mice housed individually.

2.2 Reagents and Media

1. PMSG (5–10 IU per mouse); Dissolve 5000 IU of lyophilized PMSG in 100 mL of water, fill 100 × 1 mL syringes with PMSG solution, cap syringes with 25–27 gauge needles, label and store at −80°C.
2. HCG (5–10 IU per mouse); Dissolve 5000 IU lyophilized HCG in 100 mL of water, fill 100 × 1 mL syringes with HCG solution, cap syringes with 25–27 gauge needles, label and store at −80°C.
3. M2 embryo culture medium.
4. Hyaluronidase; 10 mg/mL in M2. Prepare 30 μL aliquots and store at −20°C.
5. Sage One-Step culture media.
6. Mineral oil is suitable for embryo culture.
7. HVJ-E (inactivated Sendai virus), reconstituted with the included suspension buffer and aliquoted into 5 μL vials. Store at −80°C, e.g., (*see* **Note 1**).

8. PVP 7% (w/v).
9. Nocodazole; 0.6 mg/mL, 2000× solution in DMSO.
10. Cytochalasin D; 2 mg/mL, DMSO, 2000× solution.
11. M2-CCD-N medium; M2 medium supplemented with 2 μg/mL cytochalasin D and 0.3 μg/mL nocodazole. Store at 4°C [38].
12. Sage-CCD-N medium; Sage 1-step supplemented with 2 μg/mL cytochalasin D and 0.3 μg/mL nocodazole. Store at 4°C.
13. Cytochalasin B; 2.5 mg/mL, DMSO, 2000× solution.
14. M2-CCB medium; M2 supplemented with 2.5 μg/mL cytochalasin B. Store at 4°C.
15. NP-40 (*see* **Note 2**).
16. NP-40/Tween; in a 50 mL tube, prepare 0.5% (v/v) NP-40 (or NP-40 Alternative, *see* **Note 2**) and 0.5% (v/v) Tween-20 in water.
17. Tween-20.

2.3 Consumables

1. 1 mL syringes.
2. 25–27 gauge needles.
3. Stripper Micropipette.
4. Lint-free tissue.
5. Dissection tools (forceps, tissue scissors).
6. 35 mm cell culture dishes.
7. 50 mm Glass-bottomed dish.
8. 50 mL tubes.
9. Eppendorf tubes.
10. Manual pipettors.
11. Filtered pipette tips.

2.4 Equipment

1. Humidified incubator gassed with 5% O_2, 6% CO_2, 89% N_2.
2. Inverted microscope housed on an antivibration table, equipped with a warm stage, micro-biopsy laser, two micromanipulators, two microinjectors, and a spindle-viewing attachment (*see* **Note 3**, Fig. 1).
3. Holding pipette (15 μm ID, 30° °foot).
4. Enucleation pipette (13 μm ID, 70° bevel, 30° °foot).
5. Heat Block.

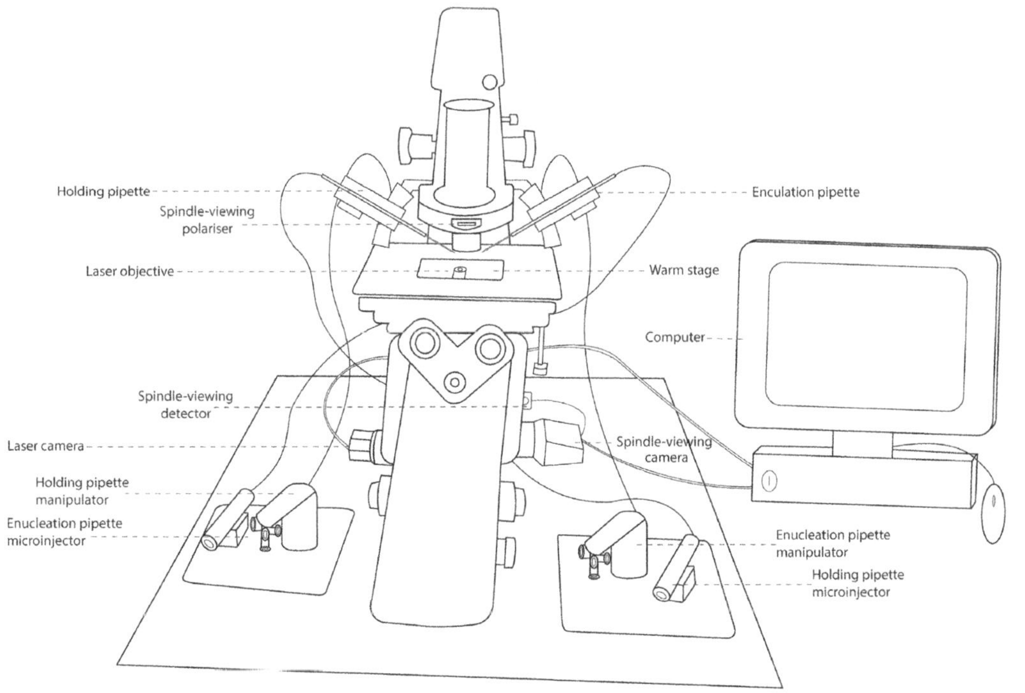

Fig. 1 Layout of the microscope and accessories used in PNT and MST. The microinjectors and micromanipulators are arranged in an "X" fashion, with the holding pipette held by the left-hand manipulator and controlled by the left-hand joystick and right-hand microinjector. The enucleation pipette is held by the right-hand manipulator and controlled by the right-hand joystick and left-hand microinjector. A spindle-viewing attachment is not needed for PNT, but advantageous for MST

3 Methods

The following describes the methods to perform pronuclear transfer and maternal spindle transfer (*see* **Note 4**).

3.1 Pronuclear Transfer

For a video demonstration of PNT, see Electronic Supplementary Video 1.

3.1.1 Collection, Isolation, and Preparation of Zygotes

1. Intraperitoneally inject 4–6 week-old female mice with 5–10 IU of PMSG followed approximately 48 h later with 5–10 IU of HCG. The volume delivered per mouse is approximately 0.1–0.2 mL using a 1 mL syringe and 25–27 gauge needle.
2. Mate the superovulated females with a male stud mouse overnight.
3. A minimum of 4 h before retrieval, prepare two 35 mm cell culture dishes with $8 \times 20\,\mu L$ drops of "Sage-CCD-N" medium under mineral oil, label "Sage-CCD-N-Donor" and "Sage-CCD-N-Recipient."

4. Also prepare three 35 mm cell culture dishes with 8 × 20 μL drops of Sage 1-step media under mineral oil and label "Culture 1-Donor," "Culture 1-Recipient," and "Culture 2."
5. Incubate dishes for a minimum of 4 h, ideally overnight, in the incubator to equilibrate.
6. On the day of zygote retrieval, prior to mouse sacrifice, prepare four 35 mm cell culture dishes filled with 2 mL of M2 media, labeled "Wash 1-4-Donor," and four 35 mm cell culture dishes filled with 2 mL of M2 media, labeled "Wash 1-4-Recipient." Also prepare two dishes, with 2 mL of M2 media supplemented with two of the prepared 30 μL hyaluronidase aliquots, to make a working solution of 0.3 mg/mL. Label these two dishes "Hyaluronidase-Donor" and "Hyaluronidase-Recipient."
7. These ten dishes can be stored at room temperature until use during zygote retrieval.
8. 16 h after HCG injection, euthanize the female mice by cervical dislocation, dissect out the oviducts, and use a 27-gauge needle to release the cumulus mass containing the zygotes into the appropriate "Wash 1" dish (*see* **Note 5**).
9. Use a micropipette to isolate the zygotes and move into the appropriate "Hyaluronidase" dish.
10. Repeatedly aspirate the zygotes up and down in the "Hyaluronidase" dish to remove all remaining cumulus cells.
11. As soon as the zygotes are free of cumulus, move them into the appropriate "Wash 2" dish and aspirate up and down in different areas of the dish to wash off the hyaluronidase. Repeat this process through appropriate "Wash 3" and "Wash 4" dishes.
12. Identify zygotes displaying two pronuclei under a microscope and transfer into the appropriate "Culture 1" dish, wash through two drops, before distributing evenly into the remaining drops, and place in the incubator (*see* **Note 6**).
13. After retrieval of zygotes, while working swiftly to minimize evaporation, prepare a glass-bottomed dish with 6 × 5 μL drops of "M2-CCD-N" media in two columns of three, a single 5 μL 7% PVP drop to one side, and an elongated 3 μL horizontal drop of HVJ-E above the two columns, before covering with mineral oil. Label as "Manipulation" and store on a 37°C heat block until use (Fig. 2).
14. After retrieval, attach an enucleation pipette to a syringe or injector, submerge the tip into "NP-40/Tween" solution, and aspirate the solution up the pipette until it reaches the bend in the foot, then expel until bubbles are visible. Repeat

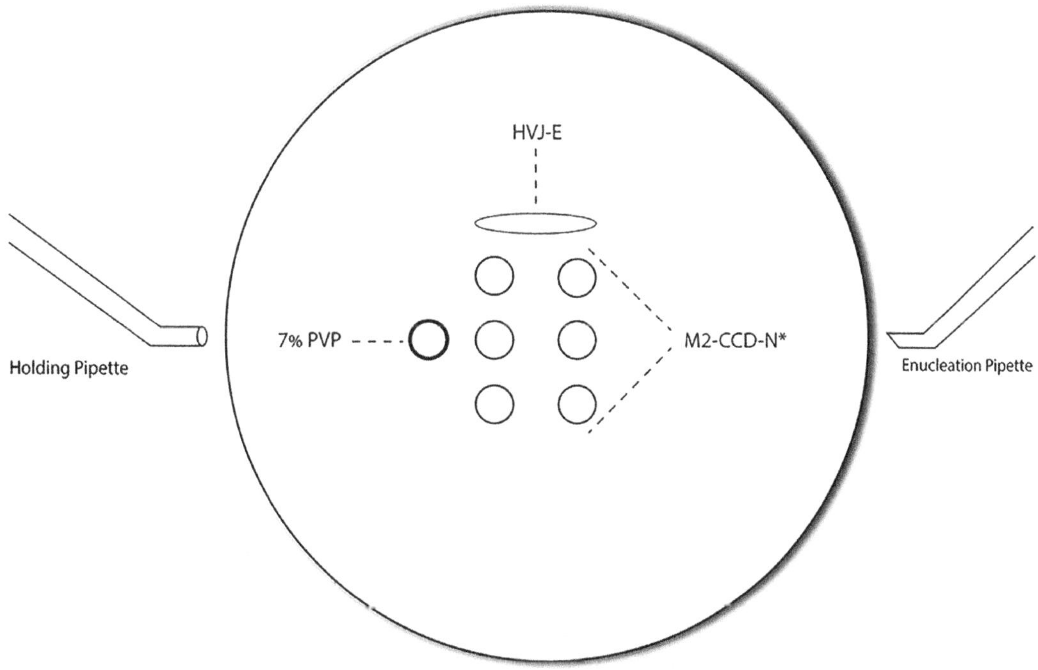

Fig. 2 Layout of "Manipulation" dish used in PNT, a 50 mm, glass-bottomed dish. 6× 5 μl volume of "M2-CCD-N" drops, 1× 3 μl volume of HVJ-E drop drawn out lengthwise with a pipette tip. The HVJ-E drop has been elongated to allow prompt access from the manipulation drops. *Note that the MST procedure will use "M2-CCD" in the 6× manipulation drops instead of "M2-CCD-N"

this three times, before withdrawing the pipette from the solution while expelling any remaining solution. Carefully wipe the pipette clean with a Kimwipe or other lint-free tissue (*see* **Note 7**).

15. Prepare the inverted microscope by ensuring the warm stage is set at 37°C. Mount and align a holding pipette and enucleation pipette on the micromanipulators.
16. At 18 h post-HCG, move the zygotes from their respective "Culture 1" dishes into their "Sage-CCD-N" dishes, wash through two drops, distribute evenly into the remaining drops, and culture for 30 min in the incubator.
17. After 30 min, wash these zygotes through the bottom two "M2-CCD-N" drops in the "Manipulation" dish, then arrange to suit in the four central drops of "M2-CCD-N" (*see* **Note 8**).
18. Move the "Manipulation" dish onto the microscope heated stage, set at 37°C.

3.1.2 Enucleating Zygotes

1. Ensure holding and enucleation pipettes remain aligned and have stable menisci (*see* **Note 9**).

2. Move the enucleation pipette into an empty drop of "M2-CCD-N."
3. Aspirate and suction the visible meniscus several times to flush any residual NP-40/Tween media from the pipette.
4. Move pipettes into a "M2-CCD-N" drop containing nuclear recipient zygote(s).
5. Position the zygote with the polar body facing towards the holding pipette and apply sufficient negative pressure (suction) to "grasp" the zygote (*see* **Note 10**, Fig. 3.1).
6. Use the laser to make a hole in the zona directly opposite the holding pipette, ensuring the laser does not contact the oocyte plasma membrane (*see* **Note 11**).

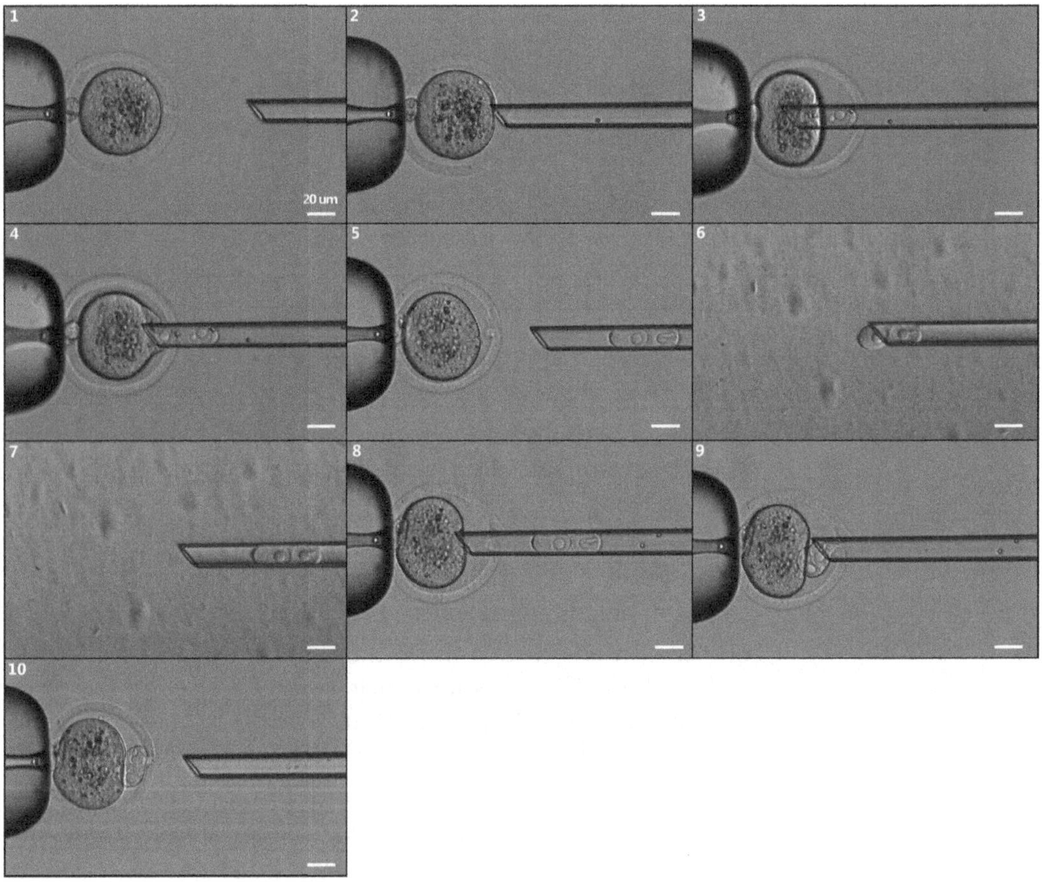

Fig. 3 Stepwise process for removing pronuclei from a zygote, before creating and exposing a karyoplast to inactivated Sendai virus and fusing it into an enucleated zygote. (1) Position the zygote and drill through the zona. (2) Insert the enucleation pipette through the zona. (3) Aspirate pronuclei into the enucleation pipette. (4) Remove enucleation pipette from zygote. (5) Secure the karyoplast in the enucleation pipette. (6) Expose karyoplast to HVJ-E. (7) Aspirate HVJ-E into the enucleation pipette. (8) Insert the enucleation pipette under the recipient zona. (9) Expel the karyoplast under the zona. 10. Remove enucleation pipette from zona

7. Gently insert the enucleation pipette through the hole in the zona and position directly next to the first pronucleus (*see* **Note 12**, Fig. 3.2).
8. Carefully apply suction to the enucleation pipette and draw the pronucleus two-thirds of the way into the pipette tip (Fig. 3.3).
9. Position the enucleation pipette tip directly next to the second pronucleus and carefully suction the second pronucleus until it is just inside the pipette tip (Fig. 3.4).
10. Gently withdraw the enucleation pipette until the plasma membrane releases, creating the enucleated zygote, and the isolated karyoplast within the pipette (*see* **Note 13**, Fig. 3.5).
11. Move the pipette containing the isolated karyoplast away from the enucleated zygote and gently expel the unneeded karyoplast (*see* **Note 14**).
12. Move the pipettes into a drop containing nuclear donor zygote(s).
13. Repeat Steps 3.1.2.5–3.1.2.10 with the nuclear donor zygote (*see* **Note 15**).
14. Ensure the nuclear donor karyoplast is stable within the enucleation pipette.

3.1.3 Transfer of Pronuclei

1. Move the enucleation pipette containing the karyoplast into the HVJ-E droplet, taking care to ensure the holding pipette does not contact the droplet.
2. Expel one-third of the karyoplast into the droplet and expose the karyoplast to the HVJ-E for 5–10 s (*see* **Note 16**, Fig. 3.6).
3. Apply suction to the karyoplast, drawing it back into the pipette along with an additional amount of HVJ-E equal to about half the length of the karyoplast (Fig. 3.7).
4. Move the pipette containing the karyoplast into the drop containing an enucleated nuclear recipient zygote.
5. Position the recipient zygote with the polar body facing towards the holding pipette, the previously laser-drilled hole in focus, and grasp the zygote (Fig. 3.8).
6. Move the enucleation pipette through the zona and slowly expel the entire karyoplast into the perivitelline space (Fig. 3.9).
7. Carefully withdraw the enucleation pipette. As you withdraw the pipette, it is possible to remove any excess cytoplasm in the karyoplast by moving the pipette in a downward direction at the point where both the pronuclei have been expelled (Fig. 3.10).

8. After completing the micromanipulation steps, move the manipulated zygote(s) into the "Culture 2" dish.
9. Wash the manipulated zygote(s) through two unused drops of medium before evenly distributing them into the remaining drops for subsequent culture.
10. Fusion of karyoplast into the recipient cytoplasm occurs after around 15 min at 37°C (*see* **Note 16**).

3.2 Maternal Spindle Transfer

For a video demonstration of MST, see Electronic Supplementary Video 2.

3.2.1 Collection, Isolation, and Preparation of Oocytes

1. Intraperitoneally inject 4–6 week-old mice with 5–10 IU of PMSG followed approximately 48 h later with 5–10 IU of HCG. The volume delivered per mouse is approximately 0.1–0.2 mL using a 1 mL syringe and 25–27 gauge needle.
2. A minimum of 4 h before retrieval, prepare three 35 mm cell culture dishes with 8 × 20 μL drops of Sage 1-step media under mineral oil and label "Culture 1-Donor," "Culture 1-Recipient," and "Culture 2."
3. Incubate dishes for a minimum of 4 h, ideally overnight, in the incubator to equilibrate.
4. On the day of oocyte retrieval, prior to mouse sacrifice, prepare four 35 mm cell culture dishes filled with 2 mL of M2 media, labeled "Wash 1-4-Donor", and four 35 mm cell culture dishes filled with 2 mL of M2 media, labeled "Wash 1-4-Recipient". Also prepare two "Hyaluronidase" dishes with 2 mL of M2 media and two of the prepared 30 μL hyaluronidase aliquots, to make a working solution of 0.3 mg/mL. Label these two dishes "Hyaluronidase-Donor" and "Hyaluronidase-Recipient."
5. These ten dishes can be stored at room temperature until use during oocyte retrieval.
6. 12 h after HCG injection, euthanize the female mice by cervical dislocation, dissect out the oviducts, and use a 27-gauge needle to release the cumulus mass containing the oocytes into the appropriate "Wash 1" dish (*see* **Note 5**).
7. Use a micropipette to isolate the oocytes and move into the appropriate "Hyaluronidase" dish.
8. Repeatedly aspirate the oocytes up and down in the "Hyaluronidase" dish to remove their cumulus cells.
9. As soon as the oocytes are free of cumulus, move them into the appropriate "Wash 2" dish and aspirate up and down in different areas of the dish to wash off the hyaluronidase. Repeat this process through appropriate "Wash 3" and "Wash 4" dishes.

10. Identify which oocytes are mature (have extruded a polar body) and move them into the appropriate "Culture 1" dish, wash through two drops, before distributing evenly into the remaining drops, and place in the incubator.
11. After retrieval of oocytes, working swiftly to minimize evaporation, prepare a glass-bottomed dish with 6×5 μL drops of "M2-CCB" media in two columns of three, a single 5 μL 7% PVP drop to one side, and an elongated 3 μL horizontal drop of HVJ-E above the two columns, before covering with mineral oil. Label as "Manipulation" and store on a 37°C heat block until use (Fig. 2).
12. After retrieval of oocytes, attach an enucleation pipette to a syringe or injector, submerge the tip into "NP-40/Tween" solution, and aspirate the solution up the pipette until it reaches the bend in the foot, then expel until bubbles are visible. Repeat this three times, before withdrawing the pipette from the solution while expelling any remaining solution. Carefully wipe the pipette clean with a Kimwipe or other lint-free tissue (*see* **Note 7**).
13. Prepare the inverted microscope by ensuring the warm stage is set at 37°C. Mount and align a holding pipette and enucleation pipette on the micromanipulators.
14. At 14 h post-HCG, move mature oocytes from their respective "Culture 1" dishes into the "Manipulation" dish and wash them through the bottom two "M2-CCB" drops, then arrange to suit in the 4 central drops of "M2-CCB" (*see* **Note 8**).
15. Move the "Manipulation" dish onto the microscope heated stage, set at 37°C and wait for 15 min for actin inhibition to take affect (commence with microscope setup during this time).

3.2.2 Enucleating Oocytes

1. Ensure the holding and enucleation pipettes are aligned and have stable menisci (*see* **Note 9**).
2. Move the enucleation pipette into an empty drop of "M2-CCB."
3. Aspirate and suction the visible meniscus several times to flush any residual NP-40/Tween media from the pipette.
4. Move pipettes into "M2-CCB" drop containing nuclear recipient oocyte(s).
5. Use the spindle-viewing attachment to visualize spindle complexes within the oocyte (*see* **Note 17**).
6. Position the oocyte with the spindle-complex on a horizontal plane, near the oocyte's plasma membrane, on the side closest to the enucleation pipette (*see* **Note 18**, Fig. 4.1).

7. Use the laser to make a hole in the zona direct opposite the holding pipette, ensuring the laser does not contact the oocyte plasma membrane or spindle-complex (*see* **Note 11**).
8. Gently insert the enucleation pipette through the hole in the zona and position it directly next to the spindle-complex (*see* **Note 19**, Fig. 4.2).
9. Carefully apply suction to the enucleation pipette and draw up the spindle until it is just inside the pipette tip (Fig. 4.3).
10. Gently withdraw the enucleation pipette until the plasma membrane releases, creating the enucleated oocyte, and the isolated spindle karyoplast within the pipette (*see* **Note 13**, Fig. 4.4).
11. Move the pipette containing the isolated karyoplast away from the enucleated oocyte, and gently expel the unneeded karyoplast (*see* **Note 14**).
12. Move the pipettes into a drop containing nuclear donor oocyte(s).
13. Repeat Steps 3.2.2.6–3.2.2.10 with the nuclear donor zygote (*see* **Note 15**).
14. Ensure the nuclear donor karyoplast is stable within the enucleation pipette.

3.2.3 Transfer of Spindle-Complex

1. Move the enucleation pipette containing the spindle karyoplast into the HVJ-E droplet, taking care to ensure the holding pipette does not contact the droplet.
2. Expel one-third of the karyoplast into the droplet and expose the karyoplast to the HVJ-E for 5–10 s (*see* **Note 16**, Fig. 4.5).

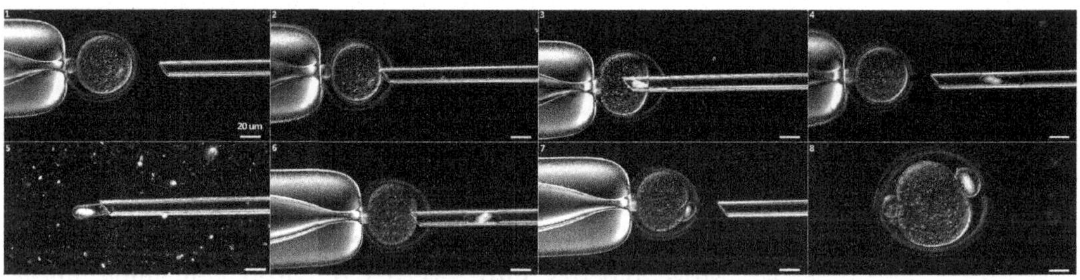

Fig. 4 Stepwise process for removing the maternal spindle from an oocyte, before creating and exposing a karyoplast to inactivated Sendai virus and fusing it into an enucleated oocyte. Contrast is generated by software associated with the Oosight Imaging System. (1) Position the oocyte and drill through the zona. (2) Insert the enucleation pipette under the zona. (3) Aspirate spindle complex. (4) Remove enucleation pipette from oocyte and zona. (5) Expose the spindle karyoplast to HVJ-E. (6) Insert the enucleation pipette through the zona of the recipient oocyte. (7) Expel the spindle karyoplast under the zona and remove the pipette. (8) Release the reconstructed oocyte from the holding pipette

3. Apply suction to the karyoplast, drawing it back into the pipette along with an additional amount of HVJ-E equal to about half the length of the karyoplast.
4. Move the pipette containing the karyoplast into the drop containing the enucleated nuclear recipient oocyte.
5. Position the oocyte with the previously laser-drilled hole in focus, and grasp the oocyte so that the hole is opposite to the holding pipette (Fig. 4.6).
6. Move the enucleation pipette through the zona and slowly expel the entire karyoplast into the perivitelline space.
7. Carefully withdraw the enucleation pipette. As you withdraw the pipette, it is possible to remove any excess cytoplasm in the karyoplast by moving the pipette in a downward direction at the point where the spindle-complex has been expelled (Figs. 4.7 and 4.8).
8. After completing the micromanipulation steps, move the manipulated oocytes into the "Culture 2" dish.
9. Wash the manipulated oocyte(s) through two unused drops of medium before evenly distributing them into the remaining drops for subsequent culture.
10. Fusion of karyoplast into the recipient cytoplasm occurs after around 15 min at 37°C (*see* **Note 16**).

4 Notes

1. GenomeONE is now discontinued. Melton Biomedical International Co. Ltd, Hong Kong, offers this product, however, the authors have not validated the product from this new supplier.
2. NP-40 is now discontinued. Sigma-Aldrich offers NP-40 Alternative, #492016, however the authors have not validated this product.
3. The equipment setup described here has been assembled with the intention of translating the murine PNT and MST techniques into use with human zygotes and oocytes in a clinic. As such, the microscope and accessories are of the highest quality to streamline the manipulation technique. Both PNT and MST can be successfully performed without the use of biopsy lasers, warm stages, and commercial pipettes.
4. Pronuclear transfer is performed on fertilized zygotes and is used to transfer both the maternal and paternal nuclear genomes into a donor, enucleated zygote. This constructs a zygote with the paternal and maternal genomes from the

nuclear donor, and the cytoplasm and zona from the nuclear recipient. This zygote is fertilized and should be developmentally competent. Maternal spindle transfer is performed on unfertilized, mature oocytes, and is used to transfer the maternal nuclear genomes into an enucleated, mature oocyte. This constructs an oocyte with the maternal genome from the nuclear donor, and the cytoplasm and zona from the nuclear recipient. This oocyte will need fertilizing or activating before any further development will occur.

5. Ensuring nuclear donors and nuclear recipient zygotes and oocytes are kept separate during their retrieval and initial culture is essential.

6. The media drops in culture and inhibition dishes can be arranged to suit the user. The authors prefer a row of two drops, which function as wash drops, and underneath these, two rows of three drops, which function as culture drops. Each wash drop can be used multiple times as long as minimal media is transferred with each oocyte or zygote. Oocytes and zygotes are generally cultured in groups of around ten per drop.

7. Flushing the enucleation pipette with NP-40/Tween can help prevent the karyoplast from sticking to the inside of the pipette. Take care to fully expel the detergents before manipulating a zygote as they may have a detrimental effect on survival rates.

8. Multiple oocytes and zygotes can be placed in each manipulation drop depending on the user's preference. The authors recommend matching the number of nuclear donors with nuclear recipients, and ensuring that total manipulation time for all zygotes or oocytes present on the dish does not exceed 1 h. The manipulation dish can be reused for multiple groups of oocytes or zygotes in a single session.

9. To reduce sensitivity of the enucleation pipette's meniscus, aspirate "columns" of alternating oil and manipulation media into the pipette. This dampens the response speed of the microinjector.

10. Try to initially position the zygote so that both pronuclei are visible at once (Fig. 3).

11. We use a Vitrolife Octax Navilase, which has a wavelength of 1.48 um and variable power between 100 mW and 250 mW. At a pulse length setting of 0.6 ms, the zona can be drilled through with two to three pulses.

12. Gently nudging the pronuclei with the enucleation pipette can help ensure the pipette is on the same Z plane as the pronuclei. This will help allow aspiration of the pronuclei with minimal excess cytoplasm.

13. Reducing the amount of excess cytoplasm in the karyoplast may help to reduce the amount of carryover of the nuclear donor's mitochondria.
14. Aspirating and expelling media from the 7% PVP drop can help to flush the enucleation pipette of debris sticking to the inside of the pipette.
15. The authors recommend only enucleating one nuclear recipient at a time, to minimalize the time the reconstructed zygote or oocyte spends without a nuclear genome.
16. HVJ-E activity can vary with reconstitution dilution factor and age. If fusion is not occurring, consider increasing exposure time or decreasing the dilution factor. If the karyoplast is lysing after exposure to HVJ-E, consider decreasing exposure time or increasing dilution factor.
17. Clear visualization of spindle complexes is essential to successful MST. We use a separate micro-biopsy laser and spindle viewing attachment on our manipulation microscope. This requires two C-mount cameras to be attached at the same time. In our setup, the MST procedure can be performed by looking down the microscope eyepieces or by looking at the computer monitor displaying either camera feed. It should be noted that the spindle-viewing software provides enhanced contrast and has a slight delay. Viewing the spindle via the eyepiece or the laser software has no delay, but no enhanced contrast.
18. Try to position the oocyte so that the spindle complex is positioned close to the right-hand side of the oocyte, and orientated horizontally along the X axis, in line with the enucleation pipette (Fig. 4).
19. Gently nudging the spindle-complex with the enucleation pipette can help ensure the pipette is on the same Z plane as the spindle. This will help allow aspiration of the spindle-complex with minimal excess cytoplasm.

Acknowledgments

This work was supported by grants from NHMRC (2001112), MRFF (2023189), and the Mito Foundation.

References

1. McGrath J, Solter D (1983) Nuclear transplantation in the mouse embryo by microsurgery and cell fusion. Science 220(4603):1300–1302. https://doi.org/10.1126/science.6857250
2. Surani MA, Barton SC, Norris ML (1984) Development of reconstituted mouse eggs suggests imprinting of the genome during gametogenesis. Nature 308(5959):548–550. https://doi.org/10.1038/308548a0
3. McGrath J, Solter D (1984) Completion of mouse embryogenesis requires both the maternal and paternal genomes. Cell 37(1):179–183. https://doi.org/10.1016/0092-8674(84)90313-1

4. Mann JR, Lovell-Badge RH (1984) Inviability of parthenogenones is determined by pronuclei, not egg cytoplasm. Nature 310(5972):66–67. https://doi.org/10.1038/310066a0
5. Barton SC, Surani MA, Norris ML (1984) Role of paternal and maternal genomes in mouse development. Nature 311(5984):374–376. https://doi.org/10.1038/311374a0
6. Howlett SK, Barton SC, Surani MA (1987) Nuclear cytoplasmic interactions following nuclear transplantation in mouse embryos. Development 101(4):915–923
7. Campbell K, Ritchie W, Wilmut I (1993) Nuclear-cytoplasmic interactions during the first cell cycle of nuclear transfer reconstructed bovine embryos: implications for deoxyribonucleic acid replication and development. Biol Reprod 49:933–933
8. Surani MAH, Barton SC, Norris ML (1986) Nuclear transplantation in the mouse: Heritable differences between parental genomes after activation of the embryonic genome. Cell 45(1):127–136. https://doi.org/10.1016/0092-8674(86)90544-1
9. Znachorova T, Dudko N, Ming H, Jiang Z, Fulka H (2024) The timing of pronuclear transfer critically affects the developmental competence and quality of embryos. Mol Hum Reprod 30(7):gaae024. https://doi.org/10.1093/molehr/gaae024
10. Heindryckx B, Rybouchkin A, Van der Elst J, Dhont M (2002) Serial pronuclear transfer increases the developmental potential of in vitro-matured oocytes in mouse cloning. Biol Reprod 67(6):1790–1795. https://doi.org/10.1095/biolreprod.102.004770
11. Christodoulaki A, He H, Zhou M, De Roo C, Baetens M, De Pretre T, Fakhar-I-Adil M, Menten B, Van Soom A, Stoop D, Boel A, Heindryckx B (2024) Pronuclear transfer rescues poor embryo development of in vitro-grown secondary mouse follicles. Hum Reprod Open 2024(1):hoae009. https://doi.org/10.1093/hropen/hoae009
12. Wang M-K, Chen D-Y, Lui J-L, Li G-P, Sun Q-Y (2001) In vitro fertilisation of mouse oocytes reconstructed by transfer of metaphase II chromosomes results in live births. Zygote 9(1):9–14
13. Kono T, Obata Y, Yoshimzu T, Nakahara T, Carroll J (1996) Epigenetic modifications during oocyte growth correlates with extended parthenogenetic development in the mouse. Nat Genet 13(1):91–94. https://doi.org/10.1038/ng0596-91
14. Tang M, Popovic M, Stamatiadis P, Van der Jeught M, Van Coster R, Deforce D, De Sutter P, Coucke P, Menten B, Stoop D (2020) Germline nuclear transfer in mice may rescue poor embryo development associated with advanced maternal age and early embryo arrest. Human Reprod 35(7):1562–1577
15. Zhang J, Zhuang G, Zeng Y, Grifo J, Acosta C, Shu Y, Liu H (2016) Pregnancy derived from human zygote pronuclear transfer in a patient who had arrested embryos after IVF. Reprod BioMed Online 33(4):529–533. https://doi.org/10.1016/j.rbmo.2016.07.008
16. Tang M, Boel A, Castelluccio N, Cardona Barberán A, Christodoulaki A, Bekaert B, Popovic M, Vanden Meerschaut F, De Sutter P, Menten B, Symoens S, Vanlander AV, Stoop D, Coucke PJ, Heindryckx B (2022) Human germline nuclear transfer to overcome mitochondrial disease and failed fertilization after ICSI. J Assist Reprod Genet 39(3):609–618. https://doi.org/10.1007/s10815-022-02401-7
17. Costa-Borges N, Nikitos E, Späth K, Miguel-Escalada I, Ma H, Rink K, Coudereau C, Darby H, Koski A, Van Dyken C, Mestres E, Papakyriakou E, De Ziegler D, Kontopoulos G, Mantzavinos T, Vasilopoulos I, Grigorakis S, Prokopakis T, Dimitropoulos K, Polyzos P, Vlachos N, Kostaras K, Mitalipov S, Calderón G, Psathas P, Wells D (2023) First pilot study of maternal spindle transfer for the treatment of repeated in vitro fertilization failures in couples with idiopathic infertility. Fertil Steril 119(6):964–973. https://doi.org/10.1016/j.fertnstert.2023.02.008
18. Hyslop LA, Blakeley P, Craven L, Richardson J, Fogarty NM, Fragouli E, Lamb M, Wamaitha SE, Prathalingam N, Zhang Q, O'Keefe H, Takeda Y, Arizzi L, Alfarawati S, Tuppen HA, Irving L, Kalleas D, Choudhary M, Wells D, Murdoch AP, Turnbull DM, Niakan KK, Herbert M (2016) Towards clinical application of pronuclear transfer to prevent mitochondrial DNA disease. Nature 534(7607):383–386. https://doi.org/10.1038/nature18303
19. Craven L, Tuppen HA, Greggains GD, Harbottle SJ, Murphy JL, Cree LM, Murdoch AP, Chinnery PF, Taylor RW, Lightowlers RN, Herbert M, Turnbull DM (2010) Pronuclear transfer in human embryos to prevent transmission of mitochondrial DNA disease. Nature 465(7294):82–85. https://doi.org/10.1038/nature08958
20. Sato A, Kono T, Nakada K, Ishikawa K, Inoue S-I, Yonekawa H, Hayashi J-I (2005) Gene therapy for progeny of mito-mice carrying pathogenic mtDNA by nuclear transplantation. Proc Natl Acad Sci U S A 102(46):16765–16770. https://doi.org/10.1073/pnas.0506197102
21. Herbert M, Turnbull D (2018) Progress in mitochondrial replacement therapies. Nat Rev Mol Cell Biol 19(2):71–72. https://doi.org/10.1038/nrm.2018.3
22. Greenfield A, Braude P, Flinter F, Lovell-Badge R, Ogilvie C, Perry ACF (2017) Assisted reproductive technologies to prevent human

mitochondrial disease transmission. Nat Biotechnol 35(11):1059–1068. https://doi.org/10.1038/nbt.3997

23. Tachibana M, Kuno T, Yaegashi N (2018) Mitochondrial replacement therapy and assisted reproductive technology: A paradigm shift toward treatment of genetic diseases in gametes or in early embryos. Reprod Med Biol 17(4):421–433
24. Zhang J, Liu H, Luo S, Chavez-Badiola A, Liu Z, Yang M, Munne S, Konstantinidis M, Wells D, Huang T (2016) First live birth using human oocytes reconstituted by spindle nuclear transfer for mitochondrial DNA mutation causing Leigh syndrome. Fertil Steril 106(3):e375–e376. https://doi.org/10.1016/j.fertnstert.2016.08.004
25. Walczak CE, Heald R (2008) Chapter three—Mechanisms of mitotic spindle assembly and function. In: Jeon KW (ed) International review of cytology, vol 265. Academic Press, San Diego, CA, pp 111–158. https://doi.org/10.1016/S0074-7696(07)65003-7
26. Baranska W, Koprowski H (1970) Fusion of unfertilized mouse eggs with somatic cells. J Exp Zool 174(1):1–14
27. Bromhall J (1975) Nuclear transplantation in the rabbit egg. Nature 258(5537):719–722
28. Willadsen SM (1986) Nuclear transplantation in sheep embryos. Nature 320(6057):63–65. https://doi.org/10.1038/320063a0
29. Costa-Borges N, Spath K, Miguel-Escalada I, Mestres E, Balmaseda R, Serafín A, Garcia-Jiménez M, Vanrell I, González J, Rink K (2020) Maternal spindle transfer overcomes embryo developmental arrest caused by ooplasmic defects in mice. Elife 9:e48591
30. McGrath J, Solter D, Mann J (2017) Pronuclear transplantation in the mouse embryo. Cold Spring Harb Protoc 2017(8):pdb.prot094417. https://doi.org/10.1101/pdb.prot094417
31. Terashita Y, Wakayama S, Yamagata K, Li C, Sato E, Wakayama T (2012) Latrunculin a can improve the birth rate of cloned mice and simplify the nuclear transfer protocol by gently inhibiting actin polymerization. Biol Reprod 86(6):180, 181–186
32. Paull D, Emmanuele V, Weiss KA, Treff N, Stewart L, Hua H, Zimmer M, Kahler DJ, Goland RS, Noggle SA, Prosser R, Hirano M, Sauer MV, Egli D (2013) Nuclear genome transfer in human oocytes eliminates mitochondrial DNA variants. Nature 493(7434):632–637. https://doi.org/10.1038/nature11800
33. Tachibana M, Sparman M, Sritanaudomchai H, Ma H, Clepper L, Woodward J, Li Y, Ramsey C, Kolotushkina O, Mitalipov S (2009) Mitochondrial gene replacement in primate offspring and embryonic stem cells. Nature 461(7262):367–372. https://doi.org/10.1038/nature08368
34. Keefe D, Liu L, Wang W, Silva C (2003) Imaging meiotic spindles by polarization light microscopy: principles and applications to IVF. Reprod Biomed Online 7(1):24–29
35. Liu L, Oldenbourg R, Trimarchi JR, Keefe DL (2000) A reliable, noninvasive technique for spindle imaging and enucleation of mammalian oocytes. Nat Biotechnol 18(2):223–225
36. Chen SU, Chao KH, Chang CY, Hsieh FJ, Ho HN, Yang YS (2004) Technical aspects of the piezo, laser-assisted, and conventional methods for nuclear transfer of mouse oocytes and their efficiency and efficacy: Piezo minimizes damage of the ooplasmic membrane at injection. J Exp Zool A Comp Exp Biol 301(4):344–351
37. Wakayama T, Yanagimachi R (1998) The first polar body can be used for the production of normal offspring in mice. Biol Reprod 59(1):100–104. https://doi.org/10.1095/biolreprod59.1.100
38. Quinn P, Barros C, Whittingham DG (1982) Preservation of hamster oocytes to assay the fertilizing capacity of human spermatozoa. Reproduction (Cambridge, England) 66(1):161–168. https://doi.org/10.1530/jrf.0.0660161

Subject Index

A

B

C

Alexandra J. Harvey and David K. Gardner (eds.), *The Preimplantation Embryo*, Methods in Molecular Biology, vol. 3038, https://doi.org/10.1007/978-1-0716-5292-3,

N

O

P

R

S

The manufacturer's authorised representative in the EU is Springer Nature Customer Service Centre GmbH, Europaplatz 3, 69115 Heidelberg, Germany. If you have any concerns regarding our products, please contact ProductSafety@springernature.com

Printed and bound by CPI Group (UK) Ltd, Croydon, CR0 4YY
18/09/2026
02219020-0006